IET ENERGY ENGINEERING 172

Lightning Interaction with Power Systems

Other volumes in this series:

Volume 1	**Power Circuit Breaker Theory and Design** C.H. Flurscheim (Editor)
Volume 4	**Industrial Microwave Heating** A.C. Metaxas and R.J. Meredith
Volume 7	**Insulators for High Voltages** J.S.T. Looms
Volume 8	**Variable Frequency AC Motor Drive Systems** D. Finney
Volume 10	**SF_6 Switchgear** H.M. Ryan and G.R. Jones
Volume 11	**Conduction and Induction Heating** E.J. Davies
Volume 13	**Statistical Techniques for High Voltage Engineering** W. Hauschild and W. Mosch
Volume 14	**Uninterruptible Power Supplies** J. Platts and J.D. St Aubyn (Editors)
Volume 15	**Digital Protection for Power Systems** A.T. Johns and S.K. Salman
Volume 16	**Electricity Economics and Planning** T.W. Berrie
Volume 18	**Vacuum Switchgear** A. Greenwood
Volume 19	**Electrical Safety: A guide to causes and prevention of hazards** J. Maxwell Adams
Volume 21	**Electricity Distribution Network Design, 2nd Edition** E. Lakervi and E.J. Holmes
Volume 22	**Artificial Intelligence Techniques in Power Systems** K. Warwick, A.O. Ekwue and R. Aggarwal (Editors)
Volume 24	**Power System Commissioning and Maintenance Practice** K. Harker
Volume 25	**Engineers' Handbook of Industrial Microwave Heating** R.J. Meredith
Volume 26	**Small Electric Motors** H. Moczala *et al.*
Volume 27	**AC-DC Power System Analysis** J. Arrillaga and B.C. Smith
Volume 29	**High Voltage Direct Current Transmission, 2nd Edition** J. Arrillaga
Volume 30	**Flexible AC Transmission Systems (FACTS)** Y.-H. Song (Editor)
Volume 31	**Embedded Generation** N. Jenkins *et al.*
Volume 32	**High Voltage Engineering and Testing, 2nd Edition** H.M. Ryan (Editor)
Volume 33	**Overvoltage Protection of Low-Voltage Systems, Revised Edition** P. Hasse
Volume 36	**Voltage Quality in Electrical Power Systems** J. Schlabbach *et al.*
Volume 37	**Electrical Steels for Rotating Machines** P. Beckley
Volume 38	**The Electric Car: Development and future of battery, hybrid and fuel-cell cars** M. Westbrook
Volume 39	**Power Systems Electromagnetic Transients Simulation** J. Arrillaga and N. Watson
Volume 40	**Advances in High Voltage Engineering** M. Haddad and D. Warne
Volume 41	**Electrical Operation of Electrostatic Precipitators** K. Parker
Volume 43	**Thermal Power Plant Simulation and Control** D. Flynn
Volume 44	**Economic Evaluation of Projects in the Electricity Supply Industry** H. Khatib
Volume 45	**Propulsion Systems for Hybrid Vehicles** J. Miller
Volume 46	**Distribution Switchgear** S. Stewart
Volume 47	**Protection of Electricity Distribution Networks, 2nd Edition** J. Gers and E. Holmes
Volume 48	**Wood Pole Overhead Lines** B. Wareing
Volume 49	**Electric Fuses, 3rd Edition** A. Wright and G. Newbery
Volume 50	**Wind Power Integration: Connection and system operational aspects** B. Fox *et al.*
Volume 51	**Short Circuit Currents** J. Schlabbach
Volume 52	**Nuclear Power** J. Wood
Volume 53	**Condition Assessment of High Voltage Insulation in Power System Equipment** R.E. James and Q. Su
Volume 55	**Local Energy: Distributed generation of heat and power** J. Wood
Volume 56	**Condition Monitoring of Rotating Electrical Machines** P. Tavner, L. Ran, J. Penman and H. Sedding
Volume 57	**The Control Techniques Drives and Controls Handbook, 2nd Edition** B. Drury
Volume 58	**Lightning Protection** V. Cooray (Editor)
Volume 59	**Ultracapacitor Applications** J.M. Miller

Volume 62 **Lightning Electromagnetics** V. Cooray
Volume 63 **Energy Storage for Power Systems, 2nd Edition** A. Ter-Gazarian
Volume 65 **Protection of Electricity Distribution Networks, 3rd Edition** J. Gers
Volume 66 **High Voltage Engineering Testing, 3rd Edition** H. Ryan (Editor)
Volume 67 **Multicore Simulation of Power System Transients** F.M. Uriate
Volume 68 **Distribution System Analysis and Automation** J. Gers
Volume 69 **The Lightening Flash, 2nd Edition** V. Cooray (Editor)
Volume 70 **Economic Evaluation of Projects in the Electricity Supply Industry, 3rd Edition** H. Khatib
Volume 72 **Control Circuits in Power Electronics: Practical issues in design and implementation** M. Castilla (Editor)
Volume 73 **Wide Area Monitoring, Protection and Control Systems: The enabler for Smarter Grids** A. Vaccaro and A. Zobaa (Editors)
Volume 74 **Power Electronic Converters and Systems: Frontiers and applications** A.M. Trzynadlowski (Editor)
Volume 75 **Power Distribution Automation** B. Das (Editor)
Volume 76 **Power System Stability: Modelling, analysis and control** B. Om P. Malik
Volume 78 **Numerical Analysis of Power System Transients and Dynamics** A. Ametani (Editor)
Volume 79 **Vehicle-to-Grid: Linking electric vehicles to the smart grid** J. Lu and J. Hossain (Editors)
Volume 81 **Cyber-Physical-Social Systems and Constructs in Electric Power Engineering** S. Suryanarayanan, R. Roche and T.M. Hansen (Editors)
Volume 82 **Periodic Control of Power Electronic Converters** F. Blaabjerg, K. Zhou, D. Wang and Y. Yang
Volume 86 **Advances in Power System Modelling, Control and Stability Analysis** F. Milano (Editor)
Volume 87 **Cogeneration: Technologies, optimisation and implementation** C.A. Frangopoulos (Editor)
Volume 88 **Smarter Energy: From Smart Metering to the Smart Grid** H. Sun, N. Hatziargyriou, H.V. Poor, L. Carpanini and M.A. Sánchez Fornié (Editors)
Volume 89 **Hydrogen Production, Separation and Purification for Energy** A. Basile, F. Dalena, J. Tong and T.N. Veziroğlu (Editors)
Volume 90 **Clean Energy Microgrids** S. Obara and J. Morel (Editors)
Volume 91 **Fuzzy Logic Control in Energy Systems with Design Applications in MATLAB®/Simulink®** İ.H. Altaş
Volume 92 **Power Quality in Future Electrical Power Systems** A.F. Zobaa and S.H.E.A. Aleem (Editors)
Volume 93 **Cogeneration and District Energy Systems: Modelling, analysis and optimization** M.A. Rosen and S. Koohi-Fayegh
Volume 94 **Introduction to the Smart Grid: Concepts, technologies and evolution** S.K. Salman
Volume 95 **Communication, Control and Security Challenges for the Smart Grid** S.M. Muyeen and S. Rahman (Editors)
Volume 96 **Industrial Power Systems with Distributed and Embedded Generation** R Belu
Volume 97 **Synchronized Phasor Measurements for Smart Grids** M.J.B. Reddy and D.K. Mohanta (Editors)
Volume 98 **Large Scale Grid Integration of Renewable Energy Sources** A. Moreno-Munoz (Editor)
Volume 100 **Modeling and Dynamic Behaviour of Hydropower Plants** N. Kishor and J. Fraile-Ardanuy (Editors)
Volume 101 **Methane and Hydrogen for Energy Storage** R. Carriveau and D.S.-K. Ting
Volume 104 **Power Transformer Condition Monitoring and Diagnosis** A. Abu-Siada (Editor)
Volume 106 **Surface Passivation of Industrial Crystalline Silicon Solar Cells** J. John (Editor)
Volume 107 **Bifacial Photovoltaics: Technology, applications and economics** J. Libal and R. Kopecek (Editors)

Volume 108 **Fault Diagnosis of Induction Motors** J. Faiz, V. Ghorbanian and G. Joksimović
Volume 110 **High Voltage Power Network Construction** K. Harker
Volume 111 **Energy Storage at Different Voltage Levels: Technology, integration, and market aspects** A.F. Zobaa, P.F. Ribeiro, S.H.A. Aleem and S.N. Afifi (Editors)
Volume 112 **Wireless Power Transfer: Theory, technology and application** N. Shinohara
Volume 115 **DC Distribution Systems and Microgrids** T. Dragičević, F. Blaabjerg and P. Wheeler
Volume 117 **Structural Control and Fault Detection of Wind Turbine Systems** H.R. Karimi
Volume 119 **Thermal Power Plant Control and Instrumentation: The control of boilers and HRSGs, 2nd Edition** D. Lindsley, J. Grist and D. Parker
Volume 120 **Fault Diagnosis for Robust Inverter Power Drives** A. Ginart (Editor)
Volume 123 **Power Systems Electromagnetic Transients Simulation, 2nd Edition** N. Watson and J. Arrillaga
Volume 124 **Power Market Transformation** B. Murray
Volume 125 **Wind Energy Modeling and Simulation, Volume 1: Atmosphere and plant** P. Veers (Editor)
Volume 126 **Diagnosis and Fault Tolerance of Electrical Machines, Power Electronics and Drives** A.J.M. Cardoso
Volume 128 **Characterization of Wide Bandgap Power Semiconductor Devices** F. Wang, Z. Zhang and E.A. Jones
Volume 129 **Renewable Energy from the Oceans: From wave, tidal and gradient systems to offshore wind and solar** D. Coiro and T. Sant (Editors)
Volume 130 **Wind and Solar Based Energy Systems for Communities** R. Carriveau and D.S.-K. Ting (Editors)
Volume 131 **Metaheuristic Optimization in Power Engineering** J. Radosavljević
Volume 132 **Power Line Communication Systems for Smart Grids** I.R.S Casella and A. Anpalagan
Volume 139 **Variability, Scalability and Stability of Microgrids** S.M. Muyeen, S.M. Islam and F. Blaabjerg (Editors)
Volume 155 **Energy Generation and Efficiency Technologies for Green Residential Buildings** D. Ting and R. Carriveau (Editors)
Volume 157 **Electrical Steels, 2 Volumes** A. Moses, K. Jenkins, Philip Anderson and H. Stanbury
Volume 905 **Power System Protection, 4 Volumes**

Lightning Interaction with Power Systems

Volume 1: Fundamentals and modelling

Edited by
Alexandre Piantini

The Institution of Engineering and Technology

Published by The Institution of Engineering and Technology, London, United Kingdom

The Institution of Engineering and Technology is registered as a Charity in England & Wales (no. 211014) and Scotland (no. SC038698).

First published 2020

The Institution of Engineering and Technology
Michael Faraday House
Six Hills Way, Stevenage
Herts, SG1 2AY, United Kingdom

www.theiet.org

British Library Cataloguing in Publication Data
A catalogue record for this product is available from the British Library

ISBN 978-1-83953-090-6 (Hardback Volume 1)
ISBN 978-1-83953-091-3 (PDF Volume 1)
ISBN 978-1-83953-092-0 (Hardback Volume 2)
ISBN 978-1-83953-093-7 (PDF Volume 2)
ISBN 978-1-83953-094-4 (Hardback Volumes 1 and 2)

Typeset in India by MPS Limited

Contents

About the editor

Alexandre Piantini graduated in Electrical Engineering from the Federal University of Paraná in 1985 and got his master's and doctoral degrees from the Polytechnic School of the University of São Paulo in 1991 and 1997, respectively. He joined the University of São Paulo in 1986 and served as Director of Technological Development of the Institute of Energy and Environment (1998–2011), where he is Associate Professor and the Head of the Lightning and High Voltage Research Centre.

He has participated in 26 research projects related mainly to lightning and electromagnetic compatibility (EMC). He coordinated 21 of these projects, of which 15 were funded mainly by power companies and national agencies for research support. IEEE Senior Member since 2004, he was the Convener of the CIGRÉ WG C4.408 'Lightning Protection of Low-Voltage Networks' and member of various IEEE and CIGRÉ working groups. He is Associate Editor of the *IEEE Trans. Electromagnetic Compatibility*, *High Voltage* (IET), *Electrical Engineering* (Springer) and member of the Editorial Advisory Panel of the *Electric Power Systems Research* (Elsevier). He is member of the Steering Committee of the Int. Project on Electromagnetic Radiation from Lightning to Tall Structures. He was Deputy Editor-in-Chief of the *Journal of Lightning Research* (2005–15) and Associate Editor of *The Open Atmospheric Science Journal* (2008–13). He has given various invited lectures and courses related to lightning in universities and international conferences organized in Brazil, Sweden, Spain, Colombia, Russia and China.

Prof. Piantini is the Chairman of the Int. Symposium on Lightning Protection (SIPDA), vice-chairman of the Int. Conf. Grounding and Earthing & Int. Conf. Lightning Physics and Effects and member of scientific committees of various conferences such as the Int. Conf. Lightning Protection (ICLP). He is a founder member of the *Institute for Lightning Protection and Safety* (ILPS), Guest Professor of the Chongqing University, China, and member of the *IEEE Award Committee of the Sun & Grzybowski Award*. In 2018, he was the recipient of the ICLP *Rudolf Heinrich Golde Award*, 'for extraordinary theoretical and experimental achievements in lightning protection of power systems'. He is the author or co-author of four book chapters and over 150 scientific papers published in prestigious peer-reviewed journals or presented at international conferences with review board. He has given over 190 interviews to national and regional TV stations, radios, newspapers, etc., in topics related mainly to lightning.

Preface

The importance of improving the reliability and robustness of power systems makes protection of transmission and distribution lines against lightning-related effects a primary concern. This situation stems mainly from the increasing emphasis on overall power system efficiency, the continuous proliferation of equipment sensitive to short-duration voltage disturbances, the increasing level of consumer demand for power quality and the high economic losses associated with power quality issues. Numerous studies have been carried out in this area with a view to a better understanding of the phenomena involved and the identification of technically and economically viable solutions that provide effective improvement of the quality of energy supplied to consumers.

Lightning is particularly noteworthy in this context, as it is often responsible for a significant number of unscheduled outages of power transmission lines and distribution networks even in regions with relatively moderate ground flash densities. Besides, renewable electricity generation capacity has been increasing rapidly all over the world. Wind turbines are growing not only in number but also in size, leading to an increasing concern for lightning protection of wind power plants. Lightning is a major source of damages to wind turbines and can cause failures either hitting the turbines directly or inducing transients on the control systems that lead to equipment failure, malfunction or degradation. Photovoltaic (PV) systems may be vulnerable to lightning transients associated with both direct and nearby strikes, which can damage sensitive electronics or weaken the dielectric strength of the PV module insulation.

Lightning is a multidisciplinary subject, and the importance of understanding the physics of the phenomenon and its interaction with various objects and materials, as well as the need to effectively protect structures, systems, people and animals against its deleterious effects, has led to the existence of several books involving different lightning-related aspects. However, the current literature lacks a comprehensive work with specific focus on the interaction between lightning and electrical power systems that addresses in depth the lightning protection of transmission and distribution networks, including smart grids and renewable energy systems. This is the aim of this book, which contains well-established information and includes the most recent advancements in the field.

This book is intended primarily for a two-semester course for undergraduate and graduate students in energy and electrical engineering, but it can be used also for a one-semester or even shorter courses. It is also useful as reference for academic scientists, researchers and engineers in the areas of electrical engineering and physics, power systems consultants, and professionals from electric power companies involved in the fields of lightning protection, electromagnetic compatibility (EMC), renewable energy systems and smart grids. The secondary readership consists of professionals from telecommunication companies and manufacturers of power equipment.

This book is divided into two volumes. The chapters in Volume 1 describe and discuss the main concepts, fundamentals and models necessary to understand and evaluate the interaction between lightning and electrical systems.

The first chapter is concerned with an assessment of how global lightning may respond to global climate change. In Chapter 2, basic lightning terminology is introduced and the main lightning processes are described. The 'classical' distributions of lightning parameters needed in engineering applications are reviewed along with the distributions based on more recent direct current measurements. Correlations between the parameters are discussed and mathematical expressions used to represent lightning current waveforms are reviewed. Chapter 3 introduces the reader to the various concepts used to construct engineering return stroke models. After describing the most important models, it provides a review of the basic features of lightning electromagnetic fields and presents methods for their calculation, including the horizontal electric field associated with return stroke over finitely conducting ground. Chapter 4 provides the basis for calculating ground flash densities, details of techniques used by modern lightning location systems (LLSs), examples of well-established LLSs in different parts of the world and methods used to validate the performance characteristics of LLSs.

In Chapter 5, the physical process and engineering models of lightning attachment to overhead power lines are described in detail, and a general procedure for the estimation of lightning incidence to overhead power lines is presented. Chapter 6 presents the coupling of lightning electromagnetic fields to overhead and underground lines based on the transmission line approximation, whereas Chapter 7 addresses the lightning response of grounding electrodes. Chapter 8, which deals with surge protective devices, presents the most common definitions, characteristics, operating mechanisms, classifications and applications of devices used in transmission and distribution networks, including low-voltage (LV) systems. Chapters 9 and 10 present and discuss models of the most important power transmission and distribution (medium-voltage (MV) and LV) system components for simulations of lightning electromagnetic transients.

The second volume, devoted to the applications, contains Chapters 1–12, which cover lightning protection of various systems, including structures and buildings, transmission and distribution networks, renewable energy systems and smart grids.

Chapter 1 is devoted to the application of the Monte Carlo method to lightning protection and insulation coordination practices and describes also the application of the stratified sampling technique to reduce the computational effort usually required. The effect of lightning on the insulation performance of substation equipment is dealt with in Chapter 2, which includes also the evaluation of the failure rates of gas-insulated switchgear and transformers. Chapter 3 organizes the lightning interactions with power transmission lines from the simple consequences of a direct stroke attachment to an unshielded line to the complex consequences of a stroke attachment to a shielded line with multiple ground wires, including the effects from phases protected with line arresters. It builds on the information in previous chapters to develop important measures in transmission line lightning performance.

Chapter 4 deals with the lightning impacts on MV power distribution systems and discusses the effects of the most important parameters on the overvoltages, as well as the effectivenesses of the main protective measures that can be applied to improve the line lightning performance. A procedure for estimating the mean annual number of line flashovers of overhead lines is presented and the lightning performances of lines with different protective measures are compared. In Chapter 5, devoted to the lightning interaction with LV power distribution networks, the major mechanisms by which lightning overvoltages can be produced are explained and the general surge characteristics are evaluated. The effectiveness of the installation of secondary arresters along the network in protecting the LV side of transformers and consumers' entrances is also discussed. Chapter 6 is dedicated to the lightning protection of common structures, including their installations and content, and persons as well. Such protection requires the combination of external and internal countermeasures, which are also discussed in the chapter.

A broad view of 'lightning protection' finds many smart grid applications of real-time lightning information in proactive protection strategies. After presenting the history of power system technologies and describing the roles that lightning research plays in successful integration of digital technologies into electric power systems, Chapter 7 discusses lightning-related digital recording technologies and addresses the lightning protection of smart grid sensors. Chapters 8 and 9 focus on the lightning protection of renewable energy systems. Chapter 8 gives an introduction to wind power generators and their components from the perspective of lightning protection, as well as an overview of lightning occurrence in relation to wind turbines. It presents the mechanisms of lightning damage to wind turbines, their classification and statistics, discusses the protection of the most sensitive components and describes the mechanisms whereby lightning surges invade a wind farm through a lightning-struck wind turbine. Chapter 9 deals with PV systems and gives a brief introduction to solar radiation, PV cells, modules and the associated effects of shading. Off-grid and grid-connected PV systems are described and the common configurations of external and internal lightning protection systems are discussed.

Chapter 10, which is about measurements of lightning currents and voltages, describes various types of sensors and discusses their application in power systems. Chapter 11 presents the fundamentals of the finite-difference time-domain method and reviews the application of the method to the analysis of lightning electromagnetic fields and lightning-caused surges in various systems. Chapter 12 describes two of the most adopted software tools for the evaluation of the lightning performance of transmission and distribution lines, namely, FLASH and LIOV-EMTP, together with some application examples.

The chapters follow a logical order and ideally should be read sequentially by a beginner reader, but they are basically independent from each other and a reader interested in a specific topic can go directly to the relevant chapter.

Alexandre Piantini

Acknowledgements

I would like to express my sincere thanks to my colleagues and friends, authors of the chapters, for their dedication and esteemed contributions. My special thanks go to Prof. Carlo Alberto Nucci, Prof. Farhad Rachidi, Prof. Marcos Rubinstein, Prof. Vernon Cooray, Prof. Vladimir A. Rakov, and Prof. William A. Chisholm, for the valuable discussions and continuous support.

I am also grateful to Dr. Christoph von Friedeburg, Senior Books Commissioning Editor at the IET, for the interesting discussions, and to Ms. Olivia Wilkins, Assistant Editor at the IET, for her kindness, sincerity and patience to deal with submission delays. Working with her was indeed a great pleasure.

I am indebted to all my former and current students, postdocs and colleagues, and specially thank Miss Michele N.N. Santos, PhD student, for her precious help during the organization of the book.

Finally, I would like to thank my parents, Farley and Elza, my 102-year-old grandmother Nair, my sisters Andrea and Barbara, and my nieces Angel, Farly and Isabella, for their unlimited love and support and for always bringing joy to my life.

Alexandre Piantini

About the authors

Marina Bernardi was born in Milan, Italy, in 1966. She received the doctoral degree in High Energy Physics at the University of Milan, Italy, in 1992. Her first experiences saw her engaged in research at the CERN (Center Européen pour la Recherche Nucleare) in Geneva for the LEP-Large Electron Collider. She then worked for CESI spa, as researcher and project manager for various activities of lightning impact assessment on medium and high voltage power lines, on laboratory experiments for atmospheric discharges, on model simulations of lightning propagation and on models of induction of the electromagnetic field generated by lightning. In the years from 2000 to 2005, she was responsible for various Public research activities, related to the lightning impact on the National Electricity Grid. Since 1999, she has been the technical manager for the Italian Lightning Detection Network (SIRF) owned by CESI. She had numerous collaborations with Italian Universities for research and educational activities; numerous international publications on the subject of lightning, lightning physics and lightning detection. She has been an Invited Scientific Lecturer for post-graduate courses and national public conferences on atmospheric science, lightning detection and lightning physics, and a member of the board of EUCLID – European Cooperation for Lightning Detection. Over the years, she has been a member of several WGs in CIGRÉ, IEC and CENELEC on lightning various aspects. Currently, she is Secretary of CENELEC-TC81X and IEC-TC81 for 'Protection against Lightning'.

Vernon Cooray (IEEE Fellow) is the professor responsible for the postgraduate studies in the field of Atmospheric Electrical Discharges at Uppsala University. He is also the professor-in-charge of the High Voltage Laboratory at Uppsala University. From 1999 until 2003, he was the head of the Division for Electricity.

He has authored and co-authored about 375 scientific papers. He is the editor of four books, *The Lightning Flash* (2003), *Lightning Protection* (2009), *Lightning Electromagnetics* (2012) and *The Lightning Flash* (second and expanded edition, 2014) published by the Institution of Engineering and Technology (former IEE), London, UK. He is also the author of the book *Introduction to Lightning*, published by Springer in 2014. For this book, he received the best book award from the association of librarians in 2015.

Prof. Vernon Cooray is the president of ICLP and the recipient of the Berger award for his contributions to lightning research. In 2015, he was awarded a DSc degree from the University of Colombo, Sri Lanka, for his contributions to lightning research.

Alberto De Conti was born in Belo Horizonte, Brazil. He received the BSc, MSc and doctoral degrees in electrical engineering from the Federal University of Minas Gerais (UFMG), Brazil, in 2000, 2001 and 2006, respectively.

In 2006, he was a guest researcher in the Division for Electricity and Lightning Research, Uppsala University, Sweden. Currently, he is Associate Professor of the Department of Electrical Engineering of UFMG. His research interests include electromagnetic transients, modelling of power system components for transient studies, lightning interaction with electrical systems and electromagnetic compatibility.

Jinliang He was born in Changsha, China, in 1966. He received the BSc degree in electrical engineering from Wuhan University of Hydraulic and Electrical Engineering, Wuhan, China, in 1988, the MSc degree in electrical engineering from Chongqing University, Chongqing, China, in 1991, and the PhD degree in electrical engineering from Tsinghua University, Beijing, China, in 1994. He became Lecturer in 1994 and Associate Professor in 1996 in the Department of Electrical Engineering, Tsinghua University. From 1997 to 1998, he was a Visiting Scientist with the Korea Electrotechnology Research Institute, Changwon, Korea. In 2001, he was promoted to Professor at Tsinghua University. During January 2014 to January 2015, he was a visiting professor in Department of Electrical Engineering, Stanford University, Stanford, CA, USA. Currently, he is the Chair of the High Voltage Research Institute, Tsinghua University. His research interests include advanced power transmission technology, nanodielectrics, electromagnetic sensors and big data application. He is the author of 8 books and 500 technical papers. He was selected as an IEEE Fellow in 2007, and he is also the recipient of the Technical Achievement Award from IEEE EMC Society in 2010, the Rudolf Heinrich Golde Award from International Conference on Lightning Protection in 2016 and 2018 IEEE Herman Halperin Electric Transmission and Distribution Award.

Holger Heckler was born in the city of Bückeburg (Germany, Lower Saxony) in 1965. He graduated from the University of Applied Sciences of Bielefeld (Germany) with a diploma (Dipl-Ing '92) in Electrical Engineering/Power Engineering and spent his career working for industrial control companies such as Hartmann & Braun, Elsag Bailey, ABB Utility Automation and Phoenix Contact Germany. He has written scientific papers for the International Conference on Lightning Protection (ICLP 2006, 2008, 2010, 2018), for the International Symposium on Lightning Protection (SIPDA 2007, 2011), for the Australian Earthing, Lightning & Surge Protection Forum (2011), for the South African Earthing, Lightning & Surge Protection Conference (2011), for the International Conference & Expo on Emerging Technologies for a Smarter World (CEWIT 2011), for the Australian Earthing, Lightning & Surge Protection Forum (2013) and for German VDE/ABB conference on Lightning Protection (2007, 2008, 2017). He frequently writes technical papers on lightning and surge protection and on general electrical engineering. His technical papers have been published in 16 different countries and he has authored several bilingual dictionaries on Electrical Engineering. In addition, he authored several comprehensive bilingual vocabularies for a piece of well-known German vocabulary learning software (Langenscheidt Vokabeltrainer). Holger Heckler has carried out training seminars and lectures on lightning and surge protection, on general electrical engineering, on electromagnetic compatibility, on electrical safety and on computer-related topics (network administration, desktop publishing, digital image processing) in 25 different countries in Europe, North America, Africa, Oceania and Asia. He is a member of standardization groups on 'DC in Low-Voltage Power Distribution Systems' [VDE, German Association for Electrical, Electronic & Information Technologies] and he supports VDE, CENELEC and IEC standardization groups on lightning and surge protection.

Pantelis N. Mikropoulos was born in Kavala, Greece, in 1967. He received the MEng and PhD degrees in electrical and computer engineering from Aristotle University of Thessaloniki (AUTh), Thessaloniki, Greece, in 1991 and 1995, respectively. He held postdoctoral positions at AUTh and the University of Manchester, Manchester, UK. He was a Senior Engineer with Public Power Corporation SA, Athens, Greece. In 2003, he joined the faculty of AUTh, where currently he serves as Professor of High Voltage Engineering and the Director of the High Voltage Laboratory. His research interests include the broad area of high-voltage engineering with emphasis given to air and surface discharges, electric breakdown, lightning protection and earthing, as well as insulation coordination for power systems.

Amitabh Nag received the MS and PhD degrees in electrical and computer engineering in 2007 and 2010, respectively, from the University of Florida (UF), Gainesville. From 2010 to 2011, he was employed as a postdoctoral research associate at the International Center for Lightning Research and Testing at UF. From 2011 to 2016, he worked as a scientist and product manager at Vaisala Inc., Boulder, Colorado. Currently, Dr Nag is an assistant professor in the Aerospace, Physics and Space Sciences Department at Florida Institute of Technology, Melbourne, Florida. Dr Nag is the author or co-author of over 90 papers and technical reports on various aspects of lightning electromagnetics. Dr Nag is a Senior Member of the IEEE, and a member of IEC and CIGRÉ working groups, the American Meteorological Society Scientific and Technological Activities Commission on Atmospheric Electricity and the American Geophysical Union.

Acácio Silva Neto was born in Ribeirão Preto, Brazil, in 1977. He graduated in Electrical Engineering from the São Paulo State University (UNESP) in 2000 and obtained a master's degree in Electrical Engineering from the Polytechnic School of the University of São Paulo (USP) in 2004. In 2012, he received the PhD degree in Science from USP. He joined the University of São Paulo in 2001 as a researcher at the Lightning and High Voltage Center (CENDAT) of the Institute of Energy and Environment (IEE/USP). His interests involve the effects of lightning strikes on electrical systems, lightning protection and lightning attachment processes.

Carlo Alberto Nucci graduated with honours in electrical engineering from the University of Bologna, Bologna, Italy, in 1982. He is a Full Professor and the Head of the Power Systems Laboratory of the Department of Electrical, Electronic and Information Engineering 'Guglielmo Marconi', University of Bologna. He is the author or co-author of over 370 scientific papers published in peer-reviewed journals or in proceedings of international conferences. Prof. Nucci is a Fellow of the IEEE and of the International Council on Large Electric Systems

(CIGRÉ), of which he is also an Honorary member, and has received some best paper/technical international awards, including the CIGRÉ Technical Committee Award and the ICLP Golde Award. From January 2006 to September 2012, he served as Chairman of the CIGRÉ Study Committee C4 (System Technical Performance). He has served as IEEE PES Region 8 Rep in 2009 and 2010. Since January 2010, he has served as the Editor-in-Chief of the *Electric Power Systems Research* journal (Elsevier). He has served as the President of the Italian Group of the University Professors of Electrical Power Systems (GUSEE) from 2012 to 2015. He is an Advisor of the Global Resource Management Program of Doshisha University, Kyoto, Japan, supported by the Japanese Ministry of Education and Science, and has represented PES in the IEEE Smart City Initiatives Program since 2014. Prof. Nucci is Doctor *Honoris Causa* of the University Politehnica of Bucharest and a member of the Academy of Science of the Institute of Bologna.

Alexandre Piantini graduated in Electrical Engineering from the Federal University of Paraná in 1985 and got his master's and doctoral degrees from the Polytechnic School of the University of São Paulo in 1991 and 1997, respectively. He joined the University of São Paulo in 1986 and served as Director of Technological Development of the Institute of Energy and Environment (1998–2011), where he is Associate Professor and the Head of the Lightning and High Voltage Research Centre.

He has participated in 26 research projects related mainly to lightning and EMC. He coordinated 21 of these projects, of which 15 were funded mainly by power companies and national agencies for research support. IEEE Senior Member since 2004, he was the Convener of the CIGRÉ WG C4.408 'Lightning Protection of Low-Voltage Networks' and member of various IEEE and CIGRÉ working groups. He is Associate Editor of the *IEEE Trans. Electromagnetic Compatibility*, *High Voltage* (IET), *Electrical Engineering* (Springer) and member of the Editorial Advisory Panel of the *Electric Power Systems Research* (Elsevier). He is member of the Steering Committee of the Int. Project on Electromagnetic Radiation from Lightning to Tall Structures. He was Deputy Editor-in-Chief of the *Journal of Lightning Research* (2005–15) and Associate Editor of *The Open Atmospheric Science Journal* (2008–13). He has given various invited lectures and courses related to lightning in universities and international conferences organized in Brazil, Sweden, Spain, Colombia, Russia and China.

Prof. Piantini is the Chairman of the Int. Symposium on Lightning Protection (SIPDA), vice-chairman of the Int. Conf. Grounding and Earthing & Int. Conf. Lightning Physics and Effects and member of scientific committees of various conferences such as the Int. Conf. Lightning Protection (ICLP). He is a founder member of the *Institute for Lightning Protection and Safety* (ILPS), Guest

Professor of the Chongqing University, China, and member of the *IEEE Award Committee of the Sun & Grzybowski Award.* In 2018, he was the recipient of the ICLP *Rudolf Heinrich Golde Award*, 'for extraordinary theoretical and experimental achievements in lightning protection of power systems'. He is the author or co-author of 4 book chapters and over 150 scientific papers published in prestigious peer-reviewed journals or presented at international conferences with review board. He has given over 190 interviews to national and regional TV stations, radios, newspapers, etc., in topics related mainly to lightning.

Georgij V. Podporkin was born on 26 August 1950. He received his BS and PhD degrees in Electrical Engineering from the St. Petersburg Technical University in 1973 and 1977, respectively. In 1989, he received degree of Doctor of Science for thesis 'Compact extra high voltage overhead transmission lines with increased natural capacity'.

From 1973 until 1991, he worked as a research scientist at the Extra High Voltage Laboratory of the St. Petersburg Technical University. During 1992–95, he was a Scientific Consultant of CEPEL in Rio de Janeiro, Brazil.

With his co-workers, he invented and developed lightning arresters of new types: the so-called Long Flashover Arresters and Multi Chamber Arresters. His fields of interest are lightning protection and insulation of overhead transmission and distribution lines and compact lines with increased natural capacity. He is co-owner and Director of Research of the Streamer Electric Company and a part-time professor at the St. Petersburg Technical University.

Dr Podporkin is also a Senior member of IEEE.

Farhad Rachidi (M'93–SM'02–F'10) received the MS degree in electrical engineering and the PhD degree from the Swiss Federal Institute of Technology, Lausanne, Switzerland, in 1986 and 1991, respectively. He was with the Power Systems Laboratory, Swiss Federal Institute of Technology, until 1996. In 1997, he joined the Lightning Research Laboratory, University of Toronto, Toronto, ON, Canada. From 1998 to 1999, he was with Montena EMC, Rossens, Switzerland. He is currently a Titular Professor and the Head of the EMC Laboratory with the Swiss Federal Institute of Technology, Lausanne,

Switzerland. He has authored or co-authored over 190 scientific papers published in peer-reviewed journals and over 400 papers presented at international conferences.

Dr Rachidi is currently a member of the Advisory Board of the IEEE Transactions on Electromagnetic Compatibility and the President of the Swiss National Committee of the International Union of Radio Science. He has received numerous awards including the 2005 IEEE EMC Technical Achievement Award, the 2005 CIGRÉ Technical Committee Award, the 2006 Blondel Medal from the French Association of Electrical Engineering, Electronics, Information Technology and Communication (SEE), the 2016 Berger Award from the International Conference on Lightning Protection, the 2016 Best Paper Award of the IEEE Transactions on EMC and the 2017 Motohisa Kanda Award for the most cited paper of the IEEE Transactions on EMC (2012–16). In 2014, he was conferred the title of Honorary Professor of the Xi'an Jiaotong University in China. He served as the Vice-Chair of the European COST Action on the Physics of Lightning Flash and its Effects from 2005 to 2009, the Chairman of the 2008 European Electromagnetics International Symposium, the President of the International Conference on Lightning Protection from 2008 to 2014, the Editor-in-Chief of the Open Atmospheric Science Journal (2010–12) and the Editor-in-Chief of the IEEE Transactions on Electromagnetic Compatibility from 2013 to 2015. He is a Fellow of the IEEE and of the SUMMA Foundation, and a member of the Swiss Academy of Sciences.

Vladimir A. Rakov received the MS and PhD degrees in electrical engineering from the Tomsk Polytechnical University, Russia, in 1977 and 1983, respectively. He is currently Professor in the Department of Electrical and Computer Engineering, University of Florida, Gainesville, and Co-Director of the International Center for Lightning Research and Testing (ICLRT). He is the author or co-author of 4 books and over 700 other publications on various aspects of lightning, with about 300 papers being published in peer-reviewed journals. Dr Rakov is a Fellow of four major professional societies, the IEEE, the American Meteorological Society, the American Geophysical Union and the Institution of Engineering and Technology (formerly IEE). He is also a recipient of Karl Berger Award for distinguished achievements in lightning research, developing new fields in theory and practice, modelling and measurements (2012) and Toshio Takeuti Award for outstanding contribution to worldwide recognition of winter lightning (2017). In 2015, he was awarded Honorary Doctoral degree by the Institute of Applied Physics of the Russian Academy of Sciences (IAP RAS).

Marcos Rubinstein (M'84–SM'11–F'14) received the master's and PhD degrees in electrical engineering from the University of Florida, Gainesville, FL, USA, in 1986 and 1991, respectively.

In 1992, he joined the Swiss Federal Institute of Technology, Lausanne, Switzerland, where he was involved in the fields of electromagnetic compatibility and lightning. In 1995, he was with Swisscom, where he worked in numerical electromagnetics and EMC. In 2001, he moved to the University of Applied Sciences of Western Switzerland HES-SO, Yverdon-les-Bains, where he is currently a Full Professor, the head of the advanced Communication Technologies Group and a member of the IICT Institute Team. He is the author or co-author of more than 200 scientific publications in peer-reviewed journals and international conferences. He is also the co-author of seven book chapters. He is the Chairman of the International Project on Electromagnetic Radiation from Lightning to Tall structures, served as the Editor-in-Chief of the Open Atmospheric Science Journal and currently serves as Associate Editor of the IEEE Transactions on Electromagnetic Compatibility.

Prof. Rubinstein received the best Master's Thesis award from the University of Florida. He received the IEEE achievement award and he is a co-recipient of the NASA's Recognition for Innovative Technological Work award. He is also the recipient of the ICLP Karl Berger Award. He is a Fellow of the IEEE and of the SUMMA Foundation, a member of the Swiss Academy of Sciences and of the International Union of Radio Science.

Wolfgang Schulz was born in Vienna, Austria, in 1966. He received the Dipl-Ing degree in electrical engineering and the PhD degree from the Technical University of Vienna in 1992 and 1997, respectively. In 1992, he joined the Technical University of Vienna as Assistant Professor. Currently, he is with the Austrian Lightning Detection & Information System, with a research focus on the performance of lightning location systems and E-field measurements of lightning discharges. He has authored or co-authored more than 100 scientific papers published in peer-reviewed journals and presented at international conferences.

Miltom Shigihara received the BS degree in Electrical Engineering from Polytechnic School in 2002, MSc degree in Energy in 2005 and PhD degree in Science in 2015, all from the University of São Paulo (USP), São Paulo, Brazil. He joined at the Institute of Energy and Environment of the University of São Paulo (IEE/USP) in 2002 as engineer, developing short circuit tests at the High Current Laboratory. Since 2013, he has developed researches in the lightning protection and high-voltage area. He is the author or co-author of about 30 scientific papers presented at international conferences or published in international peer-reviewed journals. Currently, he is researcher at the Lightning and High Voltage Research Center (CENDAT/USP) and engineer at the High Voltage Laboratory of IEE/USP.

Fernando H. Silveira was born in São Paulo, Brazil. He received the BSc, MSc and doctoral degrees in electrical engineering from the Federal University of Minas Gerais (UFMG), Belo Horizonte, Brazil, in 2000, 2001 and 2006, respectively.

He is with the Department of Electrical Engineering of UFMG since 2009, where he is currently an Associate Professor. He is also Associate Researcher with the Lightning Research Center (LRC). His research interests include lightning performance of transmission lines, lightning interaction with electrical systems, lightning and electromagnetic modelling of power system components.

Silverio Visacro received the BSc and MSc degrees in Electrical Engineering from Federal University of Minas Gerais (UFMG), Belo Horizonte, Brazil, in 1980 and 1984, respectively, and the PhD degree from the University of Rio de Janeiro, Brazil, in 1992.

He is an IEEE Fellow and a Full Professor of the Electrical Engineering Department of UFMG. He is also the Head of the Lightning Research Center (LRC), where he leads a team of about 30 researchers in the development of investigations and projects on electromagnetic modelling, lightning physics, grounding and lightning protection, with a special interest on the lightning performance of transmission lines.

He is the author of two books dedicated to Lightning and Grounding, respectively, and of three book chapters in the same fields. He is also the author or co-author of more than 400 scientific papers published in peer-reviewed journals and presented at international conferences.

In 2016, he was distinguished with the IEEE Motohisa Kanda Award and received the Karl Berger Award from ICLP (Int. Conference on Lightning Protection). He has been frequently invited as Lecturer and Keynote speaker of prestigious international conferences, such as ICLP2018 and APL2015 and ILDC 2016. He has been the Chairperson of GROUND & LPE Conference and Vice-Chair of SIPDA Symposium.

Prof. Visacro is an active CIGRÉ member and convened Working Groups C4.06 (Response of Grounding Electrodes to Lightning Currents) and C4.33 (Impact of Soil-Parameter Frequency Dependence on the Response of Grounding Electrodes and on the Lightning Performance of Electrical Systems), in addition to other participations as effective Member, as in WG C4.23 (Lightning Performance of Transmission Lines).

He has led more than 20 investigation projects, most of them contracted by power utilities, and is responsible for a successful relationship between his university and industry.

Prof. Visacro has been working as special consultant on Lightning Protection, Grounding and Transmission Lines.

Dr-Ing Martin Wetter was born in Münster (Westf.), Germany, in 1967. He received his diploma as Certified Engineer (Dipl-Ing) from the University of Dortmund, Germany, in 1992. In 1992, he started as a development engineer for car ignition systems. After one year as development engineer, he came back to the University of Dortmund as scientific engineer. His thesis was published in 1997 and summarizes 5 years of research on the transient behaviour of non-linear systems, especially on ferromagnetic materials. He joined the company Phoenix Contact in Blomberg, Germany, in 1997 as the head of development for Surge Protective Devices (SPDs). Since 2012, he is Vice President of Phoenix contact and the head of the Business Unit Surge Protection TRABTECH, globally responsible for the development, marketing and production of SPDs. Dr Wetter is member of several national and international committees for standardization of lightning protection and overvoltage protection.

Earle Williams is a physical meteorologist at MIT with a lifelong interest in lightning. He has led radar field experiments in all three centres of active tropical lightning: northern Australia of the 'Maritime Continent', the Amazon rainforest of South America and the semiarid Sahel of West Africa. During the initial Australian campaign in 1990, he hit on the idea to use lightning as a sensitive 'global thermometer'. This suggestion served to renew interest in the practical measurement of the global electrical circuit, a natural and inexpensive framework for monitoring global change, formed by the conductive Earth and conductive ionosphere as boundaries. The AC version of the global circuit – known as the Earth's Schumann resonances – is a global electromagnetic phenomenon maintained continuously by lightning and with a fundamental frequency of 8 Hz. Williams established the MIT field station for monitoring Schumann resonances in West Greenwich, Rhode Island, and he and his students then established that the extraordinarily energetic lightning flashes known as Q-bursts that single-handedly ring the Schumann resonances were also responsible for luminous sprites high in the mesosphere. Methods were developed for locating and characterizing such flashes on a global basis from the single receiving station in Rhode Island. In the last few years, Williams has taken on the most difficult aspect of this phenomenon: the 'background' resonances that are sustained by contributions from all lightning flashes worldwide. Multiple receiving stations are used to measure the global lightning activity on a continuous basis and in absolute units for the first time. This work constitutes an international collaboration with scientists and engineers in Antarctica, China, Czech Republic, England, France, Hungary, India, Japan, Poland, Russia, Scotland, Spain, Sweden, Tahiti and Taiwan.

Chapter 1
Lightning and climate change

Earle R. Williams[1]

Lightning is a widely recognized source of damage and disruption to electrical power systems worldwide. The climate is changing, with both natural and anthropogenic origins. This chapter is concerned with the response of lightning to changes in temperature and aerosol loading of the atmosphere that are expected to accompany climate change. In the present climate, lightning is shown to increase with both temperature and with the boundary layer populations of cloud condensation nuclei (CCN). In a future climate characterized by the continued consumption of fossil fuels, the threat from lightning is expected to increase.

1.1 Introduction

Lightning is a natural phenomenon originating in the high voltage differences encountered in thunderstorms (up to one billion volts) and exhibiting currents as large as hundreds of kiloamperes. The lightning threat to worldwide energy infrastructure is widely recognized [1]. Lightning dominates the damage to electrical/electronic equipment in homes, commercial installations, and industrial facilities. The total cost is dependent on both the total exposure and the worldwide lightning activity. Both these contributions to cost are increasing with time as a result of a growing infrastructure worldwide.

Much attention is given today to extreme events in a warmer climate (e.g., [2]). This attention serves to place lightning at center stage, to the extent that lightning is a manifestation of the extreme form of moist convection—largest clouds, strongest updrafts, and most hazardous precipitation. One can expect volatile behavior in the tail of any distribution, and for this reason alone, the recent selection of lightning as a climate variable [3] is most appropriate.

This chapter is concerned with an assessment of how global lightning may respond to global climate change. This turns out to be a difficult problem. Some understanding of this difficulty is derived from the limits of our current ability to understand the general behavior and global distribution of lightning *in the present*

[1]Parsons Laboratory, Department of Civil and Environmental Engineering, Massachusetts Institute of Technology (MIT), USA

climate. One particular challenge is that both temperature and aerosol play important roles in lightning activity in the present climate. Accordingly, this aspect shall be the point of departure in this chapter.

A global climatology for lightning measured from optical sensors on satellites in space is shown in Figure 1.1(a). This integration is based on nearly two decades of observation. The most conspicuous feature of the global distribution is the strong preference of lightning for land, with a 10- to 20-fold contrast between land and ocean (see also [4]). The leading factor in this contrast is the number of thunderstorms, with a secondary contribution from a greater flash rate per storm in the continental case [6]. Since the majority of the world's population density and infrastructure is also over land, this land dominance aggravates the lightning threat. However, since the lightning is also strongly centered on equatorial regions (for reasons that are soon to be discussed) where population and infrastructure are reduced in comparison to higher latitudes in the northern hemisphere, the overall threat is ameliorated to some extent on a global basis.

Three major continental zones straddling the equator—the Americas, Africa, and the Maritime Continent (southeast Asia, Indonesia, and northern Australia)—dominate the global lightning activity. The same three zones are also the major players in the Earth's global electrical circuit [7,8]. From a climate perspective, the three major continental zones have previously been ranked in their continentality [9], with Africa leading, followed by America, and with the Maritime Continent closest to oceanic behavior. Both the total lightning activity and the aerosol burden in these three regions follow the same order, whereas rainfall amounts follow the reverse order.

The energy involved with global lightning activity is derived from the much larger latent heat released when water vapor condenses. Rainfall is also a product of the condensation process. The global distribution of rainfall (Figure 1.1(b)), more readily measured than condensation, provides some global measure of the distribution of latent heat release. In marked contrast with the lightning distribution, rainfall and latent heat release is as prevalent over the ocean than over the land, but as will be shown by the evidence in this chapter, the vertical profile of latent heat release is markedly different between land and ocean. Both thermodynamic and aerosol effects are at play in this difference, by virtue of their impact on thunderstorm updrafts, and both are important in considerations of how lightning will respond to climate change.

The traditional explanation [10,11] for the contrast in lightning activity between land and ocean, aptly illustrated in Figure 1.1(a), is based on thermodynamics: land is hotter and more unstable to vertical motion. In recent years, a growing body of evidence [12–23] has shown that the atmospheric aerosol, and in particular the CCN that provide the embryos for cloud droplets, is also playing a key role in this contrast. The global aerosol population shown in Figure 1.1(c) also shows a prominent land–ocean contrast, with more polluted conditions over the land. Since climate change will invariably involve change in both thermodynamics and aerosol, the physical basis for both controls needs to be explored. The treatment of the thermodynamic part appears in Section 1.4 and for the aerosol part in Section 1.5. Ahead of these discussions, some attention is warranted on the workings of the thunderstorm in the next Section 1.2.

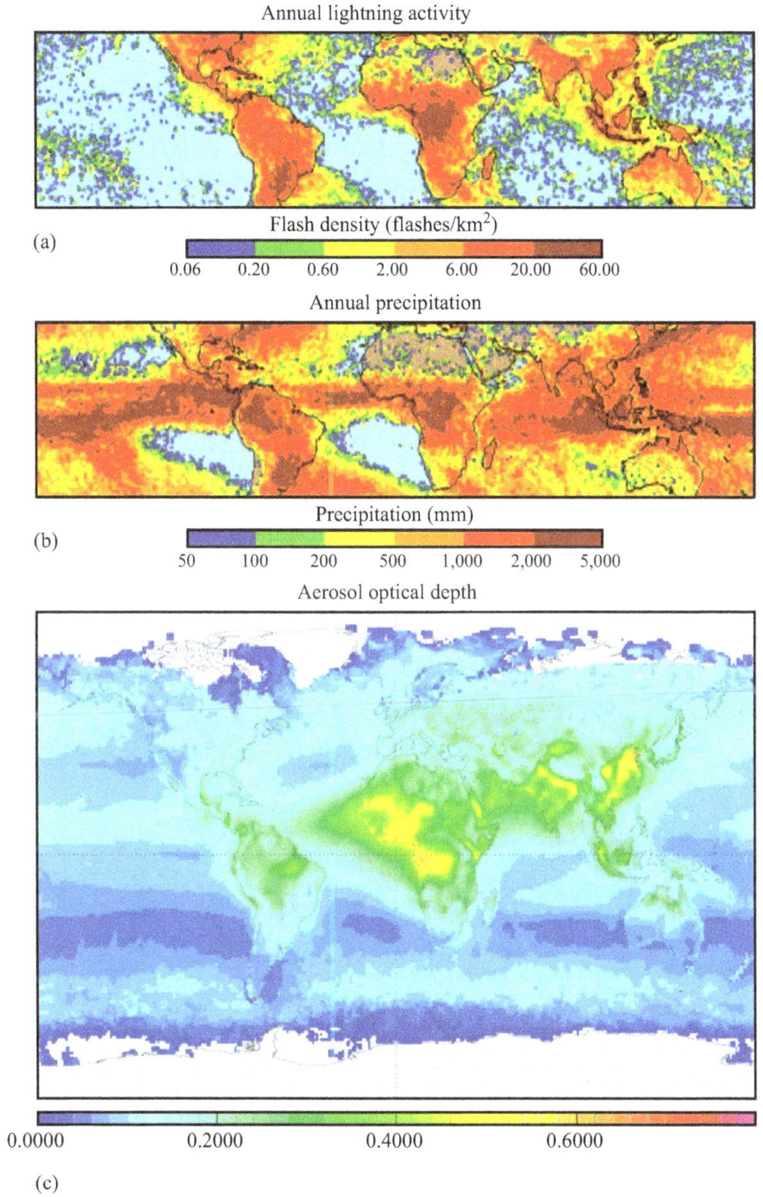

Figure 1.1 Global climatologies of (a) lightning flash density (lightning imaging sensor), (b) rainfall (NASA TRMM), and (c) aerosol concentration (as measured with satellite aerosol optical depth). Adapted from [5]

1.2 Basics of thunderstorm electrification and lightning

The deepest and most vigorous convective clouds in the atmosphere are thunderstorms, and extend deeply into the cold portion (defined here at $T < 0$ °C) of the atmosphere (Figure 1.2). Considerable evidence has accrued [24,25] that the mechanism for charging a thunderstorm and for the production of lightning flashes is based on the collisions of two kinds of particles: small ice crystals and larger graupel particles. Both ice particles are the product of mixed-phase conditions involving water substance in all three thermodynamic phases: vapor, liquid, and solid (ice). The liquid phase at "cold" temperature ($T < 0$ °C) is referred to as supercooled water. The ice crystals form by diffusion of water vapor in the so-called Bergeron process based on the asymmetry in equilibrium vapor pressure between liquid and ice. The mass of ice crystals increases at the expense of the supercooled cloud water. The graupel particles grow by the accretion of supercooled cloud droplets, which freeze in contact with the graupel surface. The collisions between graupel particles and ice crystals result in the transfer of negative charge to graupel and positive charge to the crystals, by a mechanism at the

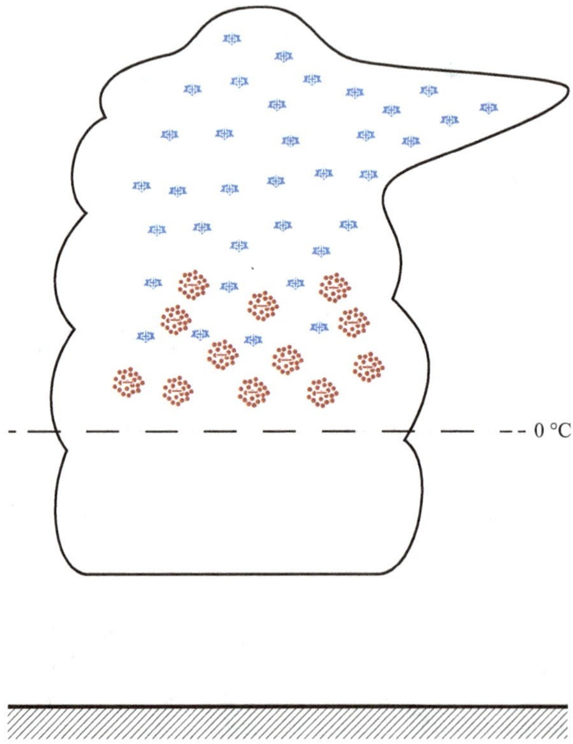

Figure 1.2 The mechanism of the thunderstorm: storm cloud with colliding ice particles and positive electric dipole

molecular scale that has long eluded scientists [25,26]. The descent of negative graupel with respect to positive ice crystals under gravity sets up the macroscopic positive dipole of the thunderstorm (Figure 1.2). This hydrometeor-based mechanism of differential charge separation (based on different fallspeeds of ice crystals and graupel) is immune to the effects of turbulence, which often shows a strong presence in thunderstorms.

1.3 Thermodynamic control on lightning activity

A number of basic thermodynamic parameters as well as relationships from physical meteorology deserve discussion when one considers possible changes in lightning in a changing climate. These items are here addressed in turn.

1.3.1 Temperature

The most commonly used thermodynamic parameter in global climate change is the temperature of surface air, typically measured at "screen level." The formal meteorological quantity is "dry bulb temperature" to distinguish it from "wet bulb temperature" and "dew point temperature" both of which involve the water vapor content of the air. Traditional estimates of the global mean temperature [27,28] and global warming involve averages of 4,000–6,000 thermometer readings of dry bulb temperature over the Earth's surface. This temperature parameter is also a key factor in other thermodynamic quantities of interest here: saturation water vapor concentration, convective available potential energy (CAPE) and cloud base height (CBH), described in greater detail below.

1.3.2 Dew point temperature

The dew point temperature T_d is a direct measure of the water vapor concentration in surface air and is typically measured by cooling a metal surface to a temperature at which condensation, or "dew," appears. For water-saturated conditions (i.e., inside a cloud), the dew point temperature is equal to the dry bulb temperature T. In contrast, in a dry desert environment, the dew point temperature can be several tens of °C lower than the dry bulb temperature.

1.3.3 Water vapor and the Clausius–Clapeyron relationship

The working substance of a thunderstorm is water vapor. Energy is released when water vapor rises and condenses to form cloud. The latent heat of condensation L_v is 2.5×10^6 J/kg of water, sufficient energy to raise the condensate 250 km against gravity if this transformation took place with perfect efficiency.

The water vapor concentration in the atmosphere in a condition of thermodynamic equilibrium is controlled by temperature in an exponential dependence known as the Clausius–Clapeyron relation. In differential form:

$$de^*(T)/dT = L_v e^*/R_v T^2 \tag{1.1}$$

where T is absolute temperature (K), $e^*(T)$ is the saturation vapor pressure of water vapor, and R_v is the gas constant for water vapor (461 J/kg/K). The integral form of this relationship (see e.g., [29]) in terms of the water vapor mixing ratio is shown graphically in Figure 1.3. As a rough rule of thumb, the equilibrium water vapor concentration $e^*(T)$ doubles for every 10 °C of temperature increase. This result has much to say about the sparsity of thunderstorms in polar regions and their predominance in tropical latitudes. A change in temperature of 30 °C amounts to nearly an order of magnitude difference in available water vapor. A quantitative consideration of the global lightning climatology shows that two of every three lightning flashes lie within ±23° of the equator [30].

The slope de^*/dT in the Clausius–Clapeyron relationship in (1.1) is a valuable benchmark for judging results on the response of lightning to temperature on various time scales (Section 1.4). This slope depends on temperature, but at the mean Earth surface temperature (~14 °C), the slope is 7% per 1 °C.

Cumulonimbus clouds are the primary agents for transporting water vapor from the planetary boundary layer in the lower troposphere to the upper troposphere.

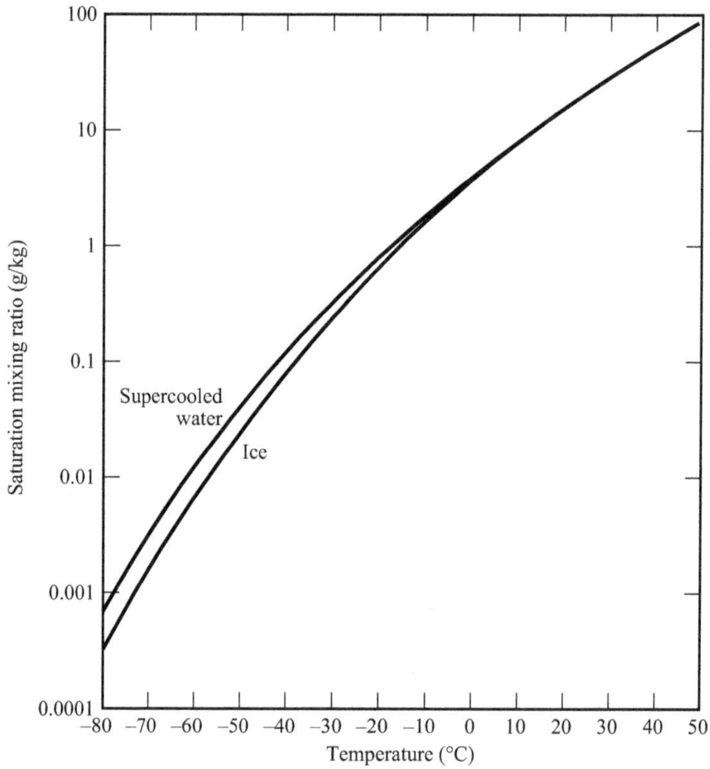

Figure 1.3 Equilibrium water vapor mixing ratio (g/kg) versus temperature: the Clausius–Clapeyron relation

Consistent with this general picture, Price [31] has shown variations in upper tropospheric water vapor correlated with lightning variations over the African continent.

1.3.4 Convective available potential energy and its temperature dependence

The maintenance of thunderstorm mixed-phase conditions and the associated "factory" for ice and electric charge requires an energy source for the updraft. That energy source is CAPE and is illustrated in Figure 1.4. CAPE is represented as the area on a thermodynamic diagram involving height (or pressure) and temperature. This area is bounded on the left by the temperature sounding in the storm

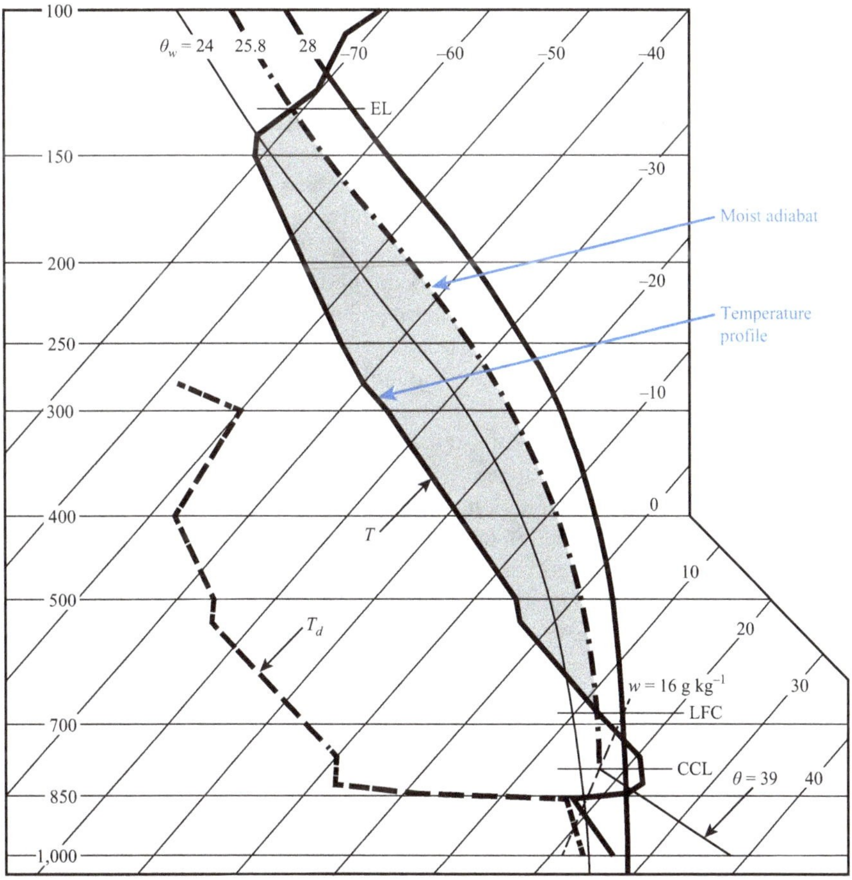

Figure 1.4 Temperature sounding, wet bulb adiabatic approximation for the updraft temperature, and the area representing convective available potential energy (CAPE)

environment and on the right by a "wet bulb adiabat" which is a theoretical prediction for the temperature of the air in an updraft parcel which is buoyant with respect to the storm's environment. At any given altitude, the different between the wet bulb adiabat and the environment is a measure of the buoyancy force acting on the updraft parcel of air. The buoyant force per unit mass at an altitude is given simply as $g(\Delta T/T)$ where ΔT is the temperature contrast between the updraft and the environment and where g is the acceleration of gravity (9.8 m/s^2). Since the wet bulb adiabat is determined by purely thermodynamic quantities temperature and dew point temperature of the surface air ingested by the storm to form the updraft, it would seem that CAPE is also a purely thermodynamic quantity. A complication arises here, however, making CAPE dependent on both thermodynamic and aerosol characteristics, and adds to the challenge of disentangling thermodynamic from aerosol influences on lightning activity. The updraft parcel buoyancy depends not only on temperature but also on the mass of condensate within the parcel. For example, if the temperature contrast is 1 °C, a typical value, the local cloud buoyancy force per unit mass is roughly 1/300 g = 0.03% of g. Since the density of surface air is 1.2 kg/m^3, a mass of condensate as small as 4 g/m^3 would completely negate the thermal buoyancy and strongly impact the dynamics of air parcels at that level. This point will be elaborated on below.

Accurate estimates of condensate mass are lacking in real thunderstorms, and theoretical CAPE calculations typically resort to two extreme assumptions, neither of which is entirely satisfied. In the most common pseudo-adiabatic (or "irreversible") approach, all the condensate is removed as the updraft parcel ascends (e.g., [13]). In this situation, only the temperature contrast (and a smaller contribution from the water vapor component) affects the parcel buoyancy. The implication is that the transformation from cloud water to precipitation is very efficient. In the context of CCN, this would be a situation with small concentration, typical of clean oceanic conditions. The observation of initial radar echoes in maritime convection at altitudes of only a few km (and in the "warm" portion of the troposphere) [32] is evidence for an efficient precipitation process in clean conditions.

The other extreme assumption in the evaluation of CAPE is that all condensate is retained as the updraft parcel ascends. The process is reversible and the wet bulb adiabat has a different mathematical form [33]. The implication is that the transformation from cloud water to precipitation has zero efficiency because the cloud droplets remain too small to coalesce. This situation is typical of rich CCN concentrations, as in polluted continental conditions. In early investigations of tropical oceanic convection [34,35] the reversible process was favored in computing CAPE, but it is now recognized that for maritime convection drawing from clean boundary layer air, the pseudoadiabatic irreversible process was a more appropriate choice. Given the negative buoyancy contribution of condensate loading, reversible CAPE is expected to be systematically less than irreversible CAPE. This expectation is consistent with numerous published results [29,36]. In early considerations of the reversible and irreversible processes [37], it was concluded that the "differences between the products of condensation as falling out or being retained, are so small as to be negligible in practice," but today this difference is acknowledged as being all important in deep moist convection.

Williams and Renno [36] also showed that CAPE in the current climate was well predicted by the wet bulb potential temperature of surface air, though different relationships were apparent for land and ocean. Global maps of wet bulb potential temperature show that maximum CAPE over land is greater than over ocean. Lucas *et al.* [38] claimed that CAPE over land was similar to values over ocean but they did not consider the diurnal variation of CAPE over land. A global climatology of CAPE has been prepared [39]. These results also show that CAPE is larger over continents than over oceans, though no consideration was given to aerosol-related effects in condensate loading. The land–ocean CAPE contrast is qualitatively consistent with the land–ocean lightning contrast, but on closer examination [11,38,40], the contrast is not sufficient to account for the order-of-magnitude contrast in lightning. This aspect will be revisited below in the context of CBH.

Given the primary role for CAPE in the charge separation and lightning activity of thunderstorms, the variation of CAPE with temperature on the long time scale of global warming is of considerable interest. This problem is nontrivial because the entire temperature profile is involved, as well as the condensate-related ambiguities of the wet bulb adiabat. In early work [41] CAPE was postulated to be a climate invariant. However, many GCM results show CAPE to increase with global warming [42,43], and Del Genio *et al.* [44] have found increases in cumulonimbus velocity in climate models in a warmer world. Furthermore, still more recent theoretical works [45,46] support a scaling of CAPE with the Clausius–Clapeyron exponential temperature dependence. On this basis alone, one expects to have more lightning in a warmer climate. However, recent model results [47] show the opposite result for the tropics. This contrast in predictions is not well understood at present.

1.3.5 Cloud base height and its influence on cloud microphysics

The contrast in physical characteristics between land and ocean surfaces exerts an important influence on the behavior of thermodynamic parameters of surface air. The contrast in heat capacity and mobility between land and ocean affect the surface air temperature, with ocean water and overlying air resisting temperature increase in response to solar heating, in comparison with dry land surfaces. The diurnal variation of ocean surface temperature is typically a fraction of 1 °C. In contrast, the diurnal variation of land surface temperature invariably exceeds 1 °C but the surface temperatures over deserts can vary by tens of °C. The contrast in available surface water between land and ocean affects the dew point temperature and relative humidity of surface air. Land surfaces are generally both hotter (larger T) and drier (smaller T_d) than oceans, and as a consequence of both of these contributions, the dew point depression of surface air ($T - T_d$), a purely thermodynamic quantity, is invariably larger over land than ocean. [Global maps of daytime dew point depression (and equivalently CBH) would show marked land–ocean contrasts as in Figure 1.1(a) and (c).]

The convenience of cloud physics is that the lifted condensation level (LCL) and CBH are both proportional to $T - T_d$. If $T = T_d$, the air is saturated (RH = 100%) and the cloud extends downward to the surface. Over oceans, typical CBHs are 500 m (corresponding to typical relative humidity of 80%), but over land can vary from 1,000 to 5,000 m (with relative humidity in the range of 20%–70%).

The contrast in CBH between continental and maritime environments is illustrated in Figure 1.5(a), for afternoon clouds at the first-radar-echo stage. The oceanic cloud achieves the first echo while still a warm cloud. In contrast, the continental cloud with systematically higher CBH is usually extended into the cold region of the atmosphere at first-echo stage. The heights of the 0 °C isotherm are similar for land and ocean, near 4,500 m MSL, but the CBHs differ markedly for thermodynamic reasons. Figure 1.5(b) shows deeper clouds for both land and ocean

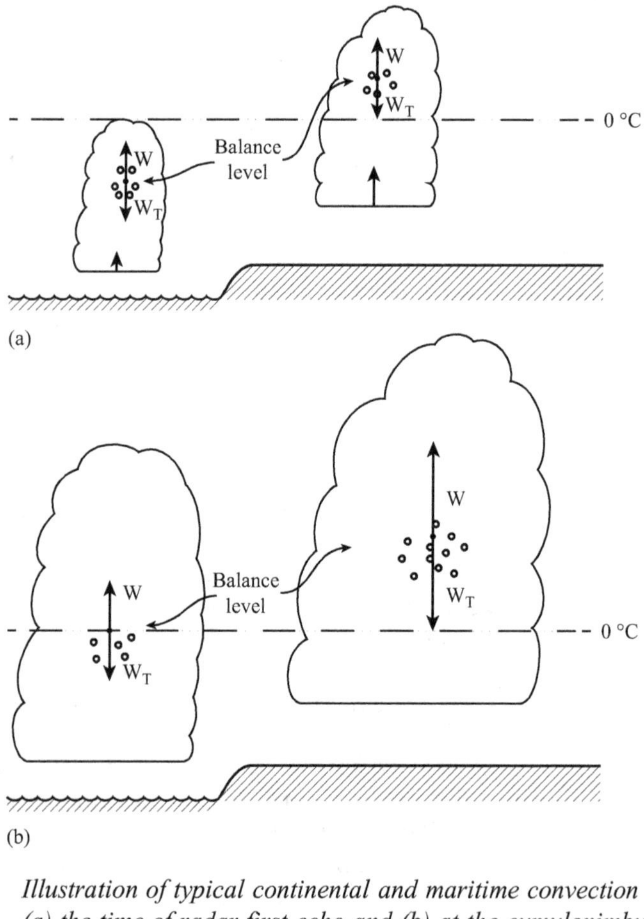

Figure 1.5 Illustration of typical continental and maritime convection at (a) the time of radar first echo and (b) at the cumulonimbus stage. The balance level heights in all cases are also indicated

that include the mixed-phase region bounded by the 0 °C and −40 °C isotherms where active charge separation can occur under appropriate conditions of cloud vertical development. The CBHs remain the same and often coincide with the top of the planetary boundary layer. The cloud widths are different based on observations showing that continental clouds are broader than maritime ones [11,48,49].

The updrafts in clouds are fundamentally important in regulating cloud microphysics and electrification. Scaling analysis indicates a sensitive fourth power relationship of lightning flash rate on updraft speed [50]. Accordingly, modest changes in CAPE can have substantial effects on lightning flash rate. Figure 1.5 also includes vertical arrows to contrast the updraft speeds in different regions (including CBH in Figure 1.5(a)) of both shallow and deep convection.

Regarding the subcloud region, Zheng and Rosenfeld [51] have found larger ascent speeds (by 50% to 100%) in the continental boundary layer than the oceanic one, and larger speeds at CBH, consistent with predictions based on thermodynamics and the contrast in surface properties in Williams and Stanfill [11]. Puzzlingly, model results on deep clouds with greater CBH do not show evidence for larger updraft speeds [52].

Earlier studies [48,49] have shown systematically larger updraft speeds in deep moist continental convection over land than over ocean. The contrast in ascent speeds is unmistakably linked with a contrast in the ice phase microphysics and lightning activity between land and ocean, but the explanation for the contrast in ascent speeds remains a controversial issue [11,40,52]. When thunderstorms over land alone are examined, lightning flash rate and CBH are positively correlated in global comparisons with the lightning imaging sensor in space and with surface thermodynamic observations of dew point depression [53]. These measurements need to be controlled for CAPE and aerosol variations to narrow down on the physical causality.

The formation of precipitation within the updraft of convective clouds is important because the precipitation can load the updraft, and ultimately reduce the updraft speed. This issue was raised initially in the context of CAPE, but here one can be more quantitative by estimating the precipitation content that will offset the effect of thermal buoyancy. Simple considerations of Archimedean buoyancy show that the force per unit mass associated with a temperature perturbation ΔT is simply $\Delta T/Tg$, where T is the ambient temperature. The negative buoyancy contribution (again force per unit total mass) from additional mass loading m in a parcel with mass of air M is simply $m/(m + M)g$. For a parcel at the 0 °C isotherm with nominal air density 0.6 kg/m^3, the condensate loading needed to offset a thermal buoyancy of 1 °C is 2.0 g/m^3.

Extensive documentation on the detection of first radar echo and the formation of precipitation in deep convection has come from radar studies (good summary in Ludlam [32], with additional examples in [54]). The D^6 dependence of the radar cross section of precipitation particles gives radar considerable sensitivity in detecting the formation of precipitation, given the runaway nature of coalescence of cloud droplets in the diameter range 25 μm [55] where the rate of droplet coalescence varies steeply with droplet size as D^5. The radar studies have shown that cloud depths (cloud top height minus CBH) in the 4,000–5,000 m range over continents and as small as 2,000 m over oceans [32,54] are needed for first echo development. These radar-based estimates are broadly consistent with aircraft in situ measurements of cloud depths needed to achieve critical cloud droplet size [55]. The relevance of these results in the context of Figure 1.5 is that the warm cloud depth in the maritime case (4,000 m), a thermodynamic effect, is large in comparison with that needed for the formation of precipitation. In the continental case this condition is not fulfilled. These comparisons are consistent with observations of radar first echoes that appear consistently in the "warm" part of the cloud over oceans, but more typically in the "cold" part of the cloud over land [32,54,56]. In a more global context, the results are also consistent with observations that warm precipitating clouds are prevalent over oceans and scarce over land [11,57]. Still the conundrum remains between a thermodynamic effect and an aerosol effect (Section 1.5). The lower CBHs over ocean overlie cleaner air with more dilute CCN concentrations [58]. Accordingly, following the discussion in Section 1.5, the cloud droplets above CBH will be larger and more prone to form precipitation and radar first echoes at lower heights. Braga *et al.* [58] give emphasis to the aerosol effect and neglect the thermodynamic effect.

The formation of precipitation in moist convection is important because it can then descend with respect to the air parcel in which it forms, and thereby load the updraft column beneath. (In barotropic conditions typical of tropical convection, the updraft is vertical.) This is sometimes called "super-adiabatic" loading because the precipitation condensate is added to the adiabatic condensate at lower levels. (In baroclinic conditions (Section 1.3.7 below) more typical of convection at higher latitudes, the updraft can be tilted and then the updraft can unload its precipitation, as is assumed in the irreversible calculation of CAPE.) Based on observations with a precipitation radar in space, warm rain clouds over oceans (where the warm cloud depths are greatest) are capable of achieving precipitation concentrations up to 2–3 g/m^3 [57]. These mass loadings are commensurate with cloud buoyancy at the 1 °C level as was shown earlier. These superadiabatic loadings represent reductions in the condensate that is delivered to the mixed-phase region by the updraft, and where the conversion of supercooled water to ice can invigorate the updraft by the latent heat of freezing. This process can strongly influence the nature of the vertical profiles of latent heat release and larger ice-phase hydrometeors and help explain the marked land–ocean contrasts in differences in lightning (Figure 1.1(a)) and rainfall (Figure 1.1(b)). In short, the warm rain cells over ocean may be substantially prevented from becoming thunderstorms by virtue of the raindrop loading they achieve. Further evidence for this suggestion is found in the next Section 1.3.6.

1.3.6 Balance level considerations in deep convection

The unloading of the updraft laden with warm rain and the delivery of condensate to the mixed-phase region are both sensitive to the updraft profile and to the location of the balance level [59,60], as illustrated in Figure 1.5. Vertically pointing Doppler radar observations of moist convection show a zero-crossing of mean Doppler velocity, with upward motions above and downward motions below. In the maritime case this balance level is located initially in the warm rain region, whereas in the continental case [60], its location is often in the mixed phase above. The precipitation particle fallspeed is balanced by the updraft w at the zero-crossing. Since the fallspeed of raindrops varies as $D^{1/2}$ at the balance level, the particle mass ($\sim D^3$) varies as W^6, and the reflectivity contribution varies as D^6 or as W^{12}. Following these sensitive dependences, a reduction in W by 40% ($\sqrt{2}$) by adiabatic loading in the maritime case relative to the continental case can cause an 8-fold reduction in mass and a 64-fold (18 dB) reduction in radar reflectivity. Eventually the supercooled raindrops will freeze and then one needs to consider the gravitational power contribution from the ice particles [61] which will scale as $D^{7/2}$, leading to an 11-fold difference in gravitational power between the land and ocean case.

Indirect evidence for the severe loading of an updraft by raindrops comes from experience in vertical mine shafts in South Africa [62–64]. Air that is nearly saturated with water at 30 °C at the bottom of a mineshaft (with vertical extent ~1500 m) is forced vertically by a powerful ventilator system. At a vertical air speed near 10 m/s, matched with the fallspeeds of the largest raindrops, the load on the ventilator system frequently exceeded its capacity and failed, allowing the suspended water to fall to the bottom of the shaft in a deluge. In a vertical shaft, no opportunity was afforded for unloading of the updraft, as in the irreversible thermodynamic process discussed in the previous section 1.3.5. In moist convection over the ocean for which the radar first echo appears below the 0 °C isotherm [32,54], this mineshaft experiment is likely applicable, and is reminiscent of suggestions by Zipser [65] that the updraft speeds in oceanic cumulonimbus clouds [49] would be limited to the fall speeds of the raindrops within them. The mineshaft experiment is also a reminder that vertical updrafts cannot unload their condensates.

In continental convection in which the radar first echo is found typically above the 0 °C isotherm [32,54], the balance level updraft loading by raindrops near 10 m/s is avoided, and larger ascent rates are possible. Now in the mixed-phase region, the main hydrometeors are graupel and so the new balance level there manifest in triple Doppler radar measurements [60] can be substantially higher. In severe storms, still larger ascent rates are possible, with a balance level accommodating the growth of hailstones that may reach softball size in updrafts approaching 100 m/s. In these situations, a BWER (bounded weak echo region) in radar observations is indicative of the balance level above, and may be 2–3 times higher than the balance level in warm rain cells. But in this strongly baroclinic situation (see Section 3.6) the updraft is strongly tilted and the hailstones can fall out of the updraft. In a special class of supercells (called LP for "low precipitation" but storms invariably productive of hail) with high CBHs (occasionally with sub-0 °C cloud base temperatures), no condensate is lost in a warm rain process and the adiabatic cloud water is available for the growth of the hail.

1.3.7 Baroclinicity

The pole-to-equator temperature difference is an important consideration for both weather and for climate [66]. The local latitudinal temperature gradient exerts a decided influence on the organization of thunderstorm activity, and may also affect the flash rate behavior of thunderstorms. This temperature contrast is set up by the latitudinal imbalance between the incoming shortwave radiation from the Sun and the outgoing longwave radiation to space. Instabilities in the prevailing westerly winds at mid-latitudes draw on the potential energy of the pole-to-equator temperature difference to produce synoptic scale weather disturbances there [67]. The condition of a latitudinal gradient in temperature is "baroclinicity" and the thermal wind equation in atmospheric dynamics [66,68] links a baroclinic atmosphere with a vertical shear in the horizontal wind. This vertical wind shear can tilt the updrafts of thunderstorms (embedded in these large scale disturbances) from the vertical, and thereby strongly influence the unloading of condensates from the updraft.

In the near equatorial zone of the tropics, the latitudinal temperature gradient nearly vanishes and the atmosphere is characterized as barotropic rather than baroclinic [68]. Without the organizing effects of vertical wind shear, air mass thunderstorms prevail, with an expectation for vertical rather than tilted updrafts. Yoshida *et al.* [69] have compared the flash rates of thunderstorm cells in continental and oceanic zones at different latitudes. In general, the mean flash rates are increasing with latitude away from the tropics. Bang and Zipser [70,71] have also found that oceanic convection is more likely to produce lightning in the presence of vertical wind shear than without it. Some of the most dramatic outbreaks of lightning over the open ocean occur in the presence of strong baroclinicity [72,73]. These findings may have explanation in baroclinicity in unloading of the updraft in the warm rain region of the storms so as to provide greater invigoration of the mixed phase. On this basis, and with all other factors the same, one might expect to have more lightning activity globally with a larger pole-to-equator temperature contrast.

Supercells are rotating thunderstorms characteristic of strongly baroclinic environments in the springtime, and to a lesser extent in the fall, in both northern and southern hemispheres [74–76]. These storms are parent to a large portion of severe weather episodes, including strong winds, large hail, and tornadoes [75,76]. Supercell thunderstorms also produce exceptional total lightning flash rates [53,77–81].

1.4 Global lightning response to temperature on different time scales

The global temperature is known to vary on a number of natural time scales: the diurnal, the semiannual, the annual, on the El Nino/La Nina time scale, and on the 11-year time scale of the solar cycle. In understanding global lightning's response to climate change on the long time scale, it is valuable to look for inconsistent patterns of response on other time scales whose physical origin is better understood. It is also possible that systematic changes in aerosol may accompany these natural variations in temperature. Here again we encounter the problem of disentangling aerosol and thermodynamic effects [82].

1.4.1 Diurnal variation

On a planet covered with ocean, the variation of global temperature over the 24-h period of the Earth's rotation in sunlight is expected to be nil. But based on the contrasting properties of land and ocean discussed in Section 1.3.5, with preferential heating of land relative to ocean in response to incoming shortwave radiation from the Sun, a consistent diurnal variation of global temperature in UT time is established. The local diurnal variation of surface air temperature typically shows a maximum value shortly after local noon, whereas the lightning activity peaks later in the afternoon, near 4 pm [7,8,83,84]. The globally integrated effect is a consistent global variation of temperature in universal time [85] that would appear to be the physical basis for the Carnegie curve of atmospheric electricity [7,8,86–88]. The regions that are sequentially heated by the Sun are the three "chimneys" of global lightning activity—the Maritime Continent, Africa, and the Americas, prominently manifest in the climatology of global lightning activity, as shown in Figure 1.6. In the global surface skin temperature [85] found a peak-to-peak diurnal variation of 3.0 °C for the globe. For a variation in global lightning activity of 60% [83,89,90], this amounts to a sensitivity of 20% per °C [9,42,91].

By using surface temperature at airports with hourly variation during a period of active measurement of ionospheric potential (V_i) of the DC global electrical circuit, Markson [92,93] established positive correlations between V_i and tropical continental air temperature on the UT diurnal time scale, with a sensitivity of 7% per °C.

The variation of aerosol and CCN concentration on the diurnal time scale has been investigated (e.g., [94]) but observations are insufficient to compile a climatology of UT variation. A climatological variation in storm flash rate in local time [84] shows a maximum between the time of maximum temperature and the time of maximum number of storms. This finding is inconclusive toward distinguishing thermodynamic and aerosol contributions to lightning activity.

1.4.2 Semiannual variation

A consistent repetition of weather twice per year is a foreign concept to mid-latitude observers but is a very consistent feature of the near-equatorial region over land [95]. The sensible semiannual variation then in a number of meteorological quantities is a direct result of the Sun's traversal of the equator twice per year at equinoxes [96]. The variation in solar insolation for the global tropics (±23° latitude) is ~7% (max-min/mean). The corresponding peak-to-peak amplitude variation in global tropical surface air temperature is about 1 °C.

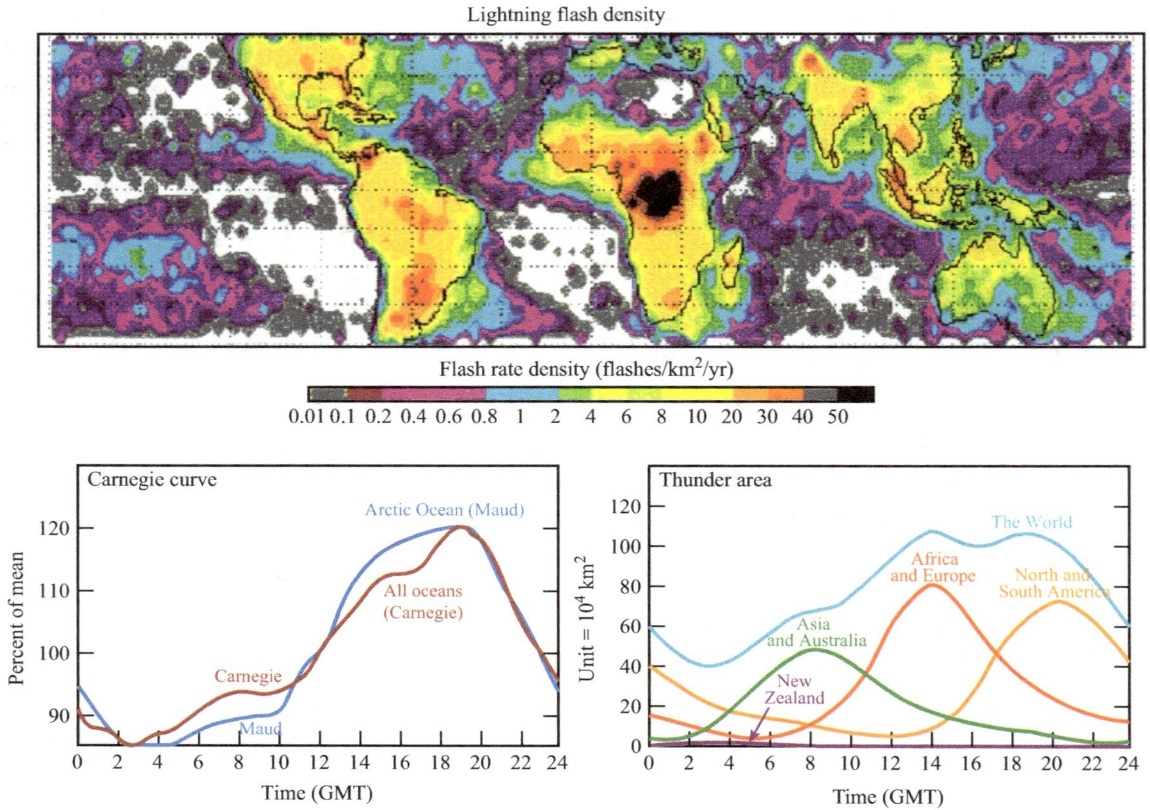

Figure 1.6 The Carnegie curve of atmospheric electricity over 24 hours of universal time and a global map showing three continental lightning chimneys

Clear evidence for a semiannual variation in global lightning with the same phase as the temperature variation has been provided by optical observations of lightning from space [97]. This signal is most conspicuous when analysis is concentrated in the near-equatorial zone. Here the sensitivity of lightning to temperature is 20%–30% per 1 °C. Rainfall observations are also consistent with double rainy seasons in northern South America and in near-equatorial Africa. The discharge record from the extensive Congo River basin, straddling the equator in central Africa, also shows a distinct semiannual variation.

Several independent observations of the global electrical circuit (both DC and AC) show evidence for semiannual signals [87,96,98–100] that are plausibly linked with the semiannual variation of temperature.

The aerosol contribution to the semiannual variation in global lightning activity has not been studied. We are presently unaware of long-term investigations of CCN concentration in the near-equatorial zone over continents that would shed further light here.

1.4.3 Annual variation

The global distribution of land/ocean area is decidedly asymmetrical in the extra-tropics [96], with a 5-fold greater land/ocean area ratio in the northern than southern hemispheres. The smaller heat capacity of land relative to ocean assures that the temperature of the surface air is greater in northern hemisphere summer than in northern hemisphere winter. This asymmetry is in large part responsible for the annual variation in mean global temperature, showing a maximum in August and with peak-to-peak amplitude variation of about 4 °C.

Many observations of global lightning have shown a seasonal variation in phase with this variation in surface air temperature, both in optical observations of lightning from space [83,89,97], but from surface-based observations of the intensity of the Earth's Schumann resonances (e.g., [101–102]). The variation of global flash rate is nearly a factor of two, and the computed sensitivity of flash rate to temperature is 11% per °C. It has also been established that the major contribution to global flash rate on the annual cycle is number of thunderstorms rather than mean flash rate per storm. If temperature is the controlling thermodynamic variable, this would imply that higher temperature is favorable for more frequent release of conditional instability rather than in increasing of CAPE. It is remarkable to have a globally integrated quantity that nearly doubles on the annual time scale. This finding is testament to the volatile nature of lightning in its apparent response to temperature.

Similar to the discussion on the semiannual time scale (Section 1.4.2), the contribution of aerosol to lightning activity on the annual time scale has not been evaluated, for lack of comprehensive observations of the global aerosol climatology. It stands to reason that substantially larger CCN concentration will be available for storm ingestion in northern hemisphere summer than winter, for the same reasons pertaining to land/ocean asymmetry, and the dramatic contrast in aerosol optical depth and CCN between land and ocean (Figure 1.1(c)). New methods now under development for the observation of CCN at CBH from satellite ([14]) will be particularly valuable in this context.

Ambiguity about the annual phase of the DC global electrical circuit began when Lord Kelvin observed [103] in the United Kingdom that the Earth's electric field was greatest in winter, contrary to the general picture based on global lightning activity and temperature. Whipple [7] identified this apparent contradiction but did not resolve it. Adlerman and Williams [104] pointed out the local effect of wintertime aerosol (in both hemispheres) in decreasing the electrical conductivity of the atmosphere and thereby increasing the electric field. A more reliable ground-based measurement of the DC global circuit is the air-earth current which represent the product of the electric field and conductivity. For example, long-term measurements of the air-earth current in Athens [105] have been shown to peak in August. Seasonal averages of the ionospheric potential [93], the preferred measure of the DC global circuit, also show a maximum in Northern Hemisphere summer.

1.4.4 ENSO

Early interest in a strong relationship between the phase of the El Nino-Southern Oscillation and lightning developed out of analysis of a single magnetic coil recording of the first resonance mode of Schumann resonances [30], for a single ENSO event (in 1973–74). A remarkably sensitive response of lightning to temperature was inferred from this analysis. Though the sign of this response has been corroborated in more recent work (e.g., [106]), the magnitude of the response in the data obtained by C. Polk is unprecedented, and may not be valid. Based on the global temperature analysis of Hansen and Lebedeff [27] and the classical analysis of rainfall variations [107] showing less rainfall over land in the warm phase, the general picture that developed was one in which all tropical land regions warmed in the El Nino warm phase, with the major upwelling in the central/eastern Pacific region causing large scale subsidence over tropical continental regions. This picture is in keeping with reductions in total river discharge in large drainage basins (Amazon and Congo), serving as continental-scale rain gauges, straddling the equatorial region [108] in the warm phase. This investigation was later extended to the Nile and the Ganges [109] River basins, with similar ENSO phasing. This picture on regional rainfall variability is also consistent with the majority of multiple ENSO events pictorialized in Allan *et al.* [110].

The regional behavior of lightning over the ENSO cycle shows somewhat less consistency overall, but a definite tendency in behavior is evident. This is most apparent in the Maritime Continent (including southeast Asia, Indonesia, and tropical Australia), where the ENSO studies are most numerous and where lightning is more prevalent in the warm El Nino phase [111–117]. Generally speaking, in this part of the world, the warm ENSO phase is also the drier (lower relative humidity and higher CBH) phase, and the more polluted phase. As noted above, the wet cool phase (La Nina) is more abundant in rainfall. So here again we have again the disentanglement issue of thermodynamics and aerosol. The opposite tendencies for rainfall and lightning on the ENSO time scale is at first peculiar, but highly variable lightning/rainfall relationships, temporarily and regionally, are now widely recognized [118,119] and are linked with differences in the vertical development of the precipitation in the cold part of the cloud. This situation of opposite ENSO phase relationships for rainfall and lightning has also led to speculation that the two global electrical circuits (DC and AC) may have opposite tendencies over ENSO cycles [120,121], though coordinated synchronous measurements are lacking to check on this prediction. In this context, it seems likely that the contribution of electrified shower clouds [122–125] will be more potent over land during the cold La Nina phase, when continental rainfall is greater [110].

In South America, both Chronis *et al.* [126] and [106] using lightning observations from space found greater amounts of lightning in northeast Brazil during the cold La Nina phase. In contrast, Pinto [127] using thunder day observations over many ENSO cycles found a conspicuous tendency for greater numbers in the warm El Nino phase. An extreme El Nino event in 1926 has been documented by Richey *et al.* [128] and by Williams *et al.* [129] in the Amazon basin, with exceptional hot and dry characteristics but no information on lightning activity is available. Chronis *et al.* [126] and Sátori *et al.* [106] agree in finding more lightning in the warm phase in southern Brazil and eastern Argentina, where the discharge of the Parana River is also maximum [108]. This region appears to be the southern component of the north-south rainfall dipole anomaly identified by Grimm and Natori [130]. This distinctly extra-tropical tendency for greater lightning in the warm phase was found earlier in the opposite hemisphere by Goodman *et al.* [131] in the Gulf of Mexico region of the United States, in a similar range of latitude.

Among three tropical lightning "chimneys" [9], Africa appears to show the weakest lightning variation on the ENSO time scale. Chronis *et al.* [126] found some enhancement in the La Nina phase, whereas Sátori *et al.* [106] reported a modest lightning increase in the warm phase. Given Africa's status as the most distant lightning chimney from the Pacific Ocean source of convective upwelling, it seems likely that global scale subsidence would have the least effect on Africa, while leaving conspicuous effects on adjacent chimneys Maritime Continent and America. Dowdy [132] has considered the effect of season on the lightning response to temperature on the ENSO time scale and this may impact the generality of a simple positive response in the warm phase. This seasonal aspect has also been discussed by LaVigne *et al.* [120].

Extreme El Nino events, such as the drought of the century in South America, can lead to so much warming and drying as to prevent both moist convection and lightning. Evidence for this situation may be found in the 1926 drought on the Amazon basin [128,129,133].

Despite the great sparsity of oceanic lightning (recall Figure 1.1(a)), easily discernible regional variations are discernible on the ENSO time scale. The general tendency for oceans is opposite to that for land: greater lightning over ocean in the cold La Nina phase [106,134,135]. This has been interpreted on a basis consistent with that for the warm phase: greater regional subsidence signifies less overall cloudiness, and hence greater surface heating and greater instability to drive moist convection.

On a worldwide basis, Satori *et al.* [106] documented greater lightning in the ENSO warm phase than the cold phase. Harrison *et al.* [136] have found evidence for inferred increases in the global electrical circuit in the warm phase.

1.4.5 Decadal time scale

Modest changes (~0.1%) in the integrated energy output of the Sun have long been recognized over the 11-year solar cycle, and the corresponding changes in global temperature have been investigated. The peak-to-peak variation of temperature from this analysis is about 0.1 °C [137–139]. Attempts to see these changes in global records of thunder day data have been mostly unsuccessful [140,141]. No solar cycle signal has been found in the LIS/OTD satellite record of global lightning. It should be noted however that solar cycle variations have been detected in the analysis of thunder day records at selected stations in Brazil [142] on the basis of wavelet analysis. Solar cycle variations in the intensity of Schumann resonances at high latitude are not plausibly explained by temperature-related variations in lightning activity [143].

Koshak *et al.* [144] studied variations in lightning incidence over the continental United States (CONUS) in the decadal period 2003–12. The trend in cloud-to-ground (CG) lightning was negative over this period, with a 12% decrease from the interval 2003–07 compared to the interval 2008–12. The trend in wet bulb temperature over the CONUS was also negative for the same period but the dry bulb temperature showed an increase. The total lightning activity measured by the lightning imaging sensor in space showed no significant trend over the same period. In retrospect, the decadal period examined in this study lay within the period now frequently referred to as the hiatus in global warming (1998–2013). It is also interesting to speculate about a possible decadal increase in CBH, as the average dry bulb temperature increased and a moisture variable decreased, together entailing an increase in the dew point depression which is proportional to the CBH. Boccippio *et al.* [145] had found a large enhancement in the IC/CG ratio in a region of the CONUS (extending from eastern Colorado northward through the Dakotas) with elevated CBH [53].

1.4.6 Multi-decadal time scale

For studies on lightning variability on time scales longer than the typical lifetimes of lightning detection networks, researchers have resorted to the use of thunder day data (and model calculations: [146]). Thunder day observations have been underway at meteorological stations and airports worldwide for more than a century. One interesting recent application here has focused on an upward lightning trend in the Sea of Japan [147]. A rough doubling in thunder day counts since 1930 has been linked with observed increases in sea surface temperature of 1.2 °C–2.2 °C.

The tendency for the current global warming to predominate at high northern latitudes is widely recognized [28]. At Fairbanks, Alaska (latitude 64.8° N) both the

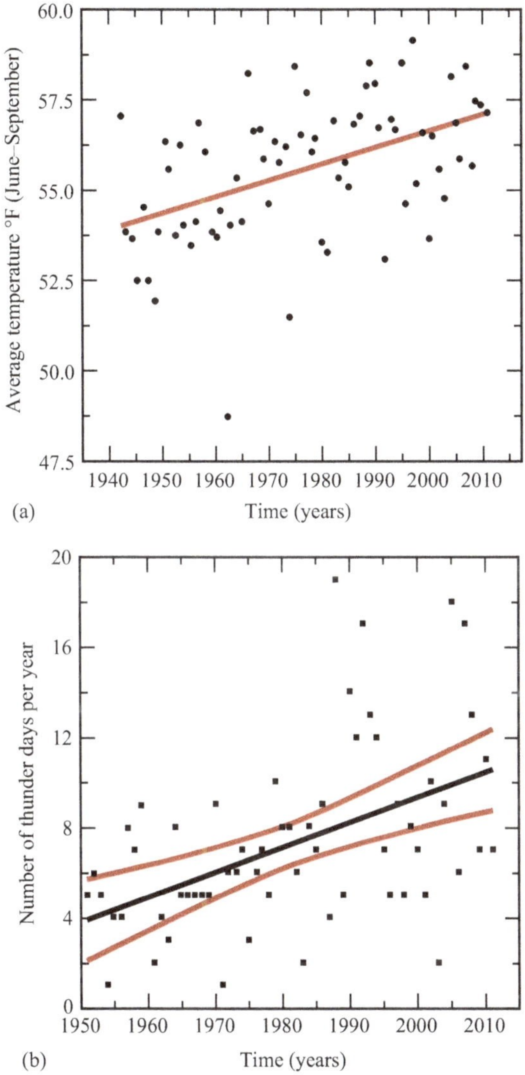

Figure 1.7 Upward trends in (a) temperature and (b) thunder days for Fairbanks, Alaska. Linear least squares fits are also included to illustrate these trends

temperature and the thunder day counts have been increasing conspicuously [42]. Figure 1.7 includes plots of both quantities with least squares fits for trend. The number of thunder days has more than doubled in 50 years. Anecdotal reports indicate that Canadian meteorological stations at the highest latitudes have noted thunder days for the first time on record.

The long-term record of mean global temperature [27,28,148] based on averaging of surface thermometers shows an increase of the order of 1 °C on the 100-year time scale but is interrupted by shorter intervals when the temperature is flat or declining with time. The two most notable intervals are the so-called "Big Hiatus" from 1940 to 1975, and the more recent (and more controversial) hiatus in global warming from 1998 to 2013. Both these intervals have been addressed recently by Williams *et al.* [149]. Appeal was made to previously published thunder day observations to address the Big Hiatus and separate analyses for both North America and Siberia show flat or declining counts of thunder days. A ~15% decline in mean annual thunder days is evident from 1940 to 1970.

For the more recent hiatus in global warming, satellite optical observations are available for nearly the entire interval from the lightning imaging sensor. Several global temperature data sets were examined and it was shown that both the temperature trend and the trend in global lightning flash rate were statistically flat during the hiatus period. The hiatus period ended with a pronounced El Nino event in 2015 and a strong resumption in the increase in global temperature [148] that had been underway between the time of the Big Hiatus and the recent hiatus. Unfortunately, the lightning imaging sensor is no longer operational in space to check the lightning behavior in this later period. Overall, the available results are not inconsistent with the hypothesis that global lightning is responsive to global temperature.

1.5 Aerosol influence on moist convection and lightning activity

1.5.1 Basic concepts

When condensation of water vapor occurs during the ascent of air parcels in the real atmosphere, every cloud droplet that forms is dependent on some nucleus to initiate its formation. The process is known as heterogeneous nucleation. The subset of the atmospheric aerosol population that serves this role is called cloud condensation nuclei (CCN) [150]. Were it not for the ubiquitous presence of CCN throughout the atmosphere, large departures in water vapor concentrations from the equilibrium predictions of the Clausius–Clapeyron relation (Section 1.3.4) would develop in a thunderstorm updraft, and these departures are not generally observed (but see recent findings in [20]). As an air parcel ascends in a thunderstorm updraft, the adiabatic cloud water content that appears (enforced by Clausius–Clapeyron) is shared roughly equally among all the available nucleation sites. This means that the cloud droplet concentration is matched with the CCN population at CBH, and when the CCN population is large (polluted conditions) the cloud droplets are smaller than they are in clean conditions. Since the tendency of cloud droplets to coalesce and form precipitation particles is strongly dependent on their size, the CCN concentration can be influential on the development of convection [12,13,15,151,152]. The contrast in conditions for convection growing over clean and polluted

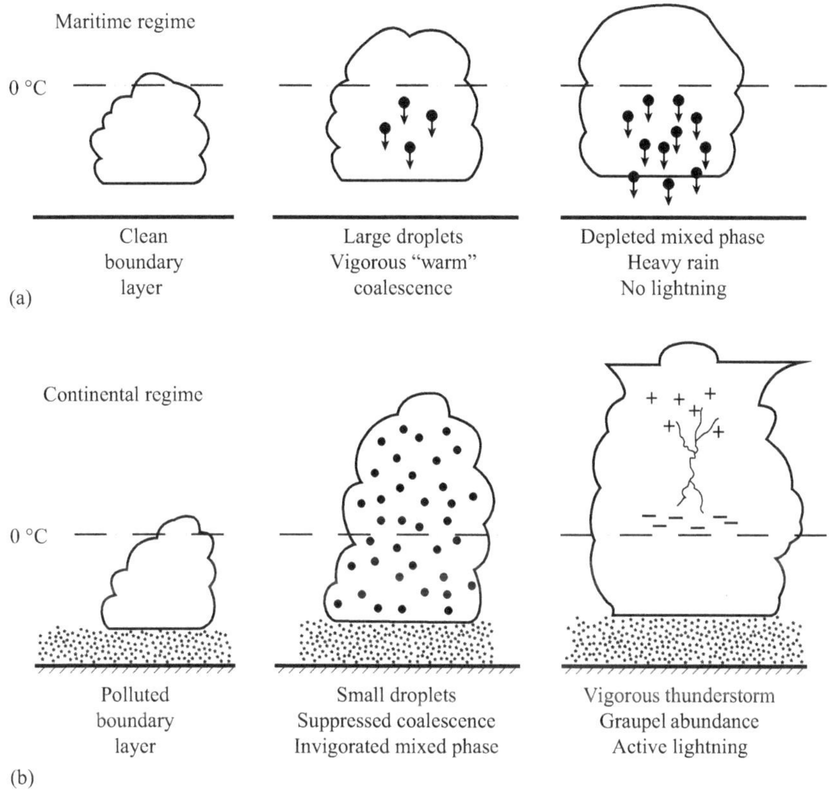

Figure 1.8 Illustration of the aerosol effect on convection and lightning in clean and polluted situations

boundary layers is illustrated in Figure 1.8. Three broad ranges of CCN population can be considered: (1) clean conditions with CCN concentrations typical of maritime air (10–100 per cc), (2) more polluted conditions (a few hundred to 1,000 per cc) typical for continents, and (3) ultra-polluted conditions with concentration exceeding several 1,000 per cc.

These regimes have been treated in cloud models [152]. Condition (1) may favor the rapid formation of precipitation and the production of rain, which may ultimately contribute to the superadiabatic loading of the air parcel (Section 1.3.5). Condition (2) may enable the retention of condensate in the updraft (consistent with the assumptions of reversible ascent ([13] and Section 1.3.5)) until the mixed-phase region is attained. The most pollution condition (3) may lead to cloud droplets so small that the formation of graupel particles in the mixed-phase region is prevented, thereby enabling the cloud water to rise as high as the −40 °C isotherm, where homogenous nucleation of the cloud water may occur.

1.5.2 Observational support

A major shortcoming in the evaluation of aerosol effects on convective vigor and lightning activity, and making detailed comparisons with competing thermodynamic contributions, has been the general absence of information on the CCN concentrations involved in specific situations. The innovative development of satellite methods, both for observing the cloud droplet sizes from space [153] and for obtaining the CCN concentration at CBH [14], is having a dramatic impact at the time of this writing on the understanding of the influence of aerosol on cloud microphysics and lightning.

Orville *et al.* [154] documented an approximate doubling of CG lightning flash density in the vicinity of oil refineries in the vicinity of Houston, Texas that was attributed to an aerosol effect. A more elaborate investigation of both thermodynamic and aerosol effects in the same general area [21] shows evidence that the lightning enhancement there is more likely due to an effect of CCN than to a heat island effect.

Even in the absence of this direct measurement of CCN, several studies have appeared that make use of global proxies for aerosol to compare with lightning activity obtained with other platforms. Stolz *et al.* [19,155] have used aerosol estimates from GEOS-chem (www.geos-chem.org) to compare with NASA TRMM lightning imaging sensor observations to show that aerosol contributions to lightning activity are comparable with thermodynamic ones. Altaratz *et al.* [156] have examined the influence of aerosol (estimated with satellite measurements of aerosol optical depth) and CAPE on lightning recorded with the world wide lightning location network (WWLLN) on a regional basis. They found that statistically significant increases in lightning activity were associated with more polluted conditions.

Published examples of perturbations in lightning activity when aerosol is introduced in maritime convection have produced the most convincing evidence for aerosol control. In the first case [157] volcanic aerosol documented by satellite was ingested by oceanic cumulonimbus clouds whose exceptional lightning activity was documented by and specific controls were placed on thermodynamic influence. In a more recent study by Thornton *et al.* [22], a rough doubling of lightning activity along sharply defined oceanic shipping lanes in Southeast Asia has been documented with the WWLLN. In this case, it can be shown that the warming of the sea by engine cooling exhaust water by the sea-going vessels is of negligible consequence, and diesel exhaust is rich in fine aerosol [158]. The introduction of rich aerosol concentrations in the pristine environment of the ocean represents a dramatic change in cloud microphysical conditions, with maximum likelihood of manifestation on lightning when none is present in the pristine state. The collection of different kinds of lightning studies by Yuan *et al.* [157], Altaratz *et al.* [156], Stolz *et al.* [18], and Thornton *et al.* [22], all over oceans, may reflect this more sensitive response of lightning to aerosol in that regime.

One may contrast the aerosol sensitivity in the maritime regime with the situation with high CBH (and shallow warm cloud depth) over land [53,159,160]. Li *et al.* [159] concluded that in this regime, there was little change in cloud microphysical behavior in response to aerosol variations. One reason for this is that the cloud droplet sizes in the mixed-phase region are already small by virtue of the proximity to cloud base [55] and the absence of a deep warm rain process to mediate with aerosol. Without the depletion by warm rain, the cloud water contents in the mixed-phase region are expected to be large in such continental storms, promoting the growth of hail and also conditions conducive to thunderstorms with inverted electrical polarity [53,79,160,161]. Lyons *et al.* [162] had earlier found that storms ingesting smoke from fire were exhibiting greater numbers of positive ground flashes.

Additional evidence that aerosol has a pervasive influence not only on lightning but other aspects of meteorology are studies that show reduced activity on weekends when the anthropogenic contribution to aerosol in industrialized regions has been shown to be reduced. Weekend effects on rainfall [163], lightning [164], hail [165], and even tornadoes [165] have been documented in recent years, with statistically verified results.

1.6 Nocturnal thunderstorms

The great majority of thunderstorms worldwide follow the diurnal variation in local time that has been discussed in Section 1.4.1, with a typical 4 pm maximum in activity when thermodynamic conditions are most favorable. This exclusive afternoon prevalence was assumed in the early analysis of global thunderstorms [166] that underlies the analysis of the global electrical circuit [7,8,125]. However, the most spectacular displays of lightning occur in the contrast of night, in nocturnal thunderstorms. Satellite-based studies at nighttime [167,168] show that the large land/ocean contrast in lightning prevalence apparent during daytime is also upheld at night. Thunderstorms that occur outside the usual afternoon diurnal cycle are of special interest in the lightning and climate context for two reasons: (1) their existence and the nocturnal land–ocean contrast provides an additional perspective on the disentanglement of thermodynamics and aerosol, and (2) the global warming signal has been dominated by nighttime temperatures over certain decadal intervals [169,170], most prominent in the period 1965 to 1985 for reasons that are not well understood.

It is appropriate first to summarize special configurations of land and water that are favorable for nocturnal thunderstorms. Nighttime lightning activity may occur in low lying areas, both lakes and valleys, adjacent to more mountainous terrain. Examples from the literature include the large tropical lakes and inland seas: Lake Victoria in Uganda, Africa [171] and Lake Maracaibo [172,173] in Venezuela, South America. Another example is when mountainous terrain lies adjacent to inland seas, as in the Mediterranean Sea [174]. Other examples of mountains adjacent to valleys productive of nocturnal thunderstorms are Phoenix, Arizona [175] and Albuquerque, New Mexico (personal observations, 1994) as well as the mountainous terrain of the Andes in Colombia [83] in South America. In all of these cases, the favored convective development in the afternoon is found in the elevated terrain where strong daytime heating of the surface is prevalent. The cold downdrafts of the ensuing thunderstorms then descend into the valleys and over lake surfaces to lift the air there and initiate deep convection where only subsidence from the adjacent mountain convection was present earlier in the day.

The most violent nocturnal convection in the Earth's atmosphere [176] occurs in two separate but conjugate regions in the Americas: in the Great Plains of North America east of the Rocky Mountains, and in Argentina, South America east of the Andes Mountains [83]. The peculiar phase of the diurnal cycle in Argentinian thunderstorms was identified in the report on worldwide thunder occurrence in Brooks [166], and Wallace [177] has systematized the local diurnal cycle in the Great Plains of North America. Prodigious nocturnal lightning activity in both areas may ultimately lead to some important reconsideration of the behavior of the global electrical circuit in universal time [7,8]. In both areas, air in the prevailing westerly flow is warmed in the afternoon in traversing the respective heated mountain ranges, and then progresses further east over low-level flow laden with moisture. The warm air aloft creates a temperature inversion in the local temperature profiles, otherwise known as a capping inversion, which serves to suppress the afternoon convection while allowing the buildup of temperature and moisture below the cap throughout the daytime that contributes to a large CAPE [29]. When the capping inversion weakens and special triggering processes occur at night, the moist convection develops with exceptional vigor [178,179]. The prominence of thunderstorms with large hail in the nighttime activity [178] is evidence for exceptional updrafts in nocturnal convection.

In all of the scenarios described above, the nocturnal thunderstorms described occur only when the usual afternoon thunderstorms are absent at the same location. This situation has special implications for the thermodynamic boundary conditions for the nighttime storms. One implication is that the nighttime air temperatures are substantially larger on the evenings with earlier afternoon storms [175,179]. There are two important reasons for this observation. One is that the customary cold downdraft air was not present. Second, the blanket of boundary layer water vapor that accumulates continuously in the daytime, and which is not reduced by its transformation to rainfall in afternoon thunderstorms, serves to reduce the outgoing longwave radiation that would otherwise serve to cool the Earth's surface.

It is important to note that these conditions contrast starkly with the usual nighttime conditions (without thunderstorms) in the tropics when the relative humidity reaches 100%, the dew point depression vanishes, and dew appears at the surface (e.g., [48,180]). In such circumstances, the contrast in CBH between land and ocean would not serve as a viable thermodynamic explanation for the land–ocean contrast in lightning activity, as discussed in Lucas *et al.* [38], Williams and Stanfill [11], and Zipser [40,65]. But an appeal to the literature shows that nocturnal thunderstorms do not occur in these circumstances. As one example study in the Great Plains [179], the development of nocturnal convection is characterized by the presence of unsaturated air at the surface and an elevated CBH (3.4–3.5 km MSL) linked with most unstable air that is elevated from the surface. CAPE values are linked with the most unstable values of wet bulb potential temperature and are also large (>3,000 J/kg) in the case of nighttime storms. The evidence that nocturnal thunderstorms are a subclass of "elevated thunderstorms" that feed on a source of most unstable air removed from the surface was established earlier by Colman [181].

Regarding the contribution of aerosol to lightning activity in the context of nocturnal thunderstorms, generally speaking one would not expect major changes in CCN from daytime to nighttime, so that a similar land/ocean contrast in CCN would pertain for daytime and nighttime convection. Further studies on this issue are warranted.

Interest in nocturnal thunderstorms is also of interest in the climate context because of the earlier idea that most of the global warming was found in the nighttime temperatures [93,169,182] in many regions worldwide. For reasons linked with the sensitivity of lightning to temperature (Section 1.4), greater lightning activity might be expected in nocturnal thunderstorms in those periods, though no analysis specific to nighttime has been undertaken to the author's knowledge. More recent global analysis of the DTR [170] shows that the most conspicuous decrease occurred in the period 1950–80, roughly coinciding with the so-called "Big Hiatus" in global warming. If the global temperature trend is flat the decline of DTR with a flat mean temperature guarantees a decline in the nighttime temperatures.

1.7 Meteorological control on lightning type

The greater danger to both mankind and infrastructure posed by CG lightning in comparison with intracloud (IC) lightning motivates some discussion on the meteorological conditions favoring the former over the latter. A related issue pertains to how a changing climate may influence the relative numbers of CG and IC lightning.

The abundant evidence that CG lightning does not occur until late in the lightning life cycle of thunderstorms [56] is strongly suggestive that clouds with substantial width of the main negative charge region are needed to provide a discharge from cloud to ground, and that multistroke CG flashes are more likely, the greater that width [183]. Narrow storms with extraordinary vertical development tend to be dominated by IC lightning in the author's personal experience. But the meteorological controls on thunderstorm width are still not well tied down.

Conspicuous variations in the IC/CG ratio over the continental United States have been shown by Boccippio *et al.* [145]. When the map of this ratio is compared with the climatology of summertime CBH, it is apparent that the largest IC/CG ratios are found in a region (eastern Colorado and extending northward into the Dakotas) with elevated CBH, and also a region where the climatology of CG lightning shows a diminished incidence of CG flash density (e.g., [144]). It has also been suggested [184] that CBH may be influencing the latitudinal variation of the IC/CG lightning ratio. Given the evidence that the surface relative humidity is quasi-invariant with climate change [185–190], it may also be true that the relative incidence of CG lightning may not change appreciably in a warmer climate. This general area of research is in need of further attention.

1.8 The global circuits as monitors for destructive lightning and climate change

The conductive Earth and the conductive ionosphere sandwich the more electrically insulating atmosphere to form two global electrical circuits. In the classical DC global circuit, a quasi-steady DC voltage of ~250 kV known as the "ionospheric potential" [93,191,192] is maintained between the Earth and ionosphere by electrical source currents from thunderstorms and electrified shower clouds [125,193–195] that together provide about 1,000 A. For the AC global circuit, otherwise known as Schumann resonances [102,195–201], the insulating space between two spherical conductors serves as a giant waveguide that supports resonant electromagnetic waves maintained continuously by the vertical charge transfers enacted by global lightning activity. The naturally occurring waveguide signals are manifest in two ways: as the "background" Schumann resonances consisting of the overlapping waveforms of ordinary lightning produced at rates of order 100 per second globally, and as the transient resonances (or "Q-bursts") produced by exceptional mesoscale lightning with global rates of only a few per minute, but which singlehandedly ring the global waveguide to amplitude levels that dominate all the other lightning contributions combined. Simple illustrations of the two global circuits are shown in Figure 1.9. The simultaneous behavior of the two global circuits has been considered recently by Williams *et al.* [202].

The two global circuits provide natural frameworks for global monitoring. In the case of the DC global circuit [93,191,192,203], the ionospheric potential is a measure of all the electrified weather underway at any time, including thunderstorms and electrified shower clouds that provide current to the ionosphere without producing lightning. For the AC circuit, the background resonances can be monitored at multiple stations (of the order of ten stations) to produce chimney-resolved measures of lightning activity in units of coul^{2}km^2/sec [201,204–206]. The AC global circuit can also be used to locate and monitor the special population of mesoscale lightning flashes worldwide that are most damaging to mankind and infrastructure, by virtue of their exceptional charge transfers and their long continuing currents [207–214]. Conventional lightning detection networks operating in the LF and VLF range do not have sufficient low frequency bandwidth to identify particularly hazardous ground flashes with long continuing current and large charge moment change.

Early attempts to evaluate the DC and AC global circuits as diagnostics for global temperature were made by Williams [30] for the AC global circuit and

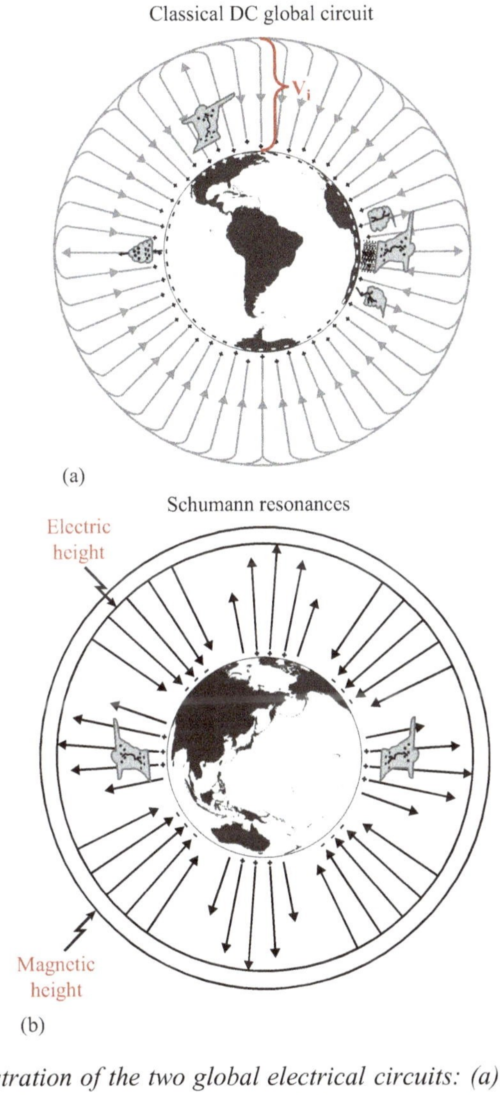

Figure 1.9 Illustration of the two global electrical circuits: (a) the classical DC global circuit, and (b) the AC global circuit, otherwise known as Schumann resonances

by Markson and Price [215] and Markson [92,93] for the DC global circuit. A substantial challenge remaining is to provide for the continuous monitoring of both global circuits over the long time scales that are relevant to climate change. When this problem is overcome, comprehensive diagnostics of global weather will be available in the electromagnetic field.

1.9 Expectations for the future

The foregoing discussion has given emphasis to the response of lightning to climate variables on a number of time scales as well as the growing body of evidence supporting an important role for aerosol in cloud electrification and lightning. On the basis of this body of evidence, one may speculate about changes in lightning and its effect on mankind and infrastructure into the future. As global warming proceeds, uncertainties remain about what quantities are invariant and what quantities are changing. Inferences about changing quantities are often based on their behavior in the current climate, but still uncertainties remain.

If total water vapor in the Earth's atmosphere follows the Clausius–Clapeyron relationship, one expects greater total water and more condensate in a warmer world, and given the need for condensates for lightning, greater lightning is expected. Global climate models [188,189,216] do show increased precipitation in a warmer world, though this change is regionally dependent. It must also be recognized that the nature of the precipitation increase is important and that an increase in warm rain alone is unlikely to be accompanied by an increase in lightning.

Given the evidence for CAPE in lightning in the present climate, more lightning is expected in a warmer climate if CAPE increases. Early speculation showed CAPE to be a climate invariant [41]. Romps *et al.* [217] have predicted increases over the CONUS in a warmer climate on the basis of both increases in CAPE and increases in precipitation. More recent global climate models show larger CAPE in warmer climates [42]. Theoretical calculations in equilibrium atmospheres [218] show CAPE scaling with the Clausius–Clapeyron relationship. Estimations of CAPE changes in the western United States based on the advection of dry desert area over the moist boundary layer with origin in the Gulf of Mexico also lead to predictions that CAPE should scale with Clausius–Clapeyron [45]. For all of these reasons, one expects greater lightning in a warmer world.

A frequent assumption in the climate community is that surface relative humidity is a climate invariant [185,190]. This assumption is based in part on the empirical evidence in the present climate for a quasi-fixed relative humidity near 80% over large areas of the tropical oceans, with an associated CBH for moist convection near 500 m. If the ocean surface temperature were to increase globally, the mean dew point temperature would also increase so as to keep the dew point depression ($T - T_d$) and the associated CBH (above local terrain) both constant. But the height of the 0 °C isotherm would increase following a presumed lifting along a dry adiabat. Accordingly, the warm cloud depth (distance between the 0 °C isotherm and the CBH) would increase. Based on findings in the current climate that the lightning flash rate is decreasing with increased warm cloud depth, in this scenario one might expect less lightning in the warmer world. However, the CAPE change in this scenario also deserves consideration. The expected increases in both T and T_d both contribute to increases in the wet bulb potential temperature of the boundary layer, a result that certainly favors greater CAPE in the current climate.

But the ultimate CAPE change depends also on the change in the overall temperature profile in a warmer climate, and this is indeterminate in the context of the foregoing assumptions.

One recent study [47] predicts less tropical lightning in a warmer world. The prediction is model-based and with the finding that less ice phase is reaching the upper portion of the troposphere where active charge separation is known to occur. More information about this model, and in the context of expected changes in warm cloud depth, is needed to understand this result. Other recent observations on extremes in tropical rainfall in a warmer climate showing enhancements exceeding the simple predictions based on Clausius–Clapeyron [219] would seem to contradict the model predictions of Finney *et al.* [47]. Further discussion on this issue and predictions for the behavior of lightning in a warmer climate can be found in Yair [220].

Given the recent evidence for an important role for CCN in increasing lightning activity in the present climate [23], one can surely speculate about changes in lightning in a more polluted world on the basis of this effect. China for example continues to undergo major industrialization and with a major reliance on coal [221] and so one can expect an increasing aerosol production from that region alone. From what is known at present, the lightning enhancements are expected over a finite range of CCN concentrations, from typical oceanic concentrations of a few per cc to up to about 1,000 per cc [21,23]. Above that level (typical of continental conditions described in Section 1.5), the lightning activity is expected to flatten and then decrease, for reasons that are evident in model calculations [152]. An assessment of where the world stands relative to this important threshold will be much aided by a new satellite method for estimation of CCN concentration at CBH [14].

References

[1] Cooray, V. (Editor): "Lightning Protection" (Series: IET Power and Energy, Institution of Engineering and Technology - IET, London, 2010)

[2] Herring, S.C., N. Christidis, A. Hoell, J.P. Kossin, C.J. Schreck III and P.A. Scott, Editors, Explaining Extreme Event of 2016 from a Climate Perspective, Special Supplement to the Bull. Am. Met. Soc., 99, No. 1, January, 2018.

[3] Aich, V., R. Holzworth, S. J. Goodman, Y. Kuleshov, C. Price and E. Williams (2018), Lightning: A new essential climate variable, *Eos*, 99, https://doi.org/10.1029/2018EO104583. Published on 07 September 2018.

[4] Virts, K. S., J. M. Wallace, M. L. Hutchins and R.H. Holzworth, Highlights of a new ground-based, hourly global lightning climatology. *Bull. Amer. Meteor. Soc.*, 94, 1381–1391, 2013. doi: http://dx.doi.org/10.1175/BAMS-D-12-00082.1

[5] Kinne, S., Remote sensing data combinations-superior global maps for aerosol optical depth, in *Satellite Aerosol Remote Sensing Over Land*, edited by A. Kokhanovsky and G. de Leeuw, Springer, chapter 12, pp. 361–380, 2009.

[6] Boccippio, D.J., S.J. Goodman and S. Heckman, Regional differences in tropical lightning distributions, J. Appl. Met., 39, 2231–2248, 1999.
[7] Whipple, F.J.W., On the association of the diurnal variation of electric potential gradient in fine weather with the distribution of thunderstorms over the globe, Quart. J. Roy. Met. Soc., 55, 1–17, 1929.
[8] Whipple, F.J.W. and F.J. Scrase, Point discharge in the electric field of the Earth, Geophys. Mem., London, 68, 1–20, 1936.
[9] Williams, E.R., Lightning and climate: A review, Atmospheric Research, 76, 272–287, 2005.
[10] Humphreys, W.J., *The Physics of the Air*, Dover Publications, New York, 1964.
[11] Williams, E. and S. Stanfill, The physical origin of the land-ocean contrast in lightning activity, Comptes Rendus—Physique, 3, 1277–1292, 2002.
[12] Rosenfeld, D. and W.L. Woodley, Closing the 5-year circle: From cloud seeding to space and back to climate change through precipitation physics, Chapter 6 of "Cloud Systems, Hurricanes, and the Tropical Rainfall Measuring Mission (TRMM)", ed., W-K. Tao and R. Adler, *Meteorol. Monogr.*, 51, 59–80, 2003.
[13] Rosenfeld, D., U. Lohmann, G.B. Raga, *et al.*, Flood or drought: How do aerosols affect precipitation?, Science, 321, 1309–1313, 2008.
[14] Rosenfeld, D., Y. Zheng, E. Hashimshoni, *et al.*, Satellite retrieval of cloud condensation nuclei concentrations by using clouds as CCN chambers, PNAS, www.pnas.org/cgi/doi/10.1073/pnas.1514044113, 2016.
[15] Venevsky, S., Importance of aerosols for annual lightning production at global scale, Atmos. Chem. Phys. Discuss., 14, 4303–4325, 2014.
[16] Yang, X., Z. Yao and Z. Li, Heavy air pollution suppresses summer thunderstorms in central China, *J. Atmos. & Solar-Terrestrial Phys.*, 28–40, doi:10.1016/j.jastp.2012.12.023, 2013.
[17] Yang, X. and Z. Li, Increases in thunderstorm activity and relationships with air pollution in southeast China, *J. Geophys. Res. Atmos.*, 119, 1835–1844, doi:10.1002/2013JD021224, 2014.
[18] Stolz, D.C., S. A. Rutledge, W. Xu and J.R. Pierce, Interactions between the MJO, aerosols and convection over the Central Indian Ocean, J. Atmos. Sci., doi: 10.1175/JAS-D-0054.1, 2017a.
[19] Stolz, D.C., S.R. Rutledge, J. Pierce and S. van den Heever, A global lightning parameterization based on statistical relationships between environmental factors, aerosols, and convective clouds in the TRMM climatology, J. Geophys. Res. Atmospheres,122, 7461–7492, 2017b.
[20] Fan, J., D. Rosenfeld, Y. Zhang, *et al.*, Substantial convection and precipitation enhancements by ultrafine aerosol particles, Science 359, 411–418, 2018.
[21] Hu, J., D. Rosenfeld, A. Ryzhkov *et al.*, Polarimetric radar convective cell tracking reveals large sensitivity of cloud precipitation and electrification properties to CCN, J. Geophys. Res. Atmos., https://doi.org/10.1029/2019JD030857, October 2019.

[22] Thornton, J.A., K.S. Virts, R.H. Holzworth and T.P. Mitchell, Lightning enhancement over major shipping lanes, Geophys. Res. Lett., doi: 10.1002/2017GL074982, 2017.

[23] Wang, Q., Z. Li, J. Guo, C. Zhao and M. Cribb, The climate impact of aerosols on lightning: Is it detectable from long-term aerosol and meteorological data? Atmos. Chem. Phys., 18, 12797–12816, 2018.

[24] Takahashi, T., Riming electrification as a charge generation mechanism in thunderstorms, J. Atmos. Sci., 35, 1536–1548, 1978.

[25] Takahashi, T., S. Sugimoto, T. Kawano and K. Suzuki, Riming electrification in Hokuriku winter clouds and comparison with laboratory observations, J. Atmos. Sci., https://doi.org/10.1175/JAS-D-16-0154.1, 2017.

[26] Reynolds, S.E., M. Brook and M. Gourley, Thunderstorm charge separation, J. Meteorol, 14, 426–436, 1957.

[27] Hansen, J.E. and S. Lebedeff, Global trends of measured surface air temperature, J. Geophys. Res., 92, 13345–13372, 1987.

[28] Hansen, J., R. Ruedy, M. Sato and K. Lo, Global surface temperature change, Rev. Geophys., 48, RG4004, doi:10.1029/2010RG000345, 2010.

[29] Emanuel, K.A., *Atmospheric Convection*, Oxford University Press, 580 pp., 1994.

[30] Williams, E.R., The Schumann resonance: A global tropical thermometer, Science, 256, 1184–1187, 1992.

[31] Price, C., Evidence for a link between global lightning activity and upper tropospheric water vapor, Nature, 406, 290–293, 2000.

[32] Ludlam, F.H., *Clouds and Storms: The Behavior and Effect of Water in the Atmosphere*, Pennsylvania State University Press, 1980, 405 pp.

[33] Iribarne, J.V. and W. L. Godson, *Atmospheric Thermodynamics*, D. Reidel, 1981 259 pp.

[34] Betts, A.K., Saturation point analysis of moist convective overturning, J. Atmos. Sci., 39, 1484–1505, 1982.

[35] Xu, K.-M. and K.A. Emanuel, Is the tropical atmosphere conditionally unstable?, Mon. Wea. Rev.,117, 1471–1479, 1989.

[36] Williams, E.R. and N.O. Renno, An analysis of the conditional instability of the tropical atmosphere, Mon. Wea. Rev.,121, 21–36, 1993.

[37] Brunt, D., *Physical and Dynamical Meteorology*, Cambridge University Press, 1952.

[38] Lucas, C., E.J. Zipser and M.A. Lemone, Convective available potential energy in the environment of oceanic continental clouds: Correction and comments, Notes and Correspondence, J. Atmos. Sci., 51, 3829–3830, 1994.

[39] Riemann-Campe, K., K. Fraedrich and F. Lunkeit, Global climatology of convective available potential energy (CAPE) in ERA-40 reanalysis, Atmos. Res., 93, 534–545, 2009.

[40] Zipser, E.J., Some views on "hot towers" after 50 year of tropical field programs and two years of TRMM data, Chapter 5 in Meteorological Monographs, American Meteorological Society, 49–58, 2003.

[41] Emanuel, K.A., J.D. Neelen and C.S. Bretherton, On large-scale circulations in convecting atmospheres, Quart. J. Roy. Met. Soc., 120, 1111–1143, 1994.

[42] Williams, E.R., Lightning and Climate, Franklin Lecture (invited), http://fallmeeting.agu.org/2012/events/franklin-lecture-ae31a-lightning-and-climate-video-on-demand/, Fall Meeting, American Geophysical Union, San Francisco, December, 2012.

[43] Ye, B., A.D. Del Genio and K.-W. Lo, CAPE variations in the current climate and in a climate change, J. Climate, 11, 1997–2015, 1998.

[44] Del Genio, A.D., Y. Mao-Sung and J. Jonas, Will moist convection be stronger in a warmer climate? Geophys. Res. Lett., 34, L16703, doi: 10.1029/2007GL030525, 2007.

[45] Agard, V. and K. Emanuel, Clausius-Clapeyron scaling of peak CAPE in continental convective storm environments, J. Atmos. Sci., doi: 10.1175/JAS-D-16-0352.1, 2017.

[46] Seeley, J. T. and D. M. Romps, Why does tropical convective available potential energy (CAPE) increase with warming?, Geophys. Res. Lett., 42, 10,429–10,437, doi:10.1002/2015GL066199, 2015.

[47] Finney, D.L., R.M. Doherty. O. Wilde, D.S. Stevenson, I.A. MacKenzie and A.M. Blyth, A projected decrease in lightning under climate change, Nature Climate Change, https://doi.org/10.1038/S41558-018-0072-6, 2018.

[48] Byers, H.R. and R.R.Braham, *The Thunderstorm Project*, U.S. Weather Bureau, U.S. Department of Commerce, Washington, D.C., 1949.

[49] Lemone, M.A.and E.J. Zipser, Cumulonimbus vertical velocity events in GATE. Part I: Diameter, intensity and mass flux, J. Atmos. Sci., 37, 2444–2457, 1980.

[50] Boccippio, D.J., Lightning scaling relations revisited, J. Atmos. Sci., 59, 1086–1104, 2001.

[51] Zheng, Y. and D. Rosenfeld, Linear relation between convective cloud base height and updrafts and application to satellite retrievals, Geophys. Res. Lett., 42, doi:10.1002/2015GL064809, 2015.

[52] Hansen, Z.R. and L.E. Back, Higher surface Bowen ratios ineffective at increasing updraft intensity, Geophys. Res. Lett., 42, doi:10.1002/2015GL066878, 2015.

[53] Williams, E.R., V.C. Mushtak, D. Rosenfeld, S.J. Goodman and D.J. Boccippio, Thermodynamic conditions favorable to superlative thunderstorm updraft, mixed phase microphysics and lightning flash rate, Atmospheric Research, 76, 288–306, 2005b.

[54] Williams, E., D. Rosenfeld, N. Madden, C. Labrada, J. Gerlach and L. Atkinson, The role of boundary layer aerosol in the vertical development of precipitation and electrification: Another look at the contrast between lightning over land and over ocean, 11th Int'l Conf. on Atmospheric Electricity, Guntersville, AL, June 7–11, 1999.

[55] Freud, E. and D. Rosenfeld, Linear relation between convective cloud drop number concentration and depth for rain initiation, J. Geophys. Res., 117, D-2207, doi:10.1029/2011JD016457, 2012.

[56] Mattos, E.V., L.A.T. Machado, E. Williams, S.J. Goodman, R.J. Blakeslee and J.C. Bailey, Electrification life cycle of incipient thunderstorms, *J. Geophys. Res.*, https://doi.org/10.1002/2016JD025772, 2017.

[57] Liu, C. and E.J. Zipser, "Warm rain" in the tropics: Seasonal and regional distributions based on 9 yr of TRMM data, J. Climate, 22, 767–779, DOI: 10.1175/2008CLI2641.1, 2009.

[58] Braga, R.C., D. Rosenfeld, R. Weigel, *et al.*, Further evidence for CCN aerosol concentrations determining the height of warm rain and ice initiation in convective clouds over the Amazon basin, Atmos. Chem. Phys., 17, 14433–14456, 2017.

[59] Atlas, D., The balance level in convective storms, J. Atmos. Sci., 23, 635–651, 1966.

[60] Lhermitte, R. and E. Williams, Thunderstorm electrification: A case study, J. Geophys. Res., 90(D4), 6071–6078, doi:10.1029/JD090iD04p06071, 1985.

[61] Williams, E. and R. Lhermitte, Radar tests of the precipitation hypothesis for thunderstorm electrification, *Journal of Geophysical Research*, 88, 10984–10991, 1983.

[62] Lambrechts, J. de V., The value of water drainage in upcast mine shafts and fan-drifts, J. Chem., Metal. and Mining Soc. South Africa, 56, 307–324 and 383–387, 1956.

[63] Carte, A.E., Mine shafts as a cloud physics facility, Proc. Int'l Conference on Cloud Physics, Toronto, Canada, August 26–30, 1968.

[64] Carte, A.E. and P.E. Lourens, Water year 1970: Convention on water for the future; South Africa (Republic) 1970, Studies of cloud and rain in mine shafts, Department of Water Affairs, 1970.

[65] Zipser, E.J., Deep cumulonimbus clouds in the tropics with and without lightning, Mon. Wea. Rev., 122, 1837–1851, 1994.

[66] Lindzen, R.S., *Dynamics in Atmospheric Physics*, Cambridge University Press, 1990.

[67] Charney, J.G., The dynamics of long waves in a baroclinic westerly current, J. Meteorology, 4, 135–162, 1947.

[68] Wallace, J.M. and P.V. Hobbs, *Atmospheric Science: An Introductory Survey*, Academic Press, 1977.

[69] Yoshida, S., T. Morimoto, T. Ushio and Z. Kawasaki, A fifth-power relationship for lightning activity from Tropical Rainfall Mission satellite observations, J. Geophys. Res., 114, D09104, doi:10.1029/2008JD010370, 2009.

[70] Bang, S.D. and E.J. Zipser, Differences in size spectra of electrified storms over land and ocean, Geophys. Res. Lett., 42, doi: 10.1002/2015GL065264, 2015.

[71] Bang, S.D. and E.J. Zipser, Seeking reasons for the differences in size spectra of electrified storms over land and ocean, J. Geophys. Res., 121, 9048–9068, 2016.

[72] Orville, R.E. , Cloud-to-ground lightning in the Blizzard of '93, Geophys. Res. Lett., 20, 1367–1370, 1993.

[73] Pessi, A.T. and S. Businger, The impact of lightning data assimilation on a winter storm simulation over the North Pacific Ocean, Mon. Wea. Rev., 137, 3177–3195, 2009.

[74] Bluestein, H.B., *Synoptic-Dynamic Meteorology in Midlatitudes, Vol. II: Observations and Theory of Weather Systems*, Oxford University Press, 1993.

[75] Doswell, C.A., III, Severe Convective Storms, *Meteorological Monographs*, Vol. 28, Number 50, American Meteorological Society, 2001.

[76] Ludlam, F.H. Severe Local Storms: A Review, in *Severe Local Storms*, Meteorol. Monograph No. 27, American Meteorological Society, 1–30, 1963.

[77] Emersic, C., P.L. Heinselman, D.R. MacGorman and E.C. Bruning, Lightning activity in a hail-producing storm observed with phased-array radar, Mon. Wea. Rev., 139, 1805–1825, 2011.

[78] Korolev, A., G. McFarquhar, P.R. Field, *et al.*, Mixed-Phase Clouds: Progress and Challenges, Meteorological Monographs, American Meteorological Society, https://doi.org/10.1175/AMSMONOGRAPHS-D-17-0001.1, 2017.

[79] Rust, W.D., D.R. MacGorman, E.C. Bruning, *et al.*, Inverted-polarity electrical structures in thunderstorms in the Severe Thunderstorm Electrification and Precipitation Study (STEPS), Atmos. Res., 76, 247–271, doi:10.1016/j.atmosres.2004.11.029, 2005.

[80] Stough, S.M., L.D. Carey, C.J. Schultz and P.M. Bitzer, Investigating the relationship between lightning and mesocyclonic rotation in supercell thunderstorms, Wea. Forecasting, 32, 2237–2259, 2017.

[81] Williams, E.R., The electrification of severe storms, 527–561, in *Severe Convective Storms*, Ed. C.A. Doswell, III, American Meteorological Society, 561 pp., 2001.

[82] Blakeslee, R.J., K.T. Driscoll, D.E. Buechler, *et al.*, 1999. Diurnal lightning distribution as observed by the Optical Transient Detector (OTD). 11th International Conf. on Atmos. Elec., Proceedings, NASA/CP–1999–209261, 742–745, Guntersville, Alabama, June 7–11, 1999.

[83] Blakeslee, R.J., D.M. Mach, M.G. Bateman and J.C. Bailey, Seasonal variations in the lightning diurnal cycle and implications for the global electric circuit, Atmos. Res., 135–136, 228–243, 2014.

[84] Williams, E.R., K. Rothkin, D. Stevenson and D. Boccippio, Global lightning variations caused by changes in thunderstorm flash rate and by changes in the number of thunderstorms, J. Appl. Met. (TRMM Special Issue), 39, 2223–2248, 2000.

[85] Price, C. Global surface temperature and the atmospheric electric circuit, Geophys. Res. Lett., 20, 1363–1366, 1993.

[86] Burns, G.B., A.V. Frank-Kamenetsky, O.A. Troschichev, E.A. Bering and B.D. Reddell, Interannual consistency of bi-monthly differences in annual variations of the ground-level, vertical electric field, J. Geophys. Res., 110, D10106, doi:10.1029/2004JD005469, 2005.

[87] Burns,G.B., A. V. Frank-Kamenetsky, B. A. Tinsley, *et al.*, Atmospheric global circuit variations from Vostok and Concordia electric field, J. Atmos. Sci., http://dx.doi.org/10.1175/JAS-D-16-0159.1, March 1, 2017.

[88] Israel, H., *Atmospheric Electricity*, Vol. II, Israel Program for Scientific Translations, Ltd, ISBN 0 7065 1129 8, 1973.

[89] Bailey, J.C., R.J. Blakeslee, D.E. Buechler and H.J. Christian, Diurnal lightning distributions as observed by the Optical Transient Detector (OTD) and the Lightning Imaging Sensor (LIS). Proceedings of the 13th International Conf. on Atmos. Elec., Vol. II, pp. 657–660. Organized by the International Commission on Atmospheric Electricity (ICAE/IAMAS/IUGG): August 13–17, 2007, Beijing Friendship Hotel, China, 2007.

[90] Cecil, D.J., D.E. Buechler and R.J. Blakeslee, Gridded lightning climatology from TRMM-LIS and OTD: Dataset description, Atmos. Res., 135–136, 404–414, 2014.

[91] Williams, E.R., Global circuit response to temperature on distinct time scales: A status report, in Atmospher*ic* and Ionospheric Phenomena *Associated* with Earthquakes, Ed., M. Hayakawa), Terra Scientific Publishing (Tokyo), 1999.

[92] Markson, R., Ionospheric potential variation from temperature change over continents, Proceedings of the 12th International Conf. on Atmos. Elec., Vol I, 283–286, Versailles, France, June 9–13, 2003.

[93] Markson, R., The global circuit intensity: Its measurement and variation over the last 50 years, Bull. Am. Met. Soc., doi:10.1175/BAMS-88-2-223, 223–241, 2007.

[94] Phillipin, S. and E.A. Betterton, Cloud condensation nuclei concentrations in Southern Arizona: Instrumentation and early observations, Atmos. Res. 43, 263–275, 1997.

[95] Kendrew, W.G., *The Climates of the Continents*, Oxford University Press, 3rd Edition, New York, 473 pp., 1942.

[96] Williams, E.R., Global circuit response to seasonal variations in global surface air temperature, Mon. Wea. Rev.,122, 1917–1929, 1994.

[97] Christian, H.J., R.J. Blakeslee, D.J. Boccippio, *et al.*, Global frequency and distribution of lightning as observed from space by the Optical Transient Detector, J. Geophys. Res., 108, 4005, doi: 10.1029/2002JD002347, 2003.

[98] Füllekrug, M. and A. Fraser-Smith, Global lightning and climate variability inferred from ELF field variations, Geophys. Res. Lett., 24, 2411–2414, 1997.

[99] Hogg, A.R., Air-earth current observations in various localities, Arch. Meteor., 3, 40–55, 1950.

[100] Sátori, G. and B. Zieger, Spectral characteristics of Schumann resonances observed in Central Europe. J Geophys Res., 101, 29663–29669, 1996.

[101] Nickolaenko A.P., G. Sátori, B. Zieger, L.M. Rabinowicz and I.G. Kudintseva, Parameters of global thunderstorm activity deduced from the long-term Schumann resonance records *J. Atmos. Sol.-Terr. Phys.*, 60, 387–399, 1998.

[102] Sátori, G., V. Mushtak and E. Williams, Schumann Resonance Signatures of Global Lightning Activity, in *Lightning: Principles, Instruments and Applications*, Ed.: H.-D Betz, U. Schumann and P. Laroche, Springer, ISBN: 978-1-4020-9078-3, pp. 347–386, 2009a.
[103] Kelvin, L., Atmospheric Electricity. Royal Institution Lecture, Papers on Electricity and Magnetism, 1860 pp. 208–226.
[104] Adlerman, E.J. and E.R. Williams, Seasonal variations of the global electrical circuit, *J.* Geophys. Res., 101, 29679–29688, 1996.
[105] Retalis, D., Study of the air-earth electrical current density in Athens, Pure and Applied Geophysics, 136, 217–233, 1991.
[106] Sátori, G., E. Williams and I. Lemperger, Variability of global lightning activity on the ENSO time scale, Atmos. Res., 91, 500–507, 2009b.
[107] Ropelewski, C.F. and M.S. Halpert, Global and Regional Scale Precipitation Patterns Associated with the El Nino/Southern Oscillation. Monthly Weather Review, 115, 1606–1626, 1987.
[108] Amerasekera, K.N., R.F. Lee, E.R. Williams and E.A.B. Eltahir, ENSO and the natural variability in the flow of tropical rivers, J. Hydrology, 200, 24, 1997.
[109] Whitaker, D.W., S.A. Wasimi and S. Islam, The El Nino-Southern Oscillation and long-range forecasting of flows in the Ganges, Int'l J. Climatology, 21, 77–87, 2001.
[110] Allan, R., J. Lindesay and D. Parker, *El Nino Southern Oscillation and Climatic Variability*, CSIRO Publishing, 1996, 405 pp.
[111] Hamid, E.Y., Z.-I. Kawasaki and R. Mardiana, Impact of the 1997–98 El Nino events on lightning activity over Indonesia, Geophys. Res. Lett., 28, 147–150, 2001.
[112] Kulkarni, M.K., J.V. Revadekar, H. Verikoden and S. Athale, Thunderstorm days and lightning activity in association with El Nino, Vayu Mandal, 41, 39–43, 2015.
[113] Kumar, P. and A.K. Kamra, Impact of the 1997–98 and 2002–03 ENSOs on lightning activity in South/Southeast Asia, Atmos. Res., 118, 84–102, 2012.
[114] Manohar, G.K, S.S. Kandalgaonkar and M.I.R. Tinmaker, Thunderstorm activity over India and Indian southwest monsoon, J. Geophys. Res., 104, 4169–4188, 1999.
[115] Saha, U., D. Siingh, S.K. Midya, R.P. Singh, A.K. Singh and S. Kumar, Spatio-temporal variability of lightning and convective activity over south southeast Asia with an emphasis during El Nino and La Nina, Atmos Res., 197, 150–166, 2017.
[116] Tinmaker, M.I.R., M.V. Aslam, S.D. Ghude and D. M. Chate, Lightning activity with rainfall during El Nino and La Nina events over India, Theor. Appl. Climatology, 130, 391–400, 2017.
[117] Yoshida, S., T. Morimoto, T. Ushio and Z. Kawasaki, ENSO and convective activities in Southeast Asia and western Pacific, Geophys. Res. Lett., 34, L21806, doi: 10.1029/2007GL030758, 2007.

[118] Petersen, W.A. and S.A. Rutledge, Regional variability in tropical convection: Observations from TRMM, J. Climate, 14, 3566–3586, 2001.

[119] Williams, E., S. Rutledge, S. Geotis, N. Renno, E. Rasmussen and T. Rickenback, A radar and electrical study of tropical 'hot towers', J. Atmos. Sci., 49, 1386–1395, 1992.

[120] LaVigne, T., C. Liu, W. Deierling and D. Mach, Relationship between the global electric circuit and electrified cloud parameters at diurnal, seasonal and interannual time scales, J. Geophys. Res., 122, doi: 10.1002/2016JD026442, 2017.

[121] Peterson, M., W. Deierling, C. Liu, D. Mach and C. Kalb, A TRMM/GPM retrieval of the total mean source current for the global electrical circuit, J. Geophys. Res., 122, 10025–10049, doi: 10.1002/2016JD026336, 2017.

[122] Liu, C., E.R. Williams, E.J. Zipser and G. Burns, Diurnal variation of global thunderstorms and electrified shower clouds and their contribution to the global electrical circuit, J. Atmos. Sci., 67, 309–323, 2010.

[123] Rycroft, M. J., A. Odzimek, N.F. Arnold, M. Fűllekrug, A. Kulak and T. Neubert, New model simulations of the global atmospheric electric circuit driven by thunderstorms and electrified shower clouds: The roles of lightning and sprites, Journal of Atmospheric and Solar-Terrestrial Physics, 69, 2485–2509, 2007.

[124] Williams, E.R. and G. Sátori, Lightning, thermodynamic and hydrological comparison of the two tropical continental chimneys, J. Atmos. Sol. Terr. Phys., 66, 1213–1231, 2004.

[125] Wilson, C.T.R., Investigations on lightning discharges and on the electric field of thunderstorms, Phil. Trans. A. 221, 73–115, 1920.

[126] Chronis, T.G., S.J. Goodman, D. Cecil, D. Buechler, F.J. Robertson and J. Pittman, Global lightning activity from the ENSO perspective, Geophys. Res. Lett., 35, L19804, doi:10,1029/2008GL034321, 2008.

[127] Pinto, O., III, Thunderstorm climatology of Brazil: ENSO and Tropical Atlantic connections, Int'l J. Climatology, 35, 871–878, 2015.

[128] Richey, J.E., C. Nobre and C. Deser, Amazon discharge and climate variability: 1903–1985, Science, 246, 101–103, 1989.

[129] Williams, E., A. Dall'Antonia, V. Dall'Antonia, *et al.*, The drought of the century in the Amazon Basin: An analysis of the regional variation of rainfall in South America in 1926. *Acta Amazonica*, 35(2), 231–238, 2005a.

[130] Grimm, A.M. and A. Natori, Climate change and interannual variability of precipitation in South America, Geophys. Res. Lett., 33,L19706, doi: 10.1029/2006GL026821, 2006.

[131] Goodman, S.J., D.E. Buechler, K. Knupp, D. Driscoll, and E.E. McCaul, Jr.,The 1997–98 El Nino event and related wintertime lightning variations in the southeastern United States, Geophys. Res. Lett., 27, 541–544, 2000.

[132] Dowdy, A.J., Seasonal forecasting of lightning and thunderstorm activity in tropical and temperature regions of the world, Nature, Scientific Reports, doi: 10.1038/srep20874,11 Feb, 2016.

[133] Marengo, J.A., E.R. Williams, L.M. Alves, W.R. Soares and D.A. Rodriguez, Extreme seasonal climate variations in the Amazon basin: Droughts and Floods, Chapter 4 in *Interactions between Biosphere, Atmosphere and Human Land Use*, Ecological Series 227, Eds. L. Nagy, B.R. Forsberg and P. Artaxo, Springer-Verlag, Berlin, 2016.
[134] Wu., Y.J., A.B. Chen, H.H. Hsu, *et al.*, Occurrence of elves and lightning during El Nino and La Nina, Geophys. Res. Lett., 39, L03106, doi: 10.1029/2011GL049831, 2012.
[135] Wu., Y.J., H.H. Hsu, A.B. Chen and K.-M. Peng, Revisiting oceanic elves and lightning occurrence rate during El Niño and La Nina episodes in a 10-year time frame, Terr. Atmos. Ocean Sci., 28, doi:10.3319/TAO.2016.07.28.01, 2016.
[136] Harrison, R.G., M Joshi and K. Pascoe, Inferring convective response to El Nino with atmospheric electricity measurements at Shetland, Environ. Res. Lett., 6, 1–3, 2011.
[137] Camp, C.D. and Tung, K.K., Surface warming by the solar cycle as revealed by the composite mean difference projection, *Geophys. Res. Lett.* 34, L14703, doi:10.1029/2007GL030207, 2007.
[138] Tung, K.K. and Camp C.D. Solar cycle warming at the Earth's surface in NCEP and ERA-40 data: A linear discriminant analysis, *J. Geophys. Res.*, 113, D05114, doi:10.1029/2007JD009164, 2008.
[139] Zhou, J. and Tung, K.K., Deducing multidecadal anthropogenic global warming trends using multiple regression analysis, J. Atmos. Sci., 70, 9–14, 2013.
[140] Brooks, C.E.P., The variation of the annual frequency of thunderstorms in relation to sunspots, Quart. J. Roy. Met. Soc., 60, 153–165, 1934.
[141] Kleymenova, E.P., On the variation of the thunderstorm activity in the solar cycle, Glav. Upirav. Gidromet. Scuzb., Met. Gidr. 8, 64–68, 1967.
[142] Pinto, O., III, O. Pinto, Jr. and I.R.C.A. Pinto, The relationship between thunderstorm and solar activity for Brazil from 1951 to 2009, J. Atmos. Sol. Terr. Phys., 98, 12–23, 2013.
[143] Williams, E.R., Comments on "11-year cycle in Schumann resonance data as observed in Antarctica" by Nickolaenko *et al.* (2015), *Sun and Geosphere*, 11, 75–76, 2016.
[144] Koshak, W.J., K. L. Cummins, D. E. Buechler, *et al.*, Variability of CONUS lightning in 2003–12 and associated impacts, J. Appl. Met. Clim., 54, 15–41, 2015.
[145] Boccippio, D.J., K. Cummins, H.J. Christian and S. J. Goodman, Combined satellite- and surface-based estimation of the intracloud–cloud-to-ground lightning ratio over the continental United States, Mon. Wea. Rev., 129, 108–122, 2001.
[146] Price, C. and D. Rind, A simple lightning parameterization for calculating global lightning distributions, J. Geophys. Res., 9919–9933, 1992.
[147] Yamamoto, K., T. Nakashima, S. Sumi and A. Ametani, About 100 years survey of the surface temperatures of Japan Sea and lightning days along

the coast, Int'l Conf. on Lightning Protection, Estoril, Portugal, 25–30 September, 2016.

[148] Hansen, J., M. Sato, R. Ruedy, G.A. Schmidt, K. Lo, and A. Persin, Global temperature in 2017, NASA GISS website, 2018.

[149] Williams, E., A. Guha, R. Boldi, H. Christian and D. Buechler, Global lightning activity and the hiatus in global warming, J. Atmos. Sol. Terr. Phys., 189, 27–34, 2019.

[150] Houghton, H.G., *Physical Meteorology*, MIT Press, 1985.

[151] Altaratz, O., I. Koren, Y. Yair and C. Price, Lightning response to smoke from Amazon fires, Geophys. Res. Lett., 37, L07801, doi:10.1029/2010GL042679, 2010.

[152] Mansell, E.R. and C. L. Ziegler, Aerosol effects on simulated storm electrification and precipitation in a two-moment bulk microphysics model, J. Atmos. Sci., 70, 2032–2050, 2013.

[153] Rosenfeld, D. and I. Lensky, Satellite-based insights into precipitation formation processes in continental and maritime convective clouds, Bull. Am. Met. Soc., 79, 2457–2476, 1998.

[154] Orville, R.E., G. Huffines, J. Nielson-Gammon, *et al.*, Enhancement of cloud-to-ground lightning over Houston, Texas, Geophys. Res. Lett., 28, 2597–2600, 2001.

[155] Stolz, D.C., S.R. Rutledge and J.R. Pierce, Simultaneous influences of thermodynamics and aerosols on deep convection and lightning in the tropics, J. Geophys. Res.: Atmospheres, 120, 12, 6207, 2015.

[156] Altaratz O., B. Kucienska, A. Kostinski, G.B. Raga and I. Koren, Global association of aerosol with flash density of intense lightning, Env. Res. Lett., 114037, 2017.

[157] Yuan, T., L.A. Remer, K.E. Pickering and H. Yu, Observational evidence of aerosol enhancement of lightning activity and convective invigoration, Geophys. Res. Lett., 38, L04701, doi:10.1029/2010GL046052, 2011.

[158] Keskinen, J. and T. Ronkko, Can real-world diesel exhaust particle size distribution be reproduced in the laboratory? A critical review, Air and Waste Management Assoc., 60, 1245–1255, doi:10.3155/1047-3289.60.10.1234, 2010.

[159] Li, Z., F. Niu, J. Fan, Y. Liu, D. Rosenfeld and Y. Ding, Long-term impacts of aerosols on the vertical development of clouds and precipitation, Nature Geoscience, 4, 888–894, 2011.

[160] Pawar, S.D., V. Gopalakrishnan, P. Murugavel, N.E. Veremey and A.A. Sinkevich, Possible role of aerosols in the charge structure of isolated thunderstorms. Atmos. Res., 183, 331–340, 2017.

[161] Qie, X., T. Zhang, G. Zhang, T. Zhang and Z. Kong, Electrical characteristics of thunderstorms in different plateau regions of China, Atmos. Res., 91, 244–249, 2009.

[162] Lyons, W.A., T.E. Nelson, E.R. Williams, J. Cramer, and T. Turner, Enhanced positive cloud-to-ground lightning in thunderstorms ingesting smoke, Science, 282, 77–81, 1998.

[163] Bell, T.L., D. Rosenfeld, K.-M. Kim, J.-M Yoo, M..-I. Lee and M. Hahnenberger, Midweek increase in U.S. summer rain and storm heights suggest air pollution invigorates rainstorms, J. Geophys. Res., 113, doi:1029/2007JD008623, 2008.

[164] Bell, T.L., D. Rosenfeld and K.-M. Kim, Weekly cycle of lightning: Evidence of storm invigoration by pollution, Geophys. Res. Lett., 36, L23805, doi:10.1029/2009GL040915, 2009.

[165] Rosenfeld, D. and T.L. Bell, Why do tornadoes and hailstorms rest on weekends?, J. Geophys. Res., 116, D20211, doi: 10.1029/2011JD016214, 2011.

[166] Brooks, C.E.P., The distribution of thunderstorms over the globe, *Geophys. Mem. London*, 24, 147–164, 1925.

[167] Kotaki, M. and C. Katoh, The global distribution of thunderstorm activity observed by the Ionospheric Sounding Satellite (ISS-b), J. Atmos. Terr. Phys., 45, 833, 1983.

[168] Orville, R.E. and R.W. Henderson, The global distribution of midnight lightning: December 1977 to August 1978, Mon. Wea. Rev., 114, 2640–2653, 1986.

[169] Kukla, G. and T.R. Karl, Nighttime warming and the greenhouse effect, Environ. Sci. Technol., 27, 1468–1474, 1993.

[170] Thorne, P.W., M.G. Donat, R.J.H. Dunn, *et al.*, Reassessing changes in diurnal temperature range: Intercomparison and evaluation of existing data set estimates, J. Geophys. Res., Atmos., 121, 5138–5158, doi:10.1002/2015JD024584, 2016.

[171] Henderson, J.P., Some aspects of climate of Uganda with special reference to rainfall, E. African Meteor. Dept. Memoirs, 2, No 5, 1–16, 1949.

[172] Albrecht, R.I., Where are the lightning hot spots on Earth?, Bull. Am. Met. Soc., https://doi.org/10.1175/BAMS-D-14-00193.1, 2016.

[173] Bürgesser, R.E., M.G. Nicora, and E.E. Avila, Characterization of the lightning activity of "Relampago del Catatumbo", J. Atmos. Sol. Terr. Phys., doi:10.1016/j.jastp.2012.01.013, 2012.

[174] Neumann, J., Land breezes and nocturnal thunderstorms, J. Meteorol., 8, 60–67, 1950.

[175] Hales, J.E., Jr., On the relationship of convective cooling to nocturnal thunderstorms at Phoenix, Mon. Wea. Rev., 105, 1609–1613, 1977.

[176] Zipser, E.J., C. Liu, D.J. Cecil, S.W. Nesbitt and D.P. Yorty, Where are the most intense thunderstorms on Earth?, Bull. Am. Meteorol. Soc., 87(8), 1057–1071, doi:10.1175/BAMS-87-8-1057, 2006.

[177] Wallace, J.M., Diurnal variations in precipitation and thunderstorm frequency over the conterminous United States, Mon. Wea. Rev., 103, 406–419, 1975.

[178] Reif, D.W. and H.N. Bluestein, A 20-year climatology of nocturnal convection initiation over the central and southern Great Plains during the warm season, Mon. Wea. Rev., 145, 1615–1639, 2017.

[179] Wilson, J.W., S.B. Trier, D.W. Reif, R.D. Roberts and T.M. Weckwerth, Nocturnal elevated convection initiation of the PECAN 4 July hailstorm, Mon. Wea. Rev., 146, 243–262, 2018.

[180] Dessens, J., Severe convective weather in the context of a nighttime global warming, Geophys. Res. Lett., 22, 1241–1244, 1995.

[181] Colman, B.R., Thunderstorms above frontal surfaces in environments with positive CAPE. Part I: A Climatology, Mon. Wea. Rev., 118, 1103, 1990.

[182] Vose, R.S., D.R. Easterling and B. Gleason, Maximum and minimum temperature trends for the globe: An update through 2004, Geophys. Res. Lett., 32, L23822, doi:10.1029/2005GL024379, 2005.

[183] Williams, E.R., E.V. Mattos and L.T. Machado, Stroke multiplicity and horizontal scale of negative charge regions in thunderclouds, *Geophys. Res. Lett.*, 43, doi:10.1002/2016GL068924, 2016.

[184] Mushtak, V.C., E.R. Williams and D.J. Boccippio, Latitudinal variations of cloud base height and lightning parameters in the tropics, Atmospheric Research, 76, 222–230, 2005.

[185] Dai, A., Recent climatology, variability, and trends in global surface humidity, J. Clim., 19, 3589–3606, 2006.

[186] Held, I.M. and B.J. Soden, Water vapor feedback and global warming, Annual Reviews Energy and Environ., 25, 441–475, 2000.

[187] O'Gorman, P.A. and C.J. Muller, How closely do changes in surface and column water vapor follow Clausius-Clapeyron scaling in climate change simulations? Env. Res. Lett., doi:10.1088/1748-9326/5/2/025207, 2010.

[188] Soden, B.J. and I.M. Held, An assessment of climate feedbacks in coupled ocean–atmosphere models, J. Clim., 19, 3354–3360, 2006a.

[189] Soden, B.J. and I.M. Held, Robust responses of the hydrological cycle to global warming. J. Clim., 19, 5686–5699, 2006b.

[190] Byrne, M.P. and P.A. O'Gorman, Trends in continental temperature and humidity directly linked to ocean warming, Proc. Nat'l Acad. Sci., 115, 4863–4868, https://doi.org/10.1073/pnas.1722312115, 2018.

[191] Markson, R., L.H. Ruhnke and E.R. Williams, Global scale comparison of simultaneous ionospheric potential measurements, Atmos. Res., 51, 315–321, 1999.

[192] Mühleisen, R., The global circuit and its parameters, 467–476, In: Dolezalek, H. and Reiter, R., (Eds.), *Electrical Processes in Atmospheres*, Steinkopff, Darmstadt, 1977.

[193] Bering, E.A., A.A. Few and J.R. Benbrook, The global electric circuit, Physics Today, 51 24–30, 1998.

[194] Chalmers, J.A., *Atmospheric Electricity*, 2nd Edition, Pergamon Press, 1967.

[195] Williams, E.R., The global electrical circuit: A Review, Atmospheric Research, 91, 140–152, 2009.

[196] Heckman, S., E. Williams and R. Boldi, Total global lightning inferred from Schumann resonance measurements, J. Geophys. Res., 103, 31775–31779, 1998.

[197] Madden, T.R. and W.B. Thompson, Low-frequency electromagnetic oscillations of the Earth-ionosphere cavity, Rev. Geophys., 3, 211, 1965.

[198] Nickolaenko, A.P. and M. Hayakawa, *Resonances in the Earth-ionosphere cavity*, Kluwer Academic Publishers, 2002.

[199] Polk C., Schumann Resonances. In *CRC Handbook of Atmospherics*, Vol. I, H. Volland, ed., CRC Press, Boca Raton, FL., pp. 55–82, 1982.

[200] Sentman, D.D., Schumann resonances, Chapter 11 in *Handbook of Atmospheric Electrodynamics*, Vol. I, Ed. H. Volland, CRC Press, 408 pp., 1995.

[201] Williams, E.R. and E.A. Mareev, Recent progress on the global electrical circuit, Atmos. Res., Volumes 135–136, 208–227, 2014.

[202] Williams, E., R. Boldi, R. Markson and M. Peterson, Comparative behavior of the DC and AC global circuits, XVI International Conference on Atmospheric Electricity, Nara city, Nara, Japan, 17–22 June, 2018.

[203] Mühleisen, R.P., New determination of the air-earth current over the ocean and measurement of ionospheric potential, PAGEOPH, 84, 112–115, 1971.

[204] Dyrda, M. A., M. Kulak, M. Mlynarczyk, *et al.*, Application of the Schumann resonance decomposition in characterizing the main African thunderstorm center, J. Geophys. Res., Atmospheres, 119, 13338–13349, 2014.

[205] Williams, E.,V. Mushtak, A. Guha, *et al.*, Inversion of multi-station Schumann resonance background records for global lightning activity in absolute units, Fall Meeting, American Geophysical Union, San Francisco, December, 2014.

[206] Prácser, E., T. Bozóki,, G. Sátori, E. Williams, A. Guha and H. Yu, Reconstruction of global lightning activity based on Schumann Resonance measurements: Model description and synthetic tests, Radio Science, 54, 254–267, 2019.

[207] Guha, A., E. Williams, R. Boldi, *et al.*, Aliasing of the Schumann resonance background signal by sprite-associated Q-bursts, J. Atmos. Sol. Terr. Phys., 165–166, 25–37, 2017.

[208] Hobara, Y., E. Williams, M. Hayakawa, R. Boldi and E. Downes, Location and electrical properties of sprite-producing lightning from a single ELF site, in *Sprites, Elves and Intense Lightning Discharges*, Ed. M. Fullekrug, E.A. Mareev and M.J. Rycroft, NATO Science Series, II, Mathematics, Physics and Chemistry- Vol. 225, Springer, 2006.

[209] Huang, E., E. Williams, R. Boldi, *et al.*, Criteria for sprites and elves based on Schumann resonance observations, J. Geophys. Res., 104, 16943–16964, 1999.

[210] Kemp, D.T., The global location of large lightning discharges from single station observations of ELF disturbances in the Earth-ionospheric cavity, *J. Atmos. Terr. Phys.*, 33, 919–928, 1971.

[211] Kemp, D.T., and D.L. Jones, A new technique for the analysis of transient ELF electromagnetic disturbances within the Earth-ionosphere cavity. *J. Atmos. Terr. Phys.*, 33, 567–572, 1971.

[212] Ogawa, T., Y. Tanaka, M. Yasuhara, A.C. Fraser-Smith and R. Gendrin, Worldwide simultaneity of occurrence of a Q-type ELF burst in the Schumann resonance frequency range, J. Geomag. Geoelectr., 19, 377–384, 1967.

[213] Williams, E.R., W.A. Lyons, Y. Hobara, *et al.*, Ground-based detection of sprites and their parent lightning flashes over Africa during the 2006 AMMA campaign, Quart. J. Roy. Met. Soc., Special Issue: Advances in understanding atmospheric processes over West Africa through the AMMA field campaign, 136, 257–271, 2010.

[214] Yamashita, K., Y. Takahashi, M. Sato and H. Kase, Improvement in lightning geolocation by time-of-arrival method using global ELF network data J. Geophys. Res.,116, A00E61, 12 PP., doi:10.1029/2009JA014792, 2011.

[215] Markson, R. and C. Price, Ionospheric potential as a proxy index for global temperature, Atmos. Res., 51, 309–314, 1999.

[216] Allen, M.R. and W.J. Ingram, Constraints on future changes in climate and the hydrologic cycle, Nature, 419, 224–232, 2002.

[217] Romps, D.M., J. T. Seeley, D. Vollaro and J. Molinari, Projected increase in lightning strikes in the United States due to global warming, Science, 346, 851–854, 2014.

[218] Romps, D.M., Clausius-Clapeyron scaling of CAPE for analytic solutions to RCE, J. Atmos. Sci., 73, 3719–3737, 2016.

[219] Guerreiro, S.B., H.J. Fowler, R. Barbero, *et al.*, Detection of continental-scale intensification of hourly rainfall extremes, Nature Climate Change, doi.org/10.1038/s41558-018-0245-3, 2018.

[220] Yair, Y., Lightning hazards to human societies in a changing climate, Env. Res. Lett., 13, 123002, 2018.

[221] Huaichuan, R., *Globalisation, Transition and Development in China: The Case of the Coal Industry*, Routledge, 2004.

Chapter 2

Lightning phenomenon and parameters for engineering application

Vladimir A. Rakov[1]

In this chapter, basic lightning terminology is introduced, different types of lightning are described, and main lightning processes are identified. For the most common negative cloud-to-ground lightning, the number of strokes per flash, interstroke intervals, flash duration, multiple channel terminations on ground, and relative stroke intensity within the flash are discussed. Traditional lightning parameters needed in engineering applications include lightning peak current, maximum current derivative, average current rate of rise, current risetime, current duration, charge transfer, and specific energy (action integral), all derivable from direct current measurements. Distributions of these parameters presently adopted by CIGRE, IEEE, and IEC are largely based on direct measurements by K. Berger and co-workers in Switzerland. Those "classical" distributions are reviewed along with the distributions based on more recent direct current measurements obtained in Austria, Brazil, Japan, and the United States (Florida). Correlations between the parameters are discussed. Additional lightning parameters, return-stroke propagation speed, and equivalent impedance of the lightning channel are considered. Mathematical expressions used to represent lightning current waveforms in models are reviewed.

2.1 Types of lightning and main lightning processes

2.1.1 Overview

Lightning can be defined as a transient, high-current (typically tens of kiloamperes) electric discharge in air whose length is measured in kilometers. As for any discharge in air, lightning channel is composed of ionized gas, that is, of plasma, whose peak temperature is typically 30,000 K, about five times higher than the temperature of the surface of the Sun. Lightning was present on Earth long before human life evolved and it may even have played a crucial role in the evolution of life on our planet. The global lightning flash rate is some tens to a hundred per

[1]Department of Electrical and Computer Engineering, University of Florida, Gainesville, FL, USA

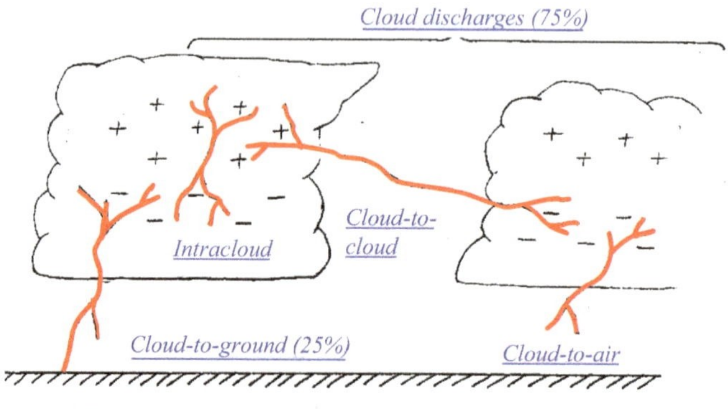

Figure 2.1 General classification of lightning discharges from cumulonimbus (thunderstorm clouds). Cloud discharges constitute 75% and cloud-to-ground discharges 25% of global lightning activity

second or so. Each year, some 25 million cloud-to-ground lightning discharges occur in the United States, and this number is expected to increase by about 50% due to global warming over the twenty-first century [1]. Lightning occurrence is not limited to the Earth's atmosphere. There exists convincing evidence for lightning or lightning-like discharges on Jupiter and Saturn.

Lightning initiates many forest fires, and over 30% of all electric power line failures are lightning related. Each commercial aircraft is struck by lightning on average once a year. A lightning strike to an unprotected object or system can be catastrophic.

The lightning discharge in its entirety, whether it strikes ground or not, is usually termed a "lightning flash" or just a "flash." A lightning discharge that involves an object on ground or in the atmosphere is referred to as a "lightning strike." A commonly used nontechnical term for a lightning discharge is a "lightning bolt." About three-quarters of lightning discharges do not involve ground. They include intracloud, intercloud, and cloud-to-air discharges and are collectively referred to as cloud discharges or flashes (see Figure 2.1) and sometimes as ICs. Lightning discharges between cloud and Earth are termed cloud-to-ground (or just ground) discharges and sometimes referred to as CGs. The latter constitute about 25% of global lightning activity. About 90% or more of global cloud-to-ground lightning is accounted for by downward (the initial process begins in the cloud and develops in the downward direction) negative (negative charge is effectively transported to the ground) lightning. The term "effectively" is used to indicate that individual charges are not transported all the way from the cloud to ground during the lightning processes. Rather the flow of electrons (the primary charge carriers) in one part of the lightning channel results in the flow of other electrons in other parts of the channel. Other types of cloud-to-ground lightning include downward positive, upward negative, and upward positive discharges (see Figure 2.2). Downward flashes exhibit downward branching, while upward flashes are branched

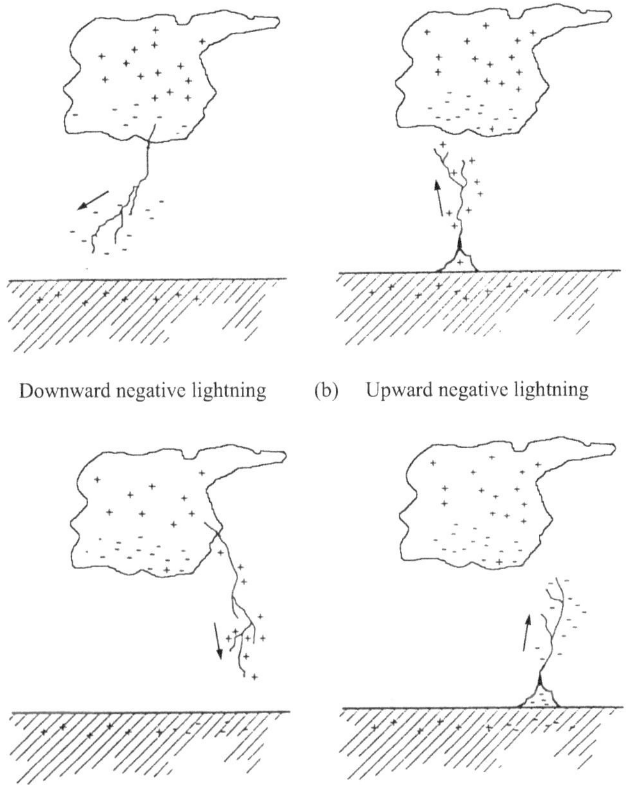

(a) Downward negative lightning (b) Upward negative lightning

(c) Downward positive lightning (d) Upward positive lightning

Figure 2.2 Four types of lightning effectively lowering cloud charge to ground. Only the initial leader is shown for each type. For each lightning-type name given below the sketch, direction (downward or upward) indicates the direction of propagation of the initial leader and polarity (negative or positive) refers to the polarity of the cloud charge effectively lowered to ground. In (a) and (c), the polarity of charge lowered to ground is the same as the leader polarity, while in (b) and (d) those polarities are opposite. Not shown in this figure are upward (object-initiated) and downward bipolar lightning flashes. Adapted from Rakov and Uman [2]

upward. Upward lightning discharges (types (b) and (d) in Figure 2.2) are thought to occur only from tall objects (higher than 100 m or so) or from objects of moderate height located on mountain tops. There are also bipolar lightning discharges sequentially transferring both positive and negative charges during the same flash. Bipolar lightning discharges are usually initiated from tall objects (are of upward type). Downward bipolar lightning discharges do exist but appear to be rare.

Cloud flashes are most likely to begin near the upper and lower boundaries of the main negative charge region, and in the former case often bridge the main negative and main positive charge regions in the cloud. Other scenarios are possible. There is a special type of cloud lightning that is thought to be the most intense natural producer of HF-VHF (3–300 MHz) radiation on Earth. It is referred to as compact intracloud discharge (CID). CIDs received their name due to their relatively small (hundreds of meters) spatial extent. They tend to occur at high altitudes (mostly above 10 km), appear to be associated with strong convection (however, even the strongest convection does not always produce CIDs), and tend to produce less light than other types of lightning discharges.

2.1.2 Downward negative lightning

We first introduce, referring to Figure 2.3(a) and (b), the basic elements of the negative downward lightning discharge, termed component strokes or just strokes. Each flash typically contains 3–5 strokes, the observed range being 1–26. Then, we will introduce, referring to Figure 2.4, the two major lightning processes comprising a stroke, the leader and the return stroke, which occur as a sequence with the leader preceding the return stroke. We will also briefly review lightning parameters, with more details on those needed for engineering applications being found in Sections 2.2–2.12.

Two photographs of a negative cloud-to-ground discharge are shown in Figure 2.3(a) and (b). The image in Figure 2.3(a) was obtained using a stationary

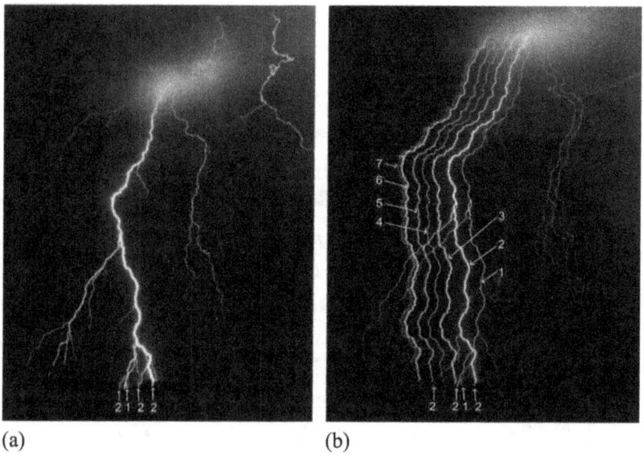

(a) (b)

Figure 2.3 Lightning flash that appears to have at least seven separate ground strike points (three of them are created by stroke 2): (a) still-camera photograph and (b) moving-camera photograph. Some of the strike points are associated with the same stroke having separate branches touching ground, while others are associated with different strokes taking different paths to ground. Adapted from Hendry [3]

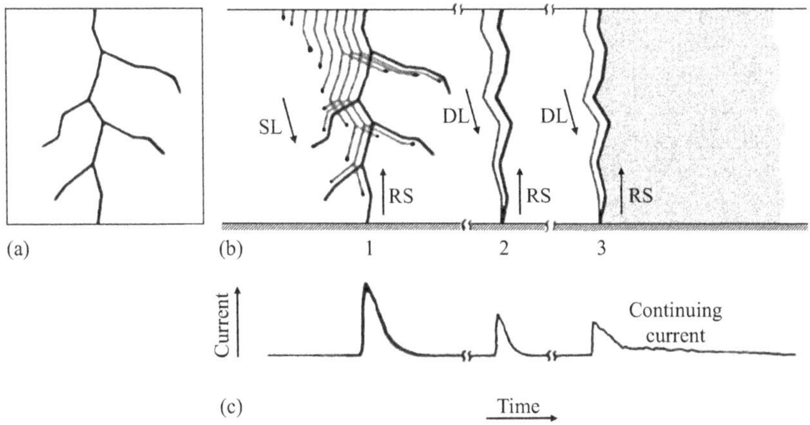

Figure 2.4 Schematic diagram showing the luminosity of a three-stroke downward negative flash and the corresponding current at the channel base: (a) still-camera image, (b) streak-camera image, and (c) channel-base current. Adapted from Rakov and Uman [2]

camera, while the image in Figure 2.3(b) was captured with a separate camera that was moved horizontally during the time of the flash. As a result, the latter image is time resolved showing several distinct luminous channels between the cloud and ground separated by dark gaps. The distinct channels are associated with individual strokes, and the time intervals corresponding to the dark gaps are typically of the order of tens of milliseconds. These dark time intervals between strokes explain why lightning often appears to the human eye to "flicker." In Figure 2.3(b), time advances from right to left, so that the first stroke is on the far right. The first two strokes are branched, and the downward direction of branches indicates that this is a downward lightning flash.

Now, we consider sketches of still and time-resolved images of the three-stroke lightning flash shown in Figure 2.4(a) and (b), respectively. A sketch of the corresponding current at the channel base is shown in Figure 2.4(c). In Figure 2.4(b), time advances from left to right, and the time scale is not continuous. Each of the three strokes in Figure 2.4(b), represented by its luminosity as a function of height above ground and time, is composed of a downward-moving process, termed a leader, and an upward-moving process, termed a return stroke. The leader creates a conducting path between the cloud charge source region and ground and distributes negative charge from the cloud source region along this path, and the return stroke traverses that path moving from ground toward the cloud charge source region and neutralizes the negative leader charge. Thus, both leader and return-stroke processes serve to effectively transport negative charge from the cloud to ground. As seen in Figure 2.4(b), the leader initiating the first return stroke differs from the leaders initiating the two subsequent return strokes (all strokes other than first are termed subsequent strokes). In particular, the first-stroke leader appears optically to be an intermittent process, hence the term stepped leader, while the tip of a subsequent-stroke leader appears to move continuously. The continuously moving subsequent-stroke leader tip appears on streak photographs as a downward-moving "dart," hence the term dart leader. The apparent difference between the two types of leaders is related to the fact that the stepped leader develops in virgin air, while the dart leader follows the "pre-conditioned" path of the preceding stroke or strokes. Sometimes, a subsequent leader exhibits stepping while propagating along a previously formed channel; it is referred to as dart-stepped leader. There are also so-called chaotic subsequent-stroke leaders.

The electric potential difference between a downward-moving stepped-leader tip and ground is probably some tens of megavolts, comparable to or a considerable fraction of that between the cloud charge source and ground. The magnitude of the potential difference between two points, one at the cloud charge source and the other on ground, is the line integral of electric field intensity between those points. The upper and lower limits for the potential difference between the lower boundary of the main negative charge region and ground can be estimated by multiplying, respectively, the typical observed electric field in the cloud, 10^5 V m^{-1}, and the expected electric field at ground under a thundercloud immediately prior to the initiation of lightning, 10^4 V m^{-1}, by the height of the lower boundary of the negative charge region above ground. The resultant range is 50–500 MV if the height is assumed to be 5 km.

When the descending stepped leader attaches to the ground, the first return stroke begins. The first return-stroke current measured at ground rises to an initial peak of about 30 kA in some microseconds and decays to half-peak value in some tens of microseconds. The return stroke effectively lowers to ground the several coulombs of charge originally deposited on the stepped-leader channel including all the branches. Once the bottom of the dart leader channel is connected to the ground, the second (or any subsequent) return-stroke wave is launched upward, which again serves to neutralize the leader charge. The subsequent return-stroke current at ground typically rises to a peak value of 10–15 kA in less than a microsecond and decays to half-peak value in a few tens of microseconds.

The high-current return-stroke wave rapidly heats the channel to a peak temperature near or above 30,000 K and creates a channel pressure of 10 atm (1 megapascal) or more, resulting in channel expansion, intense optical radiation, and an outward propagating shock wave that eventually becomes the thunder (sound wave) we hear at a distance. Each cloud-to-ground lightning flash involves an energy of roughly 10^9 to 10^{10} J (one to ten gigajoules). Lightning energy is approximately equal to the energy required to operate five 100-W light bulbs continuously for 1 month. Note that not all the lightning energy is available at the strike point, only 10^{-2}–10^{-3} of the total energy, since most of the energy is spent for producing thunder, hot air, light, and radio waves.

The impulsive component of the current in a return stroke is often followed by a continuing current, which has a magnitude of tens to hundreds of amperes and a duration up to hundreds of milliseconds (median duration is 6 ms). Continuing currents with a duration in excess of 40 ms are traditionally termed long continuing currents. Between

30% and 50% of all negative cloud-to-ground flashes contain long continuing currents. Current pulses superimposed on continuing currents, as well as the corresponding enhancements in luminosity of the lightning channel, are referred to as M-components.

2.1.3 Downward positive lightning

Positive lightning discharges are relatively rare (about 10% of global cloud-to-ground lightning activity), but there are five situations that appear to be conducive to the more frequent occurrence of positive lightning. These situations include (1) the dissipating stage of an individual thunderstorm, (2) winter (cold-season) thunderstorms, (3) trailing stratiform regions of mesoscale convective systems, (4) some severe storms, and (5) thunderclouds formed over forest fires or contaminated by smoke. Positive flashes are usually composed of a single stroke, in contrast with negative flashes about 80% of which contain two or more strokes, with three to five being typical. Multiple-stroke positive flashes do occur but they are relatively rare. Positive lightning is typically more energetic and potentially more destructive than negative lightning.

A reliable distribution of positive lightning peak currents applicable to objects of moderate height on the flat ground is presently unavailable. The sample of 26 directly measured positive lightning currents (see Table 2.1) reported by Berger *et al.* [4] is commonly used as a primary reference both in lightning research and in lightning protection studies. However, this sample is apparently based on a mix of (1) discharges initiated as a result of junction between a descending positive leader and an upward-connecting negative leader within some tens of meters of the tower top and (2) discharges initiated as a result of a very long (1–2 km) upward negative leader from the tower making contact with an oppositely charged channel inside the cloud. These two types of positive discharges, which differ by the height above the

Table 2.1 Lightning current parameters for positive flashes. Adapted from [4]

Parameters	Units	Sample size	Percent exceeding tabulated value		
			95%	50%	5%
Peak current (minimum 2 kA)	kA	26	4.6	35	250
Charge (total charge)	C	26	20	80	350
Impulse charge (excluding continuing current)	C	25	2.0	16	150
Front duration (2 kA to peak)	μs	19	3.5	22	200
Maximum dI/dt	kA/μs	21	0.20	2.4	32
Stroke duration (2 kA to half peak value on the tail)	μs	16	25	230	2000
Action integral ($\int I^2 dt$)	A^2s	26	2.5×10^4	6.5×10^5	1.5×10^7
Flash duration	ms	24	14	85	500

tower top of the junction between the upward-connecting leader and the oppositely charged overhead channel (descending positive leader or positively charged in-cloud discharge channel), are expected to produce very different current waveforms at the tower, as illustrated in Figure 2.5(a) and (b). The "microsecond-scale" current waveform shown in Figure 2.5(a) is probably a result of processes similar to those in downward negative lightning, whereas the "millisecond-scale" current waveform shown in Figure 2.5(b) is likely to be a result of the M-component mode of charge transfer to the ground, although in the latter case current peaks can be considerably higher than for ordinary M-components. It is possible that such millisecond-scale waveforms are characteristic of tall objects capable of generating very long upward-connecting leaders. The above view of the 26 positive lightning discharges documented in detail by K. Berger is an update on the previous CIGRE assumption that those events are related "principally to upward flashes" [6].

Because of the absence of other direct current measurements for positive lightning return strokes, it is still recommended (CIGRE WG C4.407 TB 549 [7]) to use the peak current distribution based on the 26 events reported by K. Berger, even though some of those 26 events are likely to be not of return-stroke type.

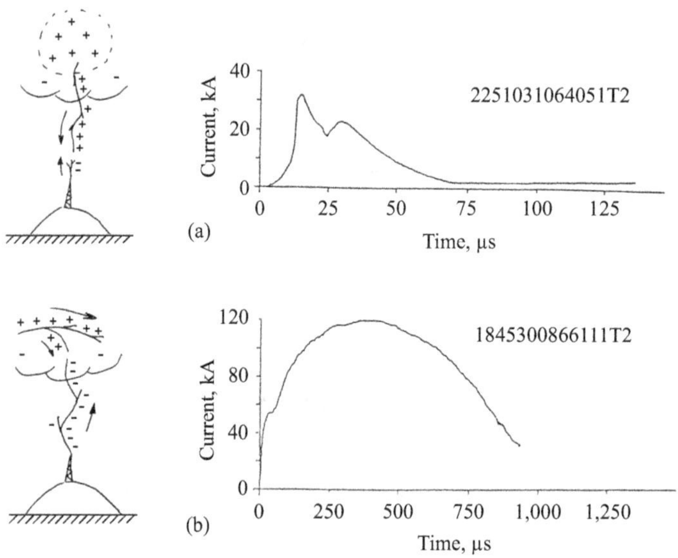

Figure 2.5 Examples of two types of positive lightning current versus time waveforms observed by K. Berger: (a) microsecond-scale waveform, similar to those produced by downward negative return strokes, and a sketch illustrating the lightning processes that might have led to the production of this waveform; and (b) millisecond-scale waveform and a sketch illustrating the lightning processes that might have led to the production of this current waveform. Arrows indicate directions of the extension of lightning channels. Adapted from Rakov [5]

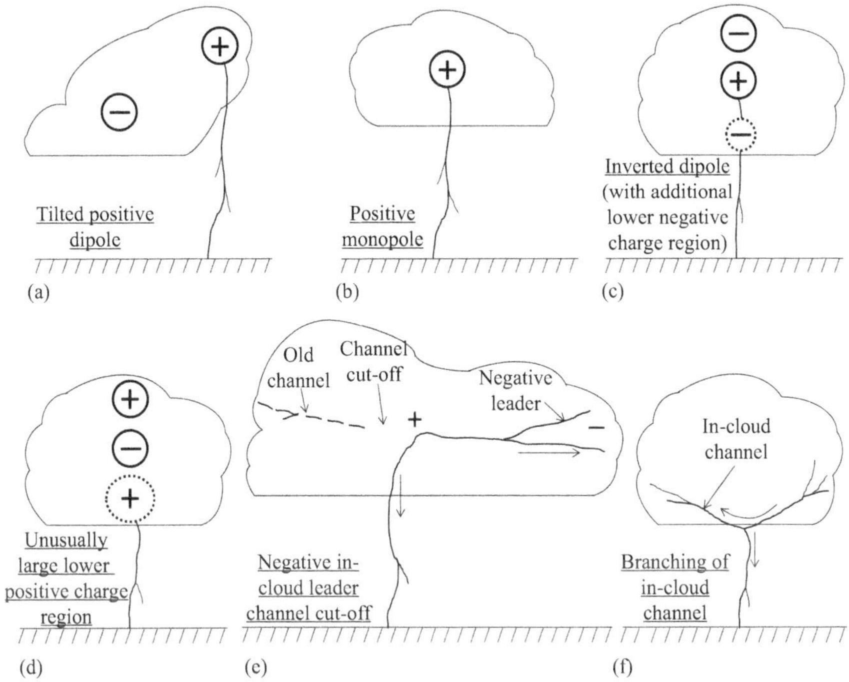

Figure 2.6 Conceptual cloud charge configurations and scenarios leading to production of downward positive lightning. Adapted from Nag and Rakov [8]

The gross charge structure of a "normal" thundercloud is often viewed as a vertical tripole consisting of three charge regions, main positive at the top, main negative in the middle, and an additional (typically smaller) positive below the main negative. Such a charge structure appears to be not conducive to production of positive cloud-to-ground lightning. Figure 2.6 illustrates six conceptual cloud charge configurations and scenarios that were observed or hypothesized to give rise to positive lightning. For four of them, tilted positive dipole, positive monopole, inverted dipole, and unusually large lower positive charge region [Figure 2.6(a)–(d)], the primary source of charge is a charged cloud region, while for the other two, negative in-cloud leader channel cutoff and branching of in-cloud channel [Figure 2.6(e) and (f), respectively], the primary source of charge is a polarized or current-carrying in-cloud lightning channel formed prior to the positive discharge to ground.

2.1.4 Artificially initiated lightning

Lightning can be artificially initiated (triggered) by launching of a small rocket trailing a thin grounded or ungrounded wire toward a charged cloud overhead. In the former case, the triggered lightning is referred to as classical and in the latter case as altitude triggered one. The sequence of processes (except for the transition from leader to return stroke stage that is referred to as the attachment process) in classical triggered lightning is schematically shown in Figure 2.7. When the rocket, ascending at about 150–200 m s^{-1}, is about 200–300 m high, the field enhancement near the rocket tip launches a positively charged leader that propagates upward toward the cloud. This upward positive leader vaporizes the trailing wire, bridges the gap between the cloud and ground, and establishes an initial continuous current with a duration of some hundreds of milliseconds that transports negative charge from cloud charge source region to the triggering facility. After the cessation of the initial continuous current, one or more downward dart-leader/upward return-stroke sequences may traverse the same path to the triggering facility. The dart leaders and the following return strokes in triggered lightning are similar to dart-leader/return-stroke sequences in natural lightning, although the initial processes in natural downward and triggered lightning are distinctly different. To date, well over 1,000 lightning discharges have been triggered by researchers in different countries using the rocket-and-wire technique, with over 450 of them at the International Center for Lightning Research and Testing (ICLRT) at Camp Blanding, Florida. The ICLRT was established in 1993 and since 2004 also includes the Lightning Observatory in Gainesville (LOG), located 45 km from the Camp Blanding facility. Photographs of two rocket-and-wire triggered lightning flashes from Camp Blanding are shown in Figure 2.8.

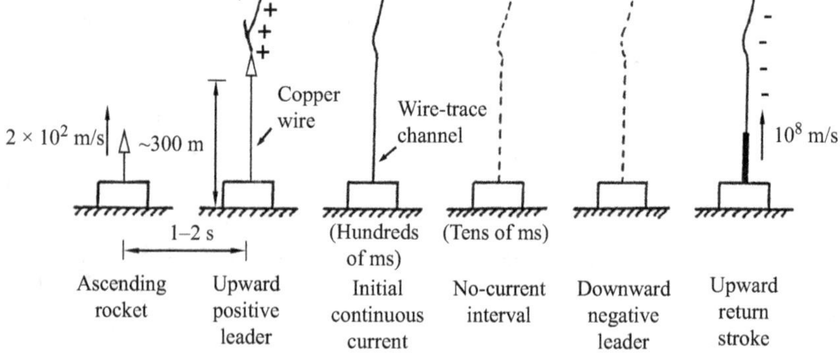

Figure 2.7 Sequence of events (except for the attachment process) in classical triggered lightning. The upward positive leader and initial continuous current constitute the initial stage of a classical triggered flash. Adapted from Rakov et al. [9]

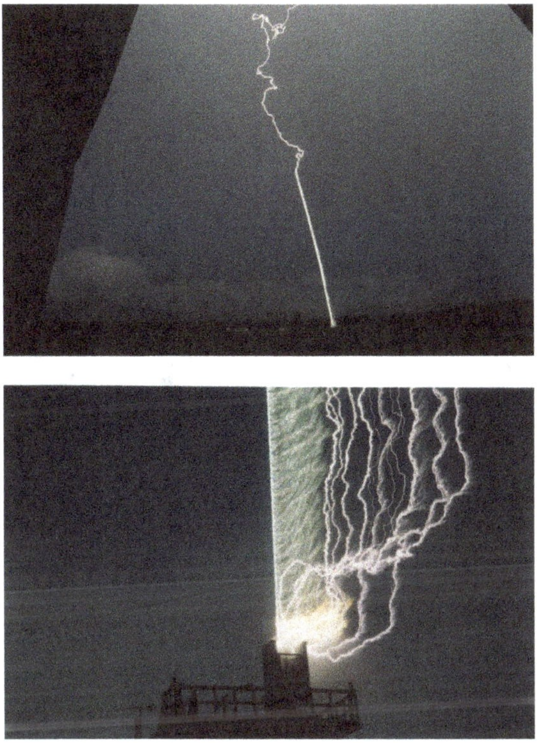

Figure 2.8 Photographs of lightning flashes triggered using the rocket-and-wire technique at Camp Blanding, Florida. Top—a distant view of a strike to the test runway; bottom—a close-up view of a strike to the test power system. Adapted from Rakov and Uman [2]

The results of triggered-lightning experiments have provided considerable insight into natural lightning processes that would not have been possible from studies of natural lightning due to its random occurrence in space and time. As an example, Figure 2.9 shows a photograph of surface arcing during a triggered-lightning flash from experiments at Fort McClellan, Alabama. The soil was red clay and a 0.3- or 1.3-m steel vertical rod was used for grounding of the rocket launcher. The surface arcing appears to be random in direction and often leaves little if any evidence on the ground. Even within the same flash, individual strokes can produce arcs developing in different directions. In one case, it was possible to estimate the current carried by one arc branch which contacted the instrumentation. That current was approximately 1 kA or 5% of the total current peak in that stroke. The observed horizontal extent of surface arcs was up to 20 m, which was the limit of the photographic coverage during the Fort McClellan experiment. These results suggest that the uniform ionization of soil, usually postulated in studies of the behavior of grounding electrodes subjected to lightning surges, may be not an adequate assumption.

Figure 2.9 *Photograph of surface arcing associated with the second stroke (current peak of 30 kA) of flash 9312 triggered at Fort McClellan, Alabama. Lightning channel is outside the field of view. One of the surface arcs approached the right edge of the photograph, a distance of 10 m from the rocket launcher. Adapted from Fisher et al. [10]*

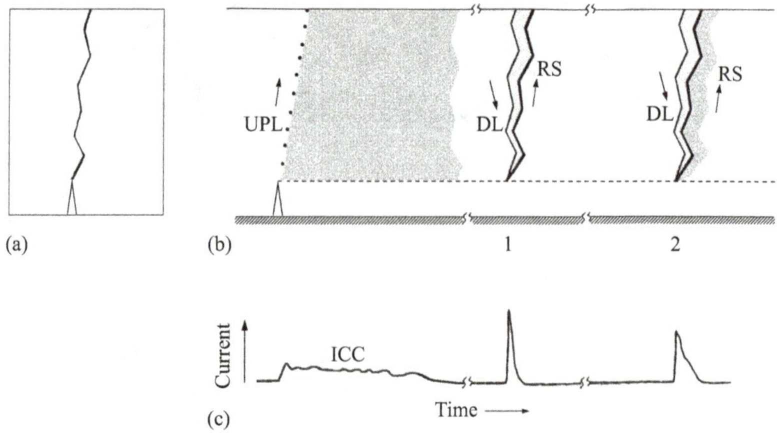

Figure 2.10 *Schematic diagram showing the luminosity of an upward negative flash and the corresponding current at the channel base: (a) still-camera image, (b) streak-camera image, and (c) current record. The flash is composed of an upward positive leader, UPL, followed by an initial continuous current, ICC, and two downward dart leader (DL)/upward return stroke (RS) sequences. The UPL and ICC constitute the initial stage of an upward negative flash. Adapted from Rakov and Uman [2]*

2.1.5 Upward lightning

The phenomenology of upward negative lightning, illustrated in Figure 2.10 by the sketches of still and time-resolved photographic records and of the corresponding

Figure 2.11 *Photograph of an upward flash initiated from a 70-m tower on Monte San Salvatore, Switzerland. Courtesy of R.E. Orville*

current record, is similar to that of negative lightning triggered using the classical rocket-and-wire technique (Section 2.1.4). In the latter case, the thin triggering wire plays the role of grounded object, one that is rapidly erected and then replaced by the plasma channel of the upward leader. Downward leader/upward return stroke sequences in upward lightning (Figure 2.10), and their counterparts in rocket-triggered lightning (Figure 2.7), are similar to subsequent strokes in natural downward lightning (Figure 2.4). For this reason, leader/return stroke sequences in upward (object-initiated) lightning and in rocket-triggered lightning are sometimes referred to as subsequent strokes. Interestingly, only 20% to 50% of object-initiated flashes contain subsequent strokes, while over 70% of rocket-and-wire triggered flashes in Florida do so. Upward flashes can be negative, positive, or bipolar. A photograph of an upward flash (note upward branching) initiated from a 70-m tower on a 640-m mountain is shown in Figure 2.11.

2.2 Number of strokes per flash

A typical negative cloud-to-ground flash is composed of 3 to 5 strokes (leader/return stroke sequences), with typical interstroke intervals of some tens of milliseconds. The largest number of strokes per flash, observed in New Mexico [11], is 26. Note that the stroke count includes both strokes developing in pre-existing channels (channel established by the first stroke in a flash) and those creating new terminations on ground (ground contact points). Parameters of new-ground-termination strokes are intermediate between first strokes and subsequent strokes developing in pre-existing channels.

The average numbers of strokes per flash and percentages of single-stroke flashes observed in different locations, using accurate stroke count methods, are

Table 2.2 Number of strokes per negative flash and percentage of single-stroke flashes

Location (Reference)	Average number of strokes per flash	Percentage of single-stroke flashes	Sample size
New Mexico [11]	6.4	13%	83
Florida [12]	4.6	17%	76
Sweden [13]	3.4	18%	137
Sri Lanka [14]	4.5	21%	81
China [15]	3.8	40%	83
Arizona [16]	3.9	19%	209
Brazil [17]	4.6	17%	883
Malaysia [18]	4.0	16%	100
Florida [19]	4.6	12%	478

summarized in Table 2.2. It follows from Table 2.2 that the percentage of single-stroke flashes previously recommended by CIGRE, 45% [6], is in most cases an overestimate by a factor of two or so. Note that the percentage of single-stroke flashes near the equator (Sri Lanka (5.9°N) and Malaysia (1°N)) is not much different from that in Sweden (60°N). However, significant variations in the average number of strokes per flash and percentage of single-stroke flashes were observed between individual storms in the same location (e.g., [19]).

2.3 Interstroke intervals and flash duration

Interstroke intervals are usually measured between the peaks of current or electromagnetic field pulses. Some interstroke intervals contain continuing currents of appreciable duration. However, these currents always vanish (become undetectable) before the next stroke [20–22]. The time interval between the end of continuing current and the beginning of the next stroke is referred to as the no-current interstroke interval. Less accurate estimates of interstroke intervals can be obtained from high-speed (interframe intervals of 1 ms or less) video observation.

From accurate-stroke-count studies of negative flashes in Florida and New Mexico, the geometric mean interstroke interval is typically about 60 ms (e.g., [2], Figure 4.4). When long continuing currents are involved, interstroke intervals can be as large as several hundreds of milliseconds. Occasionally, two leader/return stroke sequences occur in the same lightning channel with a time interval between them as short as 1 ms or less [23,24]. Interstroke intervals preceding strokes initiating long continuing currents show a clear tendency to be shorter than regular interstroke intervals [25–27]. Table 2.3 gives a summary of geometric mean interstroke intervals observed for negative lightning in different geographical locations. Additionally given are multiple-stroke flash durations.

2.4 Multiple channel terminations on ground

One-third to one-half of all lightning discharges to earth, both single- and multiple-stroke flashes, strike ground at more than one point with the spatial separation between the channel terminations being up to many kilometers. Most measurements of ground flash density do not account for multiple channel terminations on ground. When only one location per flash is recorded, while all strike points separated by distances of some hundreds of meters or more are of interest, as is the case where lightning damage is concerned, measured values of ground flash density should, in general, be increased. According to Table 2.4, the correction factor of about 1.5 to 1.7 is needed for measured values of ground flash density to account for multiple channel terminations on ground, which is considerably larger than 1.1 previously estimated by Anderson and Eriksson [6].

In most cases, multiple ground terminations within a given flash are associated not with an individual multigrounded leader but rather with the deflection of a subsequent leader from the previously formed channel. According to Thottappillil *et al.* [33], the distances between separate channel terminations in a given flash, located via TV direction finding and thunder ranging for 22 flashes in Florida, vary from 0.3 to 7.3 km with a geometric mean of 1.7 km (see Figure 2.12). The geometric mean separation between two channel terminations created by the same leader (in 7 or 32% of the 22 flashes) was also 1.7 km. The NLDN-based distances between the first stroke and 59 new-ground-termination strokes in southern Arizona had a median of 2.1 km [34].

According to Rakov and Uman [12], the percentage of subsequent leaders that create a path to ground different from that of the previous stroke path decreases rapidly with stroke order: 37% of all second leaders, 27% of all third leaders, 2% of all fourth leaders, and none of the leaders of the order of 5 and higher. Interestingly,

Table 2.3 Geometric mean interstroke intervals and flash durations (sample sizes are given in the parentheses)

Location (Reference)	Geometric mean interstroke interval, ms	Geometric mean flash duration*, ms
Florida [28]	60 (270)	—
Sri Lanka [14]	57 (284)	—
Sweden [13]	48 (568)	—
China [15]	47 (238)	—
Florida and New Mexico [2]	60 (516)	—
Arizona [16]	61 (598)	216 (169)
Brazil [16]	62 (624)	229 (179)
Malaysia [18]	67 (305)	—
Florida [19]	52 (1,710)	223 (421)

*Multiple-stroke flashes only.

Table 2.4 Number of channel terminations per flash and percentage of multigrounded flashes. Adapted from CIGRE WG C4.407 TB 549 [7]

Location (Reference)	Average number of channels per flash	Percentage of multi-grounded flashes	Sample size
New Mexico [11]	1.7	49%	72*
	1.6	42%	83**
Florida [12]	1.7	50%	76
France [29,30]	1.5	34%	2,995
Arizona [31]	1.4	35%	386
U.S. Central Great Plains [32]	1.6	33%	103
Brazil [16]	1.7	51%	138
Arizona [16]	1.7	48%	206

*Multiple-stroke flashes only;
**Including 11 single-stroke flashes assumed to each have a single channel per flash.

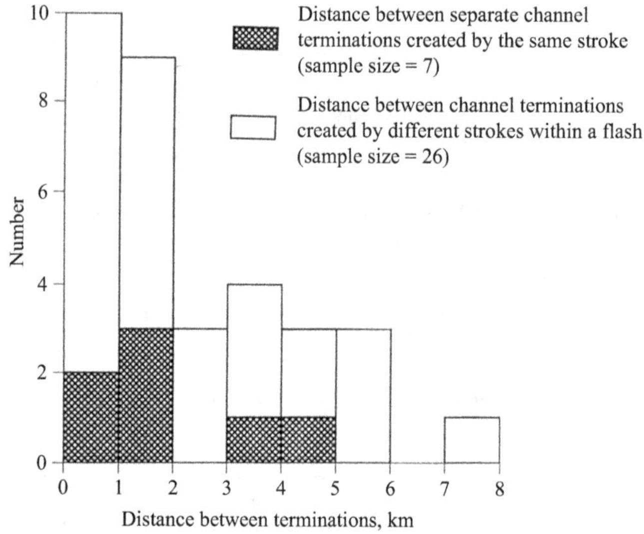

Figure 2.12 Histogram of the distances between the multiple terminations of 22 individual ground flashes in Florida. Adapted from Thottappillil et al. [33]

if only those third leaders which followed the formation of a new second-stroke path to ground (19 total) are considered, the percentage of the new terminations is 37%, the same as for second leaders, all of which are preceded by the formation of a new channel, the first-stroke channel. Rakov and Uman [12] interpreted these results as indicating that the first stroke (or even a sequence of the first two strokes) of the flash often does not create a properly conditioned channel capable of supporting the propagation of the following leader all the way to ground. An unalterable path to ground in a given flash is apparently established only after at least four (possibly more) consecutive strokes have participated in channel conditioning. Note that the behavior described above cannot be explained in terms of relatively long preceding interstroke intervals (and hence more aged channels) for strokes of the order of 2–4. In fact, the fraction of interstroke intervals lasting longer than 100 ms and not containing long continuing current for strokes from 2 to 4 is about the same as for the higher-order strokes. Further, the geometric mean preceding interstroke interval for second strokes is similar to that for strokes of the order of 5 and higher. According to Ferro *et al.* [35], the preceding interstroke interval becomes a factor after two or more strokes have used the previously created channel, with new ground terminations being more likely produced following longer interstroke intervals.

In southern Arizona, Stall *et al.* [34] observed that 59% of the time it was the second stroke that produced a new ground termination, and 27% of the time it was the third stroke (the sample size was 59). In three cases they observed new ground terminations created by the fifth stroke and in one case by the seventh stroke. Ferro *et al.* [35] reported two cases of new ground terminations created by ninth strokes (but none by eighth strokes).

The percentage of multigrounded flashes exhibits significant storm-to-storm variation. Rakov and Uman [12] reported a range of 29%–69% for three individual thunderstorm days, with a mean of 50%. Thottappillil *et al.* [33] observed up to four different strike points per flash in Florida, as did Fleenor *et al.* [32] in the U.S. Central Great Plains. Saraiva *et al.* [16] reported up to five strike points in Arizona and up to four in Brazil. Berger *et al.* [29] and Hermant [30] observed up to six strike points in France. The largest number of ground terminations in a single flash reported to date is seven [2]. Still- and moving-camera photographs of this latter flash are shown in Figure 2.3.

Kong *et al.* [36] studied multiple channel terminations created by the same negative leader in China. The percentage of flashes showing this feature varied from 11% to 20% with a mean of 15% (9 out of a total of 59 flashes). It is of interest to compare this result with observations in Florida. As noted earlier, Rakov and Uman [26] reported that 50% of 76 negative flashes in Florida created multiple channel terminations on ground. Individual channel terminations were located for 22 of their multigrounded flashes [33]. Of these 22, 7 (32%) flashes contained double-grounded leaders, which translates to 16% of all flashes, if we assume that the 22 multigrounded flashes constitute 50% of all flashes. In this case, the Florida percentage (16%) is similar to the mean percentage (15%) reported by Kong *et al.* [36].

2.5 Relative stroke intensity within the flash

Relative magnitudes of electric field peaks of first and subsequent return strokes in negative cloud-to-ground lightning flashes recorded in Florida, Austria, Brazil, and Sweden were analyzed by Nag *et al.* [37]. On average, the electric field peak of the first stroke is appreciably, 1.7 to 2.4 times, larger than the field peak of the subsequent stroke (except for studies in Austria where the ratio varies from 1.0 to 2.3, depending on methodology and instrumentation). Similar results were previously reported from electric field studies in Florida, Sweden, Sri Lanka, and China by Rakov *et al.* [28], Cooray and Perez [38], Cooray and Jayaratne [14], and Qie *et al.* [15], respectively. Directly measured peak currents for first strokes are, on average, a factor of 2.3 to 2.5 larger than those for subsequent strokes [4,6,39]. The generally larger ratio for currents than for fields possibly implies a lower average return-stroke speed for first strokes than for subsequent strokes. There appear to be some differences between first versus subsequent stroke intensities reported from different studies based on data reported by lightning locating systems (LLSs). The ratio of LLS-reported peak currents for first and subsequent strokes confirmed by video records is 1.7 to 2.1 in Brazil (for strokes followed by continuing currents with durations ranging from 4 to 350 ms), while in the United States (Arizona, Texas, Oklahoma, and the Great Plains) it varies from 1.1 to 1.6, depending on methodology used. Ratios involving arithmetic means are generally larger than those involving geometric means. The smaller ratios derived from the LLS studies are likely to be due to poor detection of relatively small subsequent strokes. The smaller values in Austria are possibly related (at least in part) to the higher percentage (about 50% vs. 24% to 38% in other studies) of flashes with at least one greater-than-the-first subsequent stroke. The effects on the ratio of excluding single-stroke flashes or subsequent strokes in newly formed channels appear to be relatively small. Additional data are needed to further clarify the issue of relative intensity of first and subsequent strokes in different geographical locations, as well as possible instrumental and methodological biases involved.

Results of Nag *et al.* [37] are summarized in Figure 2.13 and Table 2.5. Additionally given in Table 2.5 are the results of a more recent study in Florida [19].

Although first-stroke current peaks are typically a factor of 2 to 3 larger than subsequent-stroke current peaks, about one-third of cloud-to-ground flashes contain at least one subsequent stroke with electric field peak, and, by theory, current peak, greater than the first-stroke peak. This relatively high percentage suggests that such flashes are not unusual, contrary to the implication of most lightning protection and lightning test standards (e.g., [40]; Military Standard, [41]). Thottappillil *et al.* [33] observed subsequent strokes with larger field peaks both in the first-stroke channel (13 strokes) and in a different channel (12 strokes). Of the latter 12 strokes, 6 strokes created new channels to ground and the remaining 6 strokes followed a previously formed channel. Parameters of subsequent strokes with larger field peaks which followed the same channel as the first stroke are summarized in Table 2.6. Note that larger subsequent strokes are associated with relatively short leader duration (and, by inference, higher leader speed) and relatively long preceding interstroke interval. Larger subsequent strokes never followed interstroke intervals shorter than 35 ms, whereas many regular subsequent strokes did [33].

Larger-than-first subsequent strokes were also observed in direct current measurements. Five (15%) of the 33 negative downward multiple-stroke flashes striking instrumented towers in Switzerland, from the Atlas of lightning currents

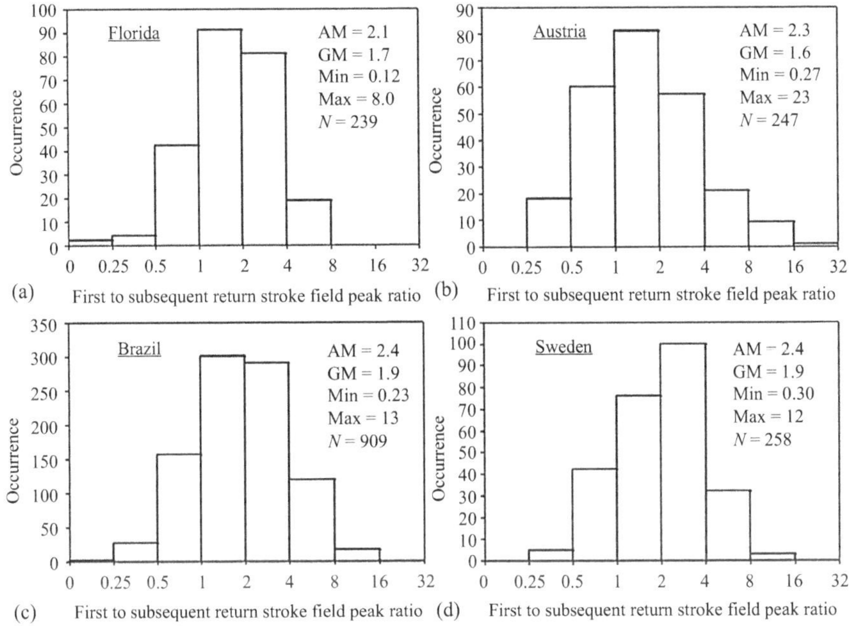

Figure 2.13 Histograms of the ratio of the first-to-subsequent-return-stroke electric field peak for multiple-stroke negative cloud-to-ground lightning flashes in (a) Florida, (b) Austria, (c) Brazil, and (d) Sweden. Adapted from Nag et al. [37]

Table 2.5 Ratio of the first to subsequent return stroke electric field peaks for multiple-stroke negative cloud-to-ground lightning flashes

Location (Reference)	Arithmetic mean	Geometric mean	Sample size
Florida [37]	2.1	1.7	239
Austria [37]	2.3	1.6	247
Brazil [37]	2.4	1.9	909
Sweden [37]	2.4	1.9	258
Florida [19]	3.1	2.4	1,693

offered by Berger [43], contained one or two subsequent strokes with initial return-stroke peak currents greater than their respective first-stroke peak currents. There were eight subsequent strokes with greater current peaks than the first-stroke peak (7% of all 115 subsequent strokes in the 33 flashes mentioned above), all of them necessarily following the same channel to the instrumented tower as the first stroke. The subsequent current peaks, greater than the first had a GM of 26 kA, which is about 1.2 times the GM of the first-stroke current peaks of those flashes and about 2.2 times the GM for all the 115 subsequent-stroke current peaks. The GM of the

Table 2.6 Geometric mean values for various parameters of larger subsequent strokes in the same channel as the first stroke versus those for all subsequent strokes in the first-stroke channel (sample sizes are given in the parentheses). Adapted from [33]

Parameter	Larger strokes	All strokes
Return-stroke field peak (at 100 km), V/m	7.7 (13)	2.6 (176)
Return-stroke current peak*, kA	−27	−8.1
Preceding interstroke interval, ms	98 (13)	53 (176)
Leader duration, ms	0.55 (8)	1.8 (117)
Ratio of subsequent to first stroke field peak	1.2 (13)	0.39 (176)

*Inferred from formula $I_p = 1.5–3.7 \times E_p$ where E_p is return-stroke initial electric field peak normalized to 100 km taken as positive and in V/m and I_p is return-stroke current peak, negative and in kA [42].

immediately preceding interstroke interval for the subsequent strokes with greater peaks was 69 ms, about 1.6 times the 43 ms GM for all the interstroke intervals.

The relative return-stroke peak and preceding interstroke interval statistics for larger subsequent strokes in the first-stroke channel derived from Florida electric field data are similar to those from direct channel-base current measurements in Switzerland.

2.6 Return-stroke peak current—"classical" distributions

Essentially all national and international lightning protection standards (e.g., IEEE Std. 1243-1997; IEEE Std. 1410-2010; IEC 62305-1) include a statistical distribution of peak currents for first strokes in negative lightning flashes (including the only strokes in single-stroke flashes). This distribution, which is one of the cornerstones of most lightning protection studies, is largely based on direct lightning current measurements conducted in Switzerland from 1963 to 1971 (e.g., [4,43]). The cumulative statistical distributions of lightning peak currents for (1) negative first strokes, (2) positive first strokes, and (3) negative subsequent strokes published by Berger *et al.* [4] are presented in Figure 2.14. The distributions are assumed to be log-normal (because they are positively skewed, that is, exhibit long "tails" extending toward higher values, so that Mean > Median > Mode) and give percent of cases exceeding abscissa value.

It is worth noting that directly measured current waveforms of either polarity found in the literature do not exhibit peaks exceeding 300 kA or so, which is consistent with the theoretically estimated upper limit for peak currents in temperate regions [13], while inferences from remotely measured electric and magnetic fields suggest the existence of considerably higher currents. It is important to note, however, that peak current estimates reported by the U.S. National Lightning

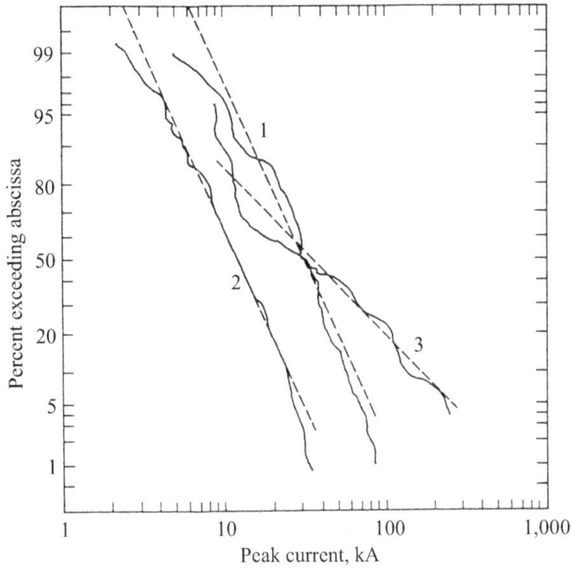

Figure 2.14 *Cumulative statistical distributions of return-stroke peak current from measurements at the tower top (solid curves) and their log-normal approximations (broken lines) for (1) negative first strokes, (2) negative subsequent strokes, and (3) positive first (and only) strokes, as reported by Berger et al. [4]*

Detection Network (NLDN) and by other similar systems are based on an empirical formula the validity of which has been tested, using triggered lightning in Florida and instrumented tower in Austria, only for negative subsequent strokes [44–49]. Remote measurements of lightning peak currents are discussed in Chapter 5 of this volume.

The log-normal probability density function for peak current I is given by

$$f(I) = \frac{1}{\sqrt{2\pi}\beta I}\exp(-z^2/2) \tag{2.1}$$

where

$$z = \frac{\ln I - \text{Mean}(\ln I)}{\beta} \tag{2.2}$$

In (2.2), ln I is the natural (base e) logarithm of I, Mean(ln I) is the mean value of ln I, and $\beta = \sigma_{\ln I}$ is the standard deviation of ln I.

For a log-normal distribution, Mean(ln I) is equal to both the logarithm of geometric mean (GM) and logarithm of median of I. It follows that the antilog of Mean(ln I) is the median (50% value) of I. Thus, a log-normal distribution is completely described by two parameters, the median and logarithmic standard deviation of the variable. Logarithmic standard deviations of lightning peak currents are often given for base 10 (base 10 logarithms are often denoted lg); those should be multiplied by ln 10 = 2.3026 in order to obtain $\beta = \sigma_{\ln I}$.

The probability for peak current to exceed a specified value I is given by

$$P(I) = \int_I^\infty \frac{1}{\sqrt{2\pi}\beta I} \exp(-z^2/2) dI \tag{2.3}$$

$P(I)$ can be evaluated as follows:

$$P(I) = 1 - \Phi(z) = \frac{1}{2} \operatorname{erfc}\left(\frac{z}{\sqrt{2}}\right) \tag{2.4}$$

where Φ is the cumulative distribution function of the standard normal distribution and erfc is the complementary error function.

Only a few percent of negative first strokes exceed 100 kA, while about 20% of positive strokes have been observed to do so. However, it is thought that less than 10% of global cloud-to-ground lightning is positive. About 95% of negative first strokes are expected to exceed 14 kA, 50% exceed 30 kA, and 5% exceed 80 kA (see Table 2.7). The corresponding values for negative subsequent strokes are 4.6, 12, and 30 kA (see Table 2.7), and 4.6, 35, and 250 kA for positive strokes (see Table 2.1). Subsequent strokes are typically less severe in terms of peak current and therefore often neglected in lightning protection studies.

Berger's peak current distribution for negative first strokes shown in Figure 2.14 is based on about 100 direct current measurements accompanied by detailed optical observations and, as of today, is thought to be the most accurate one. The minimum peak current value included in Berger's distributions shown in Figure 2.14 is 2 kA (note that no first strokes with peak currents below 5 kA were observed). Clearly, the parameters of statistical distributions can be affected by the lower and upper measurement limits. Rakov [50] showed that, for a log-normal distribution, the parameters of a measured, "truncated" distribution and knowledge of the lower measurement limit can be used to recover the parameters of the actual, "untruncated" distribution. He applied the recovery procedure to the various lightning peak current distributions found in the literature and concluded that the peak current distributions published by Berger *et al.* [4] can be viewed as practically unaffected by the effective lower measurement limit of 2 kA. Further, it has been shown by Rakov [51] that Berger's peak currents for first strokes, based on measurements at the top of 70-m towers, are not influenced by the transient process (reflections) excited in the tower. For subsequent strokes, reflections are expected to increase the tower-top current by 10% or so. The distribution of peak currents based on measurements on tall instrumented towers may be biased (relative to the ground-surface peak-current distribution) toward higher values due to the peak-current-dependent attractive effect of the tower [52,53]. Borghetti *et al.* [53], using the electrogeometric model, showed that median values of peak current based on measurements at instrumented towers should be reduced by 20% to 40% (depending on the attractive radius expression) to obtain the corresponding values for flat ground (in the absence of the tower). Interestingly, the electrogeometric model

Table 2.7 Parameters of downward negative lightning derived from channel-base current measurements. Adapted from [4]

Parameters	Units	Sample size	Percent exceeding tabulated value		
			95%	50%	5%
Peak current (minimum 2 kA)					
First strokes	kA	101	14	30	80
Subsequent strokes		135	4.6	12	30
Charge (total charge)					
First strokes	C	93	1.1	5.2	24
Subsequent strokes		122	0.2	1.4	11
Complete flash		94	1.3	7.5	40
Impulse charge (excluding continuing current)					
First strokes	C	90	1.1	4.5	20
Subsequent strokes		117	0.22	0.95	4
Front duration (2 kA to peak)					
First strokes	μs	89	1.8	5.5	18
Subsequent strokes		118	0.22	1.1	4.5
Maximum dI/dt					
First strokes	kA μs^{-1}	92	5.5	12	32
Subsequent strokes		122	12	40	120
Stroke duration (2 kA to half peak value on the tail)					
First strokes	μs	90	30	75	200
Subsequent strokes		115	6.5	32	140
Action integral ($\int I^2 dt$)					
First strokes	A^2s	91	6.0×10^3	5.5×10^4	5.5×10^5
Subsequent strokes		88	5.5×10^2	6.0×10^3	5.2×10^4
Time interval between strokes	ms	133	7	33	150
Flash duration					
All flashes	ms	94	0.15	13	1,100
Excluding single-stroke flashes		39	31	180	900

predicts that even the presence of a 5-m tall strike object appreciably alters the flat-ground peak current distribution [54], although in practice this is unlikely because of the influence of neighboring objects such as buildings and trees. As of today, there is no experimental evidence that peak current distributions for downward lightning are materially affected by the presence of the tower [55]. In fact, Popolansky [56] reported that the median negative peak currents for strike objects with heights 15–55 m ($n = 64$) and 56–65 m ($n = 81$) were 30 and 27 kA, respectively, not in support of the expected object-height dependence. For these height ranges, influence of upward lightning is usually neglected. To summarize, it appears that Berger's distributions of peak currents for first and subsequent negative strokes are not materially affected by either lower measurement limit or the presence of the tower.

In lightning protection standards, in order to increase the sample size, Berger's data are often supplemented by limited direct current measurements in South Africa and by less accurate indirect lightning current measurements obtained (in different countries) using magnetic links. There are two main distributions of lightning peak currents for negative first strokes adopted by lightning protection standards: the IEEE distribution (e.g., IEEE Std. 1243-1997 [119]; IEEE Std. 1410-2010 [120]; [40]) and CIGRE distribution (e.g., [6]). Both these "global distributions" are presented in Figure 2.15 [57].

In the coordinates of Figure 2.15 (also Figure 2.14), a cumulative log-normal distribution appears as a slanted straight line. Anderson and Eriksson [6] arbitrarily introduced two slanted lines having different slopes (implying a bimodal probability density function) and intersecting at 20 kA to approximate their "global" peak current distribution based on both direct and indirect (magnetic-link) current measurements. The same approach was adopted in the CIGRE WG 33.01 Report 63 [57].

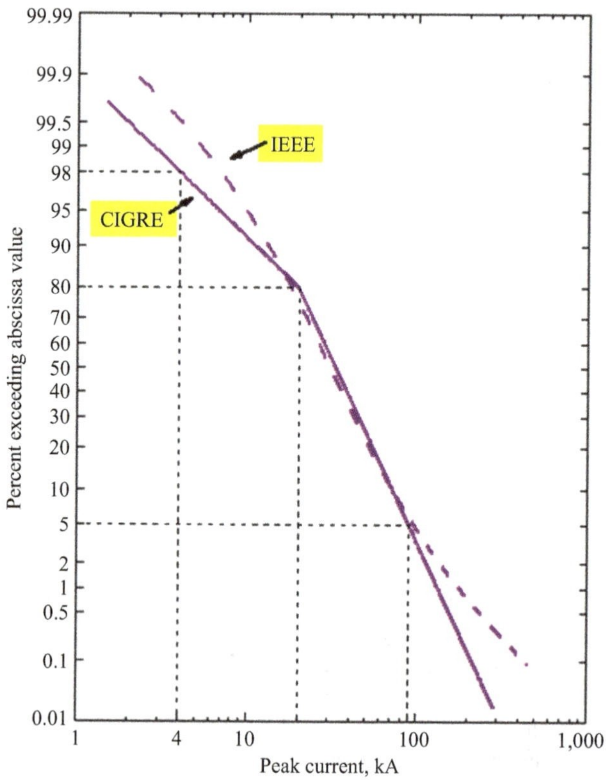

Figure 2.15 Cumulative statistical distributions of peak currents for negative first strokes adopted by IEEE and CIGRE and used in various lightning protection standards. Adapted from CIGRE WG 33.01 Report 63 [57]. Drawing by Yongfu Li

Note that IEEE Std. 1243-1997 makes reference to the two-slope CIGRE distribution as well.

For the CIGRE distribution, 98% of peak currents exceed 4 kA, 80% exceed 20 kA, and 5% exceed 90 kA.

For the IEEE distribution, the "probability to exceed" values are given by the following equation

$$P(I) = \frac{1}{1 + (I/31)^{2.6}} \tag{2.5}$$

where *P*(*I*) is in per unit and *I* is in kA. According to Hileman [58], this equation, usually assumed to be applicable to negative first strokes, is based on data for 624 strokes analyzed by Poplansky (1972), whose sample included both positive and negative strokes, as well as strokes in upward lightning. Equation (2.5) applies to values of *I* up to 200 kA. For higher peak currents, IEEE Std. 1243-1997 [119] recommends the use of the two-slope CIGRE distribution, while IEEE Std. 1410-2010 [120] apparently relies on the log-normal approximation of Berger's distribution for the global current peak (I_F) found in Table 2.8. Values of *P*(*I*) for *I* varying from 5 to 200 kA, computed using (2.5) and expressed in percent, are given in Table 2.9. The median (50%) peak current value is equal to 31 kA.

In the range of 10 to 100 kA that is well supported by experimental data, the IEEE and CIGRE distributions are very close to each other. Outside that range, the uncertainty, due to relative paucity of data, is apparently too large to allow one to favor either of the two distributions.

The peak-current distribution for subsequent strokes adopted by the IEEE (IEEE Std. 1243-1997 [119]; IEEE Std. 1410-2010 [120]) is given by

$$P(I) = \frac{1}{1 + (I/12)^{2.7}} \tag{2.6}$$

which is compared with (2.5) in Table 2.9. CIGRE recommends for negative subsequent stroke peak currents a log-normal distribution with the median of 12.3 kA and $\beta = 0.53$ (CIGRE WG 33.01 Report 63 [57]), which is also included in IEEE Std. 1410-2010 [120] (see Table 2.8).

We now further discuss the "global" distributions found in most lightning protection standards. They are not much different from the distributions based on direct current measurements by Berger *et al.* [4], which are still considered to be the most reliable ones (e.g., [59]). However, the extremely low and particularly extremely high (greater than 100 kA or so) peak current tails require much larger sample sizes (probably of the order of thousands or more) than presently available (or to be available in the foreseeable future) to bring the uncertainties within an engineering accuracy range.

In this regard, it is natural to attempt to combine as many measurements as possible to increase the sample size and, hence, reduce statistical uncertainties. One such attempt was made by Popolansky [60] who combined direct and indirect (magnetic link) current measurements made on tall objects and on power lines in

Table 2.8 Lightning current parameters based on Berger et al.'s data recommended by CIGRE WG 33.01 Report 63 [57] and IEEE Std. 1410-2010 [120]

Parameters of log-normal distribution for negative downward flashes				
Parameter	**First stroke**		**Subsequent stroke**	
	***M*, median**	**β, logarithmic (base e) standard deviation**	***M*, median**	**β, logarithmic (base e) standard deviation**
Front time (μs)				
$t_{d10/90} = T_{10/90}/0.8$	5.63	0.576	0.75	0.921
$t_{d30/90} = T_{30/90}/0.6$	3.83	0.553	0.67	1.013
$t_m = I_F/S_m$	1.28	0.611	0.308	0.708
Steepness (kA/μs)				
S_m, maximum	24.3	0.599	39.9	0.852
S_{10}, at 10%	2.6	0.921	18.9	1.404
$S_{10/90}$, 10–90%	5.0	0.645	15.4	0.944
$S_{30/90}$, 30–90%	7.2	0.622	20.1	0.967
Peak (crest) current (kA)				
I_I, initial	27.7	0.461	11.8	0.530
I_F, final	31.1	0.484	12.3	0.530
Ratio, I_I/I_F	0.9	0.230	0.9	0.207
Other relevant parameters				
Tail time to half value t_h (μs)	77.5	0.577	30.2	0.933
Number of strokes per flash	1	0	2.4	0.96 based on median $N_{total} = 3.4$
Stroke charge, Q_I (Coulomb)	4.65	0.882	0.938	0.882
$\int I^2 dt$ ((kA)^{2}s)	0.057	1.373	0.0055	1.366
Interstroke interval (ms)	—	—	35	1.066

Table 2.9 The IEEE peak current distributions for first and subsequent strokes given by (2.5) and (2.6), respectively

Peak current, *I*, kA		**5**	**10**	**20**	**40**	**60**	**80**	**100**	**200**
Percentage exceeding tabulated value, $P(I) \times 100\%$	First strokes	99	95	76	34	15	7.8	4.5	0.78
	Subsequent strokes	91	62	20	3.7	1.3	0.59	0.33	0.050

eight countries. The overall sample size was 624. Later it was realized [6] that some of the indirect measurements on taller objects could be associated with strokes in upward lightning. Since upward lightning is unlikely to occur at objects less than 60 m in height, only measurements on shorter than 60 m objects were retained for compiling the next edition of the "global" peak current distribution. Additionally, all positive current measurements were excluded and 11 current measurements from South Africa were added [6]. The overall sample size became 338. Finally, in CIGRE WG 33.01 Report 63 [57], 70 more measurements (both direct and indirect) from South Africa were added bringing the overall sample size to 408. The majority of the additional 70 currents were obtained by adding typically 4 to 5 partial currents measured with magnetic links installed on wooden poles of the test power distribution line [61].

One concern about the "global" lightning peak current distributions is the inclusion of less accurate indirect (magnetic link) measurements. Even in the case of measurements on simple lightning down-conductors or measurements at vertical strike rods mounted on the top of transmission-line towers, very significant errors are likely. Specifically, magnetic links can be saturated or demagnetized by shaking during their transportation or by incomplete discharges from the strike object top occurring in response to nearby lightning flashes. Bazelyan *et al.* [62], via modeling, showed that the collapse of charge accumulated at the tip of object (or on the unconnected upward leader) in response to a nearby downward leader can involve kiloampere-scale currents in the object at the time of return stroke initiated by that downward leader. Such induced currents usually have polarity opposite to that of direct negative strikes and, hence, can partially demagnetize the link which previously recorded current of a direct negative strike. Taller objects were found to experience higher induced currents. This effect might be responsible for the observed decrease of median peak current measured using magnetic links with strike-object height, even when objects with heights greater than 65 m (for which upward flashes could be a factor) were excluded [56]. In summary, it is probably best not to "compromise" direct current measurements by combining them with indirect measurements that may contain significant errors.

Additional concern about the "global" lightning peak current distributions is related to inhomogeneity of data coming from different sources and being combined in a single sample. Popolansky [60] noted that out of seven distributions based on indirect current measurements only two (from Czechoslovakia and Poland) were in "very good" agreement with the Swiss distribution based on direct current measurements. For one of the distributions (from the United States), the lowest measured value was 7 kA, which suggests that it might be significantly truncated (depending on logarithmic standard deviation; [50]). Nevertheless, the U.S. distribution was included in the later editions of the "global" distributions [6,57]. Further, 11 peak current values from South Africa were added by Anderson and Eriksson [6], although they suggested a quite different distribution (Median = 41 kA, Min = 10 kA). Out of the 11 values, only 8 were positively identified as corresponding to downward flashes, and 2 other values were measured with magnetic links. There has been a concern that the South African measurements, made at the bottom of the tower, might have been significantly affected by the transient process in the tower (e.g., [63]). Finally, 70 more values (including both direct and indirect measurements) from South Africa were added in CIGRE

WG 33.01 Report 63 [57], with most of the values being obtained by summing partial currents measured at multiple poles of a test distribution line. The latter data were acquired during several years for different line configurations (presence or absence of arresters, transformers, and power follow current) [64], which could have introduced additional uncertainties.

In summary, it is not clear if mixing direct current measurements with less accurate indirect ones in the "global" distributions served to build a more statistically reliable distribution; it could have actually amounted to contamination of the relatively high quality data with more numerous data of questionable quality. However, since the "global" distributions have been widely used in lightning protection studies and are not much different from that based on direct measurements only (Median = 30 kA, $\sigma_{\lg I} = 0.265$ for Berger *et al.*'s distribution for negative first strokes), continued use of these "global" distributions for representing negative first strokes is still recommended (CIGRE WG C4.407 TB 549 [7]).

2.7 Return-stroke peak current—recent direct measurements

More recently direct current measurements on instrumented towers were made in Russia, South Africa, Canada, Germany, Brazil, Japan, Austria, and again in Switzerland (on a different tower). Important results from the Brazilian, Japanese, and Austrian studies are reviewed and compared with Berger's data below. Recent direct current measurements for rocket-triggered lightning in Florida are also considered.

Brazil. Visacro *et al.* [39] presented a statistical analysis of parameters derived from lightning current measurements performed in 1985–1996 on the 60-m Morro do Cachimbo tower near Belo Horizonte, Brazil. The tower is located in the tropics (about 20° S) on the mountain (hill) top, 200 m above the surrounding terrain, and about 1,600 m above sea level.

The current measuring system included two Pearson coils, installed at the tower base, with a frequency bandwidth of 100 Hz to 10 MHz that were connected to two oscilloscopes recording with a sampling interval of 50 ns. One coil was used for measuring currents above 20 kA and the other below 20 kA. A calibrated spark gap was used to bypass the latter coil when the current attained 20 kA. Up to 16 current pulses per flash could be recorded, with the individual pulse record length being 400 μs. The trigger threshold was 800 A. The dead time between two consecutive triggers was less than 12 ms.

The current measuring system has been upgraded in 2008. Two new Pearson coils for measuring currents (of either polarity) up to 9 kA and up to 200 kA with bandwidths of 0.25 Hz to 4 MHz and 3 Hz to 1.5 MHz, respectively, were installed (again at the tower base). As of this writing, currents are recorded using a multiple-channel, 12-bit data acquisition system capable of sampling at up to 60 MHz (17-ns sampling interval). No spark gap is used. The trigger thresholds for the lower- and higher-current channels are 60 and 250 A, respectively. The record length is either

1 s with 30-ms pretrigger (33-ns sampling interval) or 0.5 s with 15-ms pretrigger (17-ns sampling interval). Thus, the entire flash current can be continuously recorded since 2008.

Before the 2008 upgrade, currents were measured for 31 first and 59 subsequent strokes in negative downward flashes, with the median peak currents being 45 kA (all values were higher than 20 kA) and 16 kA, respectively [39]. Additional measurements were obtained in 2008–2017 for 19 first and 19 subsequent strokes, with the corresponding updated median values being 43 kA ($n = 50$) and 17 kA ($n = 78$) (S. Visacro, personal communication, 2018). Clearly, these values are higher than their counterparts, 30 kA ($n = 101$) and 12 kA ($n = 135$), reported by Berger *et al.* [4]. Possible reasons for the discrepancy include: (1) a relatively small sample size in Brazil, (2) dependence of lightning parameters on geographical location (Brazil versus Switzerland), and (3) different positions of current sensors on the tower at the two locations (bottom of 60-m tower in Brazil vs. top of 70-m towers in Switzerland). For typical first strokes (longer current risetimes), the towers in question are expected to behave as electrically short objects, so that the position of current sensor should not influence measurements. However, for subsequent strokes (shorter current risetimes), the towers may exhibit a distributed-circuit behavior, in which case the peak current measured at the bottom of tower is expected to be more strongly influenced by the transient process in the tower (be higher) compared to the peak current at the top [51,63]. Visacro and Silveira [65], using a hybrid electromagnetic (HEM) model and assuming a 100-m long upward connecting leader, showed that, for typical subsequent-stroke current rise times, peak currents at the top and bottom of the Morro do Cachimbo tower should be essentially the same. Additional measurements are needed, since the Brazilian sample size is still relatively small. Interestingly, the median peak current in Japan changed from 39 kA to 29 kA as the sample size increased from 35 to 120 (see below). Similarly, the median peak current in South Africa (from measurements on the research mast) changed from 41 kA to 33 kA as the sample size increased from 11 to 29.

Another peculiarity of Brazilian measurements is nonoccurrence of upward flashes with the typical charge transfer of the order of a few tens of coulombs: no upward flashes were observed in 2008–2009 and only relatively weak ones in 2010–2017. The latter (a total of 19 with 4 of them containing leader/return stroke sequences) exhibited charge transfers as low as 0.9 to 5.9 C, with a geometric mean of 3.3 C (S. Visacro, personal communication, 2018).

Japan. Takami and Okabe [66] presented lightning return-stroke currents directly measured on 60 transmission-line towers (at the top) whose heights ranged from 40 to 140 m (90 m on average). Most of the towers were located on the mountain ridges, at altitudes ranging from 100 m to 1.5 km.

Currents were measured at 2.5-m strike rods installed on tower tops using Rogowski coils with RC external integrators connected, via short shielded cables, to 10-bit memory cards. Each memory card was connected, via a fiber optic cable, to the communication terminal at the base of the tower (data could be read out remotely). The measuring system had a frequency bandwidth of 10 Hz to 1 MHz and recorded currents on two amplitude scales: ±10 kA and ±300 kA. The record length was 3.2 ms, and the sampling interval was 100 ns. The trigger threshold was relatively high, 9 kA. The maximum number of waveforms that could be recorded was 40 (J. Takami, personal communication, 2012).

A total of 120 current waveforms for negative first strokes were obtained from 1994 to 2004. This is the largest sample size for negative first strokes as of today. The median peak current was 29 kA, which is similar to that reported by Berger *et al.* [4], although the trigger threshold in Japan (9 kA) was higher than in Switzerland. The largest measured peak current was 130 kA. Interestingly, initial data from this Japanese study (for 35 negative first strokes recorded in 1994–1997) yielded the median peak current of 39 kA [67].

Florida. Schoene *et al.* [68] presented a statistical analysis of the salient characteristics of current waveforms for 206 return strokes in 46 rocket-triggered lightning flashes. The flashes were triggered during a variety of experiments related to the interaction of lightning with power lines that were conducted from 1999 to 2004 at Camp Blanding, Florida.

Lightning channel-base currents were measured using noninductive shunts mounted at the bottom of the rocket launcher. Different shunts were used at different launchers, but in all cases the upper frequency response of the shunt exceeded 5 MHz. Shunt output signals were transmitted via fiber optic links (frequency bandwidth from dc to 15 MHz) to different digitizing oscilloscopes. The latter recorded either continuously for 1 or 2 s (at a sampling rate of 1 MHz or 2 MHz) or in a few millisecond long segments (at a sampling rate between 10 MHz and 50 MHz). The data were appropriately low-pass filtered to avoid aliasing. The lowest measured current peak was 2.8 kA, and the highest one was 42 kA.

The return-stroke current was injected into either one of two test power lines or into the earth near a power line via a grounding system of the rocket launcher. The geometric mean return-stroke peak current was found to be 12 kA, which is consistent with those reported from other triggered lightning studies (see [69]). Further, this parameter was found not to be much influenced by either strike-object geometry or level of man-made grounding, as previously reported by Rakov *et al.* [9]. Specifically, the peak current was about the same for the cases of current injection into an overhead power line conductor (impedance initially "seen" by lightning at its attachment point of about 200 Ω) and into a concentrated grounding system via a short down conductor. However, the means of the 10%–90% current risetimes were significantly different, as discussed in Section 2.8. Cooray and Rakov [70] theoretically showed that the peak current decrease is negligible as the ground conductivity decreases from infinity to 10^{-3} S/m and is about 20% lower (compared to the perfectly conducting ground case) for ground conductivity of 10^{-4} S/m. The effect of ground conductivity on the maximum rate-of-rise was much more significant (see Section 2.8).

Note that triggered-lightning strokes are initiated by continuously moving dart leaders or by dart-stepped leaders and considered to be similar to subsequent strokes in natural lightning; there is no stepped leader/first return stroke sequence in classical triggered lightning (see Section 2.1.4).

Austria. Diendorfer *et al.* [71] analyzed parameters of 457 upward negative flashes initiated from the mountain-top 100-m Gaisberg Tower in 2000–2007. After adding the data acquired in 2008–2015, the sample size for upward negative flashes increased to 775 [72]. It is worth noting that downward lightning strikes to the Gaisberg Tower are very rare, only 3 (less than 1%) of 341 flashes recorded in 2008–2015 were of downward type (Diendorfer, personal communication, 2018).

The overall current waveforms were measured at the base of the air terminal installed on the top of the tower with a current-viewing resistor (shunt) of 0.25 mΩ having a bandwidth of 0 Hz to 3.2 MHz. Fiber optic links (frequency bandwidth from dc to 15 MHz) were used for transmission of the shunt output signal to a digital recorder installed in the building next to the tower. Two separate channels of different sensitivity with current scales of ±2 kA and ±40 kA were used. The signals were recorded at a sampling rate of 20 MHz (50-ns sampling interval) by an 8-bit (presently 12-bit) digitizing board installed in a personal computer. The trigger threshold of the recording system was set to ±200 A. The record length was 800 ms with a pretrigger recording time of 15 ms. A digital low-pass filter with a cut-off frequency of 250 kHz and appropriate offset correction had been applied to the current records before the lightning peak currents were determined.

Upward flashes contain only strokes that are similar to subsequent strokes in natural downward flashes, that is, they do not contain first strokes initiated by downward stepped leaders. Many upward flashes (about two-thirds for the Gaisberg Tower [72]) contain no strokes at all, only the so-called initial-stage current with or without superimposed pulses. In 2000–2007, the median return-stroke peak current was reported to be 9.2 kA ($n = 615$) and after adding the data acquired in 2008–2015 became 9.5 kA ($n = 1{,}124$) (Diendorfer, personal communication, 2018). Both values are somewhat lower than for subsequent strokes in downward flashes and for rocket-triggered-lightning strokes. This could be due to the tall grounded object reducing the distance between the cloud charge source region and the strike point. Indeed, the lower-charge-density downward leaders that are not capable of making their way to flat ground or to a small strike object may be able to make connection to a tall tower. Note that in rocket-triggered lightning the triggering wire is destroyed during the initial stage and downward leaders have to propagate all the way to the relatively small rocket launcher. An additional factor in lowering return-stroke peak currents measured at Gaisberg Tower could be the lower height of cloud charge source region, since many measurements were obtained in cold (nonconvective) season [72].

2.8 Current waveshape parameters

Lightning parameters, other than lightning peak current, derivable from direct current measurements include the maximum current derivative, average current rate of rise, current risetime, current duration, charge transfer, and action integral (specific energy). Similar to the peak current, the most reliable and complete information on the other parameters is based on the direct current measurements of K. Berger and co-workers in Switzerland. Berger *et al.* [4] summarized the lightning current parameters for 101 downward negative cloud-to-ground lightning flashes, the types that normally strike flat terrain and structures of moderate height. This summary, which is used to a large extent as a primary reference in the literature on both lightning protection and lightning research, is reproduced in Table 2.7. The table gives the percentages (95%, 50%, and 5%) of cases exceeding the tabulated values, based on the log-normal approximations to the respective statistical distributions. A similar summary for 26 positive flashes from the same study is given in Table 2.1. The action integral represents the energy that would be dissipated in a 1-Ω resistor if the lightning current were to flow through it. All the parameters presented in Table 2.7 are estimated from current oscillograms with the shortest measurable time being 0.5 μs [73]. Anderson and Eriksson [6] digitized the return-stroke current oscillograms of Berger *et al.* [4] and determined additional wavefront parameters. Most of the current waveform parameters are illustrated in Figure 2.16. Parameters of log-normal distributions of current waveform parameters (for both first and subsequent strokes) are summarized in Table 2.8, adapted from CIGRE WG 33.01 Report 63 [57] and IEEE Std. 1410-2010.

Note from Table 2.7 that the median return-stroke current peak for first strokes is two to three times higher than that for subsequent strokes. Also, negative first strokes transfer about a factor of 4 larger total charge than do negative subsequent strokes. However, subsequent return strokes are characterized by three to four times higher maximum steepness (the current maximum rate of rise).

The median 10%–90% risetime estimated for subsequent strokes by Anderson and Eriksson [6] from Berger *et al.*'s (1975) oscillograms is 0.6 μs, comparable to the median values ranging from 0.3 to 0.6 μs for triggered-lightning strokes [21,74]. The median 10%–90% current rate of rise reported for natural subsequent strokes by Anderson and Eriksson [6] is 15 kA/μs, almost three times lower than the corresponding value of 44 kA/μs in data of Leteinturier *et al.* [74] and more than twice lower than the value of 34 kA/μs found by Fisher *et al.* [21]. The largest value of maximum rate of rise of 411 kA/μs (see Figure 2.17) was measured by Leteinturier *et al.* [74] for a triggered lightning stroke terminating on a launcher grounded to salt water. The corresponding directly measured current was greater than 60 kA, the largest value reported for summer triggered lightning. The mean value of current derivative peak reported by Leteinturier *et al.* [74] is 110 kA/μs. The higher observed values of current rate of rise for triggered-lightning return strokes than for natural-lightning return strokes are likely to be due to the use of better instrumentation (digital oscilloscopes with better upper frequency response), which implies that the current rate-of-rise parameters reported by Anderson and Eriksson [6] are underestimates. Triggered-lightning data for current rates of rise (see Figure 2.17) can be applied to subsequent strokes in natural lightning.

Schoene *et al.* [68], who presented a statistical analysis of the salient characteristics of current waveforms for 206 return strokes in 46 rocket-triggered-lightning flashes, found that the means of the 10%–90% current risetimes for strikes to the power line (geometric mean 1.2 μs) and for strikes to the ground nearby (geometric mean 0.4 μs) were significantly different. This indicates that the electrical properties of the strike object affect the risetime. This effect is likely

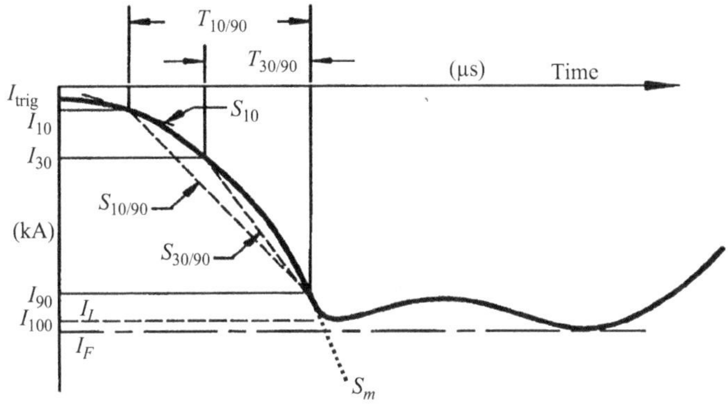

Parameter (see above)	Description
I_{10}	10% intercept along the stroke current waveshape
I_{30}	30% intercept along the stroke current waveshape
I_{90}	90% intercept along the stroke current waveshape
$I_{100} = I_I$	Initial peak of current
I_F	Final (global) peak of current (same as peak current without an adjective)
$T_{10/90}$	Time between I_{10} and I_{90} intercepts on the wavefront
$T_{30/90}$	Time between I_{30} and I_{90} intercepts on the wavefront
S_{10}	Instantaneous rate-of-rise of current at I_{10}
$S_{10/90}$	Average steepness (through I_{10} and I_{90} intercepts)
$S_{30/90}$	Average steepness (through I_{30} and I_{90} intercepts)
S_m	Maximum rate-of-rise of current along wavefront, typically at I_{90}
$t_{d\,10/90}$	Equivalent linear wavefront duration derived from $I_F/S_{10/90}$
$t_{d\,30/90}$	Equivalent linear wavefront duration derived from $I_F/S_{30/90}$
t_m	Equivalent linear wavefront duration derived from I_F/S_m
Q_I	Impulse charge (time integral of current)

Figure 2.16 Description of lightning return-stroke current waveform parameters. The waveform corresponds to the typical negative first return stroke. Adapted from CIGRE WG 33.01 Report 63 [57] and IEEE Std. 1410-2010 [120]

related to the impedance seen by lightning at the strike point and/or to reflections at impedance discontinuities within the strike object, larger effective impedances apparently resulting in larger risetimes. A dependence of the return-stroke current half-peak width on the electrical properties of the strike object was not observed. Cooray and Rakov [70] theoretically showed that the peak value of current rate-of-rise is influenced by ground conductivity: it decreases by about 40% as the ground conductivity decreases from infinity to 10^{-3} S/m and by 83% when the conductivity becomes 10^{-4} S/m. For all their data combined, Schoene *et al.* [68] reported the geometric mean values of 10%–90% current risetime and current half-peak width to be 0.9 and 1.9 μs, respectively.

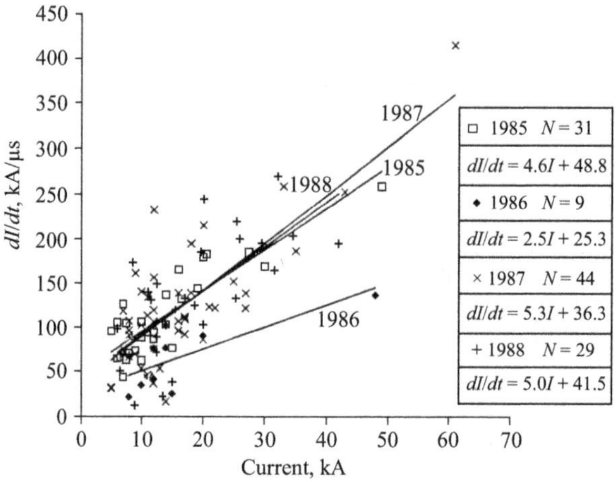

Figure 2.17 Relation between the peak value of current rate of rise and peak current from triggered-lightning experiments conducted at the NASA Kennedy Space Center, Florida, in 1985, 1987, and 1988 and in France in 1986. The regression line for each year is shown, and the sample size and the regression equation are given. Adapted from Leteinturier et al. [74]

2.9 Correlations between the parameters

Correlation coefficients for the current waveshape parameters defined in Figure 2.16 are summarized in Table 2.10. Note that there is only one current peak for subsequent strokes, which is labeled I in Table 2.10.

Anderson and Eriksson [6] gave the following relationships between S_m and $S_{30/90}$ and peak current I (I in kA and S in kA/μs) for natural lightning:

$$\text{First strokes:}\quad S_m = 3.9I^{0.55} \quad S_{30/90} = 3.2I^{0.25} \tag{2.7}$$

$$\text{Subsequent strokes:}\quad S_m = 3.8I^{0.93} \quad S_{30/90} = 6.9I^{4.2} \tag{2.8}$$

In (2.7), $I = I_I$. As noted above, the current rate-of-rise parameters estimated by Anderson and Eriksson [6] from Berger *et al.*'s (1975) oscillograms are likely to be underestimated due to limitations of the instrumentation used by Berger *et al.*

Positive correlation between the peak value of current rate-of-rise and peak current for triggered lightning is illustrated in Figure 2.17. Fisher *et al.* [21], also for triggered lightning, found a relatively strong positive correlation between the 10%–90% average steepness ($S_{10/90}$) and current peak (correlation coefficient = 0.71) and between the 30%–90% average steepness ($S_{30/90}$) and current peak (correlation coefficient = 0.74). Essentially no linear correlation was found between the current peak and 10%–90% risetime (this was also reported for triggered lightning in China;

Table 2.10 Correlation coefficients between current waveshape parameters defined in Figure 2.16. Adapted from [6]

Current peak	$T_{10/90}$	$T_{30/90}$	S_{10}	$S_{10/90}$	$S_{30/90}$	S_m
I_I (first strokes)	0.40	0.47	(0.12)	0.30	(0.19)	0.43
I_F (first strokes)	0.33	0.45	(0.06)	(0.20)	(0.17)	0.38
I (subsequent strokes)	(0.15)	(0.00)	(0.05)	0.31	0.23	0.56

Note: Values in the parentheses are not statistically significant at the 5% level.

[75]) and between the current peak and current half-peak width. Similarly, but for first strokes in natural lightning, Takami and Okabe [66] observed strong positive correlation between the current steepness characteristics and peak current and weak correlation between the peak current and front duration. Opposite trends for first strokes were reported by Visacro *et al.* [39].

According to Berger *et al.* [4], for first and subsequent negative strokes, correlation coefficients between the current peak and stroke duration (the time interval between the 2-kA point on the front and the point on the tail where the current has fallen to 50% of its peak value) are 0.56 and 0.25, respectively. Both values should be considered low, since even in the former case the determination coefficient (the square of the correlation coefficient) is as low as 0.31, which means that only 31% of the variation of one of the parameters is due to variation in the other one, while 69% is due to variation in other (unknown) factors.

All published experimental data regarding the relation between the return-stroke peak current I and charge transfer Q in natural lightning are derived from the data of K. Berger and co-workers (e.g., [4,43,73]), for lightning striking two towers in Switzerland and two towers in Italy, and have been analyzed by them and by Cooray *et al.* [76]. According to Cooray *et al.* [76], for natural negative first strokes, there is a linear regression, $Q = 0.061I$ ($R^2 = 0.88$, where R^2 is the determination coefficient), for charge transfer to 100 μs, and for natural subsequent strokes, $Q = 0.028I$ (R^2 not stated), for charge transfer to 50 μs. In the above equations, charge transfer Q is in coulombs and peak current I is in kiloamperes. Additionally, Schoene *et al.* [68] have shown that for triggered-lightning strokes (which, as noted above, are similar to natural-lightning subsequent strokes) the scatter-plot of return-stroke peak current versus charge transfer to 1 ms is surprisingly similar to the 1-ms natural-lightning first stroke data of Berger [43]. The equation for 143 triggered-lightning strokes, as given by Schoene *et al.* [68], is $I = 12.3Q^{0.54}$ ($R^2 = 0.76$) and the equation for Berger's 89 natural-lightning first strokes is $I = 10.6Q^{0.7}$ ($R^2 = 0.59$). Qie *et al.* [77] reported that $I = 18.5Q^{0.65}$ for 10 triggered-lightning strokes in China.

Schoene *et al.* [78] examined data on 117 return strokes in 31 rocket-and-wire-triggered lightning flashes acquired during experiments conducted from 1999 through 2004 at Camp Blanding, Florida, in order to relate the peak currents of the lightning return strokes to the corresponding charges transferred during various time intervals within 1 ms after return-stroke initiation. They found that the determination coefficient (R^2) for lightning return-stroke peak current versus the corresponding charge transfer decreases with increasing the duration of the charge transfer starting from return-stroke onset. For example, $R^2 = 0.91$ for a charge transfer duration of 50 μs after return-stroke onset, $R^2 = 0.83$ for a charge transfer duration of 400 μs, and $R^2 = 0.77$ for a charge transfer duration of 1 ms. Their results support the view that (1) the charge deposited on the lower portion of the leader channel determines the current peak and that (2) the charge transferred at later times is increasingly unrelated to both the current peak and the charge deposited on the lower channel section. Additionally, they found that the relation between the return-stroke peak current and charge transfer to 50 μs for triggered lightning in Florida is essentially the same as that reported by Cooray *et al.* [76] for subsequent strokes in natural lightning in Switzerland, further confirming the view that triggered-lightning strokes are very similar to subsequent strokes in natural lightning.

2.10 Return-stroke propagation speed

Lightning return stroke speed is a parameter in the models used in evaluating lightning-induced effects in power and communication lines. Further, an explicit or implicit assumption of the return-stroke speed is involved in inferring lightning currents from remotely measured electric and magnetic fields. It is known that the return-stroke speed may vary significantly along the lightning channel. As a result, optical speed measurements along the entire channel are not necessarily representative of the speed within the bottom 100 m or so; that is, at early times when the peaks of the channel-base current and remote electric and magnetic fields are formed.

The optically measured return-stroke speed probably represents the speed of the region of the upward-moving return-stroke tip where the input power is greatest. Since the power per unit length is the product of the current and the longitudinal electric field in the channel, the peak of the input power (Joule heating) wave likely occurs earlier in time than the peak of the current wave. The shape of the return-stroke light pulse changes significantly with height, so that there is always some uncertainty in tracking the propagation of such pulses for a speed measurement. For example, if the light pulse peak is tracked, then an increase in pulse risetime translates into a lower speed value than if an earlier part of the light pulse is tracked. Techniques for measuring return-stroke speed are discussed, for example, by Idone and Orville [79] and Olsen *et al.* [80].

The average propagation speed of a negative return stroke (first or subsequent) below the lower cloud boundary is typically between one-third and one-half of the speed of light. From streak-camera measurements of Idone and Orville [79], the return-stroke speed for first strokes is somewhat lower than that for subsequent strokes, although the difference is not very large (9.6×10^7 vs. 1.2×10^8 m/s). The speed within the bottom 100 m or so is expected to be between one-third and two-thirds of the speed of light. The negative return stroke speed usually decreases with height for both first and subsequent strokes. There exists some experimental evidence that the negative return stroke speed may vary nonmonotonically along the lightning channel, initially increasing and then decreasing with increasing height. For positive return strokes, the speed is of the order of 10^8 m/s, although data are very limited. More detailed information about the return-stroke speed can be found in work of Rakov [81] and in references therein.

We now discuss why the return-stroke speed is lower than the speed of light. It is well known that the propagation speed of waves on a uniform, linear, and lossless transmission line surrounded by air is equal to the speed of light, $c = (LC)^{-1/2} = (\mu_0\varepsilon_0)^{-1/2} = 3 \times 10^8$ m/s, where L is the series inductance and C is the shunt capacitance, each per unit length; μ_0 is the magnetic permeability and ε_0 is the electric permittivity of free space. However, a vertical lightning channel is a nonuniform, nonlinear, and lossy transmission line. Indeed, its L and C vary with height above ground, so that its characteristic impedance, found as $(L/C)^{1/2}$ for $R \ll \omega L$ and $G \ll \omega C$ (where R and G are the series resistance and shunt conductance, each per unit length, respectively, and ω is the angular frequency), increases with height. Thus, a return-stroke wave will suffer distortion even in the absence of losses. Further, charge cannot be confined within the narrow channel core carrying the longitudinal current; it is pushed outward via electrical breakdown forming the radial corona sheath. Finally, channel resistance per unit length ahead of the return-stroke front is relatively high (causing wave attenuation and dispersion) and decreases by two orders of magnitude or so behind the front.

Two primary reasons for the lightning return-stroke speed, v, being lower than the speed of light, c, are (1) the effect of radial corona surrounding the narrow channel core (the radius of the charge-containing corona sheath is considerably larger than the radius of the core carrying the longitudinal channel current, so that $(LC)^{-1/2} < (\mu_0\varepsilon_0)^{-1/2} = c$, and hence $v < c$) and (2) the ohmic losses in the channel core that are usually represented in lightning models by the distributed constant or current-dependent series resistance, R, of the channel (also ohmic losses in the corona sheath that are usually neglected).

The corona effect explanation is illustrated in Figure 2.18. It is based on the following assumptions:

(a) The longitudinal channel current flows only in the channel core, because the core conductivity, of the order of 10^4 S/m, is much higher than the corona sheath conductivity, of the order of 10^{-6}–10^{-5} S/m. The longitudinal resistance of channel core is expected to be about 3.5 Ω/m, while that of a 2-m radius corona sheath should be of the order of kiloohms to tens of kiloohms per meter. The corona current is radial (transverse) and hence is unrelated to the inductance of the channel.

(b) The radial voltage drop across the corona sheath is negligible compared to the potential of the lightning channel. The average radial electric field within the corona sheath is about 0.5–1.0 MV/m, which results in a radial voltage drop of 1–2 MV across a 2-m radius corona sheath (expected for subsequent strokes). The typical channel potential (relative to reference ground) is about

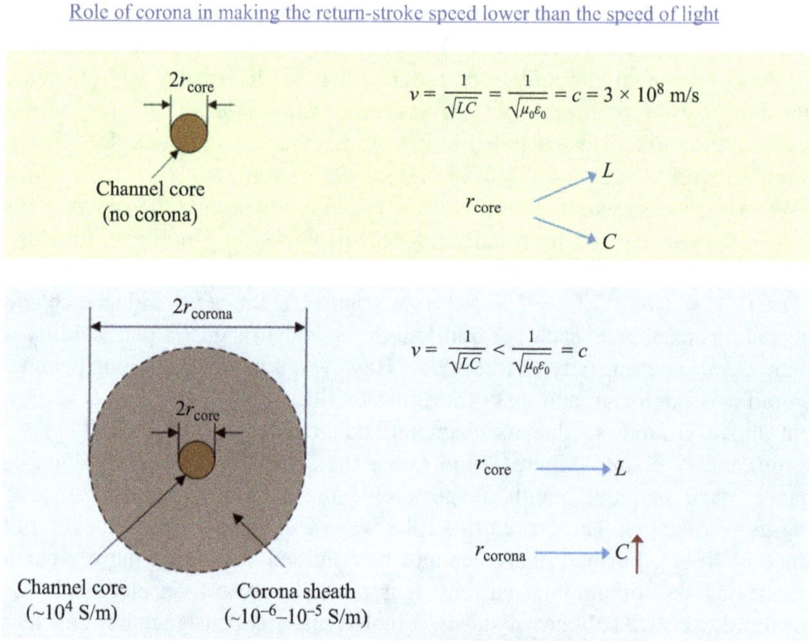

Figure 2.18 Role of corona sheath in making v < c

10–15 MV for subsequent strokes. For first strokes, both the corona sheath radius and channel potential are expected to be larger, so that about an order of magnitude difference between the corona sheath voltage drop and channel potential found for subsequent strokes should hold also for first strokes.

(c) The magnetic field due to the longitudinal current in channel core is not significantly influenced by the corona sheath. For corona sheath conductivity of 10^{-6}–10^{-5} S/m and frequency of 1 MHz, the field penetration depth is 160 to 500 m (and more for lower frequencies), which is much larger than expected radii of corona sheath of a few meters.

In summary, the corona sheath conductivity is low enough to neglect both the longitudinal current along the sheath and the shielding effect of the sheath, but high enough to disregard the radial voltage drop across the sheath. Thus, different channel radii should be used for computing L and C, which makes the return-stroke speed lower than the speed of light, even if losses are neglected.

2.11 Equivalent impedance of the lightning channel

The equivalent impedance of the lightning channel is needed for specifying the source in circuit models used in studies of either direct-strike or induced lightning effects.

When the actual current distribution in the lightning channel is of no concern (electromagnetic coupling effects are neglected), lightning is often approximated by a Norton equivalent circuit. This representation includes an ideal current source equal to the lightning return stroke current that would be injected into the ground if that ground were perfectly conducting (a short-circuit current, I_{sc}) in parallel with a lightning-channel impedance Z_{ch} assumed to be constant. In the case when the strike object can be represented by lumped grounding impedance, Z_{gr}, this impedance is a load connected in parallel with the lightning Norton equivalent [see Figure 2.19(a)]. Thus, the "short-circuit" lightning current I_{sc} effectively splits between Z_{gr} and Z_{ch} so that the current flowing from the lightning-channel base into the ground is found as $I_{gr} = I_{sc}Z_{ch}/(Z_{ch} + Z_{gr})$. Both source characteristics, I_{sc} and Z_{ch}, vary from stroke to stroke, and Z_{ch} is a function of channel current, the latter nonlinearity being in violation of the linearity requirement necessary for obtaining the Norton equivalent circuit. Nevertheless, Z_{ch}, which is usually referred to as equivalent impedance of the lightning channel, is assumed to be constant.

Equivalent impedance of the lightning channel also influences the transient process in the strike object, if that object is electrically long, that is, has dimensions that are comparable to or greater than the shortest significant wavelength of the source current, which is usually associated with the initial rising portion of the return-stroke current waveform. As an example, representation of lightning by a Norton equivalent circuit for analyzing lightning interaction with a tall object is shown in Figure 2.19(b), where Z_{ob} is the characteristic impedance of the object, and ρ_{top} and ρ_{bot} are reflection coefficients at the top and bottom of the object, respectively.

It is possible to estimate the equivalent impedance of the lightning channel from measurements of lightning current waveforms at a very tall object, if the characteristic impedance of the object and the grounding impedance are known or can be reasonably assumed. Such measurements were performed at the 540-m high Ostankino tower in Moscow, Russia. Typical lightning current waveforms, reported by Gorin and Shkilev [83], at heights of 47, 272, and 533 m above ground are shown in Figure 2.20. The median peak currents from their measurements at 47 and 533 m were 18 and 9 kA, respectively. The observed differences in peak current suggest that the effective grounding impedance of the tower is much smaller than

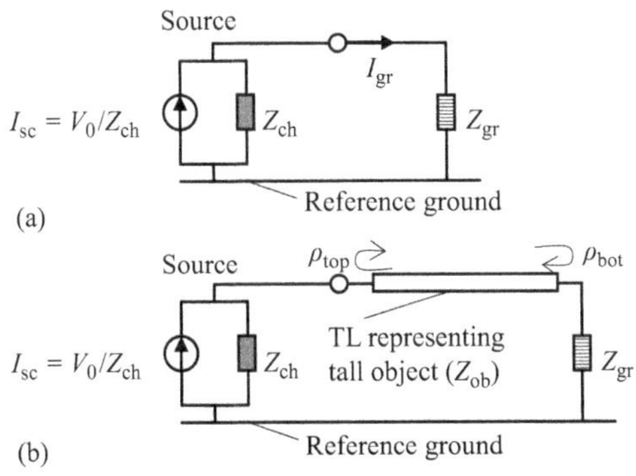

Figure 2.19 Engineering models of lightning strikes (a) to lumped grounding impedance and (b) to a tall grounded object, in which lightning is represented by the Norton equivalent circuit, labeled "Source." V_0 is the Thevenin (open-circuit) voltage of the source. Adapted from Baba and Rakov [82]

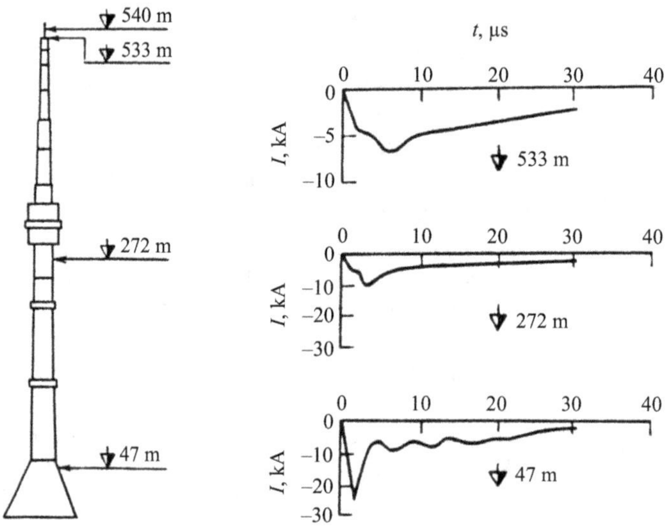

Figure 2.20 Typical return-stroke current waveforms of upward negative lightning recorded near the top (at 533 m), in the middle (at 272 m), and near the bottom (at 47 m) of the 540-m high Ostankino tower in Moscow, Russia. Differences in current waveforms at different heights are indicative of the tower behaving as a distributed circuit. Adapted from Gorin and Shkilev [83]

its characteristic impedance and that the latter impedance is appreciably lower than the equivalent impedance of lightning channel. Gorin and Shkilev [83] used current oscillograms recorded near the tower top (at 533 m), for the cases in which the current risetime was smaller than the time (about 3.5 µs) required for a current wave to travel at the speed of light from the tower top to its base and back, to estimate the equivalent impedance of the lightning channel, assumed to be a real number. Their estimates varied from 600 Ω to 2.5 kΩ when the characteristic impedance of the tower was assumed to be 300 Ω and the grounding resistance was assumed to be zero (the low-current, low-frequency value was about 0.2 Ω).

Thus, the limited estimates of the equivalent impedance of the lightning channel from experimental data suggest values ranging from several hundred ohm to a few kiloohm. In many practical situations the impedance "seen" by lightning at the strike point is some tens of ohm or less, which allows one to assume infinitely large equivalent impedance of the lightning channel. In other words, lightning in these situations can be viewed as an ideal current source. In case of direct lightning strike to an overhead conductor of a power line with 400-Ω surge impedance (effective impedance 200 Ω, since 400 Ω is "seen" in either direction), the ideal current source approximation may still be suitable. Representation of lightning by a current source with internal impedance of 400 ohm, similar to that of an overhead wire, is probably not justified (CIGRE WG C4.407 TB 549 [7]).

2.12 Mathematical expressions for the lightning current waveform

Direct effects of lightning current injection and induced effects from nearby lightning are often evaluated by numerical simulation involving assumed lightning current waveforms at the channel base. A variety of equations are used to approximate such waveworms.

Although it is well known that the return-stroke current waveform is a pulse, the step-function lightning current representation remains in use, particularly in analytical solutions (e.g., [84–88]). The Rusck formula for computing the maximum overvoltage magnitude, based on the step-function current waveform is found in IEEE Std. 1410-2010 [120]. Also, applying Duhamel's integral, the unit-step-function response of the system can be used for finding system's response to an arbitrarily-time-varying excitation (e.g., [89,90]).

A linearly rising current with linearly falling tail (triangular pulse) is a reasonable representation of lightning return-stroke current. Current waveforms of this type were used, for example, by Piantini [91–93], Andreotti *et al.* [94–96], and Rodrigues *et al.* [97]. Høidalen [98] represented the channel-base current by a single linearly rising ramp for computing the initial part (first few microseconds) of the induced-voltage waveform and Sekioka [99] employed piecewise linear approximation to the current waveform. IEEE Std. 1410-2010 [120] allows the use of "equivalent" linear front times (1.28 μs for first strokes and 0.31 μs for subsequent strokes) for induced voltage calculations.

A generalized formula for lightning channel-base current was introduced by Rakov and Dulzon [100]:

$$I(0,t) = \begin{cases} I_p(\lambda^{t/t_p} - 1)/(\lambda - 1) & t \leq t_p \\ I_p \exp\left[-\beta\left(t - t_p\right)\right] & t > t_p \end{cases} \tag{2.9}$$

where I_p is the current peak, t_p is the time of current peak (current risetime), β (not to be confused with β in (2.1)–(2.3) and in Table 2.8) is the rate of exponential current decay with time when $t > t_p$ and λ is the ratio of the steepness of the rising part of current pulse at $t = t_p$ to that at $t = 0$. For $\lambda = 1$, the rising part is a linear ramp, for $\lambda < 1$ it is convex, and for $\lambda > 1$ concave, the latter being the most realistic case. For $\lambda = 1$ and $\beta = 0$, current given by (2.9) linearly rises from 0 to t_p and then remains constant at I_p (trapezoidal waveform; e.g., [101]).

The typical subsequent-stroke current waveform at the channel base is often approximated by the Heidler function [102,103]:

$$I(0,t) = \frac{I_0}{\eta}\frac{(t/\tau_1)^n}{(t/\tau_1)^n + 1}\exp(-t/\tau_2) \quad \text{with} \quad \eta = \exp\left[-(\tau_1/\tau_2)(n\tau_1/\tau_2)^{(1/n)}\right] \tag{2.10}$$

where I_0, τ_1, n, and τ_2 are constants. This function allows one to change conveniently the current peak, maximum current derivative, and associated electrical charge transfer nearly independently by changing I_0, τ_1, and τ_2, respectively. Equation (2.10) reproduces the observed concave rising portion of a typical current waveform, as opposed to the once more commonly used double-exponential function, introduced independently by Bruce and Golde [104] and Stekolnikov [105], which is characterized by an unrealistic convex wavefront with a maximum current derivative at $t = 0$. The IEC Lightning Protection Standard (IEC 62305-1, [106]) recommends the 10/350-μs current waveform for first strokes, which is approximated by a Heidler function with $I_0 = 200$ kA (for Lightning Protection Level I), $n = 10$, $\eta = 0.93$, $\tau_1 = 19$ μs, and $\tau_2 = 485$ μs. For subsequent strokes, the 0.25/100-μs Heidler-function-approximated waveform is recommended with $I_0 = 50$ kA (for Lightning Protection Level I), $n = 10$, $\eta = 0.993$, $\tau_1 = 0.454$ μs, and $\tau_2 = 143$ μs. Sometimes the sum of two Heidler functions with different parameters is used to approximate the desired current waveshape (e.g., [90,107,108]).

One popular channel-base current waveform, which was used by Nucci *et al.* [109], Rakov and Dulzon [110], Thottappillil *et al.* [111], Moini *et al.* [112], and Romero *et al.* [113] for the calculation of subsequent return-stroke electric and magnetic fields, is represented by the sum of a Heidler function and a double-exponential function:

$$I(0,t) = \frac{I_{01}}{\eta}\frac{(t/\tau_1)^n}{(t/\tau_1)^n + 1}\exp(-t/\tau_2) + I_{02}[\exp(-t/\tau_3) - \exp(-t/\tau_4)] \quad (2.11)$$

where $n = 2$, $I_{01} = 9.9$ kA, $\eta = 0.845$, $\tau_1 = 0.072$ μs, $\tau_2 = 5.0$ μs, $I_{02} = 7.5$ kA, $\tau_3 = 100.0$ μs, and $\tau_4 = 6.0$ μs. For this latter waveform, the current peak is 11 kA, and the maximum current rate-of-rise is 105 kA/μs. It is worth reiterating that a double-exponential function alone (the last two terms in (2.11)) does not allow representation of concave current front, because the maximum time-derivative of this function occurs at $t = 0$.

De Conti and Visacro [114] employed multiple Heidler functions to reproduce with higher precision the concave front in first- and subsequent-stroke current waveforms, as well as the secondary peak typically seen in first-stroke current waveforms recorded on instrumented towers. Gamerota *et al.* [59] extended that study and other similar studies (e.g., [115–118]) of negative return-stroke current waveform characteristics to include continuing current, as well as full-flash charge transfer and action integral (specific energy), for both positive and negative flashes, and for both median (typical) and severe cases.

2.13 Summary

1. Typically 80% or more of negative cloud-to-ground lightning flashes are composed of two or more strokes. This percentage is appreciably higher than 55% previously estimated by Anderson and Eriksson [6], based on less

accurate records. The average number of strokes per flash is typically 3 to 5, with the geometric mean interstroke interval being typically about 60 ms. Roughly one-third to one-half of lightning flashes create two or more terminations on ground separated by up to several kilometers. When only one location per flash is recorded, the correction factor for measured values of ground flash density to account for multiple channel terminations on ground is about 1.5–1.7, which is considerably higher than 1.1 previously estimated by Anderson and Eriksson [6]. First-stroke current peaks are typically a factor of 2 to 3 larger than subsequent-stroke current peaks. However, about one-third of cloud-to-ground flashes contain at least one subsequent stroke with electric field peak, and, by theory, current peak, greater than the first-stroke peak.

2. From direct current measurements in different countries, the median return-stroke peak current is about 30 kA for negative first strokes and 10–15 kA for subsequent strokes, except for measurements in Brazil that yielded 43 kA ($n = 50$) and 17 kA ($n = 78$), respectively. The "global" distributions of lightning peak currents for negative first strokes currently recommended by CIGRE and IEEE (see Figure 2.15) are each based on a mix of direct current measurements and less accurate indirect measurements, some of which are of questionable quality. However, since the "global" distributions have been widely used in lightning protection studies and are not much different from that based on direct measurements only (Median = 30 kA, $\sigma_{\lg I} = 0.265$ for Berger *et al.*'s distribution for negative first strokes), continued use of these "global" distributions for representing negative first strokes is recommended (CIGRE WG C4.407 TB 549 [7]). For negative subsequent strokes, distribution 2 (Median = 12 kA, $\sigma_{\lg I} = 0.265$) in Figure 2.14 should be used. For positive lightning strokes, distribution 3 (Median = 35 kA, $\sigma_{\lg I} = 0.544$) in Figure 2.14 is recommended (CIGRE WG C4.407 TB 549 [7]), although the data are very limited and may be influenced by the presence of strike object located on the mountain top.
3. Lightning current waveshape parameters recommended by CIGRE WG C4.407 TB 549 [7] are still based on Berger *et al.*'s (1975) data (see Table 2.7), although the current rate-of-rise parameters estimated by Anderson and Eriksson [6] from Berger *et al.*'s oscillograms are likely to be underestimated, due to limitations of the instrumentation used by Berger *et al.* Triggered-lightning data for current rates of rise (see Figure 2.17) can be applied to subsequent strokes in natural lightning. Relatively strong correlation is observed between the lightning peak current and impulse charge transfer and between the current rate-of-rise characteristics and current peak, and relatively weak or no correlation between the peak and risetime.
4. The average propagation speed of a negative return stroke (first or subsequent) below the lower cloud boundary is typically between one-third and one-half of the speed of light. It appears that the return-stroke speed for first strokes is lower than that for subsequent strokes, although the difference is not very large (9.6×10^7 vs. 1.2×10^8 m/s). For positive return strokes, the speed is of the order of 10^8 m/s, although data are very limited. The negative return-stroke speed within the bottom 100 m or so (corresponding to current and field peaks)

is expected to be between one-third and two-thirds of the speed of light. The negative return stroke speed usually decreases with height for both first and subsequent strokes. There exists some experimental evidence that the negative return stroke speed may vary nonmonotonically along the lightning channel, initially increasing and then decreasing with increasing height.

5. The equivalent impedance of the lightning channel is needed for specifying the source in studies of either direct-strike or induced lightning effects. The estimates of this impedance from limited experimental data suggest values ranging from several hundred ohm to a few kiloohm. In many practical situations the impedance "seen" by lightning at the strike point is some tens of ohm or less, which allows one to assume infinitely large equivalent impedance of the lightning channel. In other words, lightning in these situations can be viewed as an ideal current source. In case of direct lightning strike to an overhead conductor of a power line with 400 Ω surge impedance (effective impedance 200 Ω, since 400 Ω is "seen" in either direction), the ideal current source approximation may still be suitable. Representation of lightning by a current source with internal impedance of 400 Ω, similar to that of an overhead wire, is probably not justified (CIGRE WG C4.407 TB 549 [7]).
6. Although positive lightning discharges account for 10% or less of global cloud-to-ground lightning activity, there are several situations, including, for example, winter storms, that appear to be conducive to the more frequent occurrence of positive lightning. The highest directly measured lightning currents (near 300 kA) and the largest charge transfers (hundreds of coulombs or more) are associated with positive lightning. Positive flashes are usually composed of a single stroke, although up to four strokes per flash were observed. Subsequent strokes in positive flashes can occur both in a new and in the previously formed channel. In spite of recent progress, our knowledge of the physics of positive lightning remains considerably poorer than that of negative lightning. Bipolar lightning discharges are usually initiated by upward leaders from tall objects. However, natural downward flashes also can be bipolar.

2.14 Future work

Suggestions for future work are given below in the form of a few presently unanswered questions.

1. Why are some very tall objects (e.g., 634-m Tokyo Skytree in Japan and buildings taller than 300 m in Guangzhou, China) often struck by downward lightning? Is the effective height a meaningless parameter?
2. Is it possible that instrumented towers, commonly used for lightning studies, interact in a unique way with the local topography, meteorological characteristics, and so on to create conditions that are not representative of "normal" lightning in the area, even when only ordinary negative downward lightning is considered and tower reflections are not an issue? Could this mask the tower-height-related bias predicted by the electrogeometrical model?

3. How to evaluate errors in peak currents reported by lightning locating systems (LLSs) for first strokes in negative lightning and for positive lightning? Rocket-and-wire triggered lightning and tower-initiated lightning have been so far not a big help for this. What do the peak currents reported by LLSs for cloud lightning discharges tell us about the source?

Acknowledgments

Much of the content of this chapter resulted from the work conducted from 2008 to 2013 by the CIGRE WG C4.407, composed of 21 members from North and South America, Europe, and Asia: V.A. Rakov, Convenor (US), A. Borghetti, Secretary (IT), C. Bouquegneau (BE), W.A. Chisholm (CA), V. Cooray (SE), K. Cummins (US), G. Diendorfer (AT), F. Heidler (DE), A. Hussein (CA), M. Ishii (JP), C.A. Nucci (IT), A. Piantini (BR), O. Pinto, Jr. (BR), X. Qie (CN), F. Rachidi (CH), M.M.F. Saba (BR), T. Shindo (JP), W. Schulz (AT), R. Thottappillil (SE), S. Visacro (BR), and W. Zischank (DE). The author would like to thank G. Diendorfer and S. Visacro for providing their unpublished data that allowed bringing this writing up to date. Y. Zhu and Z. Ding helped with preparation of the manuscript.

References

[1] Romps, D. M., J. T. Seeley, D. Vollaro, and J. Molinari. Projected increase in lightning strikes in the United States due to global warming. *Science* 346(6211): 851–4, 2014.

[2] Rakov, V. A. and M. A. Uman. *Lightning: Physics and Effects*. New York, NY: Cambridge, 2003, 687 pp.

[3] Hendry, J. Panning for lightning (including comments on the photos by M. A. Uman). *Weatherwise* 45(6): 19, 1993.

[4] Berger, K., R. B. Anderson, and H. Kroninger. Parameters of lightning flashes. *Electra* 80: 223–37, 1975.

[5] Rakov, V. A. A review of positive and bipolar lightning discharges. *Bull. Am. Meteor. Soc.* 84: 767–76, 2003.

[6] Anderson, R. B. and A. J. Eriksson. Lightning parameters for engineering application. *Electra* 69: 65–102, 1980.

[7] CIGRE WG C4.407 Technical Brochure 549, Lightning Parameters for Engineering Applications, V. A. Rakov, Convenor (US), A. Borghetti, Secretary (IT), C. Bouquegneau (BE), W. A. Chisholm (CA), V. Cooray (SE), K. Cummins (US), G. Diendorfer (AT), F. Heidler (DE), A. Hussein (CA), M. Ishii (JP), C. A. Nucci (IT), A. Piantini (BR), O. Pinto, Jr. (BR), X. Qie (CN), F. Rachidi (CH), M. M. F. Saba (BR), T. Shindo (JP), W. Schulz (AT), R. Thottappillil (SE), S. Visacro (BR), and W. Zischank (DE), 117 pp., August 2013.

[8] Nag, A. and V. A. Rakov. Positive lightning: an overview, new observations, and inferences. *J. Geophys. Res.* 117: D08109, doi:10.1029/2012JD017545, 2012.

[9] Rakov, V. A., M. A. Uman, K. J. Rambo, *et al.* New insights into lightning processes gained from triggered-lightning experiments in Florida and Alabama. *J. Geophys. Res.* 103: 14117–30, 1998.

[10] Fisher, R. J., G. H. Schnetzer, and M. E. Morris. Measured fields and earth potentials at 10 and 20 meters from the base of triggered-lightning channels. In *Proceedings of the 22nd International Conference on Lightning Protection*, Budapest, Hungary, Paper R 1c-10, 6 pp., 1994.

[11] Kitagawa, N., M. Brook, and E. J. Workman. Continuing currents in cloud-to-ground lightning discharges. *J. Geophys. Res.* 67: 637–47, 1962.

[12] Rakov, V. A. and M. A. Uman. Some properties of negative cloud-to-ground lightning flashes versus stroke order. *J. Geophys. Res.* 95: 5447–53, 1990.

[13] Cooray, V. and V. Rakov. On the upper and lower limits of peak current of first return strokes in negative lightning flashes. *Atmos. Res.* 117: 12–17, 2012.

[14] Cooray, V. and K. P. S. C. Jayaratne. Characteristics of lightning flashes observed in Sri Lanka in the tropics. *J. Geophys. Res.* 99: 21051–6, doi:10.1029/94JD01519, 1994.

[15] Qie, X., Y. Yu, D. Wang, H. Wang, and R. Chu. Characteristics of cloud-to-ground lightning in Chinese Inland Plateau. *J. Meteorol. Soc. Jpn.* 80: 745–54, 2002.

[16] Saraiva, A. C. V., M. M. F. Saba, O. Pinto Jr., K. L. Cummins, E. P. Krider, and L. Z. S. Campos. A comparative study of negative cloud-to-ground lightning characteristics in São Paulo (Brazil) and Arizona (United States) based on high-speed video observations. *J. Geophys. Res.* 115: D11102, doi:10.1029/2009JD012604, 2010.

[17] Ballarotti, M. G., C. Medeiros, M. M. F. Saba, W. Schulz, and O. Pinto Jr. Frequency distributions of some parameters of negative downward lightning flashes based on accurate-stroke-count studies. *J. Geophys. Res.* 117: D06112, doi:10.1029/2011JD017135, 2012.

[18] Baharudin, Z. A., N. A. Ahmad, J. S. Makela, M. Fernando, and V. Cooray. Negative cloud-to-ground lightning flashes in Malaysia. *J. Atmos. Sol.-Terr. Phys.* 108: 61–7, 2014.

[19] Zhu, Y., V. A. Rakov, S. Mallick, and M. D. Tran. Characterization of negative cloud-to-ground lightning in Florida. *J. Atmos. Sol.-Terr. Phys., Special Issue on Lightning* 136: 8–15, doi.org/10.1016/j.jastp.2015.08.006, 2015.

[20] Berger, K. Novel observations on lightning discharges: results of research on Mount San Salvatore. *J. Franklin Inst.* 283: 478–525, 1967.

[21] Fisher, R. J., G. H. Schnetzer, R. Thottappillil, V. A. Rakov, M. A. Uman, and J. D. Goldberg. Parameters of triggered-lightning flashes in Florida and Alabama. *J. Geophys. Res.* 98: 22887–902, 1993.

[22] McCann, E. D. The measurement of lightning currents in direct strokes. *AIEE Trans.* 63: 1157–64, 1944.

[23] Ballarotti, M. G., M. M. F. Saba, and O. Pinto Jr. High-speed camera observations of negative ground flashes on a millisecond-scale. *Geophys. Res. Lett.* 32: L23802, doi:10.1029/2005GL023889, 2005.

[24] Rakov, V. A. and M. A. Uman. Origin of lightning electric field signatures showing two return-stroke waveforms separated in time by a millisecond or less. *J. Geophys. Res.* 99: 8157–65, 1994.

[25] Shindo, T. and M. A. Uman. Continuing current in negative cloud-to-ground lightning. *J. Geophys. Res.* 94: 5189–98, 1989.

[26] Rakov, V. A. and M. A. Uman. Long continuing current in negative lightning ground flashes. *J. Geophys. Res.* 95: 5455–70, 1990.

[27] Saba, M. M. F., M. G. Ballarotti, and O. Pinto Jr. Negative cloud-to-ground lightning properties from high-speed video observations. *J. Geophys. Res.* 111: D03101, doi:10.1029/2005JD006415, 2006.

[28] Rakov, V. A., M. A. Uman, and R. Thottappillil. Review of lightning properties from electric field and TV observation. *J. Geophys. Res.* 99: 10745–50, 1994.

[29] Berger, G., A. Hermant, and A. S. Labbe. Observations of natural lightning in France. In *Proceedings of the 23rd International Conference on Lightning Protection*, Florence, Italy, Vol. 1, pp. 67–72, 1996.

[30] Hermant, A. *Traqueur d'Orages*. Paris, France: Nathan/HER, 2000.

[31] Valine, W. C. and E. P. Krider. Statistics and characteristics of cloud-to-ground lightning with multiple ground contacts. *J. Geophys. Res.* 107(D20): 4441, doi:10.1029/2001JD001360, 2002.

[32] Fleenor, S. A., C. J. Biagi, K. L. Cummins, E. P. Krider, and X. M. Shao. Characteristics of cloud-to-ground lightning in warm-season thunderstorms in the Central Great Plains. *Atmos. Res.* 91: 333–52, doi:10.1016/j.atmosres.2008.08.011, 2009.

[33] Thottappillil, R., V. A. Rakov, M. A. Uman, W. H. Beasley, M. J. Master, and D. V. Shelukhin. Lightning subsequent-stroke electric field peak greater than the first stroke peak and multiple ground terminations. *J. Geophys. Res.* 97: 7503–9, 1992.

[34] Stall, C. A., K. L. Cummins, E. P. Krider, and J. A. Cramer. Detecting multiple ground contacts in cloud-to-ground lightning flashes. *J. Atmos. Ocean. Tech.* 26: 2392–402, 2009.

[35] Ferro, M. A. S., M. M. F. Saba, and O. Pinto Jr. Time-intervals between negative lightning strokes and the creation of new ground terminations. *Atmos. Res.* 116: 130–3, doi:10.1016/j.atmosres.2012.03.010, 2012.

[36] Kong, X. Z., X. S. Qie, Y. Zhao, and T. Zhang. Characteristics of negative lightning flashes presenting multiple-ground terminations on a millisecond-scale. *Atmos. Res.* 91: 381–6, 2009.

[37] Nag, A., V. A. Rakov, W. Schulz, *et al.* First versus subsequent return-stroke current and field peaks in negative cloud-to-ground lightning discharges. *J. Geophys. Res.* 113: D19112, 2008.

[38] Cooray, V. and H. Pérez. Some features of lightning flashes observed in Sweden. *J. Geophys. Res.* 99: 10683–8, doi:10.1029/93JD02366, 1994.

[39] Visacro, S., A. Soares Jr., M. A. O. Schroeder, L. C. L. Cherchiglia, and V. J. de Sousa. Statistical analysis of lightning current parameters: measurements

at Morro do Cachimbo Station. *J. Geophys. Res.* 109: D01105, doi:10.1029/2003JD003662, 2004.

[40] Anderson, J. G. Lightning performance of transmission lines, in *Transmission Line Reference Book—345 kV and Above*, 2nd edn. Electric Power Research Institute, Palo Alto, CA, 1982, pp. 545–597.

[41] Military Standard. Lightning qualification test techniques for aerospace vehicles and hardware. MIL-STD-1757A, 1983.

[42] Rakov, V. A., R. Thottappillil, and M. A. Uman. On the empirical formula of Willett *et al.* relating lightning return-stroke peak current and peak electric field. *J. Geophys. Res.* 97: 11527–33, 1992.

[43] Berger, K. Mesungen und Resultate der Blitzforschung auf dem Monte San Salvatore bei Lugano. der Jahre 1963-1971. *Bull. SEV* 63: 1403–22, 1972.

[44] Schulz, W., G. Diendorfer, S. Pedeboy, and D. R. Poelman. The European lightning location system EUCLID—Part 1: performance validation. *Nat. Hazards Earth Syst. Sci. Discuss.* 3: 5325–55, doi:10.5194/nhessd-3-5325-2015, 2015.

[45] Zhu, Y., V. A. Rakov, M. D. Tran, and A. Nag. A study of National Lightning Detection Network responses to natural lightning based on ground-truth data acquired at LOG with emphasis on cloud discharge activity. *J. Geophys. Res. Atmos.* 121: 14,651–60, doi:10.1002/2016JD025574, 2016.

[46] Diendorfer, G., K. Cummins, V. A. Rakov, *et al.* LLS-estimated versus directly measured currents based on data from tower-initiated and rocket-triggered lightning. In *Proceedings of the 29th International Conference on Lightning Protection*, Uppsala, Sweden, June 23–26, 2008, Paper 2-1, 9 pp., 2008.

[47] Jerauld, J., V. A. Rakov, M. A. Uman, *et al.* An evaluation of the performance characteristics of the U.S. National Lightning Detection Network in Florida using rocket-triggered lightning. *J. Geophys. Res.* 110: D19106, doi:10.1029/2005JD005924, 2005.

[48] Mallick, S., V. A. Rakov, J. D. Hill, *et al.* Performance characteristics of the NLDN for return strokes and pulses superimposed on steady currents, based on rocket-triggered lightning data acquired in Florida in 2004–2012. *J. Geophys. Res. Atmos.* 119: 3825–56, doi:10.1002/2013JD021401, 2014.

[49] Nag A., S. Mallick, V. A. Rakov, *et al.* Evaluation of U.S. National Lightning Detection Network performance characteristics using rocket-triggered lightning data acquired in 2004–2009. *J. Geophys. Res.* 116: D02123. doi:10.1029/2010JD014929, 2011.

[50] Rakov, V. A. On estimating the lightning peak current distribution parameters taking into account the lower measurement limit. *Elektrichestvo* 2: 57–9, 1985.

[51] Rakov, V. A. A review of the interaction of lightning with tall objects. *Recent Res. Devel. Geophysics* 5: 57–71, Research Signpost, India, 2003.

[52] Sargent, M. A. The frequency distribution of current magnitudes of lightning strokes to tall structures. *IEEE Trans. Power Appar. Syst.* 91: 2224–9, 1972.

[53] Borghetti, A., C. A. Nucci, and M. Paolone. Estimation of the statistical distributions of lightning current parameters at ground level from the data recorded by instrumented towers. *IEEE Trans. Power Delivery* 19: 1400–9, 2004.
[54] Mata, C. T. and V. A. Rakov. Evaluation of lightning incidence to elements of a complex structure: a Monte Carlo approach. In *Proceedings of the 3rd International Conference on Lightning Physics and Effects (LPE) and GROUND' 2008*, Florianopolis, Brazil, November 16–20, 2008, pp. 351–4, 2008.
[55] CIGRE TF 33.01.03, Report 118. Lightning exposure of structures and interception efficiency of air terminals, 86 pp., 1997.
[56] Popolansky, F. Lightning current measurement on high objects in Czechoslovakia. In *Proceedings of 20th* International Conference *on Lightning Protection*, Interlaken, Switzerland, paper 1.3, 7 pp., 1990.
[57] CIGRE WG 33.01, Report 63. Guide to Procedures for Estimating the Lightning Performance of Transmission Lines, 61 pp., 1991.
[58] Hileman, A. R. *Insulation Coordination for Power Systems*. New York, NY: Marcel Dekker, 1999, 767 pp.
[59] Gamerota, W. R., J. O. Elismé, M. A. Uman, and V. A. Rakov. Current waveforms for lightning simulation. *IEEE Trans. Electromagn. Compat.* 54: 880–8, 2012.
[60] Popolansky, F. Frequency distribution of amplitudes of lightning currents. *Electra* 22: 139–47, 1972.
[61] Eriksson, A. J. and D. V. Meal. The incidence of direct lightning strikes to structures and overhead lines. In *Lightning and Power Systems*, London: IEE Conference Publication No. 236, pp. 67–71, 1984.
[62] Bazelyan, E. M., N. L. Aleksandrov, R. B. Carpenter, and Yu. P. Raizer. Reverse discharges near grounded objects during the return stroke of branched lightning flashes. In *Proceedings of the 28th International Conference on Lightning Protection*, Kanazawa, Japan, pp. 187–92, 2006.
[63] Melander, B. G. Effects of tower characteristics on lightning arc measurements. In *Proceedings of the 1984* International Conference *on Lightning and Static Electricity*, Orlando, FL, 1984, pp. 34/1–34/12, 1984.
[64] Eriksson, A. J., C. L. Penman, and C. L. Meal. A review of five years' lightning research on an 11 kV test-line. In *Lightning and Power Systems*. London: IEE Conference Publication No. 236, pp. 62–6, 1984.
[65] Visacro, S. and F. H. Silveira. Lightning current waves measured at short instrumented towers: the influence of sensor position. *Geophys. Res. Lett.* 32: L18804-1–5, doi:10.1029/2005GL023255, 2005.
[66] Takami, J. and S. Okabe. Observational results of lightning current on transmission towers. *IEEE Trans. Power Delivery* 22: 547–56, 2007.
[67] Narita, T., T. Yamada, A. Mochizuki, E. Zaima, and M. Ishii. Observation of current waveshapes of lightning strokes on transmission towers. *IEEE Trans. Power Delivery* 15: 429–35, 2000.

[68] Schoene, J., M. A. Uman, V. A. Rakov, *et al.* Characterization of return-stroke currents in rocket-triggered lightning. *J. Geophys. Res.* 114: D03106, doi:10.1029/2008JD009873, 2009.

[69] Schoene, J., M. A. Uman, V. A. Rakov, V. Kodali, K. J. Rambo, and G. H. Schnetzer. Statistical characteristics of the electric and magnetic fields and their time derivatives 15 m and 30 m from triggered lightning. *J. Geophys. Res.* 108: 4192, doi:10.1029/2002JD002698, 2003.

[70] Cooray, V. and V. Rakov. Engineering lightning return stroke models incorporating current reflection from ground and finitely conducting ground effects. *IEEE Trans. Electromagn. Compat.* 53: 773–81, 2011.

[71] Diendorfer, G., H. Pichler, and M. Mair. Some parameters of negative upward-initiated lightning to the Gaisberg tower (2000–2007). *IEEE Trans. Electromagn. Compat.* 51: 443–52, 2009.

[72] Diendorfer, G. Review of seasonal variations in occurrence and some current parameters of lightning measured at the Gaisberg Tower. *4th International Symposium on Winter Lightning (ISWL 2017)*, 6 pp., 2017.

[73] Berger, K. and E. Garabagnati. Lightning current parameters. Results obtained in Switzerland and in Italy: URSI Conference, Florence, Italy, 1984.

[74] Leteinturier, C., J. H. Hamelin, and A. Eybert-Berard. Submicrosecond characteristics of lightning return-stroke currents. *IEEE Trans. Electromagn. Compat.* 33: 351–7, 1991.

[75] Yang, J., X. Qie, G. Zhang, *et al.* Characteristics of channel base currents and close magnetic fields in triggered flashes in SHATLE. *J. Geophys. Res.* 115: D23102, doi:10.1029/2010JD014420, 2010.

[76] Cooray, V., V. Rakov, and N. Theethayi. The lightning striking distance—revisited. *J. Electrost.* 65: 296–306, 2007.

[77] Qie, X. S., Q. L. Zhang, Y. J. Zhou, *et al.* Artificially triggered lightning and its characteristic discharge parameters in two severe thunderstorms. *Sci. China, Ser. D: Earth Sci.* 50: 1241–50, doi:10.1007/s11430-007-0064-2, 2007.

[78] Schoene, J., M. A. Uman, and V. A. Rakov. Return stroke peak current versus charge transfer in rocket-triggered lightning. *J. Geophys. Res.* 115: D12107, doi:10.1029/2009JD013066, 2010.

[79] Idone, V. P. and R. E. Orville. Lightning return stroke velocities in the Thunderstorm Research International Program (TRIP). *J. Geophys. Res.* 87: 4903–15, 1982.

[80] Olsen, R. C., D. M. Jordan, V. A. Rakov, M. A. Uman, and N. Grimes. Observed two-dimensional return stroke propagation speeds in the bottom 170 m of a rocket-triggered lightning channel. *Geophys. Res. Lett.* 31: L16107, doi:10.1029/2004GL020187, 2004.

[81] Rakov, V. A. Lightning return stroke speed. *J. Light. Res.* 1: 80–9, 2007.

[82] Baba, Y. and V. A. Rakov. On the use of lumped sources in lightning return stroke models. *J. Geophys. Res.* 110: D03101, doi:10.1029/2004JD005202, 2005.

[83] Gorin, B. N. and A. V. Shkilev. Measurements of lightning currents at the Ostankino tower. *Elektrichestvo* 8: 64–5, 1984.

[84] Rusck, S. Induced lightning overvoltages on power transmission lines with special reference to the overvoltage protection of low-voltage networks. *Trans. R. Inst. Technol.* 120: 1–118, 1958.

[85] Barker, P. P., T. A. Short, A. Eybert-Berard, and J. B. Berlandis. Induced voltage measurements on an experimental distribution line during nearby rocket triggered lightning flashes. *IEEE Trans. Power Delivery* 11(2): 980–95, 1996.

[86] Darveniza, M. A practical extension of Rusck's formula for maximum lightning-induced voltages that accounts for ground resistivity. *IEEE Trans. Power Delivery* 22(1): 605–12, 2007.

[87] Andreotti, A., D. Assante, F. Mottola, and L. Verolino. An exact closed-form solution for lightning-induced overvoltages calculations. *IEEE Trans. Power Delivery* 24(3): 1328–43, 2009.

[88] Paulino, J. O. S., C. F. Barbosa, I. J. S. Lopes, and W. C. Boaventura. An approximate formula for the peak value of lightning-induced voltages in overhead lines. *IEEE Trans. Power Delivery* 25(2): 843–51, 2010.

[89] Thottappillil, R., V. A. Rakov, and N. Theethayi. Expressions for far electric fields produced at an arbitrary altitude by lightning return strokes. *J. Geophys. Res.* 112: D16102, doi:10.1029/2007JD008559, 2007.

[90] Piantini, A. and J. M. Janiszewski. Lightning-induced voltages on overhead lines—application of the extended Rusck model. *IEEE Trans. Electromagn. Compat.* 51(3): 548–58, 2009.

[91] Piantini, A. Lightning protection of overhead power distribution lines. *Proceedings of the 29th International Conference on Lightning Protection (ICLP)*, Uppsala, June 2008 (invited lecture).

[92] Piantini, A. Lightning transients in MV power distribution lines. *Proceedings of the V Russian Conference on Lightning Protection (RCLP)*, Saint Petersburg, Russia, May 2016 (invited lecture).

[93] Piantini, A. Analysis of the effectiveness of shield wires in mitigating lightning-induced voltages on power distribution lines. *Electr. Power Syst. Res.*, http://dx.doi.org/10.1016/j.epsr.2017.08.022, 159: 9–16, 2018.

[94] Andreotti, A., A. Pierno, and V. A. Rakov. An analytical approach to calculation of lightning induced voltages on overhead lines in case of lossy ground-Part I: model development. *IEEE Trans. Power Delivery* 28(2): 1213–23, 2013.

[95] Andreotti, A., A. Pierno, V. A. Rakov, and L. Verolino. Analytical formulations for lightning-induced voltage calculations. *IEEE Trans. Electromagn. Compat.* 55(1): 109–23, 2013.

[96] Andreotti, A., A. Pierno, and V. A. Rakov. A new tool for calculation of lightning-induced voltages in power systems—Part I: development of circuit model. *IEEE Trans. Power Delivery* 30(1): 326–33, 2015.

[97] Rodrigues, A. R., G. C. Guimarães, M. L. R. Chaves, W. C. Boaventura, D. A. Caixeta, and M. A. Tamashiro. Lightning performance of transmission lines based upon real return-stroke current waveforms and statistical variation of characteristic parameters. *Electr. Power Syst. Res.* 153: 46–59, 2017.

[98] Høidalen, H. K. Analytical formulation of lightning-induced voltages on multiconductor overhead lines above lossy ground. *IEEE Trans. Electromagn. Compat.* 45(1): 92–100, 2003.

[99] Sekioka, S. Equivalent circuit of distribution line for lightning-induced voltage analysis using analytical formulas. *Electr. Eng. Jpn.* 199(3): 2017.

[100] Rakov, V. A. and A. A. Dulzon. Calculated electromagnetic fields of lightning return stroke. *Tekh. Elektrodinam.* 1: 87–9, 1987.

[101] Paulino, J. O. S., C. F. Barbosa, I. J. S. Lopes, and W. C. Bonaventura. The peak value of lightning-induced voltages in overhead lines considering the ground resistivity and typical return stroke parameters. *IEEE Trans. Power Delivery* 26(2): 920–6, 2011.

[102] Heidler, F. Analytische Blitzstromfunktion zur LEMP- Berechnung (in German). paper 1.9, pp. 63–6, Munich, September 16–20, 1985.

[103] Heidler, F. Traveling current source model for LEMP calculation. In *Proceedings of the 6th International Zurich Symposyum on Electromagnetic Compatibility*, Zurich, Switzerland, pp. 157–62, 1985.

[104] Bruce, C. E. R. and R. H. Golde. The lightning discharge. *J. Inst. Elec. Eng.* 88: 487–520, 1941.

[105] Stekolnikov, I. S. The parameters of the lightning discharge and the calculation of the current waveform. *Elektrichestvo* 3: 63–8, 1941.

[106] IEC 62305-1. *Protection Against Lightning—Part 1: General Principles, International Electrotechnical Commission, Geneva*, 2nd edn., 2010.

[107] Diendorfer, G. and M. A. Uman. An improved return stroke model with specified channel-base current. *J. Geophys. Res.* 95: 13621–44, doi:10.1029/JD095iD09p13621, 1990.

[108] Piantini, A. Lightning protection of low-voltage networks. In *Lightning Protection*, V. Cooray (Ed.), The Institution of Engineering and Technology (IET), United Kingdom, Chapter 12, pp. 551–631, 2010.

[109] Nucci, C. A., G. Diendorfer, M. A. Uman, F. Rachidi, M. Ianoz, and C. Mazzetti. Lightning return stroke current models with specified channel-base current: a review and comparison. *J. Geophys. Res.* 95: 20395–408, 1990.

[110] Rakov, V. A. and A. A. Dulzon. A modified transmission line model for lightning return stroke field calculations. In *Proceedings of the 9th International Zurich Symposyum on Electromagnetic Compatibility*, Zurich, Switzerland, pp. 229–35, 1991.

[111] Thottappillil, R., V. A. Rakov, and M. A. Uman. Distribution of charge along the lightning channel: relation to remote electric and magnetic fields and to return-stroke models. *J. Geophys. Res.* 102: 6987–7006, 1997.

[112] Moini, R., B. Kordi, G. Z. Rafi, and V. A. Rakov. A new lightning return stroke model based on antenna theory. *J. Geophys. Res.* 105: 29693–702, 2000.

[113] Romero, F., A. Piantini, and V. Cooray. On the influence of stroke current propagation velocity on lightning horizontal electric fields. *IEEE Trans. Electromagn. Compat.* 56: 940–8, 2014.

[114] De Conti, A. and S. Visacro. Analytical representation of single- and double-peaked lightning current waveforms. *IEEE Trans. Electromagn. Compat.* 49: 448–51, 2007.

[115] Heidler, F. and J. Cvetic. A class of analytical functions to study the lightning effects associated with the current front. *Eur. Trans. Elect. Power* 12: 141–50, 2002.

[116] Andreotti, A., S. Falco, and L. Verolino. Some integrals involving Heidler's lightning return stroke current expression. *Electr. Eng.* 87: 121–8, doi:10.1007/s00202-004-0240-8, 2005.

[117] Silveira, F. H., A. De Conti, and S. Visacro. Lightning overvoltage due to first strokes considering a realistic current representation. *IEEE Trans. Electromagn. Compat.* 52: 929–35, doi:10.1109/TEMC2010.2044042, 2010.

[118] Javor, V. and P. D. Rancic. A channel-base current function for lightning return-stroke modeling. *IEEE Trans. Electromagn. Compat.* 53: 245–9, doi:10.1109/TEMC.2010.2066281, 2011.

[119] IEEE Standard 1243-1997. *IEEE Guide for Improving the Lightning Performance of Transmission Lines*, 1997.

[120] IEEE Standard 1410-2010. *IEEE Guide for Improving the Lightning Performance of Electric Power Overhead Distribution Lines*, 2010.

Chapter 3

Lightning return stroke models for electromagnetic field calculations

Vernon Cooray[1]

Return stroke models are used in the calculation of lightning-induced over-voltages in power lines because the electric and magnetic field as a function of distance and height along the line is difficult, if not impossible, to obtain experimentally. Thus the return stroke models should represent faithfully the electric and magnetic fields generated by lightning first and subsequent return strokes, the events that induce largest over-voltages in power systems.

Following an introduction and nomenclature, the reader is introduced in Sections 3.2, 3.3 and 3.4 to the various concepts used by engineers (i.e. current propagation, current generation and current dissipation) to construct engineering return stroke models. In Section 3.5, the way in which any given return stroke model could be represented as a current generation or current dissipation type is described. After showing that current propagation concept is a special case of current dissipation concept in Section 3.6, the way in which the current generation and current dissipation concepts could be combined to obtain return stroke models that are in agreement with current pulse propagation along transmission lines in the presence of corona is presented in Section 3.7. In Section 3.8, a review of the basic features of electromagnetic fields of return strokes that are being used in testing the engineering return stroke models are presented. This section is followed by a description of the evaluation of electromagnetic fields, including the horizontal electric field, from return stroke over finitely conducting ground. The chapter ends with concluding remarks.

3.1 Introduction

A model is a mathematical construct that attempts to predict the outcome of an event using a certain number of input parameters. A model may attempt to explain and predict the outcome of a phenomenon starting with fundamental principles or it may contain as inputs data obtained from experimental investigations. In the case

[1]Department of Engineering Sciences, Uppsala University, Uppsala, Sweden

of lightning return strokes, models are used to predict the electromagnetic fields from lightning at different distances and in the remote sensing of lightning currents using the measured electromagnetic fields. More sophisticated models can be used, not only to evaluate the electromagnetic fields but also to predict the temporal variation of the return stroke current along the channel. Depending on their complexity, return stroke models can be divided into five categories: physics-based models [1], electromagnetic and antenna models [2,3], transmission line models [4], waveguide models [5,6] and engineering models [1]. The physics-based models attempt to simulate the lightning return stroke using a combination of the laws of conservation, hydrodynamics, thermodynamics and electrodynamics. Unfortunately, at present we do not have successful physical models that predict the temporal variation of the return stroke current and the electromagnetic fields correctly. Electromagnetic models treat the lightning channel as a vertical conductor located above a perfectly conducting ground and the return stroke is simulated using Maxwell's equation as a current pulse propagating along this conductor. The transmission line models treat the return stroke channel as a charged transmission line and simulate the return stroke as a process that discharges the transmission line. These models can incorporate the finite conductivity of the lightning channel into the model description. The waveguide models simulate the return stroke channel as a finitely conducting waveguide and the return stroke is simulates as a current pulse propagating along this waveguide. Engineering models burrow various features from transmission line models and combine them using empirical data to create models that can predict the electromagnetic fields of lightning return stroke in space and time. At present, engineering models are the most successful in predicting the features of lightning electromagnetic fields in close agreement with the experimental data. Today, most scientists who are interested in lightning protection use engineering models in their studies. In this chapter, we will concentrate on the engineering models.

Nomenclature: Engineering models can be divided into three categories namely, current propagation, current generation and current dissipation. The current propagation models are also known in the literature as lumped current source models or transmission line type. The current generation models are also known in the literature as travelling current source type models, distributed current source models or discharge type models. Concerning the naming of these models one can add the following. Even though some of the available engineering models can be categorized as lumped source and distributed source, as we will describe in this chapter, one can create engineering models that do not fall into either lumped source or distributed source models (i.e. current dissipation models). Moreover, lumped current source engineering models available in the literature are a special case of the current dissipation models. Furthermore, the name travelling current source model is a misnomer because there are no current sources that travel along the channel. The discharge type models is also a misnomer because it gives the impression that the physics of discharge is taken into account – which is not true. In the current propagation models, how the injected current at the channel base will propagate along the return stroke channel is specified as an input parameter. In the current generation models, the return stroke current is assumed to be generated by the corona currents and in the current dissipation models the return stroke current injected at the channel base is dissipated along the channel by the corona current. For these reasons, we adopt the names, current propagation, current generation and current dissipation to describe the various engineering return stroke models available in the literature.

3.2 Basic concept of current propagation models

In these models it is assumed that the return stroke is a current pulse originating at ground level and propagating from ground to cloud along the leader channel. The model specifies how the injected current will propagate along the return stroke channel i.e. how the current will attenuate and disperse along the channel. The engineering models using this postulate as a base were constructed by Norinder [7], Bruce and Golde [8], Lundholm [9], Dennis and Pierce [10], Uman and MacLain [11], Nucci *et al.* [12], Rakov and Dulzon [13] and Cooray and Orville [14]. The models differ from each other by the way they prescribe how the return stroke current varies as it propagates along the leader channel. For example, in the model introduced by Uman and MacLain [11], popularly known as the transmission line model, the current is assumed to propagate along the channel without attenuation and with constant speed. In the model introduced by Nucci *et al.* [12] (MTLE – modified transmission line model with exponential current decay), the current amplitude decreases exponentially with height and in the one introduced by Rakov and Dulzon [13] (MTLL – modified transmission line model with linear current decay), the current amplitude decreases linearly. Cooray and Orville [14] introduced both current attenuation and dispersion into these models while at the same time allowing the return stroke speed to vary along the channel.

Let us denote the temporal variation of the current pulse injected at the channel base by $I_b(t)$. Let us assume that this current pulse propagates with variable velocity $v(z)$ along the channel and the current ahead of the return stroke front is zero. As the current pulse moves along the channel it may deposit positive charge along the leader channel and this may give rise to a decrease in the amplitude of the current. The propagating current may also suffer dispersion as it propagates along the channel and this may also lead to the change in the shape and the peak amplitude of the current with height. Thus the current at level z is given by

$$I(z,t) = A(z)F(z, t - z/v_{az}(z)) \quad t > z/v_{av}(z) \tag{3.1}$$

In the above equation $A(z)$ is a function that represents the attenuation of the current due to the deposition of positive charge along the channel and $F(z,t)$ describes the modified wave shape of the current at height z caused by dispersion, and $v_{av}(z)$ is the average return stroke velocity over the channel section whose upper end is at level z. This average velocity is given by

$$v_{av}(z) = z/\int_0^z \frac{dz'}{v(z')} \tag{3.2}$$

In the above equation $v(z')$ is the front speed of the return stroke front at z'. One can define the function $F(z,t)$ that takes care of the variation of the current shape as follows:

$$F(z,t) = \int_o^t I_b(\tau)R(z,t-\tau)d\tau \tag{3.3}$$

where $I_b(t)$ is the channel base current and $R(z,t)$ is a function that describes how the shape of the current waveform is being modified with height. This function describes how a delta impulse current function is modified as it propagates along the channel due to dispersion. That is the function $R(z,t)$ describes the current at height z in the presence of dispersion if the channel base current is given by a Dirac delta function. Since no charge is deposited when only dispersion is present along the channel, the function $R(z,t)$ should satisfy the criterion:

$$\int_o^\infty R(z,t)dt = 1 \tag{3.4}$$

Equations (3.1)–(3.4) provide the most general description of the current propagation models. In the transmission line model both $A(z)$ and $R(z,t)$ are assumed to be unity. In the MTLE model $A(z)$ is represented by an exponential function while in the MTLL model it is represented by a function that decreases linearly with height. Both MTLE and MTLL models assume $R(z,t)$ is equal to unity, that is no current dispersion. In the model introduced by Cooray and Orville [14] both current attenuation and current dispersion are taken into account.

3.3 Basic concepts of current generation models

The basic concepts of the current generation models are depicted pictorially in Figure 3.1. In these models the leader channel is treated as a charged transmission line and the return stroke current is assumed to be generated by a wave of ground potential that travels along it from ground to cloud. The arrival of the wavefront (i.e. return stroke front) at a given point on the leader channel changes its potential from cloud potential to ground potential causing the release of bound charge on the central core and the corona sheath giving rise to the current in the channel (this is called the corona current in the literature). These models postulate that as the return stroke front propagates upwards the charge stored on the leader channel collapses into the highly conducting core of the return stroke channel. Accordingly, each point on the leader channel can be treated as a current source which is turned on by the arrival of the return stroke front at that point. The corona current injected by these sources into the highly conducting return stroke channel core travels to ground with a speed denoted by v_c. In most of the return stroke models it is assumed that $v_c = c$ where c is the speed of light in free space.

The basic concept of CG models was first introduced by Wagner [16]. He assumed that the neutralization of the corona sheath takes a finite time and

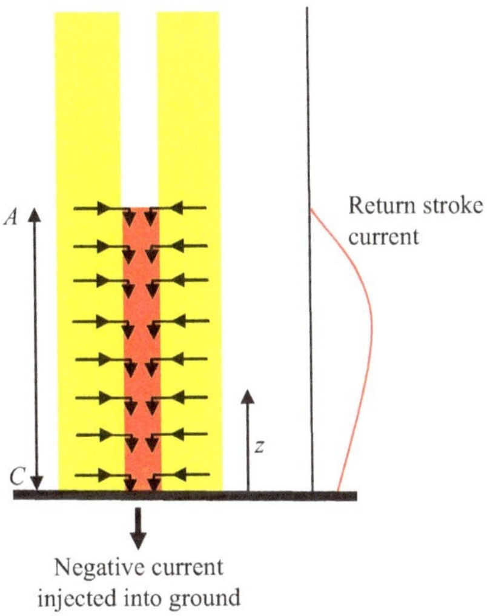

Figure 3.1 *Pictorial description of the current generation concept. According to CG concept the downward moving corona currents generate the return stroke current. Adapted from [15]*

therefore the corona current can be represented by a decaying exponential function. The decay time constant associated with this function is called the corona decay time constant. Wagner assumed however that the speed of propagation of the corona current down the return stroke channel is infinite. Lin *et al.* [17] introduced a model in which both CG and CP concepts are incorporated in the same model. In the portion of the current described by CG concept, the corona current is represented by a double exponential function. The speed of propagation of corona current down the channel is assumed to be the same as the speed of light. A modified form of this model was introduced by Master *et al.* [18] but in this modification the CG description remained intact. Heidler [19] constructed a model based on this principle in which the channel base current and the return stroke speed are assumed as input parameters. Furthermore, it was assumed that the neutralization of the corona sheath is instantaneous and hence the corona current generated by a given channel section can be represented by a Dirac Delta function. The speed of propagation of the corona current down the return stroke channel is assumed to be equal to the speed of light. This model gives rise to a current discontinuity at the return stroke front which, according to this author's understanding, is not physically reasonable. Hubert [20] constructed a current generation model rather similar to that of the Wagner's model with the exception that the downward speed of propagation of the corona current is equal to the speed of light. He utilized this model to reproduce experimental data (both current and electromagnetic fields) obtained from triggered lightning. Cooray [21,22] introduced a model in which the distribution of the charge deposited by the return stroke (i.e. sum of the positive charge necessary to neutralize the negative charge on the leader and the positive charge induced on the channel due to the action of the background electric field) and the decay time constant of the corona current are taken as input parameters with the model predicting the channel base current and return stroke speed. Moreover, he took into consideration that the process of neutralization of the corona sheath takes a finite time in reality and, as a consequence, the corona current was represented by an exponential function with a finite duration. This is the first model in which the decay time constant of the corona current (and hence the duration of the corona current) is assumed to increase with height. Since the leader channel contains a hot core surrounded by a corona sheath, he also divided the corona current into two parts, one fast and the other slow. The fast one was associated with the neutralization of the core and the slow one with the neutralization of the corona sheath. Furthermore, by treating the dart leader as an arc and assuming that the electric field at the return stroke front is equal to the electric field that exists in this arc channel, he managed to derive the speed of the return stroke. Diendorfer and Uman [23] introduced a model in which the channel base current, return stroke speed and the corona decay time constant were assumed as input parameters. They also divided the corona current into two parts: one fast and the other slow. Thottappillil *et al.* [24] and Thottappillil and Uman [25] modified this model to include variable return stroke speed and a corona decay time constant that varies with height. Cooray [26] developed the ideas introduced in [21] and [22] to create a CG model with channel base current as an input. Cooray [27] and Cooray *et al.* [28] extended this concept to include first return strokes with connecting leaders.

3.3.1 Input parameters of the CG models and the expression for the current at any height

In CG models one has the choice of selecting the channel base current, $I_b(t)$, the distribution of the charge deposited by the return stroke along the channel, $\rho(z)$, the return stroke speed, $v(z)$ and the magnitude and variation of the corona discharge time constant with height, $\tau(z)$ as input parameters.

Since the current at any given level on the channel is the cumulative effect of corona currents associated with channel elements located above that level, the return stroke current at any height in the return stroke channel $I(z,t)$ can be written as

$$I(z,t) = \int_z^{h_e} I_{cor}(\xi, t - \xi/v_{av}(\xi) - (\xi - z)/v_c) d\xi \qquad t > z/v_{av}(z) \tag{3.5}$$

$$I_{cor}(z,t) = \frac{\rho(z)}{\tau(z)} \exp\{-(t - z/v_{av}(z))/\tau(z)\} \qquad t > z/v_{av}(z) \tag{3.6}$$

Note that $I_{cor}(z)$ is the corona current per unit length associated with a channel element at height z, v_c is the speed of propagation of the corona current and $v_{av}(z)$ is the average return stroke speed over the channel section of length z with one end at ground level. The latter is given by

$$v_{av}(z) = z \Big/ \int_o^z \frac{1}{v(z')} dz' \tag{3.7}$$

The value of h_e in (3.5) can be obtained from the solution of the following equation:

$$t = \frac{h_e}{v_{av}(h_e)} + \frac{h_e - z}{v_c} \tag{3.8}$$

The current at the channel base is given by

$$I(0,t) = \int_o^{h_o} I_{cor}(\xi, t - \xi/v_{av}(\xi) - \xi/v_c)d\xi \tag{3.9}$$

with h_o given by the solution of

$$t = \frac{h_o}{v_{av}(h_o)} + \frac{h_o}{v_c} \tag{3.10}$$

As mentioned earlier, a CG model contains four model parameters and any set of three of these four input parameters will provide a complete description of the temporal and spatial variation of the return stroke current. Most of the CG models use $v(z)$ and either the $\rho(z)$ or $\tau(z)$ in combination with $I_b(t)$ as input parameters. Recently, Cooray and Rakov [29] developed a model in which $\rho(z)$, $\tau(z)$ and $I_b(t)$ are selected as input parameters. The model could generate $v(z)$ as an output. As mentioned above, a current generation model needs three input parameters which can be selected from a set of four parameters, that is $\rho(z)$, $\tau(z)$, $I_b(t)$ and $v(z)$. Once three of these parameters are specified the fourth can be evaluated either analytically or numerically. Let us now consider the mathematics necessary to do this.

3.3.2 *Evaluate $\tau(z)$ given $I_b(t)$, $\rho(z)$ and $v(z)$*

Let us start the analysis with the expression for the channel base current as given by (3.9). From this equation one can see that the channel base current, after substituting for I_{cor} from (3.6), is given by

$$I_b(0,t) = \int_o^{h_o} \frac{\rho(z)dz}{\tau(z)} [\exp\{-(t - z/v_{av}(z) - z/v_c)/\tau(z)\}] \tag{3.11}$$

where h_o can be extracted by the solution of (3.10). If we divide the channel into a large number of segments of equal length dz, the above integral can be written as a summation as follows:

$$I_b(t_m) = \sum_{n=1}^{m} \frac{\rho_n}{\tau_n} \exp\left\{-\left(t_m - \frac{(n-1)dz}{v_{av,n}} - \frac{(n-1)dz}{v_c}\right)\Big/\tau_n\right\} \tag{3.12}$$

where ρ_n is the charge deposited per unit length on the nth section, τ_n is the decay time constant of the corona current of the nth section and $v_{av,n}$ is the average return stroke speed over the channel section connecting the ground and the nth element. In this equation t_m is the time for the corona current released from the mth segment to reach the ground. This is given by the equation:

$$t_m = \left[\left(m - \frac{1}{2}\right)\Big/ v_{av,m} + \left(m - \frac{1}{2}\right)\Big/ v_c\right] dz \tag{3.13}$$

If the return stroke speed and the current at the channel base are known, then the value of the discharge time constant at different heights can be estimated progressively by moving from $m = 1$. For example when $m = 1$ the only unknown is the τ_1. Once this is found one can consider the case $m = 2$. In the resulting equation the only unknown is the value of τ_2 and it can be obtained by solving that equation. In this way the values of discharge time constants up to the mth element can be obtained sequentially [24,25].

3.3.3 Evaluate $\rho(z)$ given $I_b(t)$, $\tau(z)$ and $v(z)$

Equations (3.12) can also be used to evaluate the discharge time constant when the other parameters are given as inputs. For example, in this case when $m = 1$ the only unknown is the ρ_1. Once this is found one can consider the case $m = 2$. In the resulting equation the only unknown is the value of ρ_2 and it can be obtained by solving that equation. In this way the values of discharge time constants up to any mth element can be obtained sequentially.

3.3.4 Evaluate $v(z)$, given $I_b(t)$, $\rho(z)$ and $\tau(z)$

As before, we start with (3.12). Since $I_b(t)$, $\rho(z)$ and $\tau(z)$ are given the only unknown parameter in these equations is $v_{av,n}$, the average speed along the nth channel segment. Solving the equations as before one can observe that when $m = 1$ the only unknown is $v_{av,1}$, the average speed over the first channel segment. Once this is found the value of $v_{av,2}$ can be obtained by considering the situation of $m = 2$. In this way the average return stroke speed as a function of height can be obtained. It is important to point out that in this evaluation the value of dz in (3.12) should be selected in such a manner that it is reasonable to assume constant return stroke speed along the channel element. Once the average return stroke speed as a function of height is known the return stroke speed as a function of height can be obtained directly from it.

3.4 Basic concepts of current dissipation models

As mentioned previously, if a current pulse is propagating without corona along a transmission line, it will travel along the line without any attenuation and modification of the current waveshape. This concept is used as a base in creating current propagation models. When the current amplitude is larger than the threshold current necessary for corona generation, each element of the transmission line acts as a corona current source. Half of the corona current generated by the sources travels downward and the other half travels upwards. The upward moving corona currents interact with the front of the injected current pulse in such a way that the speed of the upward moving current pulse is reduced, and for a transmission line in air, to a value less than the speed of light [30]. In a recent publication Cooray [31] showed that the upward moving corona current concept can also be used to create return stroke models. He coined the term 'Current Dissipation Models' for the same. The basic features of the current dissipation models are illustrated in Figure 3.2. The main assumptions of the current dissipation models are the following: the return stroke is initiated by a current pulse injected into the leader channel from the grounded end. The arrival of the return stroke front at a given channel element will turn on a current source that will inject a corona current into the central core. It is important to stress here that by the statement *the arrival of the return stroke front at a given channel element* it is meant the onset of the return stroke current in that channel element (i.e. point B in Figure 3.2). Once in the core this corona current will travel upward along the channel. In the case of negative return strokes the polarity of the corona current is such that it will deposit positive charge on the corona sheath and transport negative charge along the central core. Let us now incorporate mathematics into this physical scenario.

Assume that the return stroke is initiated by a current pulse injected into the leader channel at ground level. This current pulse propagates upward along the

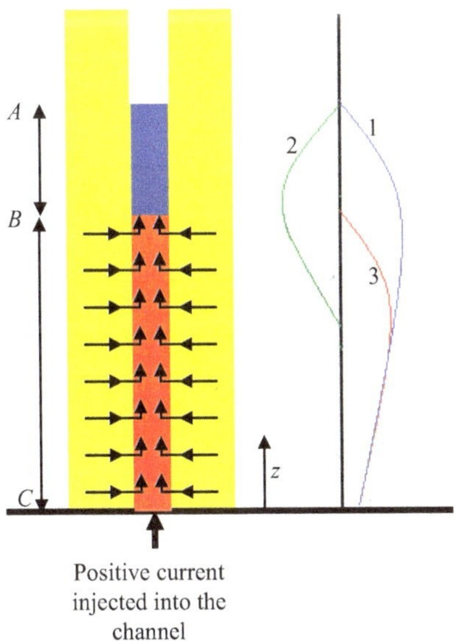

Figure 3.2 Pictorial description of the current dissipation concept. Waveforms to the right depict different current components along the channel: (1) injected current; (2) upward moving corona current and (3) net current along the channel. Adapted from [15]

channel with speed v_c (for a transmission line in air or free space $v_c = c$, the speed of light in free space). When the return stroke front (i.e. the net current front) reaches a given channel element a corona source is turned on. This source will generate a corona current that will travel upward along the central core with the same speed as the current pulse injected at the channel base (i.e. v_c). Note that the polarity of the upward moving corona current is opposite to that of the upward moving current injected at the channel base. For example, in the case of negative return stroke the current injected at the channel base carry positive charge upward whereas the corona current transports negative charge upward. According to this model the total current at a given point of the channel consists of two parts – upward moving current pulse injected at the channel base and the total contribution of the upward moving corona currents. The upward moving corona currents being of opposite polarity lead to the dissipation of the current pulse injected at the channel base. Note how the net current varies along the channel.

3.4.1 Input parameters of the CD models

As in the CG models, in CD models one has the choice of selecting the channel base current, $I_b(t)$, the distribution of the charge deposited by the return stroke along the channel, $\rho(z)$, the return stroke speed, $v(z)$, and the magnitude and variation of the corona discharge time constant with height, $\tau(z)$ as input parameters. Once three of these parameters are specified the fourth parameter can be estimated without ambiguity.

3.4.2 The connection between the channel base current (or injected current) and the corona current

In CG models we have seen that the channel base current can be described completely in terms of the corona current. For example, (3.11) gives the channel base current in terms of the corona current. The same can be done in the case of CD models.

Consider the diagram to the right in Figure 3.2. This depicts a situation at any given time t. At this time the tip of the injected current is located at point A and the return stroke front is located at point B. As in the case of a current pulse propagating along a transmission line under corona, at points above the current front or the return stroke front (i.e. points located above B), the current is zero. Of course in the case of a transmission line where a current pulse is propagating under corona emission, the current above the point B is clamped below the corona threshold. However, as pointed out by Cooray [31], since the corona in the return stroke process is caused by a neutralization process, the threshold current necessary for corona is close to zero. Now, in the CD models the sum of the corona currents completely neutralize the injected current above the return stroke front. Let us consider any height z located above the return stroke front (i.e. above point B). The net corona current at that height is given by (note that the corona current is defined as negative here because it transports negative charge towards the cloud whereas the injected current transports positive charge upwards):

$$I_{cor}(z,t) = -\int_{o}^{h_d} I_c(\xi, t - \xi/v_{ac}(\xi) - (z-\xi)/v_c)d\xi \tag{3.14}$$

In the above equation, $I_c(z,t)$ is the corona current per unit length at height z. Note that h_d is the location of the highest point on the channel whose corona current can reach the point z at time t. This can be obtained by solving the equation:

$$t = \frac{h_d}{v_{av}(h_d)} + \frac{(z-h_d)}{v_c} \tag{3.15}$$

The injected current at point z at time t is given by

$$I_{in}(z,t) = I_b(0, t - z/v_c) \tag{3.16}$$

Since the corona current annihilate the injected current at all points above the return stroke front we have

$$I_b(0, t - z/v_c) = \int_{o}^{h_d} I_c(\xi, t - \xi/v_{av}(\xi) - (z-\xi)/v_c)d\xi \tag{3.17}$$

Changing the variable we can write

$$I_b(0, t') = \int_{o}^{h_s} I_c(\xi, t' - \xi/v_{av}(\xi) + \xi/v_c)d\xi \tag{3.18}$$

with h_s given by the solution of

$$t' = \frac{h_s}{v_{av}(h_s)} - \frac{h_s}{v_c} \tag{3.19}$$

Now, a comparison of (3.18) and (3.19) with (3.9) and (3.10) shows that the only difference in the equations when moving from current generation concept to current dissipation concept is that v_c is replaced by $-v_c$. Thus any equation pertinent to CG models can be converted to those corresponding to CD models by replacing v_c by $-v_c$. Moreover, as in the case of current generation models, the input parameters of current dissipation models are the charge deposited on the channel by the return stroke, corona decay time constant, return stroke speed and the channel base current. When three of these parameters are given the fourth one can be obtained in the same manner as it was done in the case of current generation models. But in (3.11)–(3.13), v_c has to be replaced by $-v_c$ when using these equations in connection with current dissipation models.

3.5 Generalization of any model to current generation or current dissipation type

Consider any return stroke model which provides an expression for $I(z,t)$, the return stroke current at any height. Assume also that the speed of propagation of the return stroke front pertinent to that model is v. Cooray [31,32] showed that this information is enough to describe this model either as a CG model or a CD model with an equivalent corona current which depends on $I(z,t)$ and v. The equivalent corona current per unit length at any given height z that is needed to describe the model as a CG model is given by

$$I_{cg}(z,t) = -\frac{\partial I(z,t)}{\partial z} + \frac{1}{v_c}\frac{\partial I(z,t)}{\partial t} \tag{3.20}$$

In the above equation v_c is the speed of propagation of corona currents along the return stroke channel.

In a similar manner, the equivalent corona current per unit length at any given height z that is needed to describe the model as a CD model is given by

$$I_{cd}(z,t) = -\frac{\partial I(z,t)}{\partial z} - \frac{1}{v_c}\frac{\partial I(z,t)}{\partial t} \tag{3.21}$$

In the above equation v_c is the speed of propagation of corona currents along the return stroke channel. With this corona current and v as an input the variation of the return stroke current in space and time of the resulting CG or CD model would be identical to that of the original model. However, it is important to point out that the equivalent corona current of the CG model is not the same as the CD model. This shows that even though the spatial and temporal variation of the return stroke current is the same in all three models (i.e. original and the equivalent CG and CD models) the physics of charge neutralization which is the source of the corona current is different in the three models.

3.6 Current propagation models as a special case of current dissipation models

If the return stroke current associated with a current propagation model is assumed to decrease with height (as in the case of modified transmission line models [12,13]), the conservation of charge requires deposition of charge along the channel as the return stroke front propagates upward. This leakage of charge from central core to the corona sheath can be represented by a radially flowing corona current. Recently, Maslowski and Rakov [33] showed that this corona current is given by

$$I_{cp}(z,t) = -\frac{\partial I(z,t)}{\partial z} - \frac{1}{v}\frac{\partial I(z,t)}{\partial t} \tag{3.22}$$

where $I_{cp}(z,t)$ is the corona current per unit length at height z, $I(z,t)$ is the longitudinal return stroke current at the same height as predicted by the return stroke model and v is the *speed of the return stroke front*. Note that the direction of flow of the corona current is radial and, in contrast to the current generation or current dissipation models, it does not have a component flowing along the return stroke channel, that is it is a stationary corona current. Maslowski and Rakov [33] showed that any return stroke model could be reformulated as a current propagation model with an equivalent stationary corona current given by (3.22).

Let us now go back to the current dissipation models. Cooray [31] showed that in general the speed of propagation of the return stroke front in current dissipation models is less than that of the injected current (i.e. v_c). However, he also showed that one can select the parameters of the corona current in such a way that the speed of the return stroke front remains the same as that of the injected current pulse and the corona current. When such a choice is made current dissipation models reduce to MTL models. This can be illustrated mathematically as follows. Let us represent the injected current at the channel base as $I_b(0,t)$. The injected current at height z is given by

$$I_b(z,t) = I_b(0, t - z/v_c) \tag{3.23}$$

Assume that the corona current per unit length at level z is given by

$$I_{cd}(z,t) = I_b(0, t - z/v_c)A(z) \tag{3.24}$$

where $A(z)$ is some function of z. According to this equation the corona current at a given height is proportional to the injected current at that height. Substituting this expression in (3.21) one finds that

$$I_b(0, t - z/v_c)A(z) = -\frac{\partial I(z,t)}{\partial z} - \frac{1}{v_c}\frac{\partial I(z,t)}{\partial t} \tag{3.25}$$

One can easily show by substitution that the solution of this equation is given by

$$I(z,t) = A'(z)I_b(0, t - z/v_c) \tag{3.26}$$

with

$$A'(z) = -\int A(z)dz \tag{3.27}$$

Note that $I(z,t)$ in the above equation is the total current, that is sum of the corona current and the injected current. According to (3.26), the total current propagates upward with the same speed as that of the injected current and corona current. Moreover, it propagates upward without any distortion while its amplitude varies with height according to the function $A'(z)$. Indeed, (3.26) describes a MTL model. In this special case (3.21) reduces to (3.22) derived by Mazlowski and Rakov [33] because the return stroke speed v becomes equal to v_c. Thus, (3.22) is a special case of (3.21) and the latter reduces to the former in the case of MTL models. The above also demonstrates that all the current propagation models available in the literature are special cases of current dissipation models.

3.7 Physical basis of CD and CG models and a return stroke model based on their combination

The similarity and the differences between the CG and CD models and the propagation of the current pulses along a uniform transmission line in the presence of corona are described. Consider the injection of a current pulse into an ideal and uniform transmission line in air as described earlier. If however when the injected current amplitude is larger than the threshold current necessary for corona generation from the transmission line, each element of the transmission line acts as a corona current source. Half of the corona current generated by the sources travels downward and the other half travels upward. The speed of propagation of the pulses is equal to the speed of light in free space. The upward moving corona currents interact with the front of the injected current pulse in such a way that the speed of the net upward moving current (i.e. sum of the two upward moving current pulses) is reduced, and for a transmission line in air, to a value less than the speed of light in free space [30]. Thus in principle one would find three separate current waveforms along the return stroke channel. The first one is the upward moving injected current (current pulse 1). The second one is the current generated by the sum of upward moving corona currents (current pulse 2). The third one is the current generated by sum of downward moving corona currents (current pulse 3). The injected current transports positive charge upward, the current pulse due to upward moving corona current transports negative charge upward and the current pulse formed by the downward moving corona currents transports negative charge towards the ground. All the three current pulses propagate with speed of light along the return stroke channel. The net upward moving current is produced by the sum of current pulse 1 and current pulse 2.

The CG models describe the return stroke using current pulse 1 and the CD models describe it using the current pulses 1 and 2. But, as one can see from the above description a return stroke model that is in agreement with the physics of pulse propagation along transmission lines in the presence of corona should contain all the three pulses. Cooray and Diendorfer [15] was the first to introduce such a return stroke model into the literature. In this model as the injected current propagates upwards it will give rise to a corona current at each point on the return stroke channel. Half of this corona current travels downward to ground and the other half travels upward. The total currents at different level pertinent to an injected current as depicted in Figure 3.3 is shown in Figure 3.4. In this diagram the three current waveforms are depicted in the figure. The current 3 represents the CG part of the model and currents 1 and 2 represent the CD part of the model. They showed that such a model is capable of reproducing the electromagnetic fields reasonably well. However, note that the current at ground level consists of not only the injected current but also the current due to the downward moving corona current. Thus, the price that one has to pay for improving the physics is the fact that the injected current cannot be specified completely in the model.

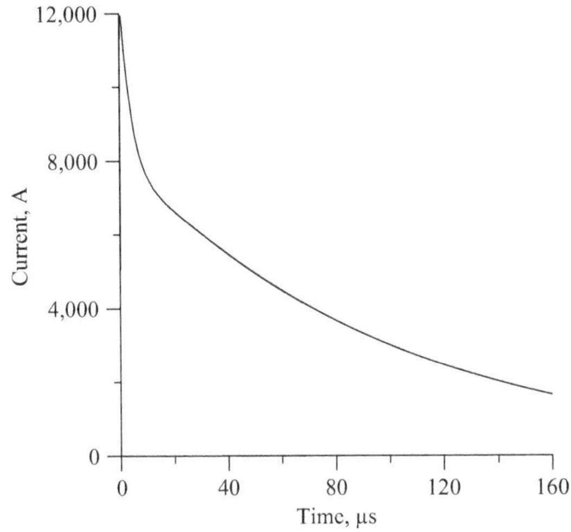

Figure 3.3 Injected current at the channel base previously by Cooray and Diendorfer [15]

3.8 Electromagnetic fields from lightning return strokes

The main goal of the engineering return stroke models is to provide a tool for the engineers to estimate the electromagnetic fields generated by lightning flashes at different distances. This requires that any engineering model is capable of generating electromagnetic fields similar to those generated by lightning flashes. In the sections to follow the features of electromagnetic fields generated by lightning flashes and how the conformity of the model generated fields with measured fields are briefly described.

In studying electromagnetic fields from lightning flashes, one can only specify a typical wave shape associated with either a first return stroke or a subsequent return stroke at a given distance. The reason for this is the following. Electromagnetic field generated by a lightning return stroke change its shape and amplitude with distance due to the presence of different field components that decrease at different rates with distance. At a given distance, there are also differences in the electromagnetic fields generated by first and subsequent return strokes. Moreover, the electromagnetic fields of either first or subsequent return strokes at a given distance may vary from one stroke to another. These changes in the electromagnetic fields of return strokes are caused probably by the different amounts of charges transported by different strokes, their channel geometry and the differences in various discharge parameters, such as charge distributions and speed of development of discharges, caused by the statistical nature of electrical breakdown. Furthermore, most of the first return strokes and some of the subsequent return strokes contain branches and the presence of these could also

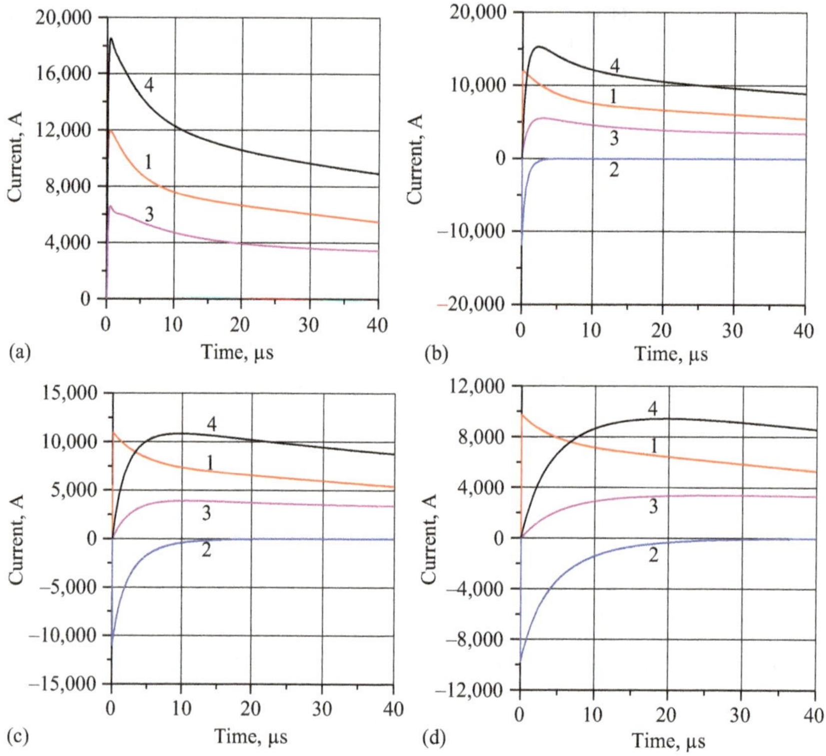

Figure 3.4 Current waveform at different heights. In each diagram the different current waveforms that exist at that level are also depicted. These waveforms are marked as follows: (1) the injected current; (2) the upward moving corona current; (3) the downward moving corona current and (4) the total current. The different diagrams correspond to different heights along the channel. They are marked as follows: (a) 0 m, (b) 100 m, (c) 500 m, (d) 1 km, (e) 2 km, (f) 3 km, (g) 4 km and (h) 5 km. Adapted from [15]

influence the structure of the electromagnetic fields [34,35]. In addition there could be differences between the electromagnetic fields generated by negative and positive return strokes [36–38]. For these reasons one can specify only the general shapes of these fields at different distances but individual strokes may produce fields that may significantly differ from the typical values and shapes specified at these distances. Due to these complexities, the engineering models can be tested only against the typical wave shape parameters extracted by studying a larger number of return stroke fields. Unfortunately, engineering models have not yet been developed to take into account the complex nature of the channel geometry or the statistical nature of lightning discharges. In the following section, we will describe the typical features of electromagnetic fields that could be utilized to test engineering models against experimental data.

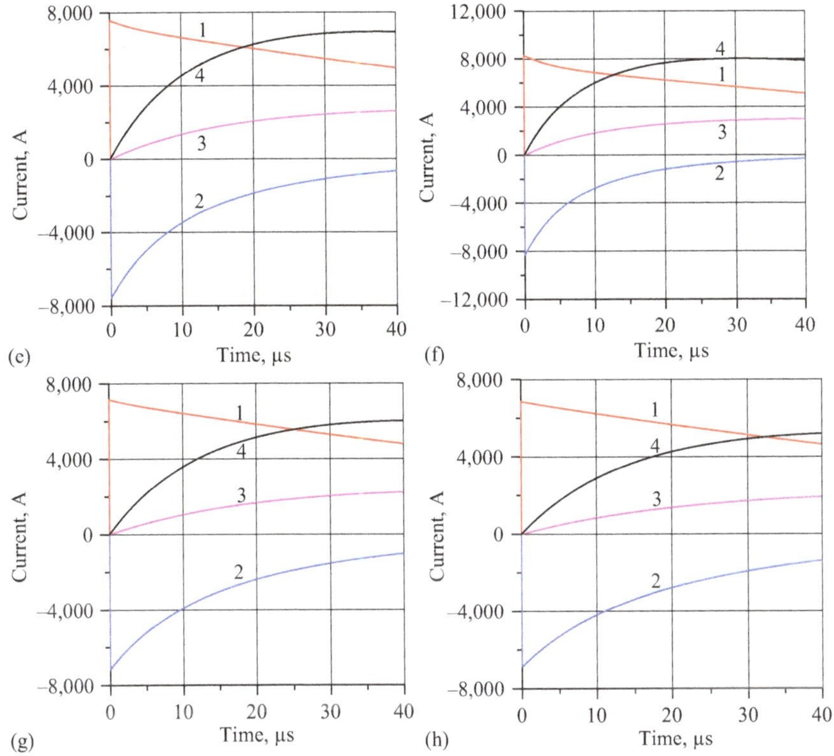

Figure 3.4 (*Continued*)

The electric field at a given distance from a lightning channel contain radiation, induction and static components while the magnetic field consists of induction and radiation terms [39]. The fields can also be divided into coulomb fields of stationary and moving charges and radiation fields [40,41]. However, in the present chapter we stick to the former division. These field components decrease at different rates with distance and at a given distance the total field may have different contributions from all the components. Moreover, it is not the distance alone but also the time that decides whether the electric field at a given distance is radiation or static. For example, the first microseconds of the close electric field at a few kilometre distance can be radiation while the tail of the electric field is dominated by the static field [42]. The reason for this is that whether a given electric field is dominatingly static or radiation is decided by the frequency content. For example, the typical wavelength associated with field variations taking place in the microsecond scale is about 300 m and for fields varying in the 100 μs scale it is about 30 km. For a field feature to be radiation the typical wavelength has to be much smaller than the distance to the source. Thus, at a few kilometre distance the fields varying in the microsecond scale are radiation while the tail of the field taking place in 50 to 100 μs scale is radiation only at distances larger than about 100 km.

But, even at these distances the electric field component of the long tails could still be static as in the case of positive return strokes [43]. Thus, it is incorrect to say that the close electric fields are static fields without indicating the time intervals over which the fields are specified. Moreover, at these distances, that is distances greater than about 100 km, the electromagnetic fields could be disturbed by ionospheric reflections [44,45] and this effect has to be taken into account in evaluating the general features of distant electromagnetic fields.

The electromagnetic fields from lightning flashes can also change due to the absorption of high frequencies by the ground as they propagate along finitely conducting ground. These effects, called propagation effects, modify mainly the electromagnetic field features such as the risetime and the initial peak value [46,47]. The exact or undistorted fields can be measured only over perfectly conducting ground. Fortunately, the sea water provides a highly conducting media and for all practical purposes the electromagnetic fields of lightning flashes striking the sea and measured in such a way that the propagation along dry land is minimal provide a rather good sample of undistorted lightning electromagnetic fields. However, one always has the question whether the lightning striking finitely conducting ground differ from those striking sea water, that is whether the fast features of lightning currents are modified by the ground conductivity [48]. All these points have to be taken into account when analysing the electromagnetic fields generated by lightning flashes.

In experimental studies with natural lightning, due to the scarcity of such strokes, it is difficult to obtain data with statistical significance when the distance to the flashes is less than about 1 km. However, close field signatures could still be obtained from triggered lightning but unfortunately such lightning flashes do not contain the first return strokes [49,50]. Thus, typical fields, assuming that the triggering process does not disturb the features of electromagnetic fields, very close to the lightning channel are available only for subsequent return strokes.

The electric fields within about 100 m obtained from triggered subsequent return strokes are shown in Figure 3.5 [49]. In this figure, the electric field to the left of the minimum is produced by the down-coming dart leader and the subsequent field recovery is due to the upward-moving return stroke. Note that the electric field rises within a few microseconds to a peak value and after that its magnitude remains more or less the same with time. This is one of the features that a return stroke model should be able to reproduce. Unfortunately, we do not have experimental evidence to confirm whether the electric fields of first return strokes, both positive and negative, also exhibit this feature. One problem of gathering such data is that the striking distance of first return strokes is on the order of 100 m and it is difficult, if not impossible, to set up an experimental station at such close distances from the striking point of first strokes [51,52]. One can show that the stabilization of the close electric field is caused by the balance of the increasing static field by the decreasing induction field with time. The radiation field does not play a significant role in generating this feature [53].

Theory and experiments show that the close magnetic field becomes closer and closer to the shape of the lightning current as the distance decreases and at very

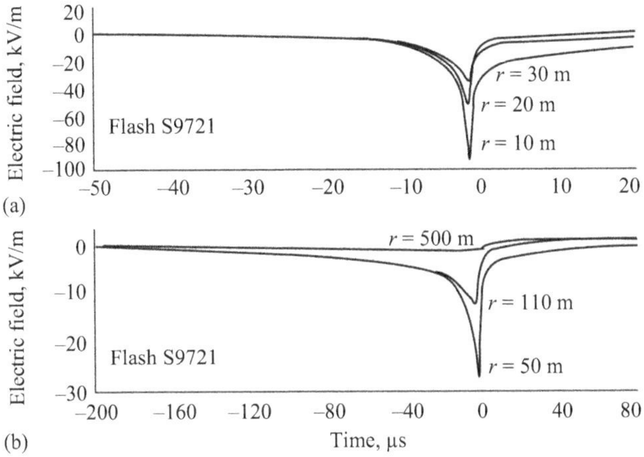

Figure 3.5 Electric field waveforms of the dart leader-return stroke sequence as recorded at different distances from a triggered lightning flash. The electric field to the left of the minimum is produced by the down-coming dart leader and the subsequent field recovery is due to the upward-moving return stroke. Adapted from [49]

close distances it resembles the channel base current completely [54]. This is what one would expect from Ampere's law. Usually, any return stroke model should have this feature in the predicted close magnetic field but exactly at what distance it resembles the channel base current may differ from one model to another due to the differences in the way in which the return stroke current is assumed to be attenuated with height in different return stroke models.

The general shape of typical electromagnetic fields of first and subsequent strokes in the distance range of 1 km to 100 km are shown in Figure 3.6 [55]. Note that the field is unipolar close to the channel and it becomes bipolar at long distances. The reason for this is the dominance of the radiation field, which is bipolar, at long distances. In the range of 1–10 km, the initial first few microseconds of the field including the initial peak is radiation. At these distances the electric field exhibits a ramp like increase with increasing time. The magnetic field is unipolar in this distance range and it shows a pronounced hump. These are the basic features that are being considered in testing the validity of engineering return stroke models. The distant radiation fields of both first and subsequent return strokes cross the zero line around 30–100 μs. The presence of the zero crossing time in this time range is also another feature that is being required in the model predictions. However, it is important to point out the following in connection with the zero crossing time.

Any discharge process that has a beginning and an end will always produce a radiation field that is bipolar [56]. In the case of return strokes, it is the initiation of the current at ground level and its termination in the cloud that decide the time of the zero crossing. Of course, part of the return stroke current may terminate along the channel due to the neutralization of the negative leader charge and this

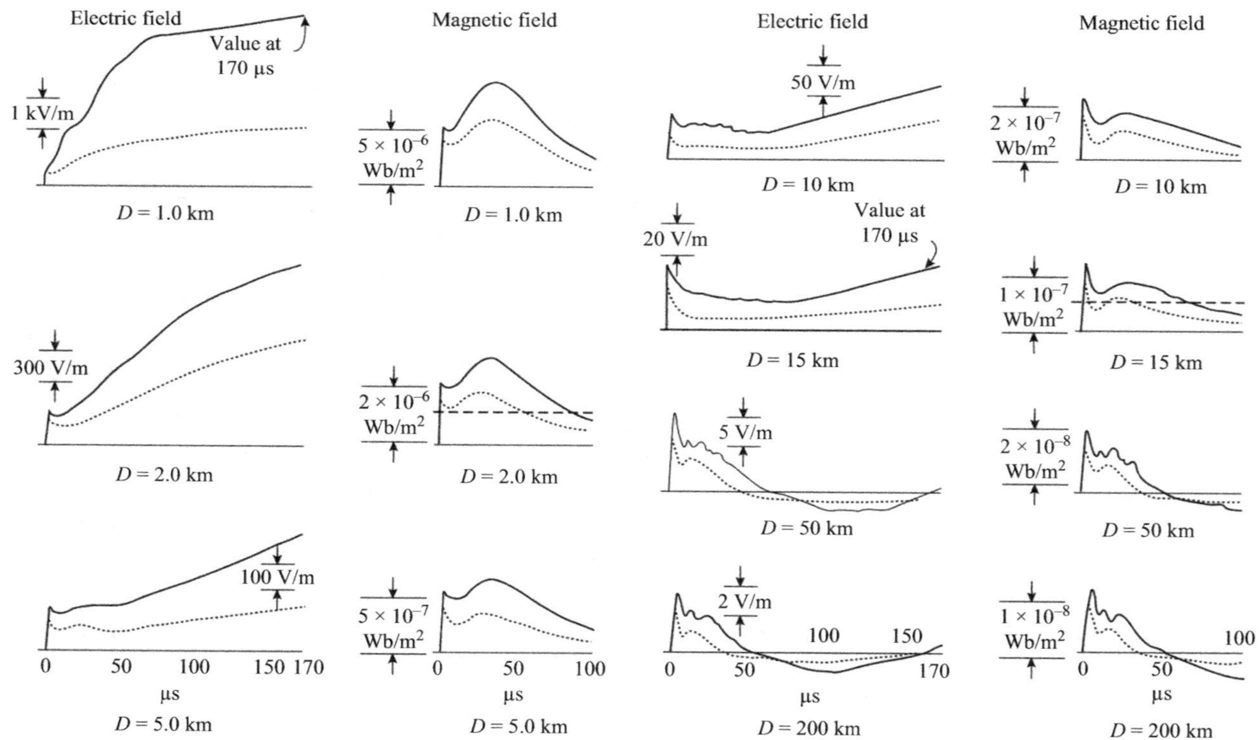

Figure 3.6 Typical electric field intensity and magnetic field density of first (solid line) and subsequent (dotted line) return strokes at different distances. Adapted from [55]

could also influence the time at which the radiation field crosses the zero line. However, some models may not adhere to this physical picture. For example, consider the TL model. In this model, if the current is allowed to propagate along an infinite channel, the radiation field will remain unipolar. It becomes bipolar only if a finite length is assumed for the channel. However, if the current is allowed to attenuate along the channel, as in the MTLE model, the radiation field will cross the zero line even if the channel length is assumed to be infinite. Similarly, a decreasing return stroke speed will also reduce the zero crossing time. Some return stroke models may have slowly decreasing current and speed and in this case if the current is allowed to propagate along an infinite vertical channel, the zero crossing time could become very large. Such a behaviour may motivate the researchers to reject the model because it cannot reproduce the observed zero crossing times. However, this observation along does not disqualify the model. The problem in this case lies not in the return stroke model but in the unphysical assumption of an infinitely long vertical channel. Note that in reality the vertical channel of return strokes are no longer than about 5–7 km. The rest of the channel located inside the cloud is horizontal. The reason for this is that there is a definite height, controlled by the ambient temperature, for the negative or positive charge centres in the cloud. Moreover, the vertical extent of the charge centres are limited and the charge distribution in a given charge centre is usually stratified horizontally [57]. Thus, the measured data should be compared with model fields calculated using return stroke channels of physically reasonable vertical extent. Since the vertical channel is the dominant radiator, the channel becoming horizontal is similar to effectively terminating the vertical channel. If one considers such a channel geometry, one can show that the electromagnetic radiation fields of all the engineering models available will cross the zero line at times appropriate for experimental data [53]. For example, if the return stroke speed is about 10^8 m/s, then any return stroke model supporting a 7 km channel will generate a zero crossing time within about 70 μs.

Figure 3.6 depicts the typical shapes of the electric and magnetic fields but it does not show the fine structure associated with them. The fine structure of radiation fields of first and subsequent return strokes are shown in Figure 3.7 [35]. Note that the radiation field starts with a small ramp-like field which is called a ‘slow front’ and subsequently rises very rapidly to a peak value. After the peak it starts to decay but, especially in the first strokes, it exhibits several peaks in the decaying part. Finally, it crosses the zero line, becomes negative and approaches the zero line again. The whole duration of the radiation field is about 100–200 μs. Almost all the engineering models, with a very few exceptions, are not capable of reproducing the fine structure exhibited by the measured electromagnetic fields. At the present time, there is no consensus among researchers as to the cause of the fine structure in the return stroke fields. Some of these features are produced by the variation of the return stroke current and the speed along the channel and the others are produced by the variation of the channel geometry. The subsequent peaks in the first return stroke radiation fields are probably generated by the current variations caused by the charges located on the channel branches [34,35]. Subsequent return strokes do not exhibit these peaks and neither do they support channel branches. Whenever the return

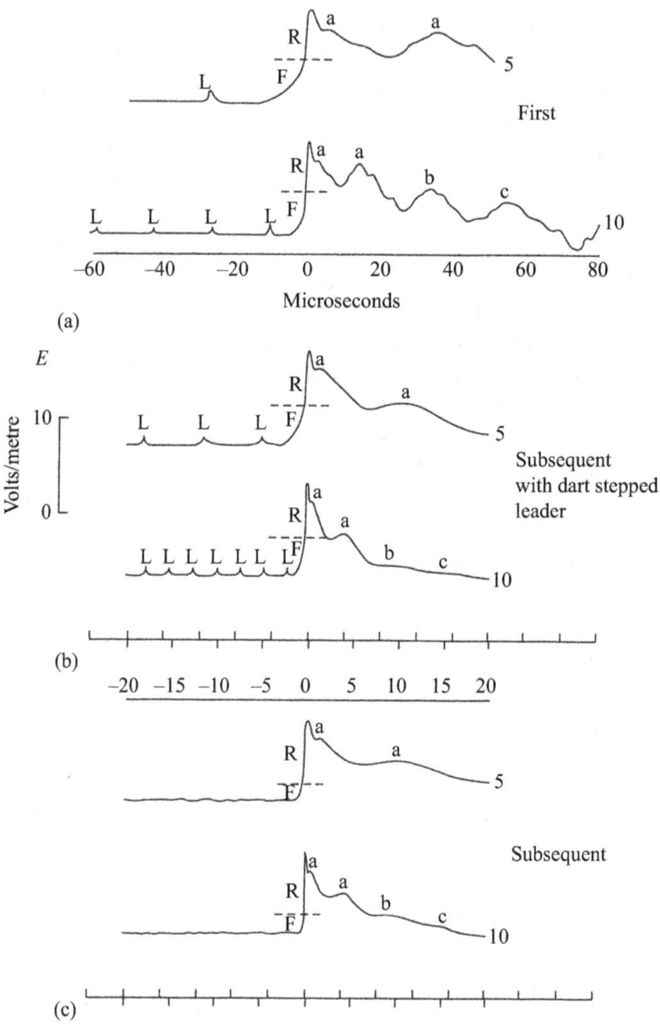

Figure 3.7 Fine structure of the radiation fields generated by first and subsequent return strokes. Adapted from [35]

stroke front reaches a branch point along the channel, both the current and the direction of propagation of the return stroke get affected (at a branch point the return stroke front travels both along the branch and along the main channel) and these changes could generate the subsequent peaks in the radiation fields. The channel tortuosity could also generate fine structure in the radiation fields due to the variation of the direction of propagation of the return stroke front and the return stroke current. In order to predict these features one has to utilize a tortuous return stroke channel with branches and make appropriate assumptions as to what happens when the return stroke front reaches a branch point. However, the slow front and the fast transition indicate what is happening at the initiation of the return stroke. Cooray [27] created an engineering model that is capable of reproducing both the slow front and the fast transition in the return stroke fields. In creating this model, he assumed that the slow front is produced by the propagation of the upward connecting leader through the streamer region of the negative leader and the fast front is generated when the hot channels of the upward connecting leader and the downward moving leader meet. Using this physical scenario he was able to reproduce the slow front and the fast transition in the radiation fields.

In conclusion one can point out that as described earlier, several typical features of electromagnetic fields from lightning can be used to test engineering return stroke models [58]. These are the following: (i) the presence of slow front and fast transition in radiation fields; (ii) the saturation of the close field within a few microseconds; (iii) the presence of a narrow initial peak in the fields at distances larger than about 1 km; (iv) the presence of a ramp like electric field at distances 2–10 km; (v) the presence of a prominent hump in the magnetic fields and (vi) the radiation field crossing the zero line around 30–100 μs. Note that the last condition has to be applied with caution when judging return stroke models without taking into account the channel geometry [53].

3.9 Calculation of lightning return stroke electromagnetic fields over ground

A return stroke model specifies the spatial and temporal variation of the return stroke current, that is $I(z,t)$. Once this information is given, the electromagnetic field can be calculated easily by dividing the channel into a large number of elementary dipoles and summing up the contribution of all the fields to obtain the total electric field. Let us first consider the electromagnetic fields generated by the lightning channel in frequency domain. It can be constructed easily by writing down the electric and magnetic fields produced at ground level by a dipole of length dz at height z and summing up the contributions of all dipoles or channel elements located along the return stroke channel. That is, the total electric and magnetic field over perfectly conducting ground can be written as (see Figure 3.8 for the relevant geometry) [38,59]:

$$e_z(j\omega,\rho) = \int_0^H \frac{I(j\omega)}{2\pi\varepsilon_o}\left(\frac{3\sin^2\theta-2}{j\omega R^3}+\frac{3\sin^2\theta-2}{cR^2}+j\omega\frac{\sin^2\theta}{c^2R}\right)e^{-j\omega R/c}dz \tag{3.28}$$

$$b_\phi(j\omega,\rho) = \int_0^H \frac{I(j\omega)\mu_0}{2\pi}\left(\frac{\sin\theta}{R^2}+j\omega\frac{\sin\theta}{cR}\right)e^{-j\omega R/c}dz \tag{3.29}$$

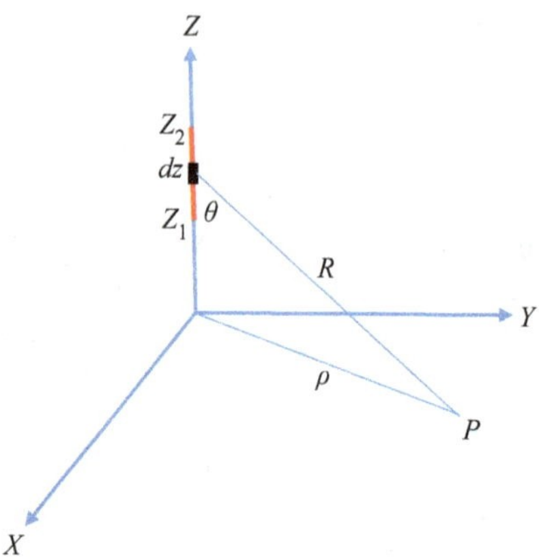

Figure 3.8 Geometry relevant to the calculation of propagation effects on electromagnetic fields generated by cloud flashes. X–Y plane represents the finitely conducting ground plane and the point of observation is located at ground level

In writing down the above equations where we have replaced the perfectly conducting ground plane by introducing an image current. The total electric and magnetic field in time domain at ground level is then given by

$$E_z(t,\rho) = \int_0^H \frac{1}{2\pi\varepsilon_o}\left(\frac{3\sin^2\theta - 2}{R^3}\int_0^t I(z,\tau - R/c)d\tau + \frac{3\sin^2\theta - 2}{cR^2} I(z,t-R/c) + \frac{\sin^2\theta}{c^2R}\frac{\partial I(z,t-R/c)}{\partial t}\right)dz \tag{3.30}$$

$$B_\phi(t,\rho) = \int_0^H \frac{\mu_0}{2\pi}\left(\frac{\sin\theta}{R^2} I(z,t-R/c) + \frac{\sin\theta}{cR}\frac{\partial I(t-R/c)}{\partial t}\right)dz \tag{3.31}$$

The exact solution for the electromagnetic fields generated by lightning return strokes over finitely conducting ground can be obtained by employing Sommerfeld's equations [60]. However, this procedure requires a significant amount of computational time and it is not suitable for estimating interaction of lightning electromagnetic fields with power lines where fields have to be calculated at larger number of places over finitely conducting ground. However, a simplified approximation based on the work of Norton [61] gives results for a good accuracy. According to Norton [61], the electric and magnetic fields in frequency domain over finitely conducting ground is given by

$$e_z(j\omega,\rho) = \int_0^H \frac{I(j\omega)}{2\pi\varepsilon_o}\left(\frac{3\sin^2\theta - 2}{j\omega R^3} + \frac{3\sin^2\theta - 2}{cR^2} + j\omega\frac{\sin^2\theta}{c^2R} a(z,j\omega,\rho)\right)e^{-j\omega R/c}dz \tag{3.32}$$

$$b_\phi(j\omega,\rho) = \int_0^H \frac{I(j\omega)\mu_0}{2\pi}\left(\frac{\sin\theta}{R^2} + j\omega\frac{\sin\theta}{cR}a(z,j\omega,\rho)\right)e^{-j\omega R/c}dz \tag{3.33}$$

In the above equation $a(z,j\omega,\rho)$ is the attenuation function which is defined in [60]. It is given by

$$a(z,j\omega,\rho) = (1+R_v) + (1-R_v)F(w,z) \tag{3.34}$$

In the above equation

$$R_v = \frac{\cos\theta - \Delta_0}{\cos\theta + \Delta_0} \tag{3.35}$$

$$\Delta_0 = \frac{\gamma_0}{\gamma}\left[1 - \frac{\gamma_0^2}{\gamma^2}\sin^2\theta\right]^{1/2} \tag{3.36}$$

$$\gamma_0 = \omega/c \quad \gamma = \gamma_0\sqrt{\varepsilon - j\sigma/\omega\varepsilon_0} \tag{3.37}$$

$$w = \frac{-j\gamma_0 R}{2\sin^2\theta_r}[\cos\theta + \Delta_0]^2 \tag{3.38}$$

$$F(w) = 1 - j(\pi w)^{1/2}e^{-w}\mathrm{erfc}(jw^{1/2}) \tag{3.39}$$

In these equations $j^2 = -1$, ε is the relative dielectric constant, σ is the conductivity of soil and erfc is the complementary error function. The angle θ is defined in Figure 3.8. The electric and magnetic fields over finitely conducting ground can be obtained by performing inverse Fourier transformation of the above equations. However, Cooray [46] showed that for all practical purposes the attenuation function can be replaced by the value corresponding to $z = 0$ and in this case an analytical expression for the electric and magnetic field can be written as

$$E_z(t,\rho) = E_{z,s}(t,\rho) + E_{z,i}(t,\rho) + \int_0^t E_{z,r}(t-\tau,\rho)A(0,\tau,\rho)d\tau \tag{3.40}$$

where the field components, $E_{z,s}(t,\rho)$, $E_{z,i}(t,\rho)$ and $E_{z,r}(t,\rho)$ are given by

$$E_{z,s}(t,\rho) = \int_0^H \frac{dz}{2\pi\varepsilon_o}\left\{\frac{3\sin^2\theta - 2}{R^3}\int_0^t i(z,\tau - R/c)d\tau\right\} \tag{3.41}$$

$$E_{z,i}(t,\rho) = \int_0^H \frac{dz}{2\pi\varepsilon_o}\frac{3\sin^2\theta - 2}{cR^2}i(z,t-R/c) \tag{3.42}$$

$$E_{z,r}(t,\rho) = \int_0^H \frac{dz}{2\pi\varepsilon_o}\frac{\sin^2\theta}{c^2R}\frac{\partial i(z,t-R/c)}{\partial t} \tag{3.43}$$

In the case of pure radiation fields (3.40) reduces to

$$E_z(t,\rho) = \int_0^t E_{z,r}(t-\tau,\rho)A(0,\tau,\rho)d\tau \tag{3.44}$$

In the above equation $A(0,t,\rho)$ is the inverse Fourier transformation of $a(0,j\omega,\rho)$.

Note in the above equation $E_{z,r}(t,\rho)$ is the radiation field that would be present if the ground is perfectly conducting. Thus, if the undistorted radiation fields of return strokes are known or estimated the effects of propagation on them can be obtained using the above equation. The accuracy of these equations have been evaluated by Cooray [62] and it was found that they can be used to evaluate the electromagnetic fields over finitely conducting ground with reasonable accuracy.

Another field component which is of interest in the calculation of voltages induced by lightning in power lines is the component of the electric field parallel to the ground. This component is usually referred to as the horizontal electric field (the component of the electric field parallel to the ground). This is the field component that drives currents along the horizontal section of the power line conductors. The horizontal electric field generated at ground level by a return stroke whose spatial and temporal variation of the current is specified can also be obtained using the Sommerfeld integrals [60]. However, a simple but a reasonably accurate procedure to obtain the horizontal electric field at ground level is given by Cooray [63]. According to this procedure, called the surface impedance method, the horizontal electric field at ground level is given by

$$E_h(\rho, 0, j\omega) = -cB_\varphi(\rho, 0, j\omega)\frac{\gamma_o}{\gamma} \tag{3.45}$$

In the above equation $E_h(\rho, 0, j\omega)$ is the horizontal electric field at the surface of the ground and $B_\varphi(\rho, 0, j\omega)$is the azimuthal magnetic field at the surface of the ground finitely conducting ground. This relationship can be used to obtain the horizontal electric field at ground to an accuracy reasonable for engineering studies if the distance to the point of observation is longer than a few hundred metres. This equation can be further simplified but by slightly compromising the accuracy of the calculated horizontal electric field by replacing the magnetic field over finitely conducting ground by the one present over perfectly conducting ground. Another and a more accurate procedure to calculate the horizontal electric field in the vicinity of the lightning channel was developed by Barbosa and Paulino [64]. Assume that the magnetic field produced by the return stroke at the point of observation located at a distance ρ from the lightning return stroke channel is a step of amplitude B_0. According to these authors, the corresponding horizontal electric field at that distance is given by

$$E_h(\rho, 0) = -\frac{Z_e B_0}{\mu_0}\left[\frac{2\varepsilon_r + a\tau(1 + 3b\varepsilon_r + 2ab\tau)}{2(1 + ab\tau)^{1/2}(\varepsilon_r + a\tau)^{3/2}}\right] \tag{3.46}$$

In the above equation $a = \pi\sigma/4\varepsilon_0$, $b = (1/Z_e\sigma\rho)^2$, Z_e is the impedance of free space, ε_r is the relative dielectric constant and τ is the time elapsed from the arrival of the electromagnetic field to the point of observation. The horizontal electric field corresponding to a magnetic field of any arbitrary shape can be obtained using the Duhamel's integral. Barbosa and Paulino [64] compared the results of their calculations with the exact results based on the integration of Sommerfeld's equations as presented by Cooray [65] and a good agreement is found between the two.

In calculating induced voltages in power lines due to lightning what is needed is the horizontal electric field at line height. A simplified procedure to obtain this field component is known in the literature as Cooray–Rubinstein approximation [63,66]. According to Cooray–Rubinstein approximation the horizontal electric field at a certain height above ground can be calculated from

$$E_\rho(\rho, z) = E_\rho(\rho, 0) + E_\infty(\rho, z) \quad (3.47)$$

where $E_\rho(\rho, z)$ is the horizontal electric field at height z at a horizontal distance ρ from the lightning channel, $E_\rho(\rho, 0)$ is the horizontal electric field at ground level at the same distance and $E_\infty(\rho, z)$ is the horizontal electric field at height z when the ground is perfectly conducting. In this formulation the horizontal electric field at ground level is given by (3.45). The validity of these procedures to calculate the horizontal electric field is described in details in Cooray [65] and Delfino *et al.* [67] among others. As mentioned previously, the errors associated with the horizontal electric field at ground level calculated using the surface impedance method increase as the distance to the point of observation decreases and when the distance is within a few hundred metres the errors could be significant, specially over poorly conducting ground [65,67]. However, this does not lead to significant errors in the induced voltages calculated using Cooray–Rubinstein formulation simply due to the fact that at close distances the horizontal electric field at line height is dominated by the second term of (3.47) and the errors associated with the first term (i.e. the horizontal electric field at ground level) playing a non-significant role in the total field.

3.10 Final comments and conclusions

In this chapter, we have given a very general description of the basic principles utilized in constructing engineering return stroke models. All the engineering models can be divided into three categories, namely, current propagation, current generation and current dissipation models. However, the currently available current propagation models, which consider only the attenuation of the current along the channel, are special cases of the current dissipation model. Any return stroke model that specifies the temporal and special variation of the return stroke current can be described either as a CG or a CD model irrespective of its origin.

Both CG and CD categories are based on simplifications and incomplete use of the physics of current pulse propagation along transmission lines in the presence of corona. If one is interested in incorporating the complete physics of corona current generation and propagation correctly into return stroke models, it is necessary to combine the two concepts into a single model.

It is also of interest to point out that both in CG and CD models, in addition to current attenuation, the current rise time increases with increasing height. However, this change in the current waveform is caused purely by the way in which the charge is deposited along the leader channel by the return stroke, that is by the features of the corona current. None of the return stroke models take into account the current dispersion along the channel but this could be a natural feature of current pulses propagating along a vertical return stroke channel. Thus, one has to consider that the signature of the corona current that provides the best predictions for the lightning return stroke electromagnetic fields is an equivalent corona current that indirectly accounts for the dispersion of the return stroke current. Thus deriving physical aspects of the neutralization of the leader corona sheath by the corona currents that best fit the engineering models should be done with caution.

References

[1] Cooray, V., Return stroke models for engineering applications, in *Lightning Protection*, edited by V. Cooray, The Institution of Engineering and Technology, London, UK, 2010.

[2] Baba, Y. and V. Rakov, Electromagnetic models of lightning return strokes, edited by V. Cooray, The Institution of Engineering and Technology, London, UK, 2012.

[3] Moini, R. and S. H. H. Sadeghi, Antenna models of lightning return stroke: an integral approach based on the method of moments, edited by V. Cooray, The Institution of Engineering and Technology, London, UK, 2012.

[4] De Conti, A., F. H. Silveira, and S. Visacro, Transmission line models of lightning return stroke, in *Lightning Electromagnetics*, edited by V. Cooray, The Institution of Engineering and Technology, London, UK, 2012.

[5] Borovsky, J. E., An electrodynamic description of lightning return strokes and dart leaders, *J. Geophys. Res.*, vol. 100, pp. 2697–2726, 1995.

[6] Cooray, V. and J. E. Borovsky, Electrodynamic model of lightning return stroke – spatial and temporal variation of the return stroke current, *2017 International Symposium on Lightning Protection (XIV SIPDA)*, Natal, Brazil, October 2–6, 2017.

[7] Norinder, H., Quelques essays recents relatifs á la determination des surten sions indirectes, *CIGRE Session 1939*, Paris, 29 June–8 July, p. 303, 1939.

[8] Bruce, C. E. R. and R. H. Golde, The lightning discharge, *J. Inst. Elect. Eng.*, vol. 88, pp. 487–520, 1941.

[9] Lundholm, Ph.D. dissertation, KTH, Stockholm, Sweden, 1957.

[10] Dennis, A. S. and E. T. Pierce, The return stroke of the lightning flash to earth as a source of atmospherics, *Radio Sci.*, pp. 777–794, 1964.

[11] Uman, M. A. and D. K. McLain, Magnetic field of lightning return stroke, *J. Geophys. Res.*, vol. 74, pp. 6899–6910, 1969.

[12] Nucci, C. A., C. Mazzetti, F. Rachidi, and M. Ianoz, On lightning return stroke models for LEMP calculations, in *19th International Conference on Lightning Protection*, Graz, Austria, 1988.

[13] Rakov, V. A. and A. A. Dulzon, A modified transmission line model for lightning return stroke field calculation, in *Proceedings of the 9th International Symposium on Electromagnetic Compatibility*, Zurich, Switzerland, 44H1, pp. 229–235, 1991.

[14] Cooray, V. and R. E. Orville, The effects of variation of current amplitude, current risetime and return stroke velocity along the return stroke channel on the electromagnetic fields generated by return strokes, *J. Geophys. Res*, vol. 95, no. D11, 1990.

[15] Cooray, V. and G. Diendorfer, Merging of current generation and current dissipation return stroke models, *Electr. Power Syst. Res.*, vol. 153, pp. 10–18, 2017.

[16] Wagner, C. F., Calculating lightning performance of transmission lines, *AIEE Trans.*, vol. 1232, p. 1985, 1956.

[17] Lin, Y. T., M. A. Uman, and R. B. Standler, Lightning return stroke models, *J. Geophys. Res.*, vol. 85, pp. 1571–1583, 1980.

[18] Master, M. J., M. A. Uman, Y. T. Lin, and R. B. Standler, Calculation of lightning return stroke electric and magnetic fields above ground, *J. Geophys. Res.*, vol. 86, pp. 12127–12132, 1981.

[19] Heidler, F., Travelling current source model for LEMP calculation, in *Proceedings of the 6th International Symposium on EMC*, Zurich, Switzerland, 29F2, pp. 157–162, 1985.

[20] Hubert, P., New model of lightning return stroke – confrontation with triggered lightning observations, in *Proceedings of the 10th International Aerospace and Ground Conference on Lightning and Static Electricity*, Paris, pp. 211–215, 1985.

[21] Cooray, V., A return stroke model, in *Proceedings of the International Conference on Lightning and Static Electricity*, U.K., pp. 6B.4.1–6B.4.6, 1989.

[22] Cooray, V., A model for subsequent return stroke, *J. Elect.*, vol. 30, pp. 343–354, 1993.

[23] Diendorfer, G. and M. A. Uman, An improved return stroke model with specified channel base current, *J. Geophys. Res.*, vol. 95, pp. 13621–13644, 1990.

[24] Thottappillil, R., D. K. Mclain, M. A. Uman, and G. Diendorfer, Extension of Diendorfer-Uman lightning return stroke model to the case of a variable upward return stroke speed and a variable downward discharge current speed, *J. Geophys. Res.*, vol. 96, pp. 17143–17150, 1991.

[25] Thottappillil, R. and M. A. Uman, Lightning return stroke model with height-variable discharge time constant, *J. Geophys. Res.*, vol. 99, pp. 22773–22780, 1994.

[26] Cooray, V., Predicting the spatial and temporal variation of the electromagnetic fields, currents, and speeds of subsequent return strokes, *IEEE Trans. Electromagn. Compat.*, vol. 40, pp. 427–435, 1998.

[27] Cooray, V., A model for first return strokes in lightning flashes, *Phys. Scripta*, vol. 55, pp. 119–128, 1996.

[28] Cooray, V., R. Montano, and V. Rakov, A model to represent first return strokes with connecting leaders, *J. Electrostat.*, vol. 40, pp. 97–109, 2004.

[29] Cooray, V. and V. Rakov, A current generation type return stroke model that predicts the return stroke velocity, *J. Lightning Res.*, vol. 1, pp. 32–39, 2007.

[30] Cooray, V. and N. Theethayi, Pulse propagation along transmission lines in the presence of corona and their implication to lightning return strokes, *IEEE Trans. Antennas Propag.*, vol. 56, no. 7, pp. 1948–1959, 2008.

[31] Cooray, V., A novel procedure to represent lightning strokes – current dissipation return stroke models, *Trans. IEEE (EMC)*, vol. 51, pp. 748–755, 2009.

[32] Cooray, V., On the concepts used in return stroke models applied in engineering practice, *Trans. IEEE (EMC)*, vol. 45, pp. 101–108, 2003.

[33] Maslowski, G. and V. A. Rakov, Equivalency of lightning return stroke models employing lumped and distributed current sources, *Trans. IEEE (EMC)*, vol. 49, pp. 123–132, 2007.

[34] Cooray, V. and M. Fernando, Effects of branches, charge irregularities and tortuosity of the stepped leader channel on the current, electromagnetic fields and hf radiation of return strokes, in *Proceedings of the International Conference on Lightning Protection*, Uppsala, 2008.

[35] Weidman, C. D. and E. P. Krider, The fine structure of lightning return stroke wave forms, *J. Geophys. Res. Lett.*, vol. 7, pp. 955–958, 1980. Correction, *J. Geophys. Res.*, vol. 87, p. 7351, 1982.

[36] Cooray, V., Further characteristics of positive radiation fields from lightning in Sweden, *J. Geophys. Res.*, vol. 84, no. 11, pp. 807–811, 815, 1984.

[37] Cooray, V., M. Fernando, C. Gomes, and T. Sorenssen, The fine structure of positive return stroke Radiation fields, *Proc. IEEE (EMC)*, vol. 46, pp. 87–95, 2004.

[38] Uman, M. A., *The Lightning Flash*, Academic Press, New York, NY, 1987.

[39] Thottappillil, R. and V. A. Rakov, On different approaches to calculating lightning electric fields, *J. Geophys. Res.*, vol. 106, pp. 14191–14205, 2001.

[40] Cooray, V. and G. Cooray, The electromagnetic fields of an accelerating charge: applications in lightning return-stroke models, *IEEE Trans. Electromagn. Compat.*, vol. 52, pp. 944–955, 2010.

[41] Cooray, V. and G. Cooray, Application of electromagnetic fields of an accelerating charge to obtain electromagnetic fields of a propagating current pulse, in *Lightning Electromagnetics*, edited by V. Cooray, The Institution of Engineering and Technology, London, UK, 2012.

[42] Cooray, V., Derivation of return stroke parameters from electric and magnetic field derivatives, *Geophys. Res. Lett.*, vol. 16, pp. 61–64, 1989.

[43] Cooray, V., R. E. Orville, and S. Lundquist, Electric fields from positive return strokes, *8th International Conference on Atmospheric Electricity*, Uppsala, pp. 472–476, 1988.

[44] Haddad, M. A., V. A. Rakov, and S. A. Cummer, New measurements of lightning electric fields in Florida: waveform characteristics, interaction with the ionosphere, and peak current estimates, *J. Geophys. Res. Atmos.*, vol. 117, no. 10, 2012.

[45] Mc Donald, T. B., M. A. Uman, J. A. Tiller, and W. H. Beasley, Lightning location and lower-ionospheric height determination from two-station magnetic field measurement, *J. Geophys. Res.*, vol. 84, no. C4, pp. 1727–1734, 1979.

[46] Cooray, V., Effects of propagation on the return stroke radiation fields, *Radio Sci.*, vol. 22, pp. 757–768, 1987.

[47] Cooray, V., M. Fernando, T. Sörensen, T. Götschl, and A. Pedersen, Propagation of lightning generated transient electromagnetic fields over finitely conducting ground, *J. Atmos. Solar-Terrestr. Phys.*, vol. 62, pp. 583–600, 2000.

[48] Cooray, V. and V. Rakov, Engineering lightning return stroke models incorporating current reflection from ground and finitely conducting ground effects, *IEEE Trans. Electromagn. Compat.*, vol. 53, pp. 773–781, 2011.

[49] Rakov, V., Lightning discharges triggered using rocket- and wire techniques, *Recent Res. Devel. Geophys.*, vol. 2, pp. 141–171, 1999.

[50] Rakov, V., Rocket and wire triggered lightning experiments, in *The Lightning Flash* (2nd edn.), edited by V. Cooray, The Institution of Engineering and Technology, London, UK, 2014.

[51] Cooray, V., V. Rakov, and N. Theethayi, The lightning striking distance – revisited, *J. Electrostat.*, vol. 65, no. 5–6, pp. 296–306, 2007.

[52] Cooray, V. and M. Becerra, Attachment of lightning flashes to grounded structures, in *Lightning Protection*, edited by V. Cooray, The Institution of Engineering and Technology, London, UK, 2010.

[53] Cooray, V., V. Rakov, F. Rachidi, and C. A. Nucci, On the relationship between the signature of close electric field and the equivalent corona current in lightning return stroke models, *IEEE Trans. Electromagn. Compat.*, vol. 50, no. 4, pp. 921–927, 2008.

[54] Fisher, R. J. and G. H. Schnetzer, 1993 triggered lightning test program: environments within 20 meters of the lightning channel and small area temporary protection concepts, Sandia National Laboratories, Albuquerque, New Mexico, SAND94-0311, 1994.

[55] Lin, Y. T., M. A. Uman, J. A. Tiller, *et al.*, Characterization of lightning return stroke electric and magnetic fields from simultaneous two-station measurements, *J. Geophys. Res.*, vol. 84, pp. 6307–6314, 1979.

[56] Yaghjian, A. D. and T. B. Hanson, Time-domain far fields, *J. Appl. Phys.*, vol. 79, no. 6, pp. 2822–2830, 1996.

[57] Krehbiel, P. R., M. Brook, and R. McCrory, An analysis of the charge structure of lightning discharges to the ground, *J. Geophys. Res.*, vol. 84, pp. 2432–2456, 1979.

[58] Nucci, C. A., G. Diendorfer, M. A. Uman, F. Rachidi, M. Ianoz, and C. Mazetti, Lightning return stroke models with specified channel base current: a review and comparison, *J. Geophys. Res.*, vol. 95, pp. 20395–20408, 1990.

[59] Cooray, V. and G. Cooray, A novel interpretation of the electromagnetic fields of lightning return strokes, *Atmosphere*, vol. 10, p. 22, 2019.

[60] Maclean, T. S. M. and Z. Wu, *Radio Wave Propagation Over Ground*, Chapman & Hall, London, 1993.

[61] Norton, K. A., Propagation of radio waves over the surface of Earth and in the upper atmosphere, II, *Proc. IEEE*, vol. 25, pp. 1203–1236, 1937.

[62] Cooray, V., On the accuracy of several approximate theories used in quantifying the propagation effects on lightning generated electromagnetic fields, *IEEE Trans. Antennas Propag.*, vol. 56, no. 7, pp. 1960–1967, 2008.
[63] Cooray, V., Horizontal fields generated by return strokes, *Radio Sci.*, vol. 27, pp. 529–537, 1992.
[64] Barbosa, C. F. and J. O. Paulino, A time domain formula for the horizontal electric field at the Earth surface in the vicinity of lightning, *IEEE Trans. (EMC)*, vol. 52, no. 3, pp. 640–645, 2010.
[65] Cooray, V., Horizontal electric field above- and underground produced by lightning flashes, *IEEE Trans. Electromagn. Compat.*, vol. 52, pp. 936–943, 2010.
[66] Rubinstein, M., An approximate formula for the calculation of the horizontal electric field from lightning at close, intermediate and long range, *IEEE Trans. (EMC)*, vol. 38, pp. 531–535, 1996.
[67] Delfino, F., R. Procopio, and M. Rossi, Lightning return stroke current radiation in presence of a conducting ground: 1. Theory and numerical evaluation of electromagnetic fields, *J. Geophys. Res.*, vol. 115, doi:10.1029/2007JD008553, 2008.

Chapter 4

Lightning geolocation information for power system analyses

Wolfgang Schulz[1] *and Amitabh Nag*[2]

In the fifties and sixties of the last century the measurements carried out by Berger [1,2] in Switzerland provided high-quality information about lightning needed for designing lightning protection systems. Those measurements were performed at two towers located on a mountain named Monte San Salvatore. All lightning flash and stroke parameters determined from those measurements, for example, return stroke peak current and flash multiplicity, are based on lightning striking the tower. Thus all strokes of each flash in Berger's dataset exhibited the tower as the ground strike point. Consequently, the flash density N_G (flashes per year per square kilometer) was defined as the risk that an object would be struck by lightning.

Today, many parts of the world are covered by accurate lightning locating systems (LLSs), and N_G can be determined using data from such systems. One of the first industries or user groups to use N_G for risk calculation were power utilities. Power utilities are interested in calculating the performance of transmission lines in the presence of lightning. For example, using the so-called electro-geometric model, the formula for computing the shielding failure rate (SFR) has been determined (CIGRE Report 63 [3]; IEEE Std. 1243 [4]):

$$\mathrm{SFR} = 2N_G L \int_{I=0}^{I=I_{\max}} D_c(I) f_1(I) dI \tag{4.1}$$

Later on, risk estimation was introduced in the lightning protection standard, IEC 62305-2 [5]. The risk of being struck by lightning (or the annual number of dangerous events, N_D) is defined [5], for example, for a building with a collection area A as

$$N_D = N_G A \tag{4.2}$$

Thottappillil *et al.* [6] and Rakov *et al.* [7] reported that negative flashes often have more than one ground strike point and that the distances between these strike

[1]Austrian Lightning Detection and Information System (ALDIS), Vienna, Austria
[2]Department of Aerospace, Physics and Space Sciences, Florida Institute of Technology, Melbourne, FL, USA

points range from 0.3 to 7.3 km, with the geometric mean being 1.7 km. Further, they found that in their data set there were no new ground strike points for strokes of order 5 or higher. The number of ground strike points reported in the literature range from 1.45 to 1.70 per flash [7–10].

The fact that all strokes in a flash often do not hit the same ground strike point is important for all risk analyses. This is because, if a flash exhibits two ground strike points on average, the risk of being struck by lightning doubles. Therefore, the risk of being struck by lightning should be estimated using the ground strike point density, N_{SG}, instead of N_G. Work is currently on-going to adopt these findings into standards. With the improving performance of modern LLSs it is possible to determine individual ground strike points with high accuracy. More details about methods that can be used to determine ground strike points from LLS data are given in Section 4.2.2.

In this chapter, we provide a general introduction to ground flash density calculation, discuss recommended methods to determine flash density from LLS data, provide details of techniques used by modern LLSs along with examples of a few well-established LLSs in different parts of the world, and discuss methods used to validate the performance characteristics of LLSs.

4.1 Introduction to ground flash density calculation

Before the availability of high-quality LLS data it was necessary to estimate N_G using other methods. Historically, there have been two main techniques that are used to determine N_G.

- The first method to do this was the calculation of N_G from the keraunic level. The keraunic level is also called thunderstorm days, T_d. A thunderstorm day is defined as a day at a given location during which thunder is heard at least once. In the past, a lot of effort was made to determine a "calibrated" N_G from thunderstorm days data resulting in the most frequently used formulas, $N_G = 0.04 * T_d^{1.25}$ [11] and $N_G = 0.1 * T_d$ (IEC 62305-2 2013 [5]). These formulas provide similar results.

  The advantage of determining N_G from the keraunic level is that these data are collected by meteorological services all over the world for many decades and, therefore, are available worldwide. The disadvantage of this method is that the keraunic level completely ignores the severity of a thunderstorm (number of flashes in a storm) and, as a result, may underestimate the risk.
- The second method to determine N_G, using lightning flash counters, was invented to address the disadvantage of using the keraunic level. A lightning flash counter registers electric field pulses produced by lightning events and reports the counts of these events at a certain location. Depending on the type of flash counter and its calibration, the effective pulse-detection range of the device is some tens of kilometers. Thus, this method to determine N_G is a local measurement. The disadvantages of using flash counters to determine N_G are as follows:

- Flash counters cannot distinguish between cloud-to-ground (CG) and intracloud (IC) lightning events, a correction factor is hence necessary to determine N_G.
- A fixed area is used to determine N_G, but large-amplitude lightning events occurring outside this area also contribute to the result.
- Flash counter measurements are only applicable locally to the regions in which such measurements are being made and many regions may not be covered.
- Flash counters often suffer from calibration issues as electric field measurements are sensitive to local electromagnetic boundary conditions.

The most commonly used flash counter is called the CIGRE counter for which there are two versions, the CIGRE 500-Hz and the CIGRE 10-kHz. More details about flash counters can be found in Rakov and Uman [12].

Nowadays, in many parts of the world LLS data are readily available (see e.g., Section 4.3.3) and in such cases N_G can be determined using data from a local LLS. Nevertheless, there are some regions in the world where no local LLSs exist. For such regions, one option is to determine N_G using data from the satellite-based OTD/LIS missions [13–15]. The optical measurements of satellites cannot distinguish between CG and IC flashes, so the lightning densities obtained from these data are total lightning flash densities, N_t (see Figure 4.1). N_t can be used to estimate N_G using the following relationship [16]:

$$N_G = 0.25N_t \tag{4.3}$$

At the time of publication of this chapter, N_t could be obtained from the website http://lightning.nsstc.nasa.gov/data/data_lis-otd-climatology.html.

Another option to determine N_G in case no local LLS data are available is to use flash densities from the GLD360 [18,19]. Figure 4.2 shows GLD360 flash densities based on 5 years (2013–17) of CG lightning data. Currently, GLD360

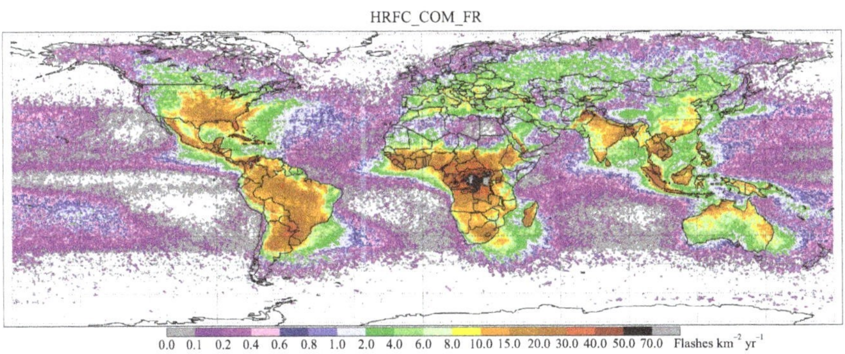

Figure 4.1 Worldwide total lightning (cloud and cloud-to-ground combined) flash densities (flashes km^{-2} yr^{-1}) obtained using combined OTD/LIS data [17]

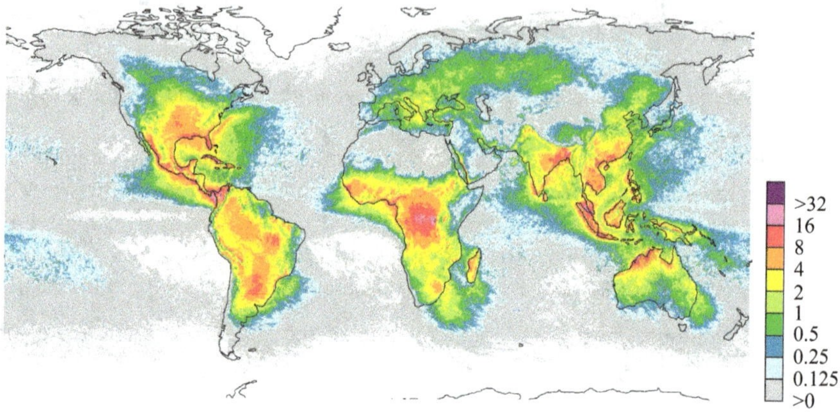

Figure 4.2 Worldwide cloud-to-ground flash densities (flashes km^{-2} yr^{-1}) based on 5 years (2013–2017) of data from the GLD360. Courtesy of Vaisala

does not distinguish between CG and IC flashes. To remove most IC pulses, only events with peak currents greater than 13 kA were included and grouped into flashes using a 20 km, 1 s space-time criteria. Note that, the data shown in Figure 4.2 are not corrected for LLS detection efficiency (DE) variations. The advantage of these data compared to satellite data is that N_G can be directly obtained and the spatial resolution of N_G is significantly better.

4.2 Standards and techniques recommended by the IEC 62858 [20]

4.2.1 Ground flash density from LLS

The ground flash density provided by LLS operators should meet certain quality criteria. This section seeks to explain what LLS operators need to take into account in order to provide reliable N_G values.

The quality of the available lightning data for calculating N_G is determined by the performance characteristics of an LLS (see Section 4.3.5; [21]). A value of N_G with an error of less than ±20% is deemed to be adequate for lightning risk assessment. For the purpose of N_G computation, data from any LLS that is able to detect CG lightning and accurately determine the ground attachment point of CG strokes can be used. The following LLS performance characteristics are desired for computation of N_G with adequate accuracy.

- Flash detection efficiency (DE): the value of the annual average flash DE of an LLS for CG lightning should be at least 80% in the region over which N_G has to be computed. This DE is usually obtained within the interior of the network. The interior of the network is defined as the region within the boundary defined by the outermost adjacent sensors of the network.

- Location accuracy (LA): the median LA of an LLS for CG strokes should be better than 500 m in the region over which N_G has to be computed. This LA is usually obtained within the interior of the network.
- Classification accuracy: if too many CG strokes are misclassified as cloud pulses, or vice versa, this may lead to erroneously low or high values of N_G. This is especially true for single-stroke CG flashes. A classification accuracy (CG flashes not misclassified as IC) of at least 85% seems to be adequate.

These performance characteristics of LLS can be determined using a variety of methods including network self-referencing and comparison against ground-truth lightning data obtained using various techniques. These methods are discussed in Section 4.3.5.

The performance parameters of an LLS depend on a few fundamental characteristics of the network. It is important for an LLS operator or data-providers to consider the following factors during the design and maintenance of the networks to ensure that the lightning data are of adequate quality for N_G computation.

- Sensor baseline distance: the distance between adjacent sensors in an LLS or sensor baseline distance is influenced by the area of desired coverage and the sensitivity of individual sensors. Sensor baseline distance is one of the factors that determine the DE and LA of an LLS. The maximum sensor baseline distance of an LLS should be such that the DE and LA of the network meet the criteria for N_G calculation described above.
- Sensor sensitivity: the sensitivity of sensors in an LLS primarily determines the ability of the network to detect lightning events of different peak currents. It is state of the art that the sensitivity of sensors in an LLS should be such that lightning events with peak currents in the range of 5 kA to 300 kA are detected and reported by the LLS. Sensor sensitivity is determined by various factors such as trigger threshold, electronic gain, sensor bandwidth, and background electromagnetic noise.
- Sensor uptime: the uptime of the individual sensors in a network influences the DE and LA of the network. The spatial and temporal variations of DE and LA are determined by the location of sensors that are up and contributing to the geolocation of lightning events. Therefore, it is important to guarantee that LLS sensors are up and communicating with the central processor with no interruption.

Because LLS normally detect and locate individual strokes it is common practice to group those individual strokes to flashes. The most widely used stroke-to-flash grouping algorithm described by Cummins *et al.* [22] uses a time and spatial criteria to group strokes to a flash. A subsequent stroke is grouped to a first return stroke if the stroke occurs less than 1 s after the first return stroke, the location is less than 10 km distant from the first return stroke and the time interval to the previous stroke is less than 500 ms.

It is further important to provide N_G only for an area over which lightning data of homogenous quality as described above are available. Network coverage normally falls off outside the boundaries of a network. In general, as a rule of thumb, lightning data within half the average sensor baseline distance (distance between adjacent sensors in the network) from the boundary of the network should be of sufficient quality for N_G calculation.

When calculating N_G it is necessary to think about the size of the area over which the flash density is averaged. The accuracy of N_G also depends on the amount of thunderstorm activity in the region of interest. If the area around a certain object is too large, N_G might be an underestimation of the real lightning density. On the other hand, if the area is too small, the accuracy with which N_G can be determined might be too low. The area over which N_G is computed can be represented as a cell in a grid. The dimensions of a cell and the observation period should be selected to comply with equation (4.4), which follows the Poisson distribution and the law of rare events, to obtain an uncertainty of less than 20% at the confidence level of 90% [23]:

$$N_G * T_{\text{obs}} * A_{\text{cell}} \geq 80 \tag{4.4}$$

where N_G is the ground flash density in km^{-2} year^{-1}; T_{obs} is the observation period in years; and A_{cell} is the area of a cell in km^2. Usually, a minimum observation period of 10 years is used. Also, the minimum cell dimension is generally selected to be larger than double the median LA of the LLS whose data is being used.

4.2.2 Ground strike point density

An updated version of the lightning protection standard IEC 62305-2 [5] will likely use the ground strike point density, N_{SG}, for calculating the risk that an object will be struck by lightning. As mentioned in the introduction of this chapter, the number of ground strike points per flash ranges from 1.45 to 1.7 based on measurements in different regions of the world. Thus, for lightning protection, the number of ground strike points per flash can be rounded to 2 (conservatively speaking) and therefore, N_{SG} can be calculated using

$$N_{\text{SG}} = 2 * N_G \tag{4.5}$$

There are different techniques or algorithms available [24-28] that provide more refined estimates of N_{SG} from LLS data directly. Each technique has its own advantages and disadvantages:

- Pedeboy [24,25] was the first to implement a reliable clustering algorithm based on the *k*-means clustering method. This approach does not employ a complete statistical treatment of the size and orientation of error ellipses and has no mechanism to deal with cases when the location uncertainties of strokes in a flash are larger than the separation distances between the strokes.
- Cummins [26] employed a discriminant analysis using a set of appropriately weighted parameters which included separation distances between strokes in a flash as well as stroke waveform parameters such as peak currents, risetimes,

and peak-to-zero times. This method can potentially identify new ground contacts (NGCs) when the location uncertainties for strokes in a flash are larger than the separation distances between the strokes, which is a weakness in Pebeboy's [24, 25] technique. However, Cummins' method relies on the availability of stroke waveform parameters such as risetime and peak-to-zero time, which some LLSs may not provide, and whose values are affected by propagation over lossy ground.

- Campos [27,28] built upon the k-means clustering approach developed by Pedeboy [24,25] by introducing in it a rigorous statistical evaluation of the location uncertainty using the size and orientation of error ellipses. This method, like Pedeboy's approach, has difficulty with cases when the stroke location-uncertainties are larger than the separation distances between strokes.

Research is ongoing to compare and refine the algorithms to determine N_{SG} from LLS data, and perhaps a new method, which combines the strengths of existing techniques, will be developed.

4.3 Lightning locating systems

4.3.1 Lightning geolocation techniques

The two most common techniques used for lightning geolocation in ground-based LLSs are time-of-arrival (TOA) and direction finding (DF) [22,29,30]. Low-Earth orbiting, satellite-based LLSs use either optical imaging (e.g., [31]) or a VHF DF technique (e.g., [32,33]) to geolocate lightning.

4.3.1.1 Time-of-arrival

The time-of-arrival (TOA) technique uses the occurrence time of a specific feature of the electromagnetic waveform (e.g., onset time, time of peak magnitude, or zero-crossing time) of a lightning event measured simultaneously at multiple sensors. A constant difference of the times at which a specific feature occurs at two different sensors defines a hyperbola, and multiple sensor pairs provide multiple hyperbolas whose intersections identify a lightning event location. Under some conditions, hyperbolas produced from only three sensors will result in two closely spaced intersections, thus leading to an ambiguous location (e.g., Figure 3 of [29]). Thus, the TOA technique requires simultaneous measurements of a lightning event from at least four sensors in order to geolocate the event uniquely, including calculation of its time, latitude, and longitude. In order to calculate the altitude of a lightning event, simultaneous measurements by at least five sensors are required. The TOA technique can be applied in ground-based LLSs operating in any frequency range. Lugrin *et al.* [34] describe the electromagnetic time reversal technique that may be used to geolocate lightning, and they mathematically demonstrate TOA to be a particular case of this technique.

In the TOA technique, a fixed random error in the time measurement results in a fixed random error in the calculated position as long as the lightning event occurs within the perimeter of the set of sensors. The position error is thus directly proportional to the timing error and independent of the distance between the sensors and the lightning event. Further, the TOA technique is not affected by directionally dependent local angle errors (known as "site errors"). The primary factors introducing timing error are variations in terrain elevation (introducing delay in the arrival time of lightning electromagnetic signals) and propagation over finite conductivity soil (introducing arrival time delay and frequency dependent attenuation of signals).

Factors such as GPS error, nonvertical lightning channels, and complexities of lightning electromagnetic wave shapes also contribute (generally to a lesser extent) to the timing error. Timing error introduced by terrain and soil conductivity variations can be addressed by applying location-specific corrections to the arrival times measured by LLS sensors [e.g., [35–37].

4.3.1.2 Direction finding

The DF technique typically employs magnetic direction finding (MDF), generally used in the VLF and LF-MF ranges, or interferometry, generally used in the VHF range (even though it has been used recently in the LF range [38]).

The MDF technique generally uses two orthogonal loop antennas to measure the magnetic field waveform produced by a lightning event. The ratio of the amplitudes of a selected waveform feature (generally the peak) measured by each loop can be used to calculate the azimuth of a lightning event location with respect to a sensor location, which gives a direction vector joining a sensor location and a lightning event location [39]. The intersections of two or more such direction vectors from two or more sensors measuring the same lightning event simultaneously identify a lightning event location. The time, latitude, and longitude of a lightning event can be calculated using this technique. In order to calculate the altitude of a lightning event, two or more three-axis magnetic field sensors are required, with each sensor having three orthogonal loop antennas and the ability to measure both the azimuth and elevation angles of a lightning event.

In VHF interferometry, each sensor consists of an array of closely spaced antennas, with the separation distance between individual antennas in the array being much smaller than the distance between the array and the lightning event location. The arrival-angle (azimuth) information can be derived from arrival time (phase) difference of the VHF signals associated with lightning events between individual antennas in a sensor. Similar to the MDF technique, simultaneous measurements of the azimuth of a lightning event location at two or more sensors can be used to determine the event location. The time, latitude, and longitude of a lightning event can be calculated using this technique. In order to add altitude to the geolocation, two or more sensors must simultaneously measure the azimuth and elevation angles of the event. VHF interferometry may be performed in narrowband (e.g., [40]) or broadband (e.g., [41–44]).

In ground-based LLSs, both MDF and interferometric techniques have to account for directionally dependent local site errors (typically due to various types of electromagnetic coupling) for each sensor in order to get accurate angle estimates. Given constant angle error, the location error associated with the estimated position of a lightning event is proportional to the distance between the sensors and the lightning event. That is, a constant angle error at the sensor translates to increasing linear error as the distance from the detector to the event increases. Additionally, when only two sensors are used to estimate position, the location error is much larger near the line joining the two sensors than it is where the sensor direction vectors intersect at a right angle.

4.3.1.3 Optical imaging

Optical emissions from CG and IC lightning discharges can be detected from above cloud top. The strongest emissions in the cloud-top lightning optical spectra (e.g., Figure 1 of [31]) are produced by the neutral oxygen (at 777.4 nm wavelength) and neutral nitrogen (at 886.3 nm wavelength) lines in the near infrared. Roughly 5% of the optical energy in a lightning flash may be radiated at and around each of these wavelengths. In satellite-based LLSs, lightning events (or groups of lightning events) are geolocated by using geometric projection of images taken from space of optical radiation from lightning within clouds. The time, latitude, and longitude of a lightning event (or a group of events) can be calculated using this technique. Altitude of lightning events, however, cannot be determined.

Detailed discussion of the factors that must be considered in satellite-based optical imaging can be found in Boccippio *et al.* [45], Suszcynsky *et al.* [46], Boccippio *et al.* [47], Christian *et al.* [13], and Mach *et al.* [14]. We do not include a detailed discussion about these factors in this chapter because satellite-based LLSs do not provide geolocation of individual ground strike points for CG discharges which is necessary for power system analyses.

4.3.1.4 Combined and special techniques

Long- and medium-range LLSs may use a combination of TOA and MDF techniques in order to utilize the combined strengths of both techniques. With direction-finding alone, a minimum of only two sensors can provide a geolocation but the position error is subject to the distance-dependent growth and it is necessary to correct the site errors affecting the angle measurements. With TOA alone, more sensors are required to determine a geolocation, but position errors are independent of distance inside the network and are generally lower than those produced from DF alone. In combined TOA-MDF LLSs, lightning events may be geolocated by as few as two contributing sensors and benefit from the position error advantage of TOA in the interior of the network, although there are larger location errors near the line joining the two sensors.

Medium range LLSs may use a combination of sensors operating in the VLF to MF and VHF ranges (e.g., [40,48]). Since different frequency ranges are suitable for measuring emissions from different types of lightning processes, such medium-range LLSs are able to geolocate lightning processes such as return strokes and cloud pulses using measurements made in the VLF to MF range and map the full spatial extent of lightning channels using measurements made in the VHF range.

4.3.2 Estimation of peak currents from measured electromagnetic fields

In the transmission line (TL) model [49] for distant lightning return strokes, the electric field peak, E_z, is proportional to the peak current, i

$$E_z = -\frac{v}{2\pi\varepsilon_0 c^2 r} i \tag{4.6}$$

where v is the return-stroke propagation speed. Equation (4.6) is valid when (a) the height above ground of the upward-moving return stroke front is much smaller than the distance r between the observation point on ground and channel base, so that all contributing channel points are essentially equidistant from the observer, (b) v is constant, (c) the return-stroke front has not reached the top of the channel, and (d) the ground conductivity is high enough, so that field propagation effects are negligible. The corresponding magnetic radiation field peak can be found from $|B_\varphi| = |E_z|/c$.

LLS sensors measure the electric or magnetic radiation field peak signal strengths (SS) for each stroke (see Figure 4.3). Additionally, the distance of each sensor to the ground strike point is determined by the LLS once the stroke has been geolocated.

The measured magnetic field signal strength, SS, is normalized to 100 km to determine a range normalized signal strength (RNSS). In order to estimate the peak current of the stroke, an empirical field-to-current conversion equation such as the following is used:

$$i_p = K * \text{Mean(RNSS)} \tag{4.7}$$

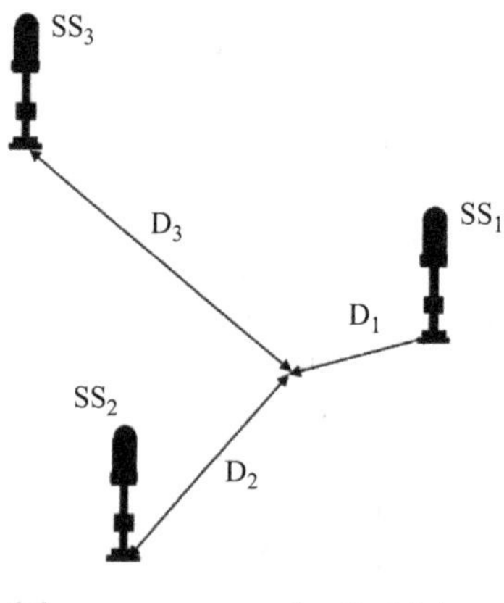

Figure 4.3 A network of three sensors each of which measures peak signal strength, SS, for a stroke occurring at a distance D. This information from each sensor is then used to calculate the range normalized signal strength and estimate the peak current for the stroke as described in Section 4.3.2

where i_p is the peak current in kA, K is an empirically determined constant, and Mean(RNSS) is the arithmetic mean of the RNSS. The LLS's central processor (which runs the geolocation and peak current algorithms) determines which sensors can participate in the peak current estimation process. Generally, contributions from sensors at distances up to several hundreds of kilometers are included. As in (4.6), (4.7) also implies that the electric or magnetic radiation field is proportional to the current.

Signal attenuation due to the propagation of lightning electromagnetic waves over lossy ground between the ground strike point and the sensors is taken into account by appropriately compensating for such propagation losses. For example, the following empirical formula has been used in the U.S. NLDN since 2004 to compensate for the field propagation effects (Cummins *et al.*, 2006 [50]):

$$\mathrm{RNSS} = \mathrm{SS}\left(\frac{r}{100}\right)\exp\left(\frac{r-100}{1{,}000}\right) \tag{4.8}$$

where r is in kilometers and SS is in the so-called LLP units. This equation assumes that the distance dependence of SS is $r^{-1}\exp(\frac{-r}{1{,}000})$. The r^{-1} term accounts for propagation over perfectly conducting ground and the $\exp(\frac{-r}{1{,}000})$ term represents additional attenuation due to the ground being lossy. The exponential function in (4.8) increases the RNSS for distances greater than 100 km to compensate for propagation effects. For example, for $r = 625$ km, this function is equal to 1.7. No compensation is provided for $r = 100$ km and for r ranging from 0 to 100 km the function varies from about 0.9 to 1.

4.3.3 Modern precision lightning locating systems

In this section, we describe several well-established continental-scale or country-wide ground-based precision LLSs that use different technologies and cover different parts of the world. Note that many other LLSs exist and we have picked LLSs that cover relatively large regions and whose characteristics are adequately described in the scientific literature. We define a precision LLS as one which provides ground-strike point location with an accuracy of better than 500 m and a flash DE of greater than 90%.

4.3.3.1 The U.S. National Lightning Detection Network (NLDN)

The NLDN has been providing real-time data since 1989 and consists of more than 100 stations separated by typically 300–350 km covering the contiguous United States (e.g., [29]). The network uses a combination of TOA and MDF techniques to geolocate lightning. The sensors in this network operate in the low-to-medium frequency range (−3 dB bandwidth of about 400 Hz to 400 kHz) and measure electric and magnetic field changes produced by lightning. Lightning events are classified into two categories: CG and IC, based upon their waveform characteristics. Additionally, prior to 2016, all positive events with estimated peak currents less than or equal to 15 kA were classified by the NLDN as IC. This peak-current-limit-based criteria for positive events are no longer applied. Peak currents for IC pulses and CG strokes are estimated from measured magnetic fields using an empirical field-to-current conversion formula based on rocket-triggered-lightning data, with the field peaks being adjusted to account for expected propagation effects (stronger than the inverse proportionality distance dependence). Further information on the evolution of the NLDN, its enabling methodology, and applications of NLDN data can be found in Cummins and Murphy [29]. The most recent network-wide sensor technology upgrade was completed in 2014 [51,52]. Technology upgrades over the years have been accompanied by performance validation studies (e.g., [53–59]) using video camera, rocket-triggered lightning, and network inter-comparison studies (e.g., Nag *et al.*, 2015 [61]). The NLDN reports CG flashes with a DE of greater than 95% and CG strokes with a DE of around 90% (e.g., [59]). For cloud flashes the DE is around 50% [57,58]. The classification accuracy is expected to be around 90% [59], even though this accuracy may vary depending upon thunderstorm type and region. The median LA of the network for CG strokes is around 250 m (e.g., [56]). Peak current estimation errors have been estimated from comparison of NLDN-reported peak currents with directly measured currents at the triggered-lightning channel base. Triggered lightning strokes are similar to negative subsequent strokes in natural lightning. The median absolute value of current estimation error is 13%–14% [55,56].

4.3.3.2 The Lightning Detection Network (LINET)

The LINET provides coverage over most of western and central Europe and uses the TOA technique to geolocate lightning. Arrival-angle information is available as well and employed as a "plausibility check" on computed locations. The sensors have a frequency bandwidth upper limit of about 200 kHz [30]. Lightning events are classified as IC or CG based upon source altitude information derived from the arrival time at the nearest reporting sensor. It is claimed that a reliable altitude estimate to separate return strokes and cloud pulses can be achieved as long as the closest sensor is within about 100 km of the lightning discharge [39]. This requires baselines (distance between adjacent sensors) of 200–250 km or less. Peak currents are estimated using a field to current conversion equation assuming direct proportionality between the peak current and peak magnetic field and inverse distance dependence of field peak. More information about LINET can be found in Betz *et al.* [60] and references therein. According to Betz *et al.* [60] a "high level of location accuracy is attained only when careful network fine-tuning is applied" and an average two-dimensional error of about 300 m can be achieved. Further, an example is shown in which 58 located strokes apparently terminated on an instrumented tower with an average location error of less than 100 m, after correcting for systematic errors that caused a location bias of about 200 m. Detection efficiencies for CG flashes/strokes and IC flashes for LINET are expected to be similar to other modern LLSs (see e.g., Nag *et al.* [61]). Peak current estimation errors for LINET are unknown (no comparison with ground-truth data has been made to date).

4.3.3.3 The Earth Networks Total Lightning Network (ENTLN)

The ENTLN covers most of North America and a number of other countries with sensors that measure electric field changes due to lightning and operate in a frequency range from 1 Hz to 12 MHz. The TOA technique is employed to geolocate IC and CG lightning. According to Heckman and Liu [62], whole electric field waveforms are used for both geolocation and classification of lightning events. The operators of the system claim greater than 50% IC flash DE across much of the United States and up to 95% in the US Midwest and East [62]. Peak currents for IC pulses and CG strokes are estimated assuming direct proportionality between the peak current and peak electric field and inverse distance dependence of field peak. Validation studies have been performed to examine the performance characteristics of the network (e.g., [63–67]). Mallick *et al.* [64], using as the ground-truth triggered-lightning data acquired at Camp Blanding, FL, in 2013, estimated the flash and stroke detection efficiencies to be 100% and 94%, respectively, for 12 flashes containing 62 negative return strokes. The median location error is 603 m, and the median absolute current estimation error was 20%. These performance characteristics were computed regardless of how the located events were classified by the system. About one-half (53%) of the triggered-lightning return strokes were misclassified as cloud discharges by the ENTLN. Rudlosky [66], who compared ENTLN detected IC and CG flashes with those detected by the lightning imaging sensor (LIS) on-board the TRMM orbital satellite over the 2011–13 period, reported that ENTLN's flash DE for IC and CG flashes combined relative to LIS varied regionally with the best performance occurring over southern CONUS where ENTLN detected 72% of LIS flashes. Recently, Zhu *et al.* [67] used 219 natural CG lightning flashes containing 608 strokes and reported the flash DE, stroke DE, and stroke classification accuracy to be 99%, 96%, and 91%, respectively. For 36 rocket-triggered lightning flashes containing 175 strokes they reported the median values of location error and absolute peak current estimation error to be 215 m and 15%, respectively.

4.3.3.4 The BrasilDAT Total Lightning Network

The BrasilDAT Total Lightning Network is an LLS that uses technology similar to the ENTLN discussed in Section 4.3.3.3. Its deployment started in December 2010 [68] and it is now composed of over 60 sensors covering the southern, central, and eastern parts of Brazil (see Figure 4.4). Like the ENTLN, BrasilDAT uses the TOA technique to geolocate IC and CG lightning. Some validation studies have been performed to examine the performance characteristics of the network. Naccarato *et al.* [69] used direct current and high-speed video camera measurements of a single flash containing seven strokes. BrasilDAT reported six out of seven strokes even though it misclassified one stroke as an IC pulse. The median absolute peak current estimation error was 13%. The location error for all strokes was greater than 500 m, with the largest location error being 1.2 km.

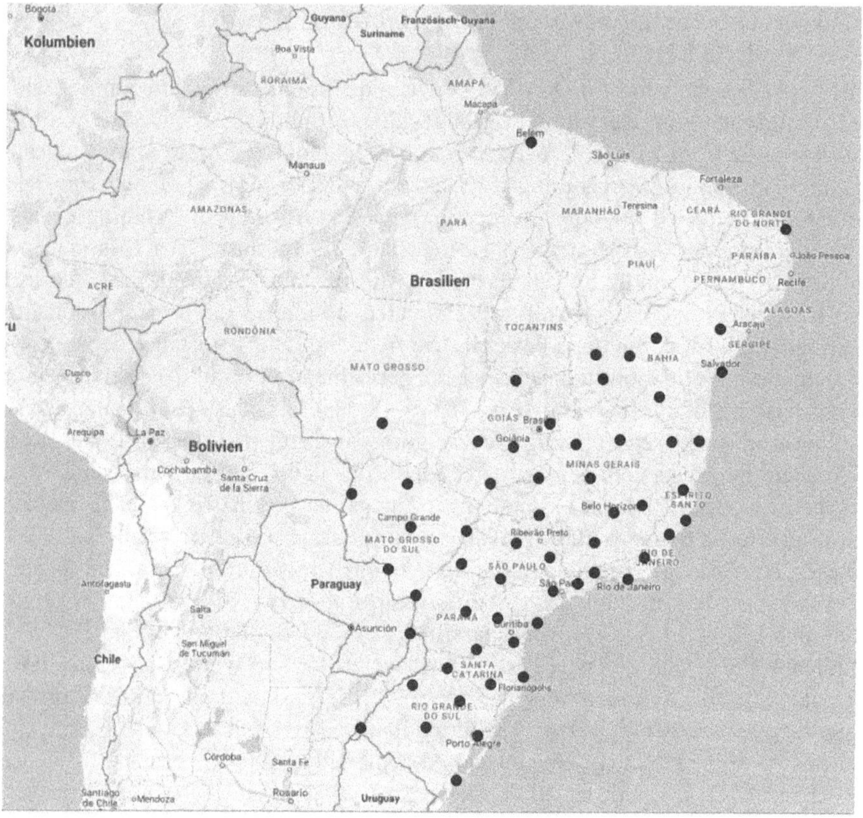

Figure 4.4 BrasilDAT network as of October 2017 (Naccarato, personal communication, October 2017). Sensor locations are shown as black dots

4.3.3.5 The Japanese Lightning Detection Network (JLDN)

The JLDN has been in operation since 1998 and has covered the main islands of Japan since 2000. This LLS uses a combination of the TOA and MDF techniques to geolocate lightning like other LLSs using Vaisala technology. Over the years, the network has undergone several technology upgrades and configuration changes. Currently, the operator of the network, Franklin Japan Corporation, is working on replacing the older-generation Vaisala sensors with newer sensors. As of January 2016, the JLDN consisted of six IMPACT-ESP, three LPATS-IV, eleven LS7001, and ten TLS200 sensors (see Figure 4.5) [70].

The LA of the JLDN has been discussed and validated several times. Matsui *et al.* [72] reported, for negative downward strokes hitting wind turbines in summer, an LA of approximately 0.44 km. Ishii *et al.* [73] reported, for upward lightning strokes developing from wind turbines in winter, an LA of 0.58 km. With the so-called propagation corrections, the LA improved from 0.44 km to 0.31 km for

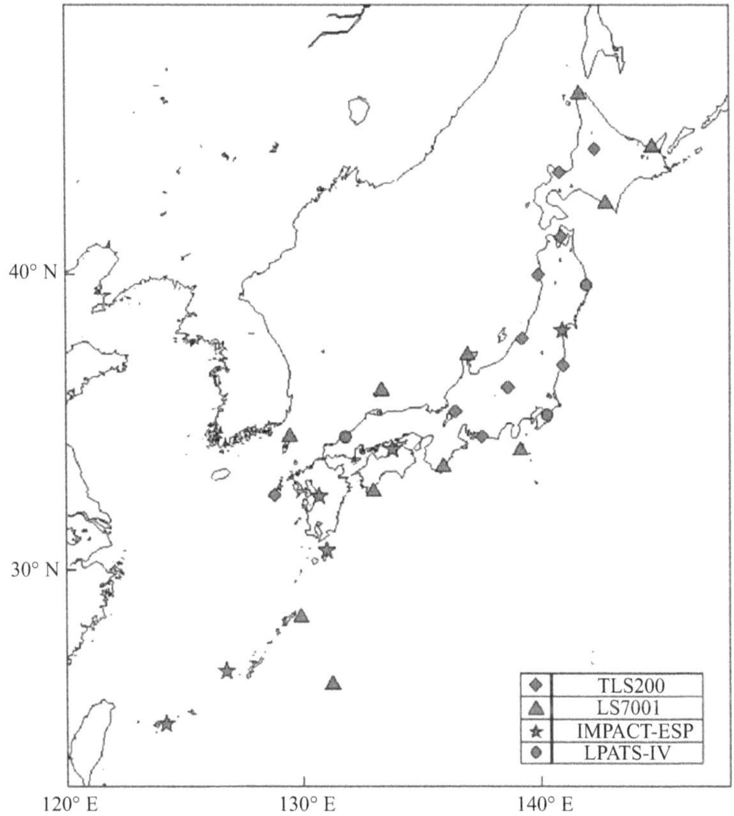

Figure 4.5 JLDN sensor types and locations as of January 2011 [71]

downward lightning strokes in summer, and it improved from 0.79 km to 0.12 km for lightning strokes in winter [68]. Further Matsui *et al.* [68] mentioned that the flash DE is greater than 90% in most areas of the four major islands of Japan. Peak current estimation errors for the JLDN are unknown (no comparison with ground-truth data has been made at the time of writing this chapter).

4.3.3.6 The European Cooperation for Lightning Detection (EUCLID)

In 2001, several countries (Austria, France, Germany, Italy, Norway, and Slovenia) started a cooperation named the European Cooperation for Lightning Detection or EUCLID. The goal of this cooperation has been to provide to end-users "European-wide" lightning data of high and nearly homogeneous quality. Since then Spain, Portugal, Finland, Sweden, and Belgium have also joined EUCLID. The EUCLID cooperation is special in the sense that it is the merger of independent national networks, and the individual partners are highly motivated to run their local networks with state-of-the-art lightning detection sensors. All the partners employ dedicated technicians to supervise and maintain the network and to react in short time in case of sensor or communication problems. As of January 2017 the EUCLID network employs 164 sensors: 34 IMPACTES/ESP, 52 LS7000/LS7001, and 78 LS7002 sensors, when listed in order from the oldest to the newest sensor version. All different sensor types, manufactured by Vaisala Inc., are operating in the same frequency range with individually calibrated sensor gains and sensitivities in order to account for any local sensor site conditions. Figure 4.6 shows the EUCLID network configuration as of 2017.

In addition to the processing carried out by each national LLS, data from all 164 sensors are processed in real-time using a central processor in Austria at

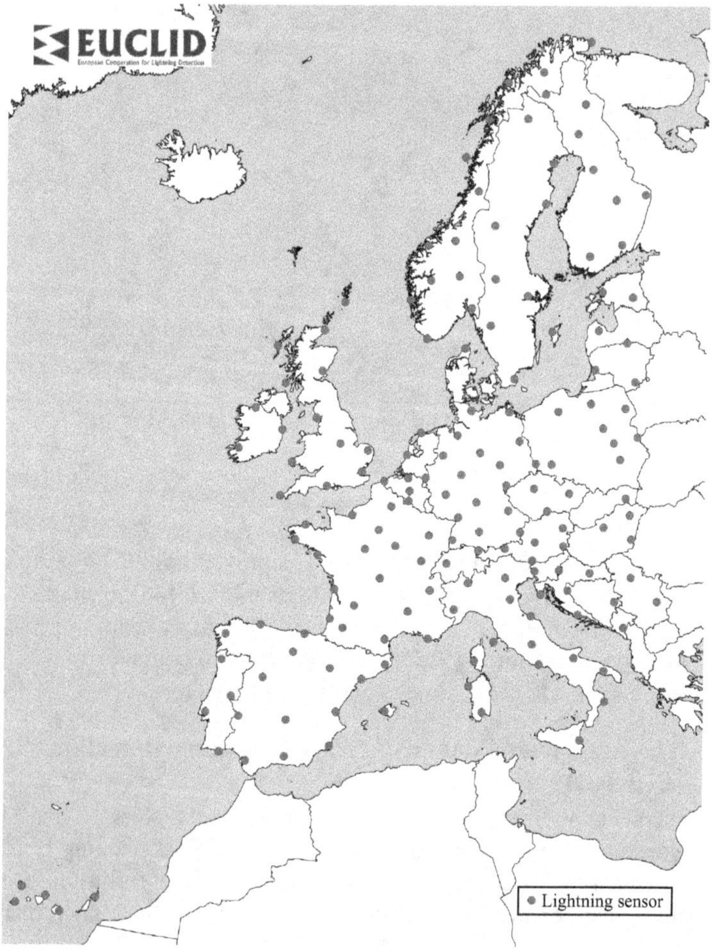

Figure 4.6 EUCLID network configuration (status 2018-01). Sensor locations are shown as gray dots

ALDIS (Austrian Lightning Detection and Information System), which also provides daily performance analyses reports for each of the sensors.

This assures that the resulting lightning data are as consistent as possible throughout Europe. In fact, the EUCLID data are frequently of higher quality than the data produced by individual national networks, being sub-networks of EUCLID. This is due to the implicit higher redundancy in EUCLID as a result of participation of additional sensors in the lightning location, which are located outside the national borders in a neighboring country. We note that there is a full backup EUCLID processing center in Germany with independent and direct data connections to all sensors. Since the beginning of the cooperation, the performance of the EUCLID network has been steadily improved. Improvements are the result of employing more advanced location algorithms, of installing state-of-the-art sensor technology, and relocating sensor positions in case of poorly performing sensor sites (e.g., local electromagnetic noise). Over the next 1–2 years, at least 10 of the remaining older type sensors are expected to be upgraded to the newest sensor type (LS7002). Detailed performance analyses of the EUCLID network are found in Schulz *et al.* [74].

4.3.3.7 Rede Integrada Nacional de Detecção de Descargas Atmosféricas (RINDAT)

The Integrated Lightning Detection Network or RINDAT was formed in 1999 by merging several regional LLSs in Brazil and covered roughly a quarter of the country including southern and south-eastern Brazil [75]. Over the years, the network grew to cover more regions in the south, north, and northeastern Brazil and by 2006 it covered approximately two-thirds of the country.

At the end of 2005, the network was upgraded with the addition of 27 more sensors (14 Impact and 13 LPATS sensors), installed in the south, north, and northeast regions. Data from the sensors are sent to six LP2000 central processors with same configuration through dedicated phone lines. The network uses a combination of older sensors (from Global Atmospherics) that use TOA only and newer ones (from Vaisala) that use a combination of TOA and MDF. Older generation sensors are being gradually replaced with newer sensors. Details of the history and configuration of RINDAT can be found in Pinto *et al.* [76], Pinot [77], and Pinto *et al.* [75]. Over the years, some validation studies have been conducted to test the performance characteristics of the RINDAT network (e.g., [78–80]). Using rocket-triggered lightning data as ground-truth, Soloranzo [78] reported a peak current estimation error of less than 20% and an LA of a few kilometers. Saba *et al.* [79] used high-speed video camera measurements of more than 400 flashes as ground-truth to report an average flash DE of 85% for RINDAT. More recently, Naccarato *et al.* [69] used direct current and high-speed video camera measurements of a single flash containing seven strokes. RINDAT reported two out of the seven strokes. The absolute peak current estimation error was about 12% for both strokes. The location error was less than 500 m for one stroke and about 600 m for the other.

4.3.4 Modern long-range lightning locating systems

Long-range LLSs mostly operate in the very low frequency or VLF (3–30 kHz) range. Lightning electromagnetic waveforms measured by sensors in long-range LLSs are affected by propagation over thousands of kilometers in the Earth-ionosphere waveguide. As a result, accounting for timing errors introduced by terrain, soil conductivity variations, and ionosphere conductivity and height variations is of fundamental importance in long-range LLSs. A recent review of long-range LLS technology is found in Said [19].

A brief summary of a few well-known long-range LLSs is provided below. Note that, the applicability of long-range LLS data for power system analyses is largely untested.

- **G**lobal **L**ightning **D**ataset (GLD360)
  GLD360 is a long-range LLS that is owned and operated by Vaisala. This network's sensors are distributed all over the world enabling GLD360 to provide lightning data on a global-scale. A combination of the TOA and MDF techniques are used by this LLS to geolocate lightning. In order to select the most appropriate time feature, a waveform correlation algorithm is used in each sensor [81]. A bank of "canonical" lightning waveforms corresponding to propagation over different distance ranges are used for this purpose. The GLD360 provides the time, latitude, longitude, peak current, and polarity of lightning. Since its inception in 2011, the performance of GLD360 has gradually improved with implementation of better location algorithms (e.g., [82]). The performance characteristics of this long-range LLS have been evaluated by comparing with "ground-truth" lightning data (e.g., [83,84]) using methods discussed in Section 4.3.5.
- **W**orld **W**ide **L**ightning **L**ocation **N**etwork (WWLLN)
  WWLLN, operated by the University of Washington in Seattle, consists of a network of sensors operating in the VLF range. WWLLN sensors are distributed all over the world and therefore it can geolocate lightning globally. The WWLLN uses a technique called the time-of-group-arrival (TOGA) to determine the best signal arrival time for each sensor. This method is based on the fact that lightning VLF signals (sferics), propagating in the Earth-ionosphere waveguide, experience dispersion, in that the higher-frequency components arrive earlier than the lower-frequency components [85]. The time at which the derivative of phase with respect to frequency reaches zero in any sferic measured by a WWLLN sensor is used as a dispersion-corrected time and yields better results than a simple trigger time. Once this corrected time is determined at multiple sensors, the essential TOA technique is then applied in the usual way to calculate the time, latitude, and longitude of the associated lightning event. The performance of this long-range LLS has been validated using ground-truth data (see e.g., [86–88]).
- **S**ferics **T**iming **a**nd **R**anging **Net**work (STARNET)
  STARNET is a long-range LLS operating in the VLF range and was first deployed in August 2006. At that time the network consisted of two sensors in Brazil (Fortaleza and São Paulo), one in Guadeloupe (an island group in the

southern Caribbean Sea), and one each in Nigeria, Ethiopia, Tanzania, and South Africa. From 2006 to 2013 the network was improved by addition of more sensors in South America especially, in Brazil. Currently, STARNET operates with 11 VLF antennas and covers South America, part of the Atlantic Ocean, Gulf of Mexico, and the Caribbean [89]. STARNET uses the TOA technique to geolocate lightning. The arrival time differences are computed by cross-correlating sferics waveforms in a 1,024 ms time-window. Besides the discharge location STARNET also provides the polarity of lightning [90]. Performance validation studies using ground-truth lightning data are not yet available in the scientific literature for this LLS.

- Arrival **T**ime **D**ifferences NETwork (ATDnet)
  The British Meteorological service has been operating a long-range LLS since 1987. The system has evolved during those years, and the most recent version called ATDnet (Arrival Time Differencing NETwork) was introduced in 2007 [91]. The focus of ATDnet is to detect lightning strokes over Europe. Therefore, the majority of the sensors (eleven "operational outstations" and five "development outstations") are located in Europe [92]. Recent results from performance validation studies of ATDnet are available in the literature [93].

4.3.5 *Validation of LLS performance characteristics using ground-truth-data*

The two most important performance characteristics of an LLS for any application of lightning data are:

- Detection efficiency (DE) for strokes/flashes: This is the percentage of all strokes/flashes that were reported by the LLS relative to all strokes/flashes what occurred in a thunderstorm.
- Location accuracy (LA) for strokes: This is expressed as an error given by the two-dimensional difference in position between a stroke's actual location and its LLS-computed location.

In addition to DE and LA, the following performance characteristics of LLSs are often relevant for power system analyses applications:

- Peak current estimation accuracy: This is usually represented by a percentage error. It is computed by taking the absolute value of the difference between a stroke's measured peak current and the LLS-estimated peak current, which is then expressed as a percentage of the stroke's actual peak current.
- IC versus CG classification accuracy: This is the percentage of IC pulses/flashes or CG strokes/flashes that were correctly classified as such by an LLS.

These performance parameters are influenced by the LLS technology, network geometry, sensor/central analyzer configuration, and regional terrain variations.

In order to evaluate these performance characteristics, the LLS data are compared with ground-truth lightning data that are often obtained from the following sources:

- Rocket triggered lightning and lightning strikes to tall objects
- Video camera measurements

In the following, we discuss the details of validating LLS performance characteristics using ground-truth data obtained from these sources.

4.3.5.1 Rocket-triggered lightning and lightning strikes to tall objects

This method uses data from rocket-triggered lightning experiments or lightning strikes to tall objects (e.g., towers instrumented with current measurements) as ground truth to evaluate the performance characteristics of an LLS within whose coverage area the triggered lightning facility or the tall object is located. Since the location of the CG events (including return strokes, M components, and superimposed pulses) and the peak currents are independently known, it is possible to measure the DE, LA, polarity, and peak current estimation accuracy, and lightning type classification accuracy of an LLS. Examples of studies using rocket-triggered lightning for LLS performance evaluation include Jerauld *et al.* [51], Nag *et al.* [55], Chen *et al.* [73], Mallick *et al.* [56,64,84], and Mallick *et al.* [65]. Examples of studies using lightning strikes to tall objects are found in Diendorfer *et al.* [94,95], Pavanello *et al.* [96], Romero *et al.* [97], Schulz *et al.* [98], Schulz *et al.* [99], and Cramer and Cummins [100]. Data obtained from rocket-triggered lightning and lightning strikes to tall objects can be used to validate the performance characteristics of all ground-based LLSs operating at different frequency ranges.

Rocket-triggered lightning and lightning strikes to tall objects provide the best ground-truth data for performance characteristics validation for CG lightning and are the only direct ways to validate peak current estimation accuracy of LLSs. However, these studies may be very expensive and may not be practical for all regions (as there are only a few triggered lightning facilities and instrumented towers across the world). Moreover, these are valid indicators of LLS performance only for the region where the rocket-triggered lightning facility or tall object is located, especially in cases where the performance of the LLS is expected to vary significantly from region to region. In order to evaluate the CG stroke LA of an LLS with a large coverage area, multiple noninstrumented tall objects in different regions of the area covered by the LLS may be used (e.g., [100]). Note that since only 10% of all downward CG lightning is positive, almost all lightning data obtained using rocket-triggered lightning and strikes to tall objects are for negative lightning. Also, upward-initiated lightning from tall towers are often followed by downward negative strokes (e.g., [101,102]). LLS performance characteristics inferred using these methods are, therefore, applicable to negative CG lightning only.

The primary distinction between natural CG lightning and classical rocket-triggered lightning is that the stepped-leader/first-return-stroke sequence in natural lightning is replaced by the initial stage (the upward positive leader, involving destruction of the triggering wire, and initial continuous current) in classical triggered lightning. As a result, rocket-triggered lightning provides data for return strokes similar only to subsequent strokes in natural lightning and no data for strokes similar to first strokes in natural lightning can be obtained. The phenomenology of the initial stage of classical triggered lightning is similar to that of upward lightning initiated by tall grounded objects (Chapter 6 of [12]). A log-normal relationship is expected between the percentage of upward lightning and effective height (which depends upon the height of the object, local terrain, and other factors) of tall objects [103]. Strikes to objects with heights less than 100 m or so are expected to be downward, while objects with heights above 500 m height are expected to experience upward flashes only. Thus, depending upon the height of the object being used for obtaining data for LLS validation, no data for first strokes may be available. Since first strokes in natural lightning are expected to have, on average, peak fields and currents that are a factor of two larger than those for subsequent strokes (e.g., [104]), CG flash and stroke DE obtained by rocket-triggered and tall object studies may be somewhat of an underestimate.

When lightning strikes a tall metallic object, the lower part of the lightning channel to ground is replaced by a single grounded vertical metallic conductor through which the return stroke current flows. As a result the characteristics of radiation field waveforms of lightning strokes attaching to tall objects can be quite different than those of radiation field waveforms for strokes attaching to ground (e.g., [105–107]). This factor needs to be taken into account while using data from lightning strikes to tall objects as ground-truth to examine the lightning type classification of LLSs. Additionally, since in this case the radiator for the lowest several hundred meters (depending upon the height of the tower) is a vertical metallic object instead of an often tortuous and branched natural lightning channel, the return stroke radiation field waveform measured by LLS sensors is expected to have unambiguous rising portions which may be much more suitable for measuring parameters used for lightning geolocation with very small error. As a result, the LA of an LLS estimated using lightning strikes to towers as ground-truth is expected to be somewhat better than that for natural lightning. Further, since the radiated field peak is influenced by the height of the tower (e.g., [105,106,108]), the peak current estimation accuracy of an LLS should be examined using data from lightning strikes to relatively short (less than 150 m or so) towers. Finally we have to have in mind that the field enhancement due to the tower also biases (increases) the resulting stroke DE from those experiments (especially for tall towers).

4.3.5.2 Video camera measurements

Measurements of cloud and CG lightning processes using video cameras can be used as ground-truth to evaluate the performance characteristics of an LLS. The LA, DE, and lightning type classification accuracy of an LLS can be estimated using this method. Examples of studies using this method include Idone *et al.* [109,110], Ballarotti *et al.* [111], Biagi *et al.* [54], Chen *et al.* [73], Schulz *et al.* [98], Poelman *et al.* [84], Schulz *et al.* [99], Schulz *et al.* [112], and Mata *et al.* [113]. Lightning data obtained using video cameras can be used to evaluate the performance characteristics of all ground-based LLSs operating at different frequency ranges and satellite-based LLSs. The advantage of this method is that relatively large sample sizes of ground-truth lightning data obtained at different regions within the coverage area of an LLS can be obtained, rather than at a specific point within network, as is the case for rocket triggered lightning and also typically for lightning strikes to instrumented towers. Note that lightning strikes to multiple (noninstrumented) towers located in different regions within the coverage area of an LLS have been used to evaluate the LA of an LLS. Polarity and peak current estimation accuracy of an LLS, however, cannot be evaluated using data obtained by video cameras alone. A practical problem of performing video camera measurements of lightning flashes is that since the locations of lightning discharges are not known in advance, it may not be easy to capture them in the field of view of the camera. Additionally, sometimes lightning discharges occurring within the field of view of the camera may be obscured by the presence of heavy precipitation or by clouds (in case of cloud lightning).

The absolute LA for CG events (M-components and return strokes) of an LLS operating at ELF to HF can be evaluated using video camera measurements only if two or more correlated video measurements are used to triangulate the location of the lightning channel to ground. The location uncertainty in the camera system employed by Idone *et al.* [109,110] was estimated to be as good as about 500 m.

However, modern LLSs have expected median LA on the order of a few hundred meters. Thus, it is currently impractical for camera-based triangulation to be used to validate absolute LA. Using a single video camera, absolute LA can only be evaluated if the lightning channel attaches to an identifiable ground-based object in the field of view of the camera (e.g., [72]). Otherwise, it is possible to determine the relative LA of CG events in the same channel as the previous stroke in a flash (e.g., [47]). However, the ground strike points of flashes and channel branching close to ground are often not observed due to obstruction of the video camera's view by closer objects, terrain, and vegetation. Hence, the relative LA determined using this technique may be poorer than the actual LA of an LLS. In order to precisely determine the stroke DE, in addition to the CG flash DE, a high-speed video camera with frame rate faster than the shortest interstroke interval (generally of the order of few to several milliseconds) in CG flashes is needed. The LA of cloud flashes generally cannot be determined by single or multiple correlated video camera measurements because cloud discharges may extend for several tens of kilometers with horizontal branches in several directions and it is likely that significant portions of the channels are either obscured by the presence of clouds or are outside the field of view of the camera. Cloud flash DE and lightning type classification accuracy of an LLS can be determined using this technique only when the complete flash is within the field of view. Video camera measurements are often accompanied by correlated electric or magnetic field measurements. In that case, the DE, LA, lightning type classification accuracy, and polarity estimation accuracy of an LLS can be evaluated, and some ambiguous video measurements may become useable. For example, in order to obtain a reliable CG stroke DE estimate, it is essential to identify M-components and strokes occurring at intervals less than the frame rate of the video camera, and correlated electric or magnetic field measurements can be used for that purpose. Note that polarity is most easily determined if the correlated broadband electric field (rather than magnetic field) is measured.

References

[1] K. Berger, "Methoden und Resultate der Blitzforschung auf dem Monte San Salvatore bei Lugano in den Jahren 1963-1971," *SEV*, vol. 63, no. A 971, 1972.

[2] K. Berger, R. B. Anderson, and H. Kroeninger, "Parameters of lightning flashes," *Electra*, vol. 41, pp. 23–37, 1975.

[3] CIGRE Report 63, "Guide to procedures for estimating the lightning performance of transmission lines," CIGRE Report 63, 1991.

[4] IEEE Std. 1243-1997: IEEE Guide for Improving the Lightning Performance of Transmission Lines, 1997.

[5] *IEC 62305-2: Protection against Lightning – Part 2: Risk Management*, vol. 2, 2013.

[6] R. Thottappillil, V. A. Rakov, M. A. Uman, W. H. Beasley, M. J. Master, and D. V Shelukhin, "Lightning subsequent-stroke electric field peak greater than the first stroke peak and multiple ground terminations," *J. Geophys. Res. Atmos.*, vol. 97, no. D7, pp. 7503–7509, 1992.

[7] V. A. Rakov, M. A. Uman, and R. Thottappillil, "Review of lightning properties from electric field and TV observations," *J. Geophys. Res. Atmos.*, vol. 99, no. D5, pp. 10745–10750, 1994.

[8] A. Hermant, *Traqueur d'orages*, 2000.

[9] W. C. Valine and E. P. Krider, "Statistics and characteristics of cloud-to-ground lightning with multiple ground contacts," *J. Geophys. Res. Atmos.*, vol. 107, no. D20, p. 4441, 2002.

[10] M. M. F. Saba, M. G. Ballarotti, and O. J. Pinto, "Negative cloud-to-ground lightning properties from high-speed video observations," *J. Geophys. Res. Atmos.*, vol. 111, no. D3, p. D03101, 2006.

[11] R. B. Anderson, A. J. Eriksson, H. Kröninger, D. V. Meal, and M. A. Smith, "Lightning and thunderstorm parameters," *IEE*, vol. 236, 1984.

[12] V. A. Rakov and M. A. Uman, *Lightning – Physics and Effects*. Cambridge University Press, 2003.

[13] H. J. Christian, R. J. Blakeslee, D. J. Boccippio, *et al.*, "Global frequency and distribution of lightning as observed from space by the optical transient detector," *J. Geophys. Res. Atmos.*, vol. 108, no. D1, p. 4005, 2003.

[14] D. M. Mach, H. J. Christian, R. J. Blakeslee, D. J. Boccippio, S. J. Goodman, and W. L. Boeck, "Performance assessment of the optical transient detector and lightning imaging sensor," *J. Geophys. Res. Atmos.*, vol. 112, no. D9, p. D09210, 2007.

[15] S. Beirle, W. J. Koshak, R. J. Blakeslee, and T. Wagner, "Global patterns of lightning properties derived by OTD and LIS," *Nat. Hazards Earth Syst. Sci.*, vol. 14, no. 10, pp. 2715–2726, 2014.

[16] W. A. Chisholm, "Estimates of lightning ground flash density using optical transient density," in *IEEE PES Transmission and Distribution Conference and Exposition (IEEE Cat. No.03CH37495)*, 2003, vol. 3, pp. 1068–1071.

[17] D. J. Cecil, "LIS/OTD 0.5 degree high resolution full climatology (HRFC), Dataset available online from the NASA Global Hydrology Center DAAC, Huntsville, Alabama, U.S.A.," 2001. [Online]. Available: http://dx.doi.org/10.5067/LIS/LIS-OTD/DATA302.

[18] R. K. Said, M. B. Cohen, and U. S. Inan, "Highly intense lightning over the oceans: Estimated peak currents from global GLD360 observations," *J. Geophys. Res. Atmos.*, vol. 118, no. 13, pp. 6905–6915, 2013.

[19] R. Said, "Towards a global lightning locating system," *Weather*, vol. 72, no. 2, pp. 36–40, 2017.

[20] IEC 62858, "Lightning density based on lightning location systems (LLS) – General principles," 2015.

[21] CIGRE Report 376, "Cloud-to-ground lightning parameters derived from lightning location systems – the effects of system performance," CIGRE Report 376, 2009.

[22] K. L. Cummins, M. J. Murphy, E. A. Bardo, W. L. Hiscox, R. B. Pyle, and A. E. Pifer, "A combined TOA/MDF technology upgrade of the U.S. National Lightning Detection Network," *J. Geophys. Res.*, vol. 103, no. D8, p. 9035, 1998.

[23] G. Diendorfer, "Some comments on the achievable accuracy of local ground flash density values," in *Lightning Protection (ICLP), 2008 International Conference on*, 2008, no. June, pp. 1–6.

[24] S. Pedeboy, "Identification of the multiple ground contacts flashes with lightning location systems," in *22nd International Lightning Detection Conference and 4th International Lightning Meteorology Conference (ILDC/ILMC)*, 2012.

[25] S. Pédeboy and W. Schulz, "Validation of a ground strike point identification algorithm based on ground truth data," in *International Lightning Detection Conference and International Lightning Meteorology Conference (ILDC/ILMC)*, 2014.

[26] K. L. Cummins, "Analysis of multiple ground contacts in cloud-to-ground flashes using LLS data: the impact of complex terrain," in *International Lightning Detection Conference and International Lightning Meteorology Conference (ILDC/ILMC)*, 2012.

[27] L. Z. S. Campos, K. L. Cummins, and O. J. Pinto, "An algorithm for identifying ground strike points from return stroke data provided by lightning location systems," in *9th Asia-Pacific International Conference on Lightning (APL)*, 2015, pp. 475–478.

[28] L. Z. S. Campos, "On the mechanisms that lead to multiple ground contacts in lightning," PhD thesis, INPE, Brazil, 2016.

[29] K. L. Cummins and M. J. Murphy, "An overview of lightning locating systems: history, techniques, and data uses, with an in-depth look at the U.S. NLDN," *Electromagn. Compat. IEEE Trans.*, vol. 51, no. 3, pp. 499–518, 2009.

[30] V. A. Rakov, "Electromagnetic methods of lightning detection," *Surv. Geophys.*, no. July, 2013.

[31] H. J. Christian, R. J. Blakeslee, and S. J. Goodman, "The detection of lightning from geostationary orbit," *J. Geophys. Res. Atmos.*, vol. 94, no. D11, pp. 13329–13337, 1989.

[32] A. R. Jacobson, X. Shao, and R. H. Holzworth, "Satellite triangulation of thunderstorms, from fading radio fields synchronously recorded on two orthogonal antennas," *Radio Sci.*, vol. RS6012, no. 46, 2011.

[33] T. Morimoto, H. Kikuchi, M. Sato, M. Suzuki, A. Yamazaki, and T. Ushio, "VHF lightning observations on JEM-GLIMS mission: gradual approach to realize space-borne VHF broadband digital interferometer," *IEEJ Trans. Fundam. Mater.*, vol. 131, no. 12, pp. 977–982, 2011.

[34] G. Lugrin, N. M. Parra, F. Rachidi, M. Rubinstein, and G. Diendorfer, "On the location of lightning discharges using time reversal of electromagnetic fields," *Electromagn. Compat. IEEE Trans.*, vol. 56, no. 1, pp. 149–158, 2014.

[35] V. Cooray, "Effects of propagation on the return stroke radiation fields," *Radio Sci.*, vol. 22, no. 5, pp. 757–768, 1987.

[36] K. L. Cummins, M. J. Murphy, J. A. Cramer, W. D. Scheftic, N. W. S. Demetriades, and A. Nag, "Location accuracy improvements using propagation corrections: a case study of the U.S. National Lightning Detection Network," in *Lightning Protection (ICLP), 2010 International Conference on*, 2010.

[37] N. Honma, M. J. Murphy, K. L. Cummins, A. E. Pifer, and T. Rogers, "Improved lightning locations in the Tohoku Region of Japan using propagation and waveform onset corrections," *IEEJ Trans. Power Energy*, vol. 133, no. 2, pp. 195–202, 2012.

[38] F. Lyu, S. A. Cummer, R. Solanki, *et al.*, "A low-frequency near-field interferometric-TOA 3-D lightning mapping array," *Geophys. Res. Lett.*, vol. 41, no. 22, p. 2014GL061963, 2014.

[39] E. P. Krider, C. R. Noggle, and M. A. Uman, "A gated, wideband magnetic direction finder for lightning return strokes," *J. Appl. Meteorol.*, vol. 15, no. 3, pp. 301–306, 1976.

[40] J.-Y. Lojou, M. J. Murphy, R. L. Holle, and N. W. S. Demetriades, "Nowcasting of thunderstorms using VHF measurements," in H. D. Betz, U. Schumann, P. Laroche (eds.), *Lightning: Principles, Instruments, and Application, Review of Modern Lightning Research*, pp. 253–270, 2009 (Chapter 11).

[41] X.-M. Shao and P. R. Krehbiel, "The spatial and temporal development of intracloud lightning," *J. Geophys. Res. Atmos.*, vol. 101, no. D21, pp. 26641–26668, 1996.

[42] R. Mardiana and Z.-I. Kawasaki, "Broadband radio interferometer utilizing a sequential triggering technique for locating fast-moving electromagnetic sources emitted from lightning," *Instrum. Meas. IEEE Trans.*, vol. 49, no. 2, pp. 376–381, 2000.

[43] Z. Sun, X. Qie, M. Liu, D. Cao, and D. Wang, "Lightning VHF radiation location system based on short-baseline TDOA technique – validation in rocket-triggered lightning," *Atmos. Res.*, vol. 129–130, no. 0, pp. 58–66, 2013.

[44] M. G. Stock, M. Akita, P. R. Krehbiel, *et al.*, "Continuous broadband digital interferometry of lightning using a generalized cross-correlation algorithm," *J. Geophys. Res. Atmos.*, vol. 119, no. 6, pp. 3134–3165, 2014.

[45] D. J. Boccippio, W. J. Koshak, R. J. Blakeslee, *et al.*, "The optical transient detector (OTD): instrument characteristics and cross-sensor validation," *J. Atmos. Ocean. Technol.*, pp. 441–458, 2000.

[46] D. M. Suszcynsky, T. E. L. Light, S. Davis, J. L. Green, J. L. L. Guillen, and W. Myre, "Coordinated observations of optical lightning from space using the FORTE photodiode detector and CCD imager," *J. Geophys. Res. Atmos.*, vol. 106, no. D16, pp. 17897–17906, 2001.

[47] D. J. Boccippio, W. J. Koshak, and R. J. Blakeslee, "Performance assessment of the optical transient detector and lightning imaging sensor. Part I: predicted diurnal variability," *J. Atmos. Ocean. Technol.*, 2002.

[48] M. J. Murphy, A. Nag, J.-Y. Lojou, and R. K. Said, "Preliminary analysis of the Vaisala TLS200 network deployed during the CHUVA campaign," in *6th Conference on the Meteorological Applications of Lightning Data*, American Meteorological Society, Austin, 2013.

[49] M. A. Uman and D. K. McLain, "Magnetic field of lightning return stroke," *J. Geophys. Res.*, vol. 74, no. 28, pp. 6899–6910, 1969.

[50] K. L. Cummins, J. A. Cramer, C. J. Biagi, E. P. Krider, J. Jerauld, M. A. Uman, and V. A. Rakov, "The U.S. National Lightning Detection Network: Post-Upgrade Status," *Am. Meteorol. Soc.*, 2006.

[51] A. Nag, M. J. Murphy, K. L. Cummins, A. E. Pifer, and J. A. Cramer, "Recent evolution of the U.S. National Lightning Detection Network," in *23rd International Lightning Detection Conference and 5th International Lightning Meteorology Conference (ILDC/ILMC)*, 2014.

[52] A. Nag, M. J. Murphy, and J. A. Cramer, "Update to the U.S. National Lightning Detection Network," in *24th International Lightning Detection Conference and 6th International Lightning Meteorology Conference (ILDC/ILMC)*, 2016.

[53] J. E. Jerauld, V. A. Rakov, M. A. Uman, *et al.*, "An evaluation of the performance characteristics of the U.S. National Lightning Detection Network in Florida using rocket-triggered lightning," *J. Geophys. Res. Atmos.*, vol. 110, no. D19, p. D19106, 2005.

[54] C. J. Biagi, K. L. Cummins, K. E. Kehoe, and E. P. Krider, "National lightning detection network (NLDN) performance in southern Arizona, Texas, and Oklahoma in 2003–2004," *J. Geophys. Res. Atmos.*, vol. 112, no. D5, p. D05208, 2007.

[55] A. Nag, S. Mallick, V. A. Rakov, *et al.*, "Evaluation of U.S. National Lightning Detection Network performance characteristics using rocket-triggered lightning data acquired in 2004–2009," *J. Geophys. Res. Atmos.*, vol. 116, no. D2, p. D02123, 2011.

[56] S. Mallick, V. A. Rakov, J. D. Hill, *et al.*, "Performance characteristics of the NLDN for return strokes and pulses superimposed on steady currents, based on rocket-triggered lightning data acquired in Florida in 2004–2012," *J. Geophys. Res. Atmos.*, vol. 119, no. 7, p. 2013JD021401, 2014.

[57] M. J. Murphy and A. Nag, "Cloud lightning performance and climatology of the U.S. based on the upgraded U.S. National Lightning Detection Network," in *7th Conference on the Meteorological Applications of Lightning Data*, 2015, pp. 1–11.

[58] D. Zhang, K. L. Cummins, and A. Nag, "Assessment of cloud lightning detection by the US National Lightning Detection Network using video and lightning mapping array observations," in *95th American Meteorological Society Annual Meeting*, 2015, vol. 2013, pp. 1–6.

[59] Y. Zhu, V. A. Rakov, M. D. Tran, and A. Nag, "A study of national lightning detection network responses to natural lightning based on ground-truth data acquired at LOG with emphasis on cloud discharge activity," *J. Geophys. Res. Atmos.*, p. 2016JD025574, 2016.

[60] H.-D. Betz, K. Schmidt, P. Laroche, *et al.*, LINET – an international lightning detection network in Europe, *Atmos. Res.*, vol. 91, no. 2–4, pp. 564–573, 2009.

[61] A. Nag, M. J. Murphy, W. Schulz, and K. L. Cummins, "Lightning locating systems: insights on characteristics and validation techniques," *Earth Sp. Sci.*, no. 2, pp. 65–93, Feb. 2015.

[62] S. J. Heckman and C. Liu, "The application of total lightning detection and cell tracking for severe weather prediction," in *International Conference on Grounding and Earthing and 4th International Conference on Lightning Physics and Effects (GROUND/LPE)*, 2010, pp. 1–10.

[63] S. Mallick, V. A. Rakov, J. D. Hill, *et al.*, "Calibration of the ENTLN against rocket-triggered lightning data," in *2013 International Symposium on Lightning Protection (XII SIPDA)*, 2013, pp. 67–74.

[64] S. Mallick, V. A. Rakov, T. Ngin, *et al.*, "An update on testing the performance characteristics of the ENTLN," in *XV International Conference on Atmospheric Electricity, International Commission on Atmospheric Electricity*, 2014, no. June, pp. 15–20.

[65] S. Mallick, V. A. Rakov, J. D. Hill, *et al.*, "Performance characteristics of the ENTLN evaluated using rocket-triggered lightning data," *Electr. Power Syst. Res.*, vol. 118, no. 0, pp. 15–28, 2015.

[66] S. D. Rudlosky, "Evaluating ENTLN performance relative to TRMM/LIS," *J. Oper. Meteorol.*, vol. 3, no. 2, pp. 11–20, 2015.

[67] Y. Zhu, V. A. Rakov, M. D. Tran, *et al.*, "Evaluation of ENTLN performance characteristics based on the ground-truth natural and rocket-triggered lightning data acquired in Florida," *J. Geophys. Res. Atmos.*, p. 2017JD027270, 2017.

[68] K. P. Naccarato, A. C. V. Saraiva, M. M. F. Saba, and C. Schumann, "First performance analysis of Brasildat total lightning network in Southeastern Brazil," in *International Conference on Grounding and Earthing and 5th International Conference on Lightning Physics and Effects (GROUND/LPE)*, 2012.

[69] K. P. Naccarato, A. R. Paiva, M. M. F. Saba, C. Schumann, J. C. O. Silva, and M. A. da S. Ferro, "Preliminary comparison of direct electric current measurements in lightning rods and peak current estimates from lightning location systems," in *2017 International Symposium on Lightning Protection (XIV SIPDA)*, 2017, no. October, pp. 3–7.

[70] M. Matsui, K. Michishita, S. Kurihara, and N. Honjo, "Discussion on location accuracy improvemed by propagation delay corrections for the Japanese lightning detection network," in *2016 International Conference on Lightning Protection (ICLP)*, 2016.

[71] A. Sugita and M. Matsui, "Lightning characteristics in Japan observed by the JLDN from 2001 to 2010," in *International Lightning Detection Conference (ILDC)*, 2012.

[72] M. Matsui, K. Michishita, and S. Kurihara, "Accuracies of location and estimated peak current for negative downward return-strokes to wind turbine observed by JLDN (in Japanese)," *IEEJ Trans. Power Energy*, vol. 135, no. 10, pp. 644–645, 2015.

[73] M. Ishii, F. Fujii, M. Saito, D. Natsuno, and A. Sugita, "Detection of lightning return strokes hitting wind turbines in winter by JLDN (in Japanese)," *IEEJ Trans. Power Energy*, vol. 133, no. 12, pp. 1009–1010, 2013.

[74] W. Schulz, G. Diendorfer, S. Pedeboy, and D. R. Poelman, "The European lightning location system EUCLID – Part 1: performance analysis and validation," *Nat. Hazards Earth Syst. Sci.*, vol. 16, no. 2, pp. 595–605, 2016.

[75] O. J. Pinto, K. P. Naccarato, M. M. F. Saba, and I. R. C. A. Pinto, "Recent upgrades to the Brazilian Integrated Lightning Detection Network," in *19th International Lightning Detection Conference and 1st International Lightning Meteorology Conference (ILDC/ILMC)*, 2006.

[76] O. J. Pinto, "The Brazilian lightning detection network: a historical background and future perspectives," in *2003 International Symposium on Lightning Protection (VII SIPDA)*, 2003.

[77] O. J. Pinto, *The Art of War Against Lightning*, 2005 (in Portuguese).

[78] N. N. Solorzano, "Triggered lightning study in Brazil," PhD thesis, INPE, 2003 (in Portuguese).

[79] M. M. F. Saba, O. J. Pinto, M. G. Ballarotti, K. P. Naccarato, and G. F. Cabral, "Monitoring the performance of the lightning detection network by means of a high-speed camera," in *18th International Lightning Detection Conference (ILDC)*, 2004, no. 63, pp. 1–7.

[80] M. G. Ballarotti, "A cloud-to-ground lightning study using a high speed camera," Master thesis, INPE, 2005 (in Portuguese).

[81] R. K. Said, U. S. Inan, and K. L. Cummins, "Long-range lightning geolocation using a VLF radio atmospheric waveform bank," *J. Geophys. Res. Atmos.*, vol. 115, no. D23, p. D23108, 2010.

[82] R. Said and M. Murphy, "GLD360 upgrade: performance analysis and applications," in *24th International Lightning Detection Conference and 6th International Lightning Meteorology Conference (ILDC/ILMC)*, 2016, no. Ic.

[83] S. Mallick, V. A. Rakov, T. Ngin, *et al.*, "Evaluation of the GLD360 performance characteristics using rocket-and-wire triggered lightning data," *Geophys. Res. Lett.*, vol. 41, no. 10, p. 2014GL059920, 2014.

[84] D. R. Poelman, W. Schulz, and C. Vergeiner, "Performance characteristics of distinct lightning detection networks covering Belgium," *J. Atmos. Ocean. Technol.*, vol. 30, no. 5, pp. 942–951, 2013.

[85] R. L. Dowden, J. B. Brundell, and C. J. Rodger, "VLF lightning location by time of group arrival (TOGA) at multiple sites," *J. Atmos. Solar-Terrestrial Phys.*, vol. 64, no. 7, pp. 817–830, 2002.

[86] S. Mallick, V. A. Rakov, T. Ngin, *et al.*, "Evaluation of the WWLLN performance characteristics using rocket-triggered lightning data," in *International Conference on Grounding and Earthing and 6th International Conference on Lightning Physics and Effects (GROUND/LPE)*, 2014, pp. 312–316.

[87] C. J. Rodger, J. B. Brundell, R. L. Dowden, and N. R. Thomson, "Location accuracy of long distance VLF lightning location network," *Ann. Geophys.*, pp. 747–758, 2004.

[88] E. H. Lay, R. H. Holzworth, C. J. Rodger, J. N. Thomas, O. J. Pinto, and R. L. Dowden, "WWLL global lightning detection system: regional validation study in Brazil," *Geophys. Res. Lett.*, vol. 31, no. 3, p. L03102, 2004.

[89] C. A. Morales, J. R. Neves, E. Anselmo, V. Rogerio, K. S. Camara, and R. Hole, "8 years of Sferics timing and ranging NETwork-STARNET: a lightning climatology over South America," in *23rd International Lightning Detection Conference and 5th International Lightning Meteorology Conference (ILDC/ILMC)*, 2014.

[90] C. A. Morales, E. N. Anagnostou, E. R. Williams, and J. S. Kriz, "Evaluation of peak current polarity retrieved by the ZEUS long-range lightning monitoring system," *IEEE Geosci. Remote Sens. Lett.*, vol. 4, no. 1, pp. 32–36, 2007.

[91] C. Gaffard, J. Nash, N. Atkinson, *et al.*, "Observing lightning around the globe from the surface," in *International Lightning Detection Conference and International Lightning Meteorology Conference (ILDC/ILMC)*, 2008.

[92] G. Anderson and D. Klugmann, "A European lightning density analysis using 5 years of ATDnet data," *Nat. Hazards Earth Syst. Sci.*, vol. 14, no. 4, pp. 6877–6922, 2014.

[93] S.-E. Enno, G. Anderson, and J. Sugier, "ATDnet detection efficiency and cloud lightning detection characteristics from comparison with the HyLMA during HyMeX SOP1," *J. Atmos. Ocean. Technol.*, vol. 33, no. 9, pp. 1899–1911, 2016.

[94] G. Diendorfer, M. Mair, W. Schulz, and W. Hadrian, "Lightning current measurements in Austria – experimental setup and first results," in *Lightning Protection (ICLP), 2000 International Conference on*, 2000, no. September.

[95] G. Diendorfer, W. Schulz, and M. Mair, "Evaluation of a LLS based on lightning strikes to an instrumented tower," in *International Lightning Detection Conference (ILDC)*, 2000.

[96] D. Pavanello, F. Rachidi, W. Janischewskyj, *et al.*, "On the current peak estimates provided by lightning detection networks for lightning return strokes to tall towers," *Electromagn. Compat. IEEE Trans.*, vol. 51, no. 3, pp. 453–458, 2009.

[97] C. Romero, M. Paolone, F. Rachidi, *et al.*, "Preliminary comparison of data from the Säntis Tower and the EUCLID lightning location system," in *Lightning Protection (XI SIPDA), 2011 International Symposium on*, 2011, pp. 180–185.

[98] W. Schulz, C. Vergeiner, H. Pichler, G. Diendorfer, and S. Pack, "Validation of the Austrian lightning location system ALDIS for negative flashes," in *CIGRE C4 Colloquium on Power Quality and Lightning*, 2012, pp. 13–16.

[99] W. Schulz, H. Pichler, G. Diendorfer, C. Vergeiner, and S. Pack, "Validation of detection of positive flashes by the Austrian lightning location system ALDIS," in *2013 International Symposium on Lightning Protection (XII SIPDA)*, 2013, pp. 75–79.

[100] J. A. Cramer and K. L. Cummins, "Evaluating location accuracy of lightning location networks using tall towers," in *23rd International Lightning Detection Conference and 5th International Lightning Meteorology Conference (ILDC/ILMC)*, 2014.

[101] G. Diendorfer, "LLS performance validation using lightning to towers," in *International Lightning Detection Conference and International Lightning Meteorology Conference (ILDC/ILMC)*, 2010, vol. 230, no. 1, pp. 1–15.

[102] T. A. Warner, K. L. Cummins, and R. E. Orville, "Upward lightning observations from towers in Rapid City, South Dakota and comparison with National Lightning Detection Network data, 2004–2010," *J. Geophys. Res. Atmos.*, vol. 117, no. D19, p. D19109, 2012.

[103] A. J. Eriksson and D. V Meal, "The incidence of direct lightning strikes to structures and overhead lines," in *Lightning and Power Systems*, London: IEE Conference, 1984.

[104] A. Nag, V. A. Rakov, W. Schulz, *et al.*, "First versus subsequent return-stroke current and field peaks in negative cloud-to-ground lightning discharges," *J. Geophys. Res. Atmos.*, vol. 113, no. D19, p. D19112, 2008.

[105] D. Pavanello, F. Rachidi, M. Rubinstein, *et al.*, "On return stroke currents and remote electromagnetic fields associated with lightning strikes to tall structures: 1. Computational models," *J. Geophys. Res. Atmos.*, vol. 112, no. D13, p. D13101, 2007.

[106] D. Pavanello, F. Rachidi, W. Janischewskyj, *et al.*, "On return stroke currents and remote electromagnetic fields associated with lightning strikes to tall structures: 2. Experiment and model validation," *J. Geophys. Res. Atmos.*, vol. 112, no. D13, p. D13122, 2007.

[107] H. Pichler, G. Diendorfer, and M. Mair, "Some parameters of correlated current and radiated field pulses from lightning to the Gaisberg tower," *IEEJ Trans. Electr. Electron. Eng.*, vol. 5, no. 1, pp. 8–13, 2010.

[108] J.-L. Bermúdez, F. Rachidi, W. Janischewskyj, *et al.*, "Determination of lightning currents from far electromagnetic fields: Effect of a strike object," *J. Electrostat.*, vol. 65, no. 5–6, pp. 289–295, 2007.

[109] V. P. Idone, D. A. Davis, P. K. Moore, *et al.*, "Performance evaluation of the U.S. National Lightning Detection Network in eastern New York: 1. Detection efficiency," *J. Geophys. Res. Atmos.*, vol. 103, no. D8, pp. 9045–9055, 1998.

[110] V. P. Idone, D. A. Davis, P. K. Moore, *et al.*, "Performance evaluation of the U.S. National Lightning Detection Network in eastern New York: 2.

Location accuracy," *J. Geophys. Res. Atmos.*, vol. 103, no. D8, pp. 9057–9069, 1998.

[111] M. G. Ballarotti, M. M. F. Saba, and O. J. Pinto, "A new performance evaluation of the Brazilian Lightning Location System (RINDAT) based on high-speed camera observations of natural negative ground flashes," in *International Lightning Detection Conference and International Lightning Meteorology Conference (ILDC/ILMC)*, 2006, vol. 122, pp. 1–4.

[112] W. Schulz, S. Pedeboy, C. Vergeiner, E. Defer, and W. Rison, "Validation of the EUCLID LLS during HyMeX SOP1," in *International Lightning Detection Conference and International Lightning Meteorology Conference (ILDC/ILMC)*, 2014.

[113] C. T. Mata, A. G. Mata, A. Nag, V. A. Rakov, and J. Saul, "Evaluation of the performance characteristics of CGLSS II and U.S. NLDN using ground-truth data from Launch Complex 39B, Kennedy Space Center, Florida," in *International Lightning Detection Conference and International Lightning Meteorology Conference (ILDC/ILMC)*, 2014.

Chapter 5

Lightning attachment to overhead power lines

Pantelis N. Mikropoulos[1], *Jinliang He*[2] *and Marina Bernardi*[3]

Lightning is the main cause of unscheduled interruptions in overhead power lines, affecting reliability of power supply and thus, consequently, resulting in economic losses. Lightning-caused insulation flashover in overhead power lines is associated with the fast-front overvoltages across line insulation, arising due to direct lightning strokes or induced by nearby lightning.

Shielding against direct lightning strokes to phase conductors of overhead power lines is provided by shield wires. The latter are metallic elements that are able to, by physical means, launch a connecting upward discharge that intercepts the descending lightning leader from a distance, called striking distance, commonly also called attractive radius or lateral distance. Lightning leaders intercepted by shield wires, increasing the potential of the transmission-line tower, may result in power-line outages due to backflashover, that is, insulation flashover between tower and phase conductors. However, some of the less intense lightning strokes, not being intercepted by shield wires terminating thus to the phase conductors, may cause power-line outages due to shielding failure. In addition, descending lightning leaders which are not intercepted by the line conductors, striking to ground nearby the power line or to adjacent structures may result in power-line outages due to induced voltages on line conductors causing insulation flashover. As the line operation voltage increases, a higher line insulation level is utilized, and the lightning performance of overhead power lines becomes increasingly determined by the direct stroke flashover rate.

In this chapter, the physical process of lightning attachment to overhead power lines is presented and discussed. Engineering models of lightning attachment are described in detail. Finally, a general procedure for the estimation of lightning incidence to overhead power lines is presented.

[1]Aristotle University of Thessaloniki, School of Electrical and Computer Engineering, Thessaloniki, Greece
[2]Tsinghua University, Department of Electrical Engineering, Beijing, China
[3]Centro Elettrotecnico Sperimentale Italiano (CESI S.p.A.), Engineering and Environment Division, Milan, Italy

5.1 Lightning attachment

The lightning performance of overhead power lines is generally assessed by considering solely negative downward flashes. This is justified by the facts that the vast majority of downward lightning flashes are of negative polarity in that they lower a negative charge to earth and that very tall objects (higher than 100 m or so) or objects of moderate height located on mountain tops experience primarily upward lightning flashes.

A negative downward lightning flash (cloud-to-ground lightning) normally traverses some kilometers before completing its long path to earth surface. The negative leader initiates from a cloud charge region and progresses toward the earth surface in a noncontinuous way, through a series of steps each one elongating the discharge tortuous channel (stepped-leader), the latter accompanied by extensive downward branching. As the descending stepped leader approaches the earth surface, the electric field at ground especially in the proximity of ground objects increases. When the electric field at the upper extremities of ground objects attains a critical value, upward positive discharges emanate from them developing toward the descending lightning leader. Depending primarily on the characteristics of the negative stepped leader as well as on the local geometry of the ground object, the upward positive discharge intercepts the descending negative leader from a distance, when the average potential gradient along their interception path reaches a critical value, around 500 kV/m, sufficient to sustain positive streamer stable propagation. The moment of establishment of the interception phase, which, as generally accepted, is qualitatively similar to that of the final jump in long electric laboratory sparks, the process of rapid leader charge neutralization begins, resulting in an upward-propagating return stroke. For the case of overhead power lines, lightning attachment to line conductors occurs through a connecting positive upward leader that, emerging from a conductor due to the local electric field enhancement caused by the approaching stepped leader, intercepts the negative descending lightning leader from a distance. The radial distance from the downward leader tip to the struck conductor and the lateral distance between them the moment the upward connecting leader initiates from the conductor are called striking distance and interception radius, respectively.

From this brief description of lightning attachment to overhead line conductors it should be evident that the evaluation of the lightning performance of an overhead power line, under either design or operation, requires a lightning attachment model that considers as accurately as possible the physical processes involved, appropriately aimed for engineering applications. The following Section 5.2 describes several engineering models representing to some extent the final stage of the complex phenomenon of lightning attachment to overhead line conductors. The models are categorized into two main categories:

1. *Electrogeometric models*, which employ the striking distance as the basic lightning interception parameter. These models recognize that the final phase of lightning attachment is dominated by the average potential gradient in the

gap between the tip of the descending leader and the ground object and have been widely used for shielding analysis of overhead power lines since 1960s.
2. *Leader propagation models*, which employ as basic lightning interception parameter the attractive radius or lateral distance; the latter is defined as the maximum interception radius at which an upward positive leader from the conductor intercepts the descending negative lightning leader. These models recognize that the inception and subsequent propagation of a positive leader from the structure dominate lightning attachment. It must be mentioned however, that, besides inception and progression of an upward positive leader from the structure, the interception phase is a necessary condition that must be met for a lightning flash to occur, contributing, in its turn, to the stochastic nature of lightning attachment.

Both electrogeometric and leader propagation models can be analytically applied for evaluating the lightning performance of overhead transmission lines. Alternative approaches for this task are computations employing leader propagation models or fractal models of lightning attachment.

5.2 Lightning attachment models

The word "model" generally stands for a conceptual representation of the physical processes of interest, aiming at reproducing the natural behavior under scrutiny. Often the model is implemented in a software code, or simulation model (also "model" for short), that will give simulation results to be compared with experimental and field data. Such comparisons give information on how much the model is a good representation of the reality and in which aspects. An accurate evaluation of the possible point of impact of lightning on a power line could be done with simulation models that reproduce the lightning behavior in the presence of power lines or more generally of tall objects; ideally this is achieved considering all the physics involved, namely the streamer development from conductors, towers, and shielding wires, the streamer transition to upward leader, and the possible final jump with connection of the upward leader to the downward stepped leader.

5.2.1 Electrogeometric models

In an effort to assess the protection afforded by lightning conductors and to conduct calculations of lightning incidence to power lines, R.H. Golde [1] developed, as early as 1945, the electrogeometric theory for lightning strokes, in the sense that he provided, under some assumptions, a numerical approach for estimating the electric gradient underneath the downward lightning leader as a function of the charge on the leader channel. In his later studies [2,3], by considering also that this charge is related to the return-stroke current, a relationship between striking distance and return-stroke current was shown graphically. Subsequently, several electrogeometric models were developed, mainly in 1960s, primarily aiming to provide an engineering tool for the evaluation of the effectiveness of shielding of overhead transmission lines against direct lightning strokes to phase conductors. Although the motivation behind these models arose from the need to remedy shielding design deficiencies, the latter as detailed in [4,5], electrogeometric models have contributed significantly to our understanding of the mechanism of lightning attachment to line conductors.

In their early development, by considering that in many respects the final stage of attachment of a descending lightning leader to a ground structure is qualitatively similar to that of the final jump phase in long laboratory spark discharges, electrogeometric models were based on two basic assumptions:

(i) The lightning stepped leader descents vertically, approaching the earth surface unperturbed by ground objects until it reaches a point of discrimination where the distance from the most likely ground object to be struck equals the striking distance. The motion of the descending stepped leader is unaffected by the ground object until within the lightning collection width defined by the interception radius (Figure 5.1) of that object. This assumption, in other words

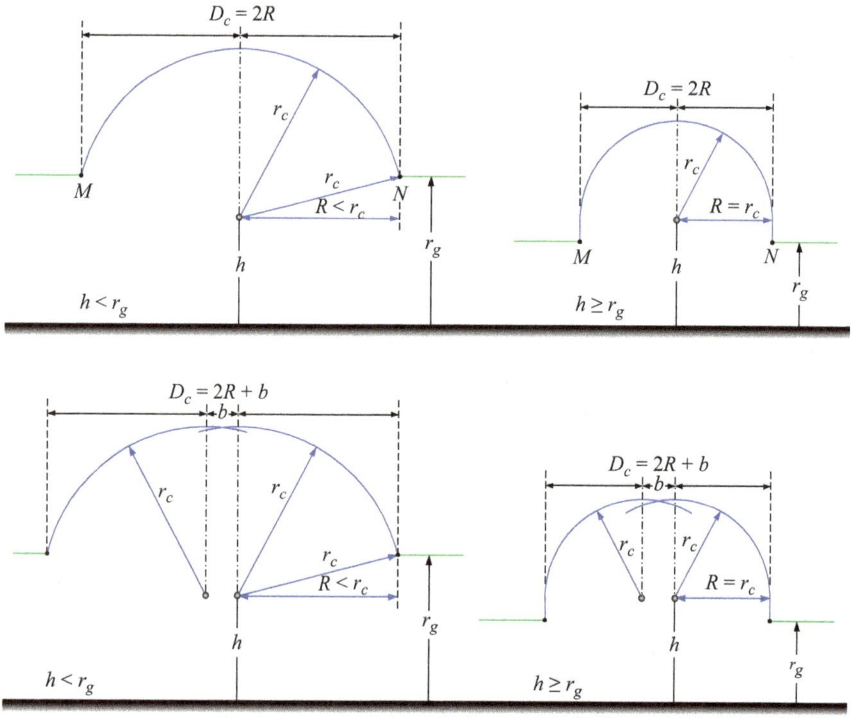

Figure 5.1 Lightning incidence to line conductors: r_c is the striking distance to conductor; r_g is the striking distance to earth surface; R is the interception radius; h is the height of the conductor; and D_c is the collection width (strokes to conductor)

by neglecting the physical processes preceding the final phase of lightning attachment, despite its simplicity, allows for a straightforward estimation of the collection width of a ground object.

(ii) The striking distance to a ground object is solely a function of the amplitude of the prospective lightning stroke current, and, in most models, independent of the local geometry of the ground structure. In making this assumption it was recognized that the collection width of a ground object depends on lightning stroke current amplitude, therefore it differs from that defined by geometrical approaches, which are based solely on the height of the ground object.

According to the electrogeometric models, the descending lightning leader is intercepted by an upward discharge from the ground object when the final jump condition is fulfilled, that is, when the average potential gradient in the gap between the tip of the descending leader and the ground object (striking distance), attains a critical value. This condition corresponds to the final phase of sparkover (final jump) of long laboratory spark discharges, which starts when the downward leader corona streamers meet the upward connecting discharge from an earthed electrode. Thus, striking distance can be considered as a function of the potential of the tip of the descending lightning leader. Since this potential depends on the charge of the leader, it follows that striking distance is a function of the leader charge; the greater the charge on the leader channel the longer the striking distance. By considering that the return stroke neutralizes the charge deposited along the leader channel, it follows that the amplitude of the return-stroke current is also a function of the leader charge; a greater leader charge corresponds to a higher return-stroke current. Thus, the striking distance is an increasing function of the amplitude of the prospective lightning stroke current.

In electrogeometric models, the relationship between striking distance and prospective lightning stroke current is represented in equation form by a general power function, which in the case of overhead power lines is given as

$$r_c = AI^B = \gamma r_g \tag{5.1}$$

where I in kA is the prospective lightning stroke current and r_c, r_g are, respectively, the striking distance to line conductors and earth surface in meters. Factors A, B as well as the proportionality factor γ have been shown to vary among electrogeometric models. Generally, the striking distances to line conductors and earth surface differ in length, because of the different electric field distributions along the corresponding interception paths (from the tip of the downward leader to conductor or earth surface) resulting in different breakdown gradients. This is accounted for in (5.1) by the factor γ, which, having the notion of a "gap factor" to increase the conductor striking distance slightly relative to striking distance to earth, takes values equal or slightly higher than unity, increasing with conductor height. It must be noted, however, that factor γ should also vary depending on prospective lightning stroke current and lateral displacement of the downward leader from the line conductor.

To illustrate the final stage of lightning attachment to overhead line conductors, let us first consider the simple case of a single line conductor positioned at a height h above ground as shown in Figure 5.1. A lightning leader that descends vertically within the collection width $D_c = 2R$, as defined by the arc between M and N, where R is the interception radius of the conductor, is more likely to terminate to the conductor; otherwise the stroke will terminate to earth surface. The line parallel to earth surface at a height r_g and the arc of radius r_c, constructed with center at the conductor and continued until the horizontal line is reached, are drawn according to (5.1) for a prospective lightning stroke current.

From Figure 5.1, the interception radius, R, of the conductor is given as

$$R = \sqrt{r_c^2 - (r_g - h)^2} = \sqrt{\left(1 - \frac{1}{\gamma^2}\right) r_c^2 + \frac{2}{\gamma} r_c h - h^2}, \text{ for } h < r_g \tag{5.2a}$$

$$R = r_c, \quad \text{for } h \geq r_g. \tag{5.2b}$$

In an analogous way, in the case of two line conductors separated by a distance, b (Figure 5.1) approximating, for example, the case of an overhead transmission line utilizing two shield wires, the collection width of the set of conductors becomes $D_c = 2R + b$.

Figure 5.2 illustrates the final stage of lightning attachment to overhead transmission line conductors by considering the simplified case of a single shield wire and one phase conductor and by using as lightning attachment parameter the striking distance. The shield wire, by intercepting the downward lightning leader,

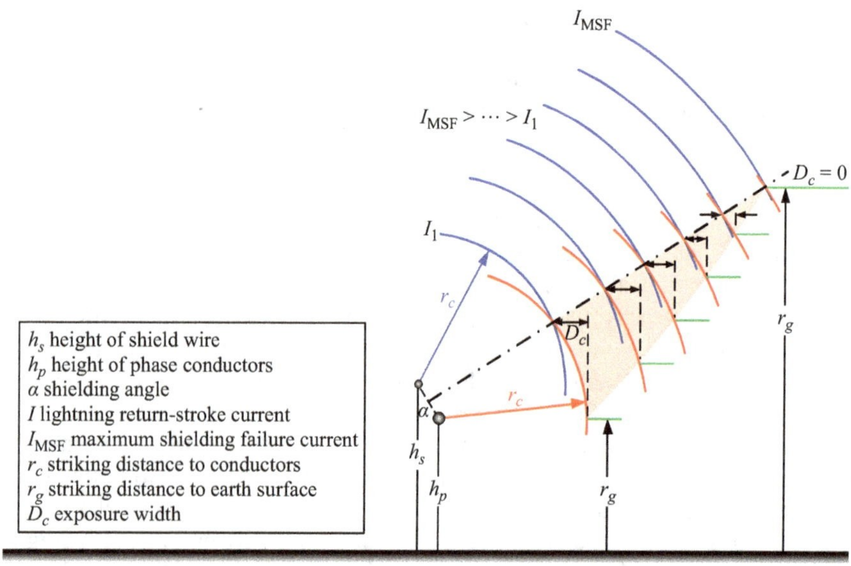

Figure 5.2 Lightning attachment to overhead line conductors according to electrogeometric models

provides shielding against direct lightning strokes to the phase conductor. However, some of the less intense strokes may not be intercepted and, thus, terminate on the phase conductor (shielding failure). A lightning leader that descends vertically within the exposure width, D_c, is more likely to terminate on the phase conductor. Both striking distance and interception radius are increasing functions of the prospective lightning stroke current, I. Thus, as I increases D_c decreases (Figure 5.2), becoming zero for an upper limit of stroke currents, denoted as I_{MSF}, that is, the maximum peak current of all possible lightning strokes that may terminate on the phase conductor.

5.2.1.1 C.F. Wagner and A.R. Hileman model

C.F. Wagner [6] in his return-stroke model derived a relation between the velocity of the return stroke and the stroke current, as graphically depicted in Figure 5.3, using as basis the energy required to establish a spark; this energy was determined from laboratory tests on rod–rod gaps. Under a number of simplifying assumptions made to develop the model, he showed that the potential of the downward leader, V_s, is a function of the velocity of the return stroke, v, in per unit of that of light, according to the following expression:

$$V_s = 120\frac{v}{1 - 2.2v^2}\ (\mathrm{MV, p.u.}) \tag{5.3}$$

Thus, to estimate the striking distance for a prospective lightning stroke current, given the latter the velocity of the return stroke is estimated from Figure 5.3. Then, with the aid of expression (5.3) the potential of the descending lightning

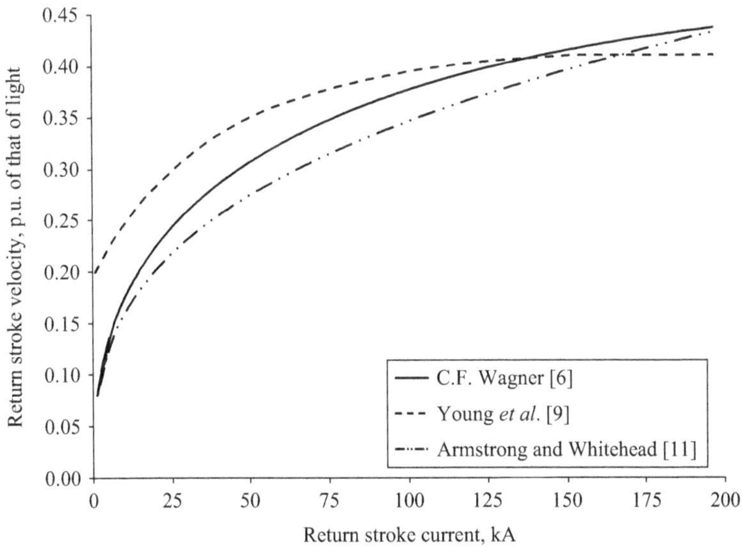

Figure 5.3 Velocity of return stroke as a function of return-stroke current

leader is calculated. The striking distance is equal to this potential divided by the "critical breakdown gradient," that is, by the average potential gradient in the gap between the tip of the downward leader and the ground object. By assuming that the final stage of lightning attachment is governed by the lightning impulse breakdown characteristics of long air gaps, the value of 600 kV/m was suggested as appropriate for the "critical breakdown gradient" [7]. The figure of 600 kV/m was considered as the upper limit of breakdown gradient, resulting in conservative shielding analysis results; the shorter the striking distance, the smaller the shielding angle required to provide adequate shielding against direct strokes to line phase conductors, as can also be deduced form Figure 5.2.

A.R. Hilleman [8] placed this shielding analysis method in the equation form of (5.1) by introducing the following relationship between striking distance and prospective stroke current:

$$r_s = 14.2I^{0.42}\ (\text{m}, \text{kA}) \tag{5.4}$$

applicable to lightning strokes to both line conductors and earth surface.

5.2.1.2 P.S. Young, J.M. Clayton, and A.R. Hileman model

Young *et al.* [9] modified the Wagner and Hileman model [8] by calibrating it on the basis of field data on lightning performance of overhead transmission lines. They proposed a new return stroke velocity-current relationship, as shown in Figure 5.3, and a variable critical breakdown field between 550 and 600 kV/m, decreasing for shield wire heights greater than 18 m. Such a variation appeared reasonable when considering that the final stage of attachment of a vertically descending lightning leader to towers of relatively higher heights would be similar to that of negative breakdown of relatively long rod–rod gaps, characterized by a breakdown gradient lower that of rod–plane gaps.

This pioneering and comprehensive study on lightning performance of overhead transmission lines clearly stressed the need for utilizing in power lines a decreasing shielding angle with increasing tower height. Young *et al.* [9] model employs the following expressions for the striking distances to line conductors, r_c, and earth, r_g, as introduced by A.R. Hilleman [8]:

$$r_g = 27I^{0.32}\ (\text{m}, \text{kA}) \tag{5.5a}$$

$$r_c = \gamma r_g \tag{5.5b}$$

$$\gamma = \begin{cases} 1 & \text{for } h \leq 18\ \text{m} \\ \dfrac{444}{462 - h} & \text{for } h > 18\ \text{m} \end{cases} \tag{5.5c}$$

where I is the prospective lightning stroke current and h is the shield wire height.

5.2.1.3 Armstrong and Whitehead model

The Armstrong and Whitehead electrogeometric model was developed based on the field experience gained by the Pathfinder research program [5], which was designed to investigate the mechanism of lightning strokes to overhead transmission lines. The experience with the first extra-high-voltage (EHV) lines, which were designed with shielding angles of about 30° in accordance with an AIEE Working Group report [10], revealed a poor shielding performance for these lines, especially when compared to that of lower voltage lines, substantially dependent upon the height of the lines.

To address these problems, Armstrong and Whitehead [11] introduced a new relationship between striking distance and prospective lightning stroke current. For the derivation of this relationship, first, the return-stroke current was associated with the potential of the downward leader channel and, then, the latter was associated with the striking distance. They adopted the Wagner's [8] return-stroke model after some simplifications made to adapt it to their model, as follows:

$$V_s = 60(I/v_l)\ln\left(2r_g/d\right) \tag{5.6}$$

where V_s is the leader voltage, I is the prospective lightning stroke current to earth, v_l is the velocity of the return-stroke channel in per unit velocity of light, r_g is the striking distance to earth, and d is the expected corona radius of the downward leader at heights well above r_g.

A fixed value of 4.6 was tentatively selected for the logarithmic factor in (5.6), by considering that it varies very slowly as both r_g and d increase with V_s. Then, by adopting, as a useful compromise, the following expression between return-stroke velocity and stroke current, as depicted in Figure 5.3:

$$I = 2400v_l^3 \text{ (kA, p.u.)} \tag{5.7}$$

they derived an approximate relationship between the potential of the downward leader channel and the prospective lightning stroke current as

$$V_s = 3.7I^{2/3} \text{ (MV, kA)} \tag{5.8}$$

By extrapolating data on negative switching impulse breakdown of the rod–rod electrode configuration with spacings in the range between 1.5 and 5 m, based on the comprehensive studies by Y. Watanabe [12] and L. Paris [13], the striking distance was associated with the leader voltage to ground by the following empirical formula:

$$r_c = 1.4V_s^{1.2} \text{ (m, MV)} \tag{5.9}$$

which employs a nonlinear dependence of the critical breakdown gradient on striking distance.

Combining (5.8) and (5.9) the relationship between striking distance and prospective lightning stroke current was obtained as

$$r_c = 6.72I^{0.8} \text{ (m, kA)} \tag{5.10}$$

Expression (5.10) refers to lightning strokes to shield wires; for the striking distance to earth, r_g, a proportionality factor of 0.9 [$1/\gamma$ in (5.1)] was proposed.

It must be mentioned that Armstrong and Whitehead were the first to introduce the general power function representing in equation form (5.1) the relationship between striking distance and prospective lightning stroke. They were also the first to introduce in shielding analysis of overhead transmission lines a stroke angle distribution, although recognizing that more field data were needed to accurately assess the leader angle effects on the shielding performance of overhead lines.

5.2.1.4 Whitehead and colleagues electrogeometric models

Whitehead and colleagues [14–20] have contributed further to the electrogeometric method of shielding of overhead transmission lines in terms of both assessing field experience and conducting analytical studies. The latter led to refinements in the power function relating striking distance with prospective lightning stroke current, with suggested values for factors A, B, and γ in (5.1) as listed in Table 5.1. These studies have provided valuable discussions on lightning attachment to overhead line conductors and presented detailed guidelines for designing effectively shielded overhead transmission lines by considering the geometrical and electrical parameters of the line, distribution of the lightning stroke current, terrain topography, and the statistical nature of striking distance.

5.2.1.5 Concluding remarks on conventional electrogeometric models

Figure 5.4 shows the variation of striking distance to earth surface with prospective return-stroke current as predicted by the electrogeometric models listed in Table 5.1. Although there is significant variability in striking distance estimates

Table 5.1 Factors A, B, and γ to be used in striking distance (5.1)

Electrogeometric model	$r_c = AI^B = \gamma r_g$		
	A	B	γ
Wagner and Hileman [7]	14.2	0.42	1
Young *et al.* [9]	27γ	0.32	1 for $h < 18$ m $\frac{444}{462-h}$ for $h > 18$ m h: shield wire height
Armstrong and Whitehead [11]	6.72	0.80	1.11
Brown and Whitehead [14]	7.1	0.75	1.11
E.R. Love [15]	10	0.65	1
E.R. Whitehead [17]	9.4	0.67	1
J.G. Anderson [24]	8	0.65	$1/\beta^*$
IEEE Std. 1410 [22]	10	0.65	1.11
IEEE Std. 1243 [23]	10	0.65	$1/\beta^{**}$

Note: r_c and r_g are the striking distances to line conductors and earth, respectively.
$^*\beta = 0.64$ for UHV lines, 0.8 for EHV lines, and 1 for other lines.
$^{**}\beta = 0.36 + 0.17\ln(43 - h)$ for $h < 40$ m, $\beta = 0.55$ for $h > 40$ m where h is the phase conductor height.

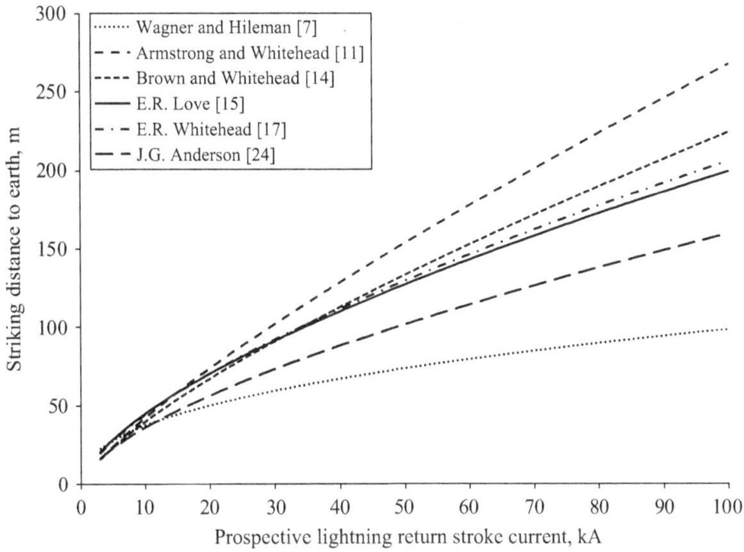

Figure 5.4 Striking distance to earth surface as a function of the prospective return-stroke current

among models, especially as stroke current increases, it must be recognized that this has not prevented the design and operation of overhead power lines with acceptable lightning performance over the years.

Also, E.R. Love's expression [15] ($r_c = 10I^{0.65}$) for striking distance is widely applied to the design of lightning protection systems of common structures according to the IEC Std. 62305 [21]. This expression is used for relating the "rolling sphere radius" to certain values of return-stroke current that correspond to acceptable protection levels. The same expression has been adopted by the IEEE Standards 1410 [22] and 1243 [23], providing guidelines for an improved performance of overhead power distribution and transmission lines, respectively.

The studies on conventional electrogeometric models presented earlier, including those by J.G. Anderson [24] and Mousa and Srivastava [25,26], have added valuable knowledge to the understanding of the mechanism of lightning attachment to a ground structure while primarily aiming to improve the design of effectively shielded overhead transmission lines. However, in that direction, as early recognized by Armstrong and Whitehead [11], an electrogeometric model should be considered as a tentative step, which is continuously subject to modifications as new field data become available. Such is the case of ultra-high-voltage (UHV) overhead power lines, where the model of Armstrong and Whitehead [11] had to be appropriately revised to mitigate the discrepancies between its predictions and the observed field data.

5.2.1.6 Revised electrogeometric models for large-scale EHV and UHV power lines

Based on field experience on the lightning performance of 500 kV and UHV designed large-scale overhead transmission lines [27,28], Taniguchi *et al.* [29] modified the Armstrong and Whitehead [11] electrogeometric model by introducing an application of the relationship between striking distance and return-stroke current that considers the distribution of the return-stroke velocity.

Taniguchi *et al.* [29] adopted the Wagner's [8] return-stroke model and the simplifications made by Armstrong and Whitehead [11] for the leader voltage, as

$$V_s = 0.276(I/v_l) \text{ (MV, kA)} \tag{5.11}$$

where v_l is the velocity of the return-stroke channel in per unit velocity of light. Then, based on extrapolation of data on negative switching impulse breakdown of rod–rod gaps with spacings in the range between 1 and 6 m [30], the striking distance was related to the leader voltage to ground as

$$r_c = 1.34{V_s}^{1.43} \text{ (m, MV)} \tag{5.12}$$

By combining (5.11) and (5.12), the striking distance is expressed in equation form as

$$r_c = (0.338/v_l)^{1.43} I^{1.43} \text{ (m, kA)} \tag{5.13}$$

Then, by using the distribution of measured return-stroke velocities reported by Idone and Orville [31], striking distance is computed from (5.13) as a weighted average with respect to probability of occurrence of the return-stroke velocity. Equation (5.13) refers to striking distance to line conductors (shield wires and phase conductors). For the striking distance to earth a proportionality factor of 1.1 [$1/\gamma$ in (5.1)] in (5.3) was suggested, accounting for a longer striking distance to earth than to line conductors; this value higher than unity was justified by the higher breakdown voltage of rod–rod than rod–plane air gaps as experimentally obtained for spacings longer than about 4 m [30].

Generally, the presented revised model for large-scale power lines yields striking distances longer than those estimated by the conventional electrogeometric models. By using also, a probability density function of the lightning stroke current waveform that considers the correlation between the wavefront duration and amplitude, obtained based on observation, a satisfactory agreement between model predictions and field data on lightning incidence and shielding failure of 500 kV and UHV designed large-scale overhead transmission lines was obtained.

The latest detailed analysis on engineering models of lightning attachment to line conductors of EHV and UHV DC and AC overhead transmission lines is presented in the Technical Brochure No. 704 based on the report prepared by CIGRE WG C4.26 [32].

5.2.1.7 A.J. Eriksson's modified electrogeometric model

A.J. Eriksson [33,34], in an effort to analytically describe lightning attractiveness of ground structures based on more physical grounds, modified the electrogeometric model. He introduced as lightning interception parameter the attractive radius of the structure, defined as the "capture" radius at which an upward positive leader from the structure intercepts the downward negative leader. Thus, depending upon the relative positions of the two leaders and their relative velocities of approach, interception occurs when the downward leader approaches to within the attractive radius of the structure.

In Eriksson's model the negative lightning stepped leader approaching a ground structure is represented by a vertically descending leader possessing along its channel a uniform charge distribution; the latter is related to the prospective stroke current though an empirical power function relationship, derived based on K. Berger's measurements [35]. By considering the electric field enhancement experienced by the structure as the descending leader channel approaches the ground, Eriksson employed as criterion for upward connecting leader inception from the ground structure the Carrara and Thione [36] critical radius concept, applying to the estimation of the minimum switching impulse sparkover voltage of large air gaps. The downward leader approaches the structure with a propagation velocity equal to that of the upward leader, being unaffected by the presence of the latter. Interception occurs when the tips of the leader channels meet each other, within the volume above the structure as defined by the attractive radius of the structure.

Based on model results and on lightning flash incidence data collected during some 10,000 structure years, Eriksson developed the following approximate expression for the attractive radius of overhead line conductors:

$$R_a = 0.67h^{0.6}I^{0.74} \ (\text{m, kA}) \tag{5.14}$$

It is noted that according to (5.14), the attractive radius, R_a, depends not only on prospective lightning stroke current, I, but also on conductor height, h.

A schematic representation of Eriksson's model as applied to line conductors is shown in Figure 5.5. For the case of a single conductor above ground, a horizontal line at the height of conductor is first drawn. Then, (5.14) is used to construct for a prospective stroke current an arc with a center at the conductor and continued until the horizontal line is reached. A downward leader that enters within the volume between points M and N under the arc determined by R_a will terminate on the conductor; otherwise it results in a stroke to earth surface. Thus, the collection width of the single conductor becomes $D_c = 2R_a$. By comparing Figure 5.5 and Figure 5.1, it can be seen that in Eriksson's model there exists no striking distance to earth; in principle, the latter represents a default condition once the downward leader has passed beyond the attractive range of the conductor.

Eriksson suggested a shielding analysis for power lines similar to that of electrogeometric models, however, using instead of striking distance the attractive radius of the conductors. For the simplified case of a single shield wire and one phase conductor (Figure 5.5), using (5.14), for a prospective return-stroke current arcs of attractive radii R_s and R_p are constructed around the shield wire and conductor, respectively; the latter arc is continued until the horizontal line drawn at the

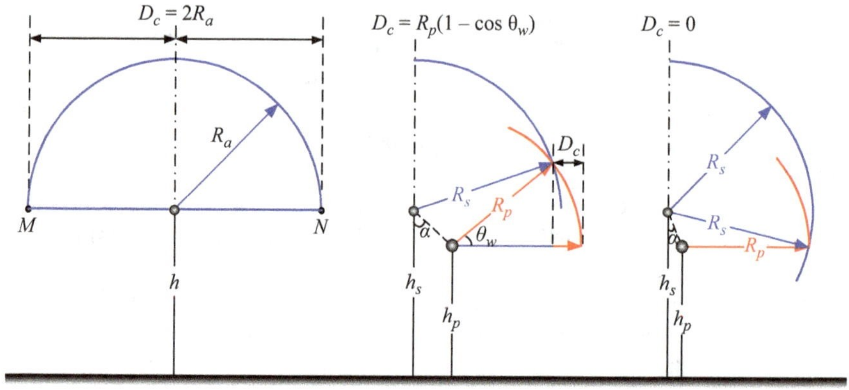

Figure 5.5 Lightning attachment to line conductors according to A.J. Eriksson modified electrogeometric model: R_a is the attractive radius of line conductor; α is the shielding angle; h_s and h_p are the height of shield wire and phase conductor, respectively; R_s and R_p are the attractive radius of shield wire and phase conductor, respectively; and D_c is the exposure width

height of phase conductor is reached. The exposure of the phase conductor to direct lightning strokes (shielding failure) is determined by the arc $\widehat{D}_c = R_p\theta_w$; a lightning leader that descends vertically within the exposure width $D_c = R_p(1 - \cos \theta_w)$ will terminate on the phase conductor. As the lightning stroke current increases, the exposure arc decreases and becomes zero at the maximum shielding failure current, I_{MSF}, that is, the maximum peak current of all possible lightning strokes that may terminate on the phase conductor. For a given lightning stroke current D_c can also be reduced to zero by decreasing the shielding angle utilized in the overhead power line (Figure 5.5).

Eriksson's modified electrogeometric model is generally accepted as the first attempt to simulate lightning attachment to ground structures by considering that the final jump phase depends on the interaction between the downward lightning leader and the upward connecting leader from the structure. Generally, as was also shown by A.R. Hilleman [8], Eriksson's model results on critical shielding angle, when compared to those obtained by the conventional electrogeometric models, are rather insensitive to line height, due to the strong dependence of attractive radius on conductor height in (5.14). It must be noted, however, that the IEEE Standards 1410 [22] and 1243 [23] recommend the use of Eriksson's method for calculations of lightning incidence to overhead power lines.

5.2.1.8 Statistical model

Lightning flash is a stochastic phenomenon. Lightning attachment to a ground structure depends on the probability for a certain amplitude of the return-stroke current and on the probability for an upward connecting discharge from the structure. Thus, for a given lightning return-stroke current, the basic parameters of engineering models representing lightning attachment to a ground structure, namely striking distance and interception radius, are statistical quantities; they vary within a range depending upon the lightning interception probability of the structure, that is, the probability for an upward connecting discharge from the structure. The fact that striking distance is a statistical quantity was first noted by Armstrong and Whitehead in [11] and was accounted for in shielding design by using the "effective" striking distance, taken as one standard deviation (10%) lower than the mean of the striking distance distribution [16–18].

Investigations on the lightning interception probability of ground structures through scale model experiments (Figure 5.6) made possible to obtain probability distributions for striking distance and interception radius; thus, a statistical instead of a deterministic approach to the problem of lightning attachment to ground structures was introduced by Mikropoulos and Tsovilis [37–39]. The lightning interception probability of a ground structure is dominated by the relative extent of development of the positive upward discharge from the structure along the interception path, and is affected by possible competing upward discharges from neighboring structures. As the extent of development of the upward discharge from a structure increases, striking distance, thus also interception radius, increases, but the probability of lightning interception reduces.

Experimental data, including those relevant from literature [40–42], were associated with the final stage of natural lightning attachment to a ground structure by using the striking distance to earth surface as a reference. Based also on data from field observations of lightning incidence to ground structures and of shielding performance of power lines reported in literature, a statistical lightning attachment model was developed [37–39]. The statistical model considers, in addition to the prospective lightning stroke current and structure height, the lightning interception probability of the structure (probability of the structure to be struck by lightning) and proximity effects. The distribution of interception radius of a ground structure

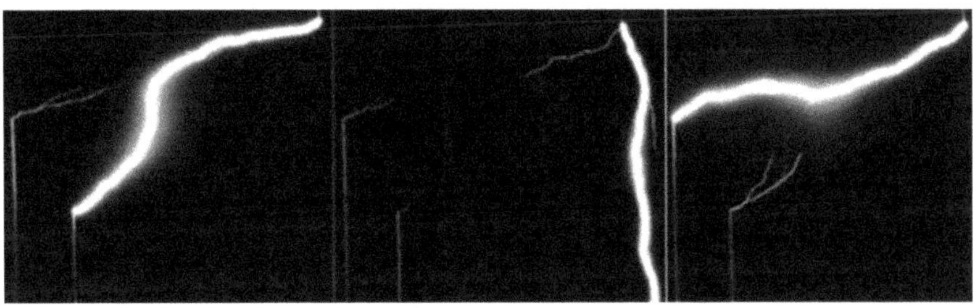

(a) (b) (c)

Figure 5.6 Scale models experiments on lightning interception; breakdown to: (a) protected rod, (b) earthed plane, (c) protective rod (adapted from [39]) is well approximated by a normal distribution, with mean, R_{ic}, called hereafter critical interception radius, and standard deviation, σ, given by the following set of equations, as was derived by adopting the widely applied striking distance expression $10I^{0.65}$:

$$R_{ic} = 6.2h^{0.3}I^{0.455} \text{ (m, kA)} \tag{5.15a}$$

$$\sigma\,(\%) = 13.5h^{-0.43}I^{0.28} \tag{5.15b}$$

Application examples of these expressions are shown in Figure 5.7 for conductor heights of 10 and 40 m, which are commonly utilized in overhead distribution and transmission lines, respectively. The dispersion of the lightning interception probability with respect to the mean increases with increasing return-stroke current, but it decreases markedly with increasing conductor height. For a given lightning return-stroke current, interception may occur within a range of interception radii, which with respect to the mean radius is larger for lower conductor heights.

A schematic representation of the statistical model as applied to line conductors is shown in Figure 5.8. For the case of a single conductor above ground, a vertically descending lightning leader that approaches to within the range of interception radii of the conductor, as estimated using (5.15a) and (5.15b) for the conductor height and prospective return-stroke current, terminates on the

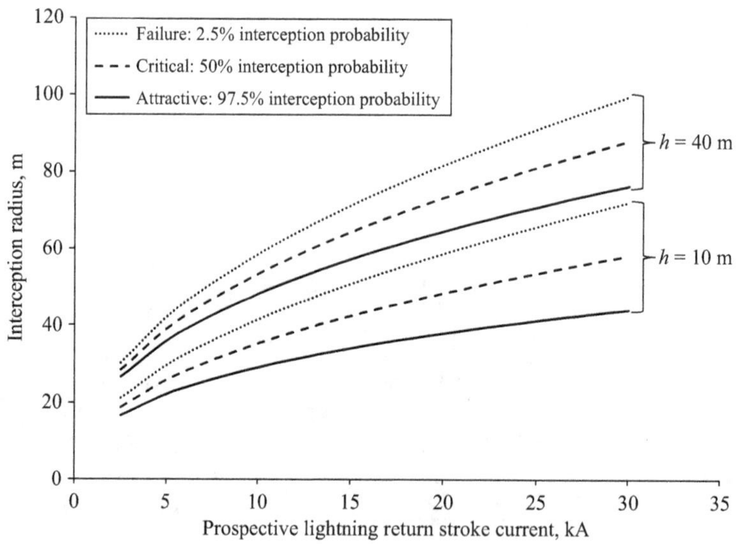

Figure 5.7 Interception radius as a function of prospective lightning stroke current; curves are drawn according to the set of expressions (5.15); and h is the conductor height

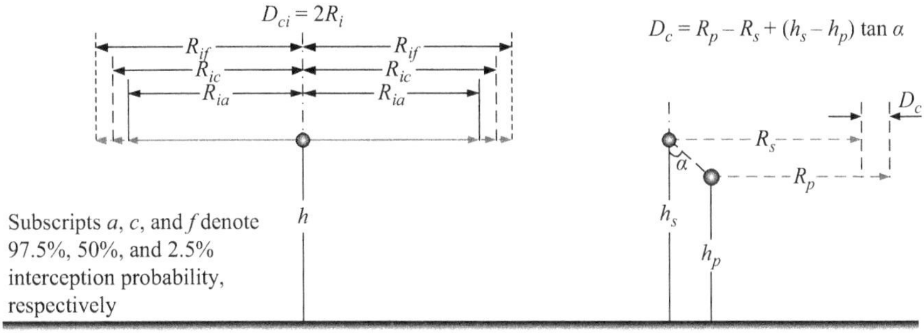

Figure 5.8 *Lightning attachment to line conductors according to statistical model. R_i interception radius of conductor; α shielding angle; h_s and h_p height of shield wire and phase conductor, respectively; R_s and R_p critical interception radius (50% interception probability) of shield wire and phase conductor, respectively; D_{ci} collection width; and D_c exposure width*

conductor; otherwise the stroke will terminate to earth surface. Thus, a range of collection widths, D_{ci}, is obtained depending on the lightning interception probability of the conductor. For the simplified case of a single shield wire and one phase conductor (Figure 5.8), using (5.15a) and (5.15b) for a prospective return-stroke current a range of exposure widths of the phase conductor is obtained depending on the lightning interception probability of the conductors. As the lightning stroke current increases, the exposure width D_c decreases, becoming zero at the maximum shielding failure current, I_{MSF}. Figure 5.8 shows the critical exposure width of the phase conductor determined by using the critical interception radii of the conductors corresponding to 50% interception probability. This exposure width (critical) should be used in studies aimed at evaluating the shielding performance of overhead power lines; for the design of effectively shielded lines a wider exposure width should be used, obtained by using the attractive radius of the shield wire corresponding to 2.5% lightning interception probability.

5.2.2 Leader propagation models

One of the first attempts to model the physical processes involved in lightning attachment to overhead power lines was the modified electrogeometric model of Eriksson [33,34] (Section 5.3.1.7). This model is still recommended for lightning incidence calculations in IEEE Standards [22,23]. Since Eriksson's model though, lightning research developed a deeper understanding of the processes involved.

Many studies concentrated on the modelling of single parts of the lightning formation process, from streamer zone ([43–45] and others) to the downward stepped leader, return stroke and attachment process to a structure ([36,46–50], and others).

The following models were and are applying all the available scientific information to date, in order to simulate lightning attachment to structures, and can be used to improve the evaluation of power line lightning performance. These simulation models are generally named leader progression models (LPM), after the first to use the name—the LPM of Dellera and Garbagnati [51,52], or alternatively "leader inception models" or "leader propagation models."

All leader propagation models make assumptions to simulate processes that lead to the attachment of a downward lightning to an earthed object:

- Thunderstorm cloud height
- Thunderstorm cloud charge distribution and/or cloud-generated electric field at ground
- Channel space-charge distribution
- Relationship between lightning current amplitude and charge along the channel
- Direction of the downward channel
- Inception of an upward streamer/leader
- Direction of the upward leader
- Speed of downward leader and upward leader
- Average voltage gradient between downward and upward tips

Parameters and assumptions are often derived from high voltage switching impulses experiments in long air gaps, or from rocket and natural lightning experiments. Some of the most widely known models will be shortly described here, as examples of the category, namely models from:

- Dellera–Garbagnati [53–56] and its recent evolution LPM-FEM [57]
- F.A.M. Rizk [58–63]
- Cooray and Becerra [64–71]

Many others have been published and can be found in literature (see [72–76]).

A number of researchers developed software codes to reproduce the same concept of the three mentioned models, making explicit the involved image charge equations (see [77]), choosing among the main assumptions of the classical models or introducing new hypothesis like the influence of the AC power voltage in the conductors [78]; all with the aim to calculate the shielding failure and back-flashover quantities for power lines.

The named representative models are implemented in software codes that iterate over all the statistical spectrum of flash currents and proceed step by step in the progression of the downward and upward leaders at different time and space values. Depending on the owners choices and the calculation power, the number of steps and the granulation of space-time are more or less refined, as much as graphical outputs.

Here we will limit the description of parameters used for the negative downward—positive upward configuration.

All these models were compared with experimental results, on power lines, towers, rocket lightning, and so on, to verify the predictions. All models presented here were written to include simulation for towers (rod–plane and rod–cloud configurations), for horizontal conductors (conductor–plane and conductor–cloud configurations), and some of them also for buildings of any shape and concurrent upward leaders (Dellera–Garbagnati and Cooray–Becerra); for sake of simplicity we will describe here only the main aspects relevant to tower configuration.

5.2.2.1 Upward leader inception criteria, ambient electric field, and channel space charge in leader inception models

Usually in the presence of a thunderstorm electric field tall objects act as a distortion to the local field and the induced potential gradient near their tip becomes high enough to originate corona and streamer ionization; in the enhanced transit electric field due to a near descending lightning channel, streamers may evolve in upward leaders (inception) that may further become stable and propagate upward, often being chosen by the downward stepped leader to connect to ground.

The inception of an upward leader is a complex event, ruled by the local field, as mentioned, but also by the wind, the humidity, and the electric configuration of the structure.

Rizk's model

Rizk's model [60–62] represents the positive leader initiation from earthed objects under the influence of a downward channel, applying the high voltage laboratory experimental results, in particular the mechanism of breakdown in long air gaps for long positive front pulses [58,59,63] and configurations for which corona inception voltage is lower than leader inception voltage: a continuous and stable positive leader starts from a streamer zone in front of the electrode when the external ambient potential (potential in the area of study in structure absence) reaches a given critical field value, U_{lc}. The critical potential value for a sustaining leader inception is evaluated experimentally in HV laboratory for rod–plane configuration and conductor–plane configuration when adapted to the downward lightning—tower tip configuration becomes:

$$U_{lc} = 1556/[1 + (3.89/h)]\ (\mathrm{kV, m}) \tag{5.16}$$

where h is the height of the structure (vertical tower).

The model considers the external or ambient field as constant everywhere, with the additional effect, only at the tip of the tall object at ground potential, of the field generated by the downward leader proceeding toward ground.

To calculate the electric field and the induced potential on the tip of the tall object, the model uses the charge simulation method [79], where the simulation of the cloud, grounded structure, lightning downward channel, upward leader, and ground plane is done with image charges.

The presence of the downward channel is represented with a linear decaying charge distribution along the channel path, from the highest value at the structure top to zero at the cloud bottom.

A relationship between the peak lightning current to be simulated and the total charge in the downward channel is derived, calculating the integral of the current waveshapes measured by [80] up to the first peak, for negative first strokes.

Dellera–Garbagnati model

Dellera–Garbagnati model keeps into account the results of high voltage laboratory tests on large air gaps stressed with long front impulses [56,81–85]. These experiments give the conditions for a stable leader inception on electrodes. In particular, it was realized [36] that the breakdown voltage for spherical electrodes was independent from the electrode radius, up to a certain "critical radius"—R_c; the critical radius is then the minimum radius for a spherical tip that will produce directly a stable leader at the first corona. This concept is introduced in the model, as the rule for the inception of upward sustaining leader from any particular point of the simulated structure, when the electric field at the critical radius distance from the point reaches the breakdown value.

For an electrode radius R_c, the necessary field for the inception of the leader E_i is, from experimental results:

$$E_i\ (\mathrm{kV/cm}) = 23 * \left(1 + 1.22/R_c^{0.37}\right) \tag{5.17}$$

The structures at ground are simulated with ring, line or point charges, and the ambient electric field is calculated by means of the charge simulation method [79]. The ambient electric field, calculated step by step at each point in space and time, is originated by the contributions of the cloud-charge, the downward channel charge, the downward streamer zone charge, the upward leader charge, the upward streamer zone charge, and the ground. In particular, each step sees the addition of a few meters of downward/upward channel to the previous ones, in form of linear charges. The potential gradient is calculated also on all structures per each step.

The cloud charge is simulated as a unipolar charge (equivalent to a bipolar vertical distribution) uniformly distributed per horizontal rings, over a 10 km horizontal radius and at a height of 2 km from ground.

The equivalent unipolar charge in the cloud is chosen in each simulation in order to get a specific lightning peak current value. The correlation between the cloud charge and the lightning peak current was obtained assuming the same probability values in the relevant distributions; the distribution of the probability of the cloud charge was obtained calculating the equivalent cloud charge necessary to cause an upward discharge from critical radius electrodes at different electrode heights and matching these values to the frequency of occurrence for upward flashes on different tower heights [86,87]. The leader channel is simulated by an equivalent leader channel, where the charge per unit length represents the total charge per length of all the lightning branches. Analyzing lightning current waveshapes from tower experiments [88–91] authors identified a common behavior, with an initial increase attributed to the thermalization of the channel, a plateau due to the flowing of current in the channel after the thermalization, a first slope variation explained as the reflection of the current at the bottom of the cloud, a following trend due to the neutralization of cloud charges; the quantity of charge in the lightning channel is then obtained by the integration of the current waveshape up to the first slope change after the plateau. This was done for all recorded available data. To evaluate the charge per unit length of the channel (q), it was assumed that the total charge of the channel is uniformly distributed along the channel and that the return stroke reaches an altitude of 2 km in the time interval between the starting of the current waveshape and the reflection at the channel end.

The interpolation of above analytical results gave the formula of the charge per unit length in the downward channel:

$$q = 38 * I^{0.68} \text{ (m/µs, kA)} \tag{5.18}$$

This formula is implemented in the model for each descending channel step, considering all branches contributing to the charge, while for the last leader step, closer to ground, only the contribution of the relevant branch is considered, assuming $q = 100$ µC/m from laboratory tests, consistent with the lowest charge in the lowest current for downward flashes [92]. A streamer zone is simulated at the tip of the leader; the extension of such a zone is assumed equal to the region in front of the downward leader tip in which the electric field is above or equal 300 kV/m.

The upward leader density charge is evaluated from experimental laboratory tests equal to 50 µC/m. At the tip of the upward leader there is an area of streamer ionization, that is where the electric field is above or equal 500 kV/m.

Borghetti et al. LPM evolution

Borghetti *et al.* introduced in 2010 [93] an evolution of the Dellera–Garbagnati model, developing a code with the finite element method (FEM) in substitution of the charge simulation method. All other general assumptions used in the original LPM were maintained in the LPM-FEM simulation.

Upward and downward leader propagations are determined with the solution of the steady-state electrostatic problem. In the domain, leader charges represent the source while the earthed structure, together with the ground zero-potential and the cloud charge distribution, are the boundary conditions. The problem consists in the calculation of the electrostatic field into a dielectric material region (air) $\Omega \in \Re 3$ with the Poisson's equation:

$$-\nabla(\varepsilon_0 \nabla \mathbf{V} - \mathbf{P}) = \rho \tag{5.19}$$

where V is the electric scalar potential, ε_0 is the permittivity of vacuum, P is the electric polarization, and ρ is the space charge density.

The adopted FEM model solves (5.19) as a function of the dependent variables, that is, the components Vx, Vy, and Vz of the electric scalar potential V. In correspondence of the exterior boundaries $\partial\Omega$ that define Ω, the following different conditions are set for the LPM solution:

- Electric potential: $V\partial\Omega$
- Ground, $V\partial\Omega = 0$
- Surface charge density: $\rho\partial\Omega$
- Zero charge: $\rho\partial\Omega = 0$

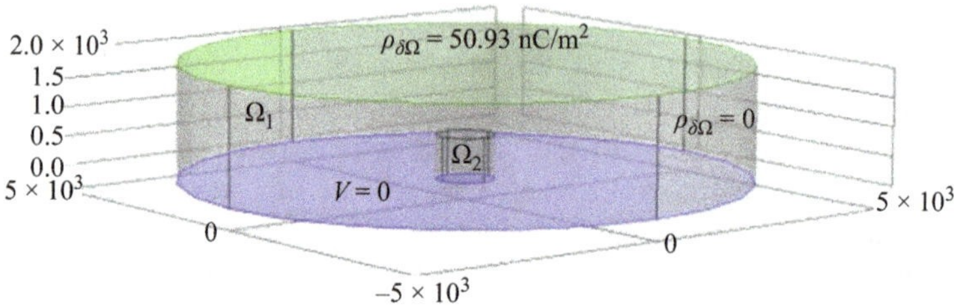

Figure 5.9 Representation of domains Ω_1 and Ω_2 and external boundaries. Adapted from [57]

The domain $\Omega \in \Re 3$ (see Figure 5.9) is divided into two sub-domains: the first one, Ω_1, consists of a cylindrical region of air characterized by 5 km radius and 2 km height; the second one, Ω_2, is an included cylindrical region of air of a reduced size, surrounding the earthed structure. As both upward leader and final jump take place into a region of space close to the earthed structure, the aim of Ω_2 is to provide a control volume in which a more refined mesh is used.

The cloud charge source is represented with a constant surface charge density, in correspondence of the top circular boundary of Ω_1. For the side surfaces of Ω_1, a boundary condition is assumed, $\rho\partial\Omega = 0$.

The earthed object is represented by a vertical mast that is an interior boundary of Ω_2, which is imposed as a grounded ($V = 0$) edge. It is worth observing that, compared to the charge simulation method, the direct adoption of such a boundary condition avoids the use of the so-called virtual charges, used in the original LPM, in order to obtain a profile of the zero-level electric potential as close as possible to the earthed structure.

The downward leader is represented by means of a linear charge distribution which value is determined as the ratio between the total charge correlated to the lightning current, Q_{fp}, and the average total leader length of 2 km.

The propagation direction of the downward leader and negative streamers follow the streamlines of the electric field and, like in the original LPM, it is assumed that the direction of the downward leader corresponds to the one of the maximum gradient of the electric potential, with an additional streamer zone equal to a region in front of the downward leader tip in which the electric field is above 300 kV/m. Thus, the estimation of the maximum gradient of the electric potential is performed in correspondence of a cap, centered on the leader tip, with a radius that can spread from few tens to few meters as a function of the streamer zone extension. The critical radius, Rc, defines a region of space surrounding the earthed structure, which calculates the electric field in order to determine the upward leader inception; in this case implemented with a so-called "control-surface" Σ that surrounds the earthed structure.

Cooray and Becerra model

The self-consistent leader inception and propagation model—SLIM—of Cooray and Becerra [64–68,70,71] simulates the development of successive corona bursts with multiple streamers and the consequent transition to unstable leader, the possible quenching of the leader by corona, and the eventual transition to stable leader at the tip of a grounded structure. Previous models delegate all this part of the process to assumptions from the high voltage laboratory results, deciding the inception of a stable leader with the surpassing of an experimental field threshold.

The charge simulation method [79] is applied and point, line, and ring charges are used to simulate objects and to calculate the electric field in any time and space point.

The corona streamer zone at the tip of the structure is simulated with ring charges, while the upward stem/leader channel is represented by finite line charges with increasing length and a point charge at the tip [64].

The downward channel, simulated with line charges, is vertical and unmoved by the presence of the upward leader. The charge density of the downward leader follows Cooray equation [94].

The field at the object tip varies with the descending of the downward leader steps (improvement after a first "static" approach with the ambient field considered unperturbed by the lowering of the downward leader) [65,66]. As soon as the downward channel tip is low enough to give a sufficient electric field on the object tip, the first corona is initiated.

In particular, the corona inception happens when the local field at the object tip satisfies the streamer inception criterion [52,72], assuming that enough free electrons are always available. The corona, represented here by a succession of ring charges, simulates the multiple streamer fan originating at the grounded rod tip as documented in natural flashes [95,96].

The corona zone simulation follows the electrostatic representation as proposed by Goelian *et al.* [76] and considering this zone as characterized by a constant potential gradient, here the total charge is calculated with the charge simulation method.

The first corona shields the electric field at the structure tip and a dark period happens, when no streamers-corona are created, which ends when the streamer condition is satisfied again due to the descending leader effect. When the streamer criterion is reached again, more corona streamers develop, up to when the total corona charge is equal to 1 μC (condition of transition from streamer to leader) and a first unstable leader step is created; in the following round of simulation the total charge of a new corona/streamer together with the first leader is calculated, and if it satisfies the threshold value of 40 nC per unit length it will create a second step of the leader, this becoming a stable leader.

Once the first leader charge segment is calculated, the length of the following leader steps uses a relation between leader current and velocity as proposed by Bondiou–Clergie and Gallimberti [74] where the velocity is directly proportional to the current and inversely proportional to the charge per unit length necessary to get from streamer to leader.

The upward leader tip potential is calculated using the thermo-hydrodynamical model for the leader channel as proposed by Gallimberti [72] and specific parameters [66,97].

5.2.2.2 Leader continuous propagation and final jump

As well known in literature, downward negative flashes are created via a dark process known as stepped leader, guided by a stem/streamer ionization area in front of each step. This process develops an ionized channel travelling toward Earth. On the opposite side, an initiation of an upward leader is possible from tall objects, under the presence of a strong induced potential, due to the downward leader end/or to the thunderstorm cloud. In the presence of a strong enough potential gradient at the object tip, the upward leader becomes stable, that is, has the energy to proceed in length and velocity without extinguishing itself. When the upward leader is stable, depending on distance, direction, and field intensity, the downward and upward leaders can connect, creating the final jump.

Rizk's model

In Rizk's model [58–63] the negative downward channel continues its vertical downward motion unperturbed, up to when the critical conditions for a final jump are fulfilled. After the inception of the stable upward leader, this proceeds by steps of linear charge and the upward leader tip potential is evaluated as the voltage drop corresponding to the length of the upward leader already created and from experimental results [59]. The vector direction of the positive leader tip always aims at the downward channel tip and the two leaders have a velocity of propagation with ratio equal to 1. When the mean potential gradient between the two tips is higher than 500 kV/m the final jump occurs and the upward leader connects with the downward one. If the two leader tips reach the same height from ground without fulfilling the jump condition, the downward leader descends to hit the ground.

Dellera–Garbagnati model

Dellera and Garbagnati model [51–56] simulates each step of the downward channel with a linear charge, and the direction of propagation is, per each step, evaluated looking for the direction of maximum electric field along an equipotential surface in front of the streamer zone. The same rule is used for the step-by-step direction of the upward leader.

The model avoids using absolute downward and upward velocities but only uses the ratio between them, varying from 4 at the upward leader inception to 1 just before the final jump, with the relationship:

$$v-/v+ = 1/\exp[1.5 * (\Delta E/\Delta E_c - 1)] \tag{5.20}$$

where ΔE_c is the average gradient between the two leader tips at final jump and ΔE is the average gradient between the two leader tips at any other instant, as in laboratory data [53]. The final jump occurs when downward and upward streamers meet.

Due to the fact that downward and upward leader direction of propagation is dependent only on the electric field values, and the leaders are not forced to proceed one toward the other, the model allows the occurrence of lateral impacts on structures or of "near miss" leaders with the downward closing at ground or even the simulation of concurrent upward. The existence of upward flashes is possible and taken into account.

Borghetti et al. LPM evolution

In the Borghetti *et al.* LPM-FEM evolution, the motion of downward and upward leaders can become a final jump when and if, in the area between the two leader tips is available an electric streamline where, along the overall streamline, the voltage gradient is larger than 500 kV/m (Figure 5.10).

Cooray and Becerra model

In Cooray and Becerra model, the downward channel proceeds vertically while the thermo-hydrodynamic leader channel model of Gallimberti is applied [72] to calculate potential gradient, channel radius, current, and propagation velocity of the upward leader, thus deciding if this particular leader is stable and will develop in further steps. When a stable leader is formed, this propagates toward the downward vertical channel accelerating. The step's length of the stable leader are evaluated with the Bondiou–Clergie [74] formula and the model calculates the velocity of the upward leader from the amount of corona charge necessary to create a next leader step. The upward leader charge per unit length is derived as well from Gallimberti and the final jump condition is satisfied when between the two leader tips the potential gradient is 500 kV/m [69].

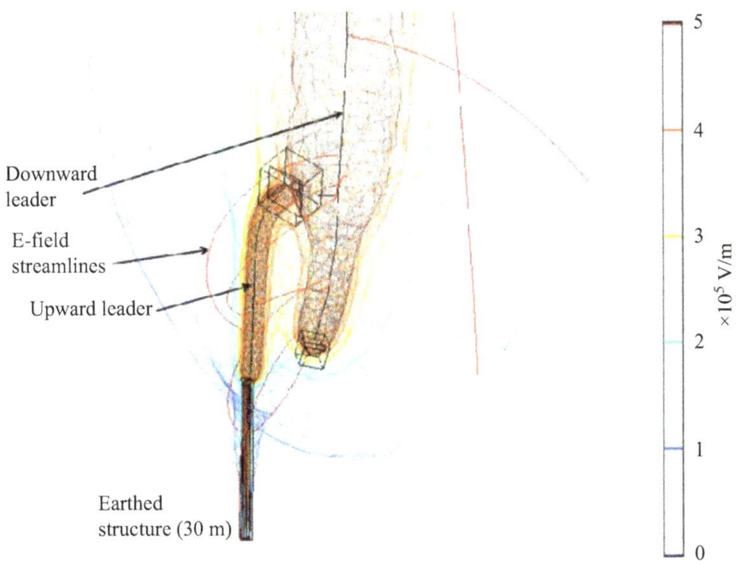

Figure 5.10 Example of final jump structure in a LPM-FEM simulation. Adapted from [57]

5.2.2.3 Lightning interception and maximum intercepting radius

In nature a downward lightning can proceed to ground thanks to the ionized channel propagation and the interception of the last leader-streamer part to the ground. Alternatively, upward streamer and leader can develop from tall grounded objects (for instance trees, chimneys, towers, and buildings) under the effect of the electric field generated by the cloud and eventually by a descending channel. If the downward and upward leader tips are close enough and the electric path between them is the easiest way to close the circuit to ground, the two will join in the final jump stage and a return stroke will happen in the channel between the cloud and the object at ground. Generally speaking, the further the downward channel is, in horizontal distance, from the object, the unlikelier is that the downward channel will connect to the upward leader from the object. At least in theory, there is a maximum distance between the downward direction of the channel and the object, after which the channel will go to ground instead of connecting to the upward leader and the object.

This distance can be called maximum intercepting distance, maximum attractive radius (Rizk), maximum lateral distance (Dellera–Garbagnati), and others, with slightly different geometric meaning in different models (see Figure 5.11).

This distance is usually represented as a function of the object height and of the lightning current. Some papers were published on the comparison for different model maximum intercepting effects (see [64,98,99]).

Rizk's model

In Rizk's model [62], the interception on a tall structure is simulated having the downward channel proceeding straight vertical up to when the condition for the final jump is satisfied. At this point the two leaders connect and the object is considered hit. The maximum distance between the downward leader and the structure, for which a final jump connects the two leaders, is called "maximum

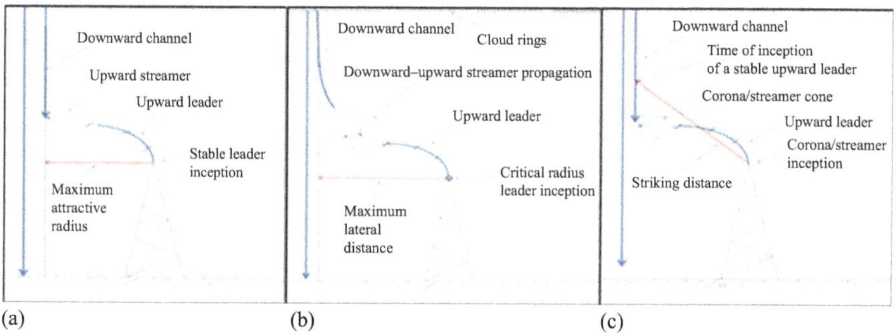

Figure 5.11 Maximum intercepting distance (in red) for various models: (a) Rizk model; (b) Dellera–Garbagnati model; (c) Cooray–Becerra model—author's design

attractive radius." Various studies have been done with different ambient fields, different currents, and object heights [60,61,63]; also, the presence of a mountain or hill under the structure was evaluated. With a median negative stroke current of −31 kA and an ambient ground field of 3 kV/m a regression analysis of model results was used to calculate an overall attractive radius, R_{ao}, of a vertical structure [100]:

$$R_{ao}(h) = 26.6 * h^{0.48}\ [\mathrm{m,m}] \tag{5.21}$$

for structures in the range 10–200 m of height.

Dellera–Garbagnati model

In Dellera and Garbagnati model [51–55], the downward moving channel is proceeding from the cloud toward earth with steps of a few meters, each step deciding in which direction to proceed based on the maximum total field in the area in front of the streamer zone.

When the total field, due to cloud, downward channel, and downward streamers, reaches the critical electric field gradient ([36,56]) at the grounded object (at the critical radius distance from the tip), an upward leader inception occurs. As mentioned before the upward leader-streamer proceeds step by step in the direction of the maximum electric field gradient in front to the streamer area along an equipotential line, up to the meeting of the two streamer zones, that is the final jump.

The parameter of interest is here called "maximum lateral distance" and represents the horizontal distance between the vertical direction of the downward channel and the object tip, for the furthest downward channel that will get a final jump (see Figure 5.11). This parameter was calculated for different height of the structure up to 200 m, for all the lightning current probability curve from 0 to 120 kA. Multiple configurations of tall slim structures (towers) horizontal conductors in elevation, buildings were studied, considering also lateral impacts or impacts on lateral facades, and studying the effect of structure position relevant to mountain and hilly terrain. The model was extensively applied to the study of power lines, in particular shielding failure and backflashover, giving shielding tripout rate for different line heights and basic insulation levels [51,53–56,93,100,101]. A regression formula was derived for calculation of the maximum lateral distance L_d as a function of tower height h_t and lightning current I_f:

$$L_d = 0.03 * h_t * I_f + 3 * h_t^{0.7}\ (\mathrm{m,kA}). \tag{5.22}$$

Cooray and Becerra model

In Cooray and Becerra model [64–68,70,71] the "striking distance" is defined as the distance between the tip of the downward channel and the tip of the high object at ground at the moment when a stable upward leader is born. The maximum striking distance is evaluated for different object heights, for different object tip radii, for structures with asymmetry (building corners). The object tip radius affects the striking distance results of a 10%, for radius between 0.002 and 0.04 m. Simulations that changes the background electric field accordingly to downward channel propagation give values of striking distances doubled with respect to simulations with a static background electric field.

5.2.2.4 Concluding remarks on leader propagation models

A schematic representation of the leader propagation models as applied to power line conductors is shown in Figure 5.12. For the case of a single conductor above ground, a vertically descending lightning leader that approaches to within the attractive radius of the conductor, R_a, the latter as dependent upon conductor height and prospective return-stroke current, terminates on the conductor; otherwise the stroke will terminate to earth surface. Thus, the collection width of the single conductor becomes $D_{ca} = 2R_a$. For the simplified case of a single shield wire and one phase conductor, by adopting Rizk's shielding analysis [60], the exposure of the phase conductor to direct lightning strokes (shielding failure) is determined by the exposure width, D_c, which according to Figure 5.12 is estimated as

$$D_c = R_{pa} - R_{sa} + \left(h_s - h_p\right)\tan a \tag{5.23}$$

The exposure width decreases as the lightning stroke current increases, becoming zero at the maximum shielding failure current, I_{MSF}.

The following general expression can be used to estimate the attractive radius of horizontal conductors, R_a, defined as the longest lateral distance from the conductor at lightning interception, according to the leader propagation models:

$$R_a = \xi h^E I^F + \zeta h^G \tag{5.24}$$

where R_a is in meters, I (kA) is the prospective lightning peak current, h (m) is the conductor height, and factors ξ, E, F, ζ, and G are listed in Table 5.2.

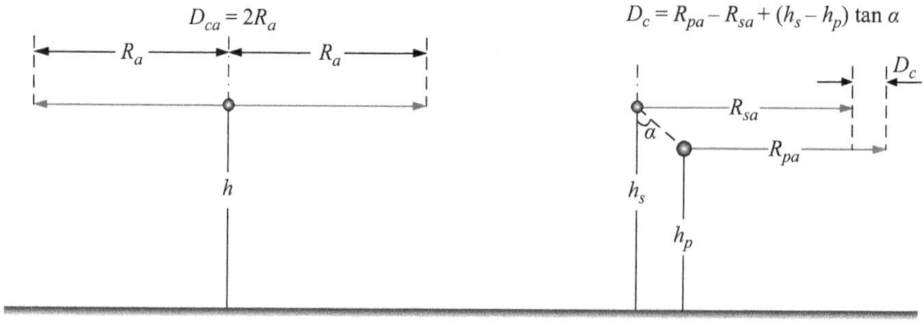

Figure 5.12 Lightning attachment to line conductors according to leader propagation models. R_a is the attractive radius of conductor; α is the shielding angle; h_s and h_p are the height of shield wire and phase conductor, respectively; R_{sa} and R_{pa} attractive radius of shield wire and phase conductor, respectively; D_{ca} is the collection width; and D_c is the exposure width

Table 5.2 Factors ξ, E, F, ζ, and G to be used in (5.24)

Leader propagation model	$R = \xi h^E I^F + \zeta h^G$				
	ξ	E	F	ζ	G
F.A.M. Rizk [60]	1.57	0.45	0.69	0	0
Borghetti *et al.* [99] from Dellera and Garbagnati [52,55]	0.028	1	1	3	0.6
Cooray and Becerra [70]	2.17	0.5	$0.5h^{0.04}$	0	0
h conductor height					

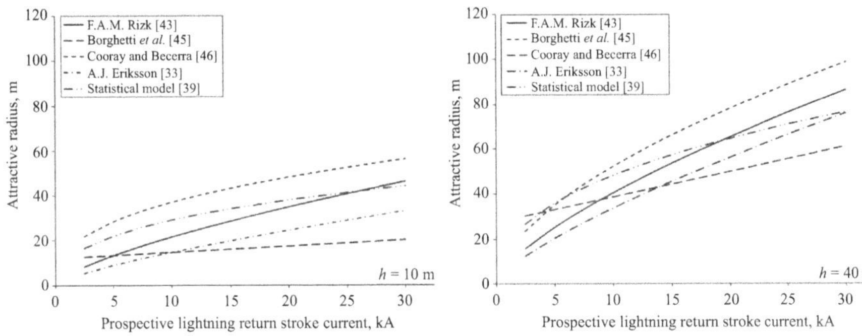

Figure 5.13 Attractive radius as a function of lightning stroke current; h is the conductor height

Application examples of these attractive radius expressions are shown in Figure 5.13 for horizontal conductor heights of 10 and 40 m, which are commonly utilized in overhead distribution and transmission lines, respectively. For comparison purposes Figure 5.13 also includes the results of the application of the attractive radius expressions employed in Eriksson's (Section 5.3.1.7) and Statistical (Section 5.3.1.8) models. There is a great variation in attractive radius among lightning attachment models, both in terms of value and rate of increase with increasing lightning stroke current and conductor height. Under such variations of attractive radius, lightning incidence to overhead power lines should also vary among lightning attachment models.

5.3 Lightning incidence due to direct lightning strokes

The lightning performance of overhead power lines is expressed as the annual number of lightning-caused flashovers of line insulation on a tower-line length basis; therefore, it is related to the fast-front overvoltages across line insulation, arising due to direct lightning strokes or induced by nearby lightning. With increasing system voltage, a higher line insulation level is utilized, and the lightning performance of power lines becomes increasingly determined by the direct lightning stroke flashover rate. A lightning flash to ground consists of a first return stroke that may be followed by subsequent strokes and other impulsive or continuing currents. In what follows, we will limit the analysis on lightning incidence to power lines only to negative first strokes.

5.3.1 Definitions and terminology

In overhead transmission lines direct lightning strokes terminate predominantly on the shield wire(s); the annual number of these events on a tower-line length basis is called lightning stroke collection rate of the shield wire(s), N_s, and is commonly given in units of strikes per 100 km per year. Direct strokes to shield wire(s) may cause backflashover, a term that is used for insulation flashover when the lightning struck component of the line is normally at ground potential; the annual number of line power outages caused by backflashover on a tower-line length basis is called "backflashover rate," BFR (flashovers/100 km/yr). Less intense lightning strokes may bypass the shield wire(s) and terminate directly on the phase conductors (shielding failures); the annual number of these events on a tower-line length basis is called shielding failure rate, SFR (strikes/100 km/yr). Not all direct strokes to phase conductors result in line insulation flashover; as also in the case of backflashover, the arising overvoltage should exceed the line insulation level for a flashover to occur. The annual number of line power outages caused by shielding failures on a tower-line length basis is called "shielding failure flashover rate," SFFOR (flashovers/100 km/yr). In practice, an effective shielding of transmission lines against direct lightning strokes to phase conductors is realized based on an acceptable SFFOR. It is noted that as the line operation voltage increases, thus a higher line insulation level is utilized, the lightning performance of overhead transmission lines becomes increasingly determined by SFFOR.

In overhead distribution lines the vast majority of direct lightning strokes to line conductors result in flashover, because of the relatively low line insulation levels utilized. The annual number of line power outages caused by direct lightning strokes resulting in flashover on a line length basis is called "direct stroke flashover rate," DSFR (flashovers/100 km/yr). In addition, line power outages may occur due to insulation flashover caused by induced voltages on line conductors, originating from lightning strokes terminating to ground nearby the power line or on adjacent structures. The annual number of line power outages caused by induced voltages resulting in flashover on a line length basis is called induced-voltage flashover rate, IVFR (flashovers/100 km/yr). It is noted that as the line operation voltage decreases, thus a lower line insulation level is utilized, the lightning performance of overhead distribution lines becomes increasingly determined by IVFR.

From the above it should be evident that the lightning performance of overhead power lines is essentially affected by, in addition to line electrical and geometrical parameters, the latter with respect to the surrounding terrain, the lightning activity along the routing of the line. Therefore, estimating the lightning performance of power lines primarily requires knowledge of the quantity of ground flash density,

N_g (flashes/km^2/yr), that is, the average number of lightning flashes to ground per unit area per unit time at a particular location. Reliable data for N_g, which varies seasonally and geographically, are available by direct measurements using lightning location systems or can be obtained after appropriate conversions from thunderstorm data. Following, a model of lightning attachment to overhead power lines has to be employed, so as to estimate the number of lightning strokes terminating directly on the line conductors, and, thus, by considering also the line insulation level, to obtain the associated lightning-caused flashover rates of the power line.

5.3.2 Lightning stroke collection rate of shield wire(s)

From the conceptual representation of the lightning attachment to a single horizontal conductor presented in Section 5.2 it can be deduced that the lightning stroke collection rate of the shield wire(s) per 100 km line length, N_s (strikes/100 km/yr), can be given as

$$N_S = N_g A_s = N_g\left[0.1\left(2R_{eq} + b\right)\right] \tag{5.25}$$

where N_g (flashes/km^2/yr) is the ground flash density, A_s (km^2) is the equivalent collection area of the shield wire(s) per unit length of the line, b (m) is the separation distance between the shield wires (zero in the case of a single shield wire), and R_{eq} (m) is the equivalent interception radius of the shield wire.

The fundamental quantity of R_{eq} is introduced in (5.25) to account for the fact that the shield wire interception radius, R, the latter as defined according to the adopted lightning attachment model, is dependent upon the peak of the lightning stroke current. In this regard, R_{eq} is calculated by integrating R weighted by the probability density function of the lightning peak current distribution, $f(I)$:

$$R_{eq} = \int_0^{\infty} R(I) f(I) dI \tag{5.26}$$

where I is the peak of the prospective lightning stroke current.

The probability density function of the lightning peak current is assumed lognormally distributed as

$$f(I) = \frac{1}{\sqrt{2\pi}\sigma_{\ln} I} \exp\left[-\frac{(\ln I - \ln \bar{I})^2}{2\sigma_{\ln}^2}\right] \tag{5.27}$$

where $\bar{I}$ and $\sigma_{\ln}$ are the median value and the standard deviation of the natural logarithm of the lightning peak current, respectively.

From (5.25)–(5.27) it is evident that, given a reasonably long-term value for N_g, the estimation of lightning incidence to shield wire(s) of an overhead power line depends on the adopted lightning attachment model and statistical distribution of the lightning peak current, both affecting the estimation of the equivalent interception radius (5.26).

5.3.2.1 Effects of lightning attachment models

According to electrogeometric models (Section 5.3.1), the interception radius of a single line conductor, R, at height h above ground is given by (5.2). Thus, by combining (5.2) and (5.26) the equivalent interception radius based on electrogeometric models is given as

$$R_{eq} = \int_0^{I_{rg}} r_c f(I)dI + \int_{I_{rg}}^{\infty} \left[\sqrt{\left(1 - \frac{1}{\gamma^2}\right) r_c{}^2 + \frac{2}{\gamma} r_c h - h^2} \right] f(I)dI \tag{5.28}$$

where r_c and r_g are respectively the striking distance to line conductors and earth, $I_{r_g} = \left(\frac{\gamma h}{A}\right)^{1/B}$ is the lightning peak current at which $r_g = h$ and factors A, B, γ, as defined by (5.1), are given in Table 5.1.

On the basis of leader propagation models (see Section 5.3.2), the attractive radius of a horizontal conductor, R_a, positioned at height h above ground is given by the general expression (5.24), then (5.26) is obtained as

$$R_{eq} = \int_0^{\infty} (\xi\, h^E I^F + \zeta h^G) f(I)dI \tag{5.29}$$

where I is the prospective lightning stroke current and factors ξ, E, F, ζ, and G are given in Table 5.2.

Expression (5.29) can be applied also on the basis of Eriksson's (Section 5.3.1.7) and Statistical (Section 5.3.1.8) models, after placing the corresponding expressions for attractive radius (5.14) and interception radius distribution (5.15) in the general form of (5.24).

When numerically integrating (5.28) and (5.29) using a statistical distribution for lightning peak current, the equivalent interception radius of a horizontal conductor at height h above ground is well approximated by the following general expression:

$$R'_{eq} = \xi' h^E \tag{5.30}$$

which should be regarded as yielding average values of R_{eq} for the entire distribution of return-stroke peak currents. By using the probability density function of the lightning peak current distribution with a median value $\bar{I} = 30.1$ kA and $\sigma_{\ln} = 0.76$, as suggested by IEEE TF 15.09 [102] for negative first strokes, Table 5.3 summarizes expressions in the form of (5.30) appropriate to lightning attachment models.

Figure 5.14 shows comparisons between the lightning incidence estimates obtained by the lightning attachment models listed in Table 5.3 and the field data provided by A.J. Eriksson [34]. Based on the tendency of the curves shown in this figure, it is evident that, with the exception of the electrical shadow [24] and Eriksson's [33] methods, all lightning attachment models predict within the evaluated range of shield wire heights a slower increase of *Req* with *h* than the observation data. It is noted that, as also recognized by Eriksson [34], these data show considerable dispersion, mainly related to topographic effects and to the

Table 5.3 Estimates of lightning incidence to shield wires, N_s (strikes/100 km/yr), of typical 115 ÷ 735 kV overhead transmission lines; $N_g = 1$ flash/km^2/yr

Lightning attachment model	R'_{eq} (m)		Overhead transmission line*						
			115/138 kV	**150 kV**	**230 kV**	**345 kV**	**400 kV**	**500 kV**	**735 kV**
Wagner and Hileman [7]	$11h^{0.43}$		9.1	9.9	10.6	11.4	11.9	11.8	13.7
Young *et al.* [9]	$14.3h^{0.44}$		12.1	13.3	14.2	15.2	15.8	15.3	17.8
Armstrong and Whitehead [11]	$33.7h^{0.29}$		17.5	18.6	19.4	20.3	20.9	20.5	22.8
Brown and Whitehead [14]	$30.1h^{0.29}$		15.7	16.6	17.4	18.2	18.7	18.5	20.6
E.R. Love [15]	$13.9h^{0.46}$		12.6	13.9	14.8	15.9	16.6	15.9	18.6
E.R. Whitehead [17]	$14.2h^{0.46}$		12.8	14.2	15.1	16.2	16.9	16.2	19.0
J.G. Anderson [24]	i. $58.3h^{0.08}$								17.8
	ii. $44.3h^{0.14}$					15.5	15.7	16.4	
	iii. $12.5h^{0.45}$		11.0	12.1	12.9				
Electrical shadow method [24]	$2h^{1.09}$		13.7	18.1	19.8	23.0	25.9	20.3	30.3
IEEE Std. 1243 [23]	$17.5h^{0.47}$		16.2	18.1	19.1	20.5	20.6	20.1	20.6
A.J. Eriksson [33]	$9.7h^{0.60}$		13.7	15.8	16.9	18.4	19.6	17.8	22.1
	iv. $14h^{0.60}$		19.7	22.8	24.1	26.3	28.0	24.9	31.1
Statistical model [39]	$R'_{eq} = 31h^{0.30}$	a.	10.7	11.3	11.7	12.1	12.4	12.1	13.2
	$\sigma(\%) = 38.2h^{-0.43}$	c.	13.3	13.7	14.0	14.4	14.6	14.5	15.4
		f.	15.8	16.0	16.3	16.6	16.8	17.0	17.5
F.A.M. Rizk [60]	$18.9h^{0.45}$		16.4	18.2	19.2	20.6	21.5	20.2	23.8
Borghetti *et al.* [99] from Dellera and Garbagnati [52,55]	$3.38h^{0.83}$		10.1	12.3	13.4	15.1	16.5	14.4	19.2
Cooray and Becerra [70]	$12.56h^{0.59}$		17.1	19.8	21.0	22.8	24.3	21.8	27.1

Note: h shield wire height (m) at tower.
*Parameters of lines as given in [103] and [104]; i., ii., and iii.: for UHV, EHV, and other lines, respectively; iv.: as proposed by A.J. Eriksson [33] and adopted in IEEE Std. 1243 [23] and IEEE Std. 1410 [22]; and a, c, and f: 97.5%, 50%, and 2.5% interception probability.

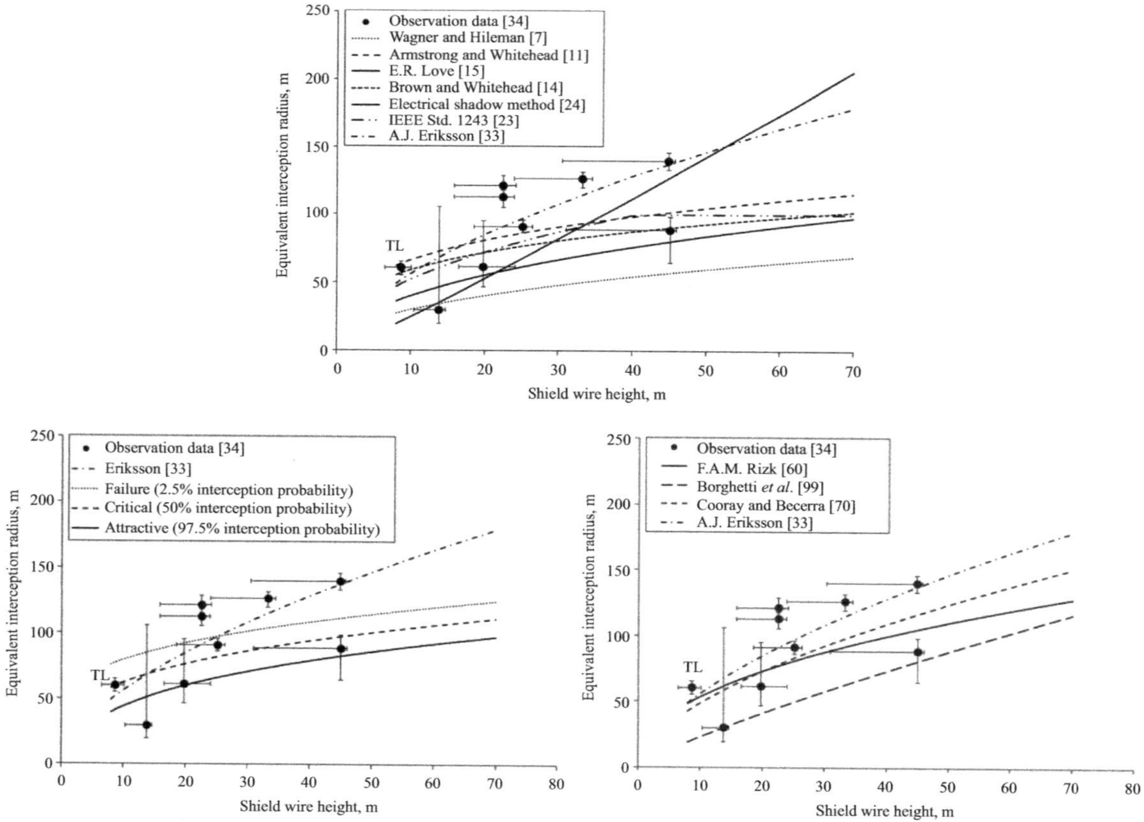

Figure 5.14 Equivalent interception radius as a function of shield wire height; curves were drawn according to the general (5.30) and Table 5.3

relationship between keraunic level and ground flash density. Nevertheless, the best agreement between lightning incidence estimates and the data point designated TL in Figure 5.14, which was considered by Eriksson as the most reliable of the listed field observations, is obtained, together with Eriksson's method, by the electrogeometric models [11] and [14], the statistical model [39], and Rizk's [60] leader propagation model.

Table 5.3 shows the expected lightning stroke collection rate of the shield wire(s), N_s (strikes/100 km/yr), of typical 115 kV up to 735 kV overhead transmission lines with tower height between 25 and 50 m. These estimates refer to $N_g = 1$ flash/km^2/yr and were obtained using in (5.28) and (5.29) the probability density function of the lightning peak current distribution with a median value $\bar{I} = 30.1$ kA and $\sigma_{\ln} = 0.76$ [102]. There is a great variability in the lightning incidence estimates among lightning attachment models. Generally, electrogeometric models yield N_s values smaller than those obtained by LPM; Eriksson's method [33], adopted in IEEE Std. 1243 [23] and IEEE Std. 1410 [22], yields the greatest N_s values.

5.3.2.2 Effects of lightning peak current distribution

Equations (5.28) and (5.29) were solved by performing numerical integration with the aid of a mathematical software package and the following approximate general expression for the equivalent interception radius of line conductors was derived [105]:

$$R_{eq} = \xi \exp\left(F \ln \bar{I} + \frac{F^2 \sigma_{\ln}^2}{2}\right) h^E + \zeta h^G \tag{5.31}$$

where $\bar{I}$ (kA) and $\sigma_{\ln}$ are the median value and the standard deviation of the natural logarithm of the lightning peak current, respectively, and factors ξ, E, F, ζ, and G are given in Table 5.4.

With the aid of (5.31) the effects of conductor height and lightning peak current distribution on R_{eq}, thus also N_s, can be quantified. Moreover, this expression is valid for all models describing lightning attachment; also, by employing the statistical model a statistical distribution instead of a deterministic value for R_{eq}, thus also N_s, can be obtained.

Figure 5.15 shows the variation of N_s of typical transmission lines as obtained by using in (5.31) the "global distributions" of lightning peak currents for negative first strokes recommended for engineering applications by the IEEE TF 15.09 ($\bar{I} = 30.1$ kA and $\sigma_{\ln} = 0.76$) [102] and CIGRE WG C4.407 ($\bar{I} = 30$ kA and $\sigma_{\ln} = 0.61$) [106], as well as the distribution with $\bar{I} = 45$ kA and $\sigma_{\ln} = 0.46$ derived from field observations in Brazil [107]. Evidently, N_s increases with increasing $\bar{I}$ and/or $\sigma_{\ln}$; the intensity of this effect varies with line geometry and among lightning attachment models. This dependence of N_s upon lightning peak current distribution is not accounted for in the expression for lightning incidence calculation suggested by IEEE Std. 1243 [23], as evidenced by the dashed line in Figure 5.15.

Table 5.4 Factors ξ, E, F, ζ, and G to be used in (5.31)

Lightning attachment model		$R_{eq} = \xi \exp\left(F \ln \bar{I} + \frac{F^2\sigma_{\ln}^2}{2}\right) h^E + \zeta h^G$				
		ξ	E	F	ζ	G
Electrogeometric models*	(i)	$\gamma\sqrt{2A}$	$0.36A^{0.16B}(4.1\gamma^2 - 12\gamma + 8.9)$	$0.5B(-6\gamma^2 + 17\gamma - 10)$	0	0
	(ii)	A	0	B	0	0
A.J. Eriksson [33]		0.67	0.6	0.74	0	0
Statistical model [39]**		6.21	0.30	0.455	0	0
F.A.M. Rizk [60]		1.57	0.45	0.69	0	0
Borghetti *et al.* [99] from Dellera and Garbagnati [52,55]		0.028	1	1	3	0.6
Cooray and Becerra [70]		2.17	0.5	$0.5h^{0.04}$	0	0

Note: h (m) is the line conductor height, $\bar{I}$ (kA) and $\sigma_{\ln}$ are the median value and the standard deviation of the natural logarithm of the lightning peak current, respectively.

*Factors A, B, and γ, as defined in (5.1), are given in Table 5.1; (i): for $h < 70$ m and (ii): for $h > 70$ m.

**The standard deviation of the equivalent interception radius distribution is $\sigma\,(\%) = 13.3e^{(0.18\sigma_{\ln})}\bar{I}^{0.27}h^{-0.43}$ [105].

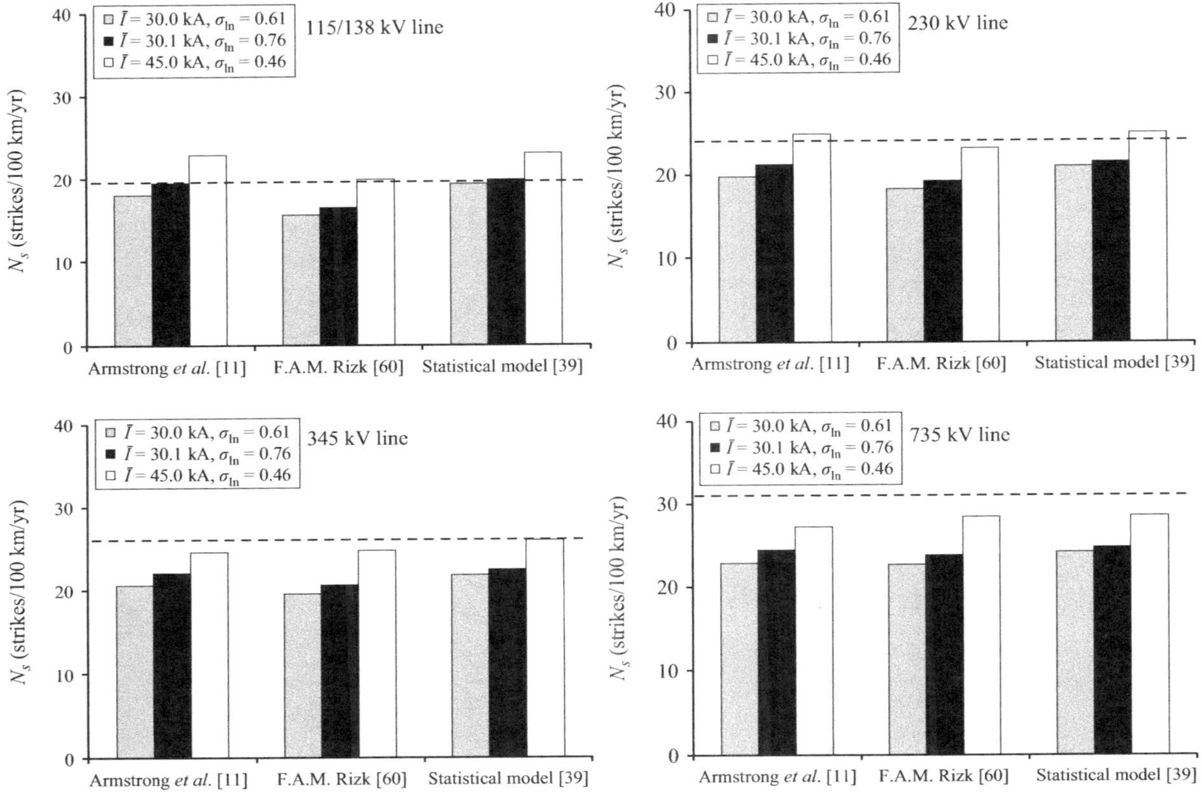

Figure 5.15 Estimates of lightning incidence to shield wires of typical 115 ÷ 735 kV overhead transmission lines by considering the lightning peak current distribution as parameter; N_g = 1 flash/km^2/yr. Dashed line according to IEEE Std. 1243 [23]

5.3.3 Lightning stroke collection rate of phase conductors (shielding failure rate)

From the conceptual representation of lightning attachment to a phase conductor when shielded against direct lightning strokes by a single shield wire as presented in Section 5.2, it can be deduced that, by considering shielding failures on both sides of an overhead line, the lightning stroke collection rate of the phase conductors of a shielded overhead power line, that is, the shielding failure rate, SFR (strikes/100 km/yr), as defined in Section 5.3.1, is given as

$$SFR = 0.2N_g \int_0^{I_{\text{MSF}}} D_c(I) f(I) dI \tag{5.32}$$

where N_g (flashes/km^2/yr) is the ground flash density and D_c (m) is the exposure width, that is, the lateral width along the shielded power line within which a descending lightning leader may terminate on the phase conductors. I_{MSF} (kA) is the maximum shielding failure current, defined as the maximum peak current of all possible lightning strokes that may terminate on the phase conductors, and $f(I)$ is the probability density function of the lightning peak current distribution as given by (5.27).

From (5.32) it is evident that, given a reasonably long-term value for N_g, the estimation of *SFR* of an overhead line depends on the adopted statistical distribution of the lightning peak current. More important, it requires the current-dependent formulation of the exposure width and an estimate for the upper integration limit, I_{MSF}. Both these tasks can be accomplished based on Table 5.5, adapted from [108], on the basis of the lightning attachment model adopted for shielding failure analysis. It is noted that the estimation of SFR based on (5.32) and Table 5.5 refers to flat terrain and to vertically descending lightning leaders, in accordance with the shielding analysis method in IEEE Std. 1243 [23].

5.3.3.1 Effects of lightning attachment models

Figure 5.16 shows for the case of an 220 kV (single-circuit) overhead power line the variation of the exposure width, D_c, of the outer phase conductor with lightning peak current, I, as obtained by employing in shielding analysis several lightning attachment models. There is significant variability in D_c, both in terms of value and rate of reduction with I, among lightning attachment models. This is also the case for the maximum shielding failure current, I_{MSF} (Figure 5.17) which, corresponding to $D_c = 0$, is given by the intersections of the D_c curves with the x-axis in Figure 5.16. Such variations in I_{MSF} and D_c affect considerably the estimated SFR of overhead power lines, as demonstrated for the evaluated overhead transmission lines in Table 5.6. Generally, the application of electrogeometric models in shielding analysis, thus also adopted in IEEE Std. 1243 [23], results in the highest SFR values.

The I_{MSF} and SFR estimates in Table 5.6 refer to the outer and upper phase conductor of the evaluated single-circuit and double-circuit power lines,

Table 5.5 *Exposure width, D_c (m), and maximum shielding failure current, I_{MSF} (kA) expressions to be used in shielding failure analysis on the basis of lightning attachment models*

Electrogeometric models: factors A, B, and γ, as defined in (5.1), are given in Table 5.1

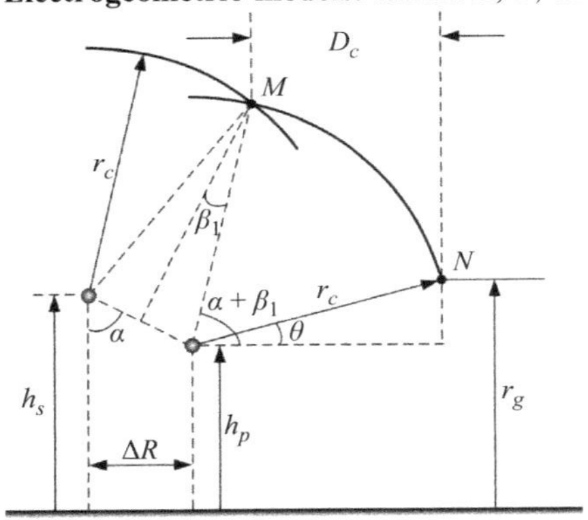

$$D_c = \begin{cases} AI^B[\cos\theta - \cos(\alpha+\beta_1)] \text{ for } I > (\gamma h_p/A)^{1/B} \\ AI^B[1 - \cos(\alpha+\beta_1)] \text{ for } I \leq (\gamma h_p/A)^{1/B} \end{cases}$$

$$\alpha = \tan^{-1}\Delta R/(h_s - h_p)$$

$$\theta = \sin^{-1}\frac{D - h_p}{S} = \sin^{-1}\left(\frac{1}{\gamma} - \frac{h_p}{AI^B}\right)$$

$$\beta_1 = \sin^{-1}\frac{\sqrt{\Delta R^2 + (h_s - h_p)^2}}{2AI^B}$$

$$I_{\text{MSF}} = \left[\frac{\gamma(h_s + h_p)/2}{A(1 - \gamma\sin\alpha)}\right]^{\frac{1}{B}}$$

Erikson's method

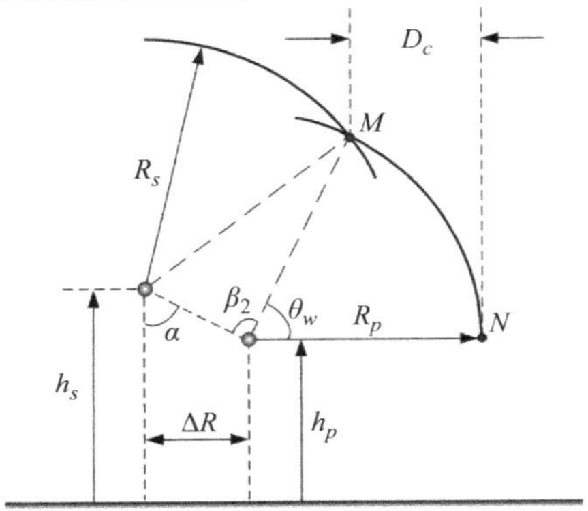

$$D_c = 0.67h_p^{0.6}I^{0.74}\left[1 - \cos\left(\alpha - \beta_2 + \frac{\pi}{2}\right)\right]$$

$$\beta_2 = \cos^{-1}\frac{R_p^2 - R_s^2 + \Psi^2}{2\Psi R_p}$$

$$\Psi = \sqrt{(h_s - h_p)^2 + \Delta R^2}$$

$$I_{\text{MSF}} = \left[\frac{\Delta R + \sqrt{\Delta R^2 + \Psi^2(\Gamma^2 - 1)}}{0.67h_p^{0.6}(\Gamma^2 - 1)}\right]^{\frac{1}{0.74}}$$

$$\Gamma = (h_s/h_p)^{0.6}$$

(Continues)

Table 5.5 (Continued)

Statistical model

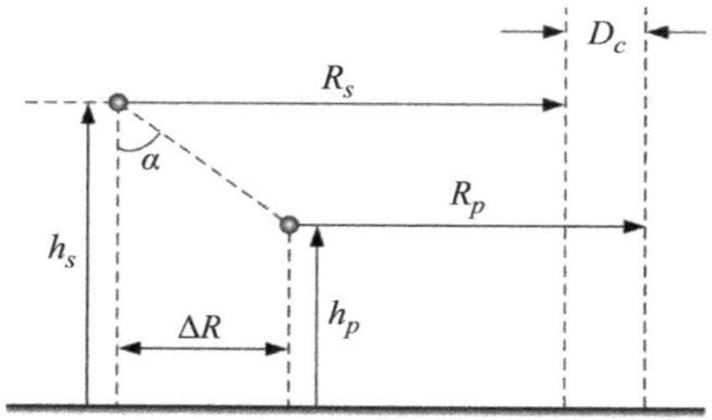

$$D_c = 2.72I^{0.65}\ln(h_p/h_s) + 0.01h_s^{1.3} + (h_s - h_p)\tan\alpha$$

$$I_{\mathrm{MSF}} = \left[\frac{(h_s - h_p)\tan\alpha + 0.01h_s^{1.3}}{2.72\ln(h_s/h_p)}\right]^{\frac{1}{0.65}}$$

Leader propagation models: factors ξ, E, F, ζ, and G, as defined in (5.24), are given in Table 5.2

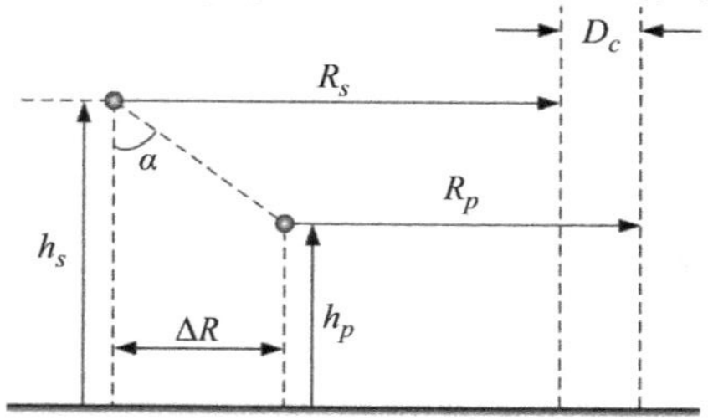

$$D_c = \xi\left(h_p^E - h_s^E\right)I^F + \zeta\left(h_p^G - h_s^G\right) + (h_s - h_p)\tan\alpha$$

$$I_{\mathrm{MSF}} = \left[\frac{(h_s - h_p)\tan\alpha - \zeta\left(h_s^G - h_p^G\right)}{\xi\left(h_s^E - h_p^E\right)}\right]^{\frac{1}{F}}$$

Note: D_c is the exposure width.
h_s and h_p height of shield wire and phase conductor, respectively.
α is the shielding angle; ΔR is the horizontal separation distance between shield wire and phase conductor.
r_c is the striking distance to conductors; r_g is the striking distance to earth surface.
R_s and R_p attractive radius of shield wire and phase conductor, respectively.

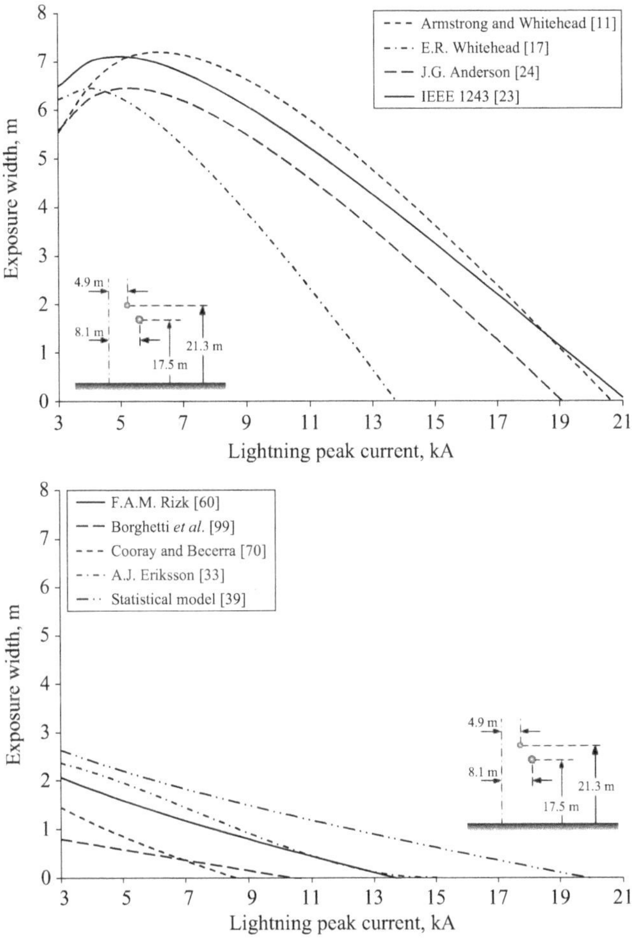

Figure 5.16 Variation of exposure width with lightning peak current as obtained by employing in shielding analysis of a 220 kV power line several lightning attachment models. Maximum shielding failure current is given by the intersection of the curves with x-axis

respectively. They also refer to the average line height, that is, height at the tower minus two-thirds of the sag of the conductors; using the height of the shield wire and phase conductor at the tower results in larger I_{MSF} and SFR values due to the larger shielding angle at the tower. A more complete shielding analysis requires the estimation of SFR of all phase conductors by considering the shielding effect provided to the lower phase conductors by, besides shield wire(s), the higher phase conductors.

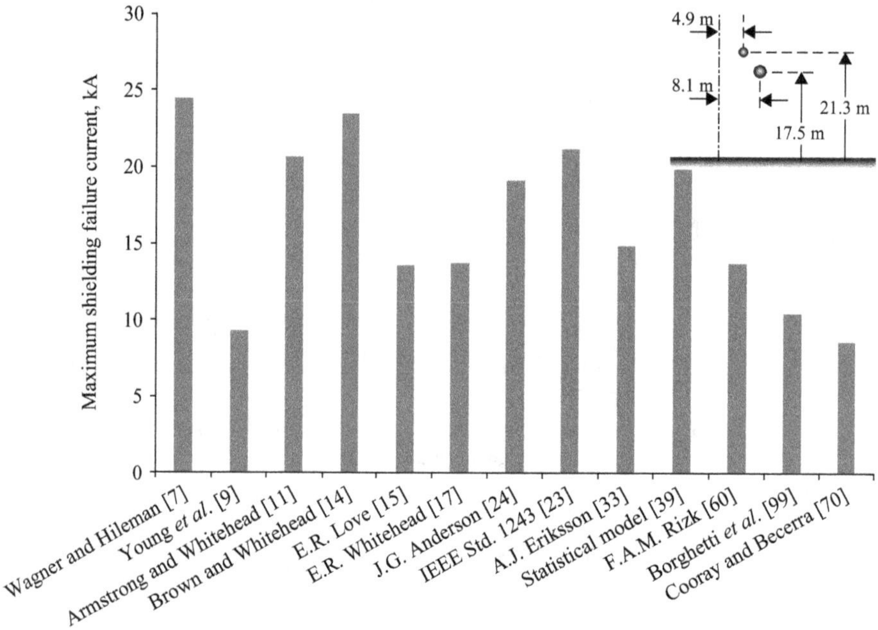

Figure 5.17 Maximum shielding failure current as obtained by employing in shielding analysis of an 220 kV power line several lightning attachment models

5.3.3.2 Effects of lightning peak current distribution

The SFR estimates in Table 5.6 were obtained by using in (5.32) the two-slope CIGRE distribution ($\bar{I} = 61.1$ kA and $\sigma_{\ln} = 1.34$ for $I \leq 20$ kA, else $\bar{I} = 33.3$ kA and $\sigma_{\ln} = 0.61$) [103], which is also recommended by the IEEE Std. 1243 [23]; the use of this lightning peak current distribution generally results in conservative SFR estimates; this can be deduced from (5.32) together with Figure 5.18, and has also been shown in [108].

5.3.4 Concluding remarks on lightning incidence due to direct lightning strokes

General procedures and closed-form expressions for the estimation of lightning incidence to overhead power lines on the basis of several engineering models of lightning attachment has been presented. It has been shown that estimating the lightning stroke collection rate of shield wire(s) and phase conductors of overhead power lines depends upon the lightning attachment model used for incidence and shielding analyses. Experimental and theoretical studies on lightning attachment to overhead power lines will continue; although procedures for evaluating the lightning incidence and shielding performance of overhead power lines are well established, under the lack of sufficient reliable field data consensus on the use of

Table 5.6 Maximum shielding failure current, I_{MSF} (kA), and shielding failure rate, SFR (strikes/100 km/yr), of typical 115 ÷ 735 kV overhead transmission lines; $N_g = 1$ flash/km^2/yr

Lightning attachment model	Overhead transmission line*													
	115/138 kV		150 kV		230 kV		345 kV		400 kV		500 kV		735 kV	
	I_{MSF}	SFR	I_{MSF}	SFR	I_{MSF}	SFR	I_{MSF}	SFR	I_{MSF}	SFR	I_{MSF}	SFR	I_{MSF}	SFR
Wagner and Hileman [7]	5.5	0.008	17.6	0.092	10.9	0.027	13.0	0.037	19.3	0.084	4.6	0.006	13.2	0.057
Young *et al.* [9]	1.3	0	6.2	0.017	3.2	0.003	4.1	0.005	7.0	0.018	1.0	0	4.2	0.007
Armstrong and Whitehead [11]	7.6	0.017	14.6	0.091	10.8	0.032	11.7	0.038	14.5	0.072	6.7	0.014	12.1	0.064
Brown and Whitehead [14]	8.1	0.018	16.3	0.102	11.7	0.036	12.8	0.043	16.1	0.080	7.1	0.015	13.2	0.071
E.R. Love [15]	5.1	0.008	11.0	0.057	8.0	0.020	9.0	0.025	11.6	0.051	4.6	0.007	9.1	0.040
E.R. Whitehead [17]	5.4	0.009	11.2	0.059	8.3	0.021	9.2	0.026	11.9	0.052	4.8	0.008	9.3	0.042
J.G. Anderson [24]	7.2	0.015	15.4	0.090	11.3	0.032	20.1	0.072	26.9	0.157	10.0	0.023	41.1	0.339
IEEE Std. 1243 [23]	6.4	0.012	16.7	0.104	11.5	0.034	13.4	0.044	19.8	0.099	5.5	0.009	12.7	0.064
A.J. Eriksson [33]	6.4	0.003	10.1	0.014	6.7	0.004	7.0	0.006	8.4	0.012	6.2	0.007	9.1	0.016
Statistical model [39]	2.1	0.001	15.6	0.040	10.1	0.014	9.1	0.014	12.8	0.032	3.0	0.001	10.0	0.024
F.A.M. Rizk [60]	3.7	0.002	7.9	0.012	4.0	0.002	3.2	0.001	4.6	0.004	1.6	0	5.0	0.006
Cooray and Becerra [70]	1.8	0	4.3	0.003	1.9	0	1.4	0	2.1	0.001	0.6	0	2.4	0.001

*Parameters of lines as given in [103] and [104].

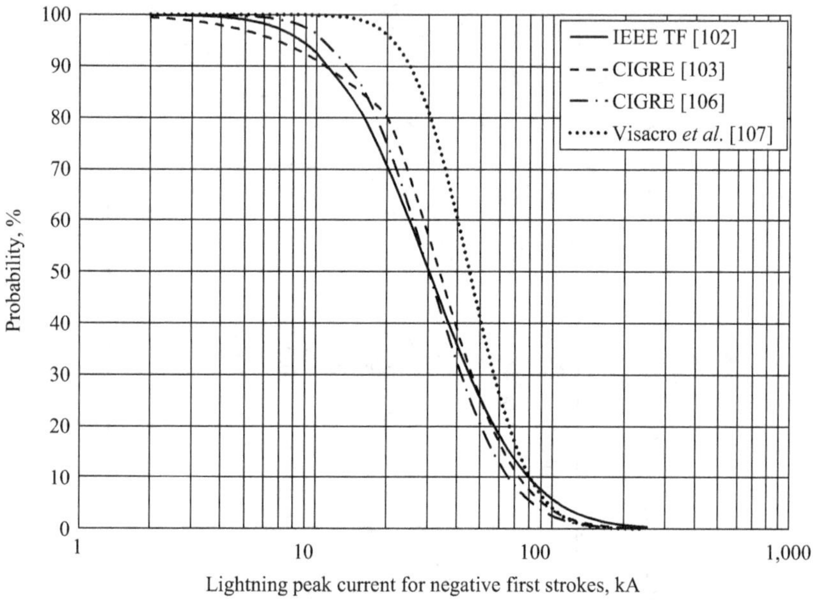

Figure 5.18 Cumulative statistical distributions of peak currents (percent values) for negative first strokes

an engineering model of lightning attachment is difficult to be reached. The estimation of the lightning stroke collection rate of line conductors also depends on the adopted statistical distribution of the lightning peak current. As the latter varies seasonally and geographically, using, instead of a "global," a "local" distribution of lightning peak current obtained from direct current measurements on instrumented towers would certainly result in an more accurate evaluation of the lightning performance of overhead power lines.

References

[1] R.H. Golde, "The frequency of occurrence and the distribution of lightning flashes to transmission lines," *AIEE Trans.*, vol. 64, pp. 902–910, 1945.

[2] R.H. Golde, "The lightning conductor," *J. Franklin Inst.*, vol. 283, no. 6, pp. 451–477, 1967.

[3] R.H. Golde, "Lightning and tall structures," *Proc. IEE*, vol. 125, no. 4, pp. 347–351, 1978.

[4] W.S. Price, S.C. Bartlett, and E.S. Zobel, "Lightning and corona performance of 330-kV lines of the American Gas and Electric and the Ohio Valley Electric Corporation Systems," *AIEE Trans.*, Part. III, vol. 75, pp. 583–597, 1956.

[5] H.R. Armstrong, and E.R. Whitehead, "A lightning stroke pathfinder," *IEEE Trans. PA&S*, vol. 83, no. 12, pp. 1223–1227, 1964.
[6] C.F. Wagner, "The relation between stroke current and the velocity of the return stroke," *IEEE Trans. PA&S*, pp. 609–617, 1963.
[7] C.F. Wagner, and A.R. Hileman, "The lightning stroke-II," *AIEE Trans. PA&S*, pp. 622–642, 1961.
[8] A.R. Hileman, in Discussion of ref. [28].
[9] F.S. Young, J.M. Clayton, and A.R. Hileman, "Shielding of transmission lines," *IEEE Trans. Power App. Syst.*, vol. S82, no. 4, pp. 132–154, 1963.
[10] AIEE Committee Report, "A method of estimating lightning performance of transmission lines," *AIEE Trans.*, vol. 69, pp. 1187–96, 1950.
[11] H.R. Armstrong, and E.R. Whitehead, "Field and analytical studies of transmission line shielding," *IEEE Trans. Power App. Syst.*, vol. 87, pp. 270–281, 1968.
[12] Y. Watanabe, "Switching surge flashover characteristics of extremely long air gaps," Paper 31 TP 66-420, presented at the *IEEE Summer Power Meeting*, New Orleans, LA, July 10–15, 1966.
[13] L. Paris, "Influence of air gap characteristics on line to ground switching surge strength," Paper 31 TP 66–421 presented at the *IEEE Summer Power Meeting*, New Orleans, LA, July 10–15, 1966.
[14] G.W. Brown, and E.R. Whitehead, "Field and analytical studies of transmission line shielding: Part II," *IEEE Trans. PA&S*, vol. PAS-88, pp. 617–626, 1969.
[15] E.R. Love, "Improvements in lightning stroke modeling and applications to design of EHV and UHV transmission lines," M.Sc. thesis, University of Colorado, Denver, CO, 1973.
[16] D.W. Gilman, and E.R. Whitehead, "The mechanism of lightning flashover on high-voltage and extra-high-voltage transmission lines," *Electra*, vol. 27, pp. 65–96, 1973.
[17] E.R. Whitehead, "CIGRE survey of the lightning performance of EHV transmission lines," *Electra*, vol. 33, pp. 63–89, 1974.
[18] L.D. Darveniza, F. Popolansky, and E.R. Whitehead, "Lightning protection of UHV lines," *Electra*, vol. 41, pp. 39–69, 1975.
[19] G.W. Brown, "Lightning performance: I. Shielding failures simplified," *IEEE Trans. PA&S*, vol. PAS-97, no. 1, pp. 33–38, 1978.
[20] G.W. Brown, "Lightning performance: II. Updating backflash calculations," *IEEE Trans. PA&S*, vol. PAS-97, no. 1, pp. 39–52, 1978.
[21] IEC 62305, "Protection against lightning," 2010
[22] IEEE Guide for Improving the Lightning Performance of Electric Power Overhead Distribution Lines, IEEE Std. 1410–2011, January 2011.
[23] IEEE Guide for Improving the Lightning performance of Transmission Lines, IEEE Std. 1243–1997, December 1997.
[24] J.G. Anderson, *Transmission Line Reference Book – 345 kV and Above*, 2nd edn, 1982, Chapter 12, Electric Power Research Institute, Palo Alto, CA.

[25] A.M. Mousa, and K.D. Srivastava, "The Implications of the electrogeometric model regarding the effect of height of structure on the median amplitude of collected lightning strokes," *IEEE Trans. Power Delivery*, vol. 4, no. 2, pp. 1450–1460, 1989.

[26] A.M. Mousa, and K.D. Srivastava, "Modelling of power lines in lightning incidence calculations," *IEEE Trans. Power Delivery*, vol. 5, no. 1, pp. 303–310, 1990.

[27] J. Takami, and S. Okabe, "Characteristics of direct lightning strokes to phase conductors of UHV transmission lines," *IEEE Trans. Power Delivery*, vol. 22, pp. 537–546, 2007.

[28] S. Taniguchi, T. Tsuboi, and S. Okabe, "Observation results of lightning shielding for large-scale transmission lines," *IEEE Trans. Dielectr. Electr. Insul.*, vol. 16, pp. 552–559, 2009.

[29] S. Taniguchi, T. Tsuboi, S. Okabe, Y. Nagaraki, J. Takami, and H. Ota, "Improved method of calculating lightning stroke rate to large-sized transmission lines based on electric geometry model," *IEEE Trans. Dielectr. Electr. Insul.*, vol. 17, no. 1, pp. 53–62, 2010.

[30] S. Taniguchi, S. Okabe, A. Asakawa, and T. Shindo, "Flashover characteristics of long air gaps with negative switching impulses," *IEEE Trans. Dielectr. Electr. Insul.*, vol. 15, pp. 399–406, 2008.

[31] V.P. Idone, and R.E. Orville, "Lightning return stroke velocities in the Thunderstorm Research International Program (TRIP)," *J. Geophys. Res.*, vol. 87, no. C7, pp. 4903–4916, 1982.

[32] J.L. He, R. Zeng, C.J. Zhuang, *et al.*, "Evaluation of lightning shielding analysis methods for EHV and UHV DC and AC transmission-lines," *Electra*, no. 295, pp. 39–43, 2017. Technical Brochure no. 704, WG C4.26.

[33] A.J. Eriksson, "The incidence of lightning strikes to power lines," *IEEE Trans. Power Delivery*, vol. PWRD-2, no. 3, pp. 859–870, 1987.

[34] A.J. Eriksson, "An improved electrogeometric model for transmission line shielding analysis," *IEEE Trans. Power Delivery*, vol. PWRD-2, no. 3, pp. 871–886, 1987.

[35] K. Berger, "Method and results of research on lightning on Mount San Salvatore, 1963–1971," *Bull. SEV*, vol. 63, no. 24, pp. 1403–1422, 1972.

[36] G. Carrara, and L. Thione, "Switching surge strength of large air gaps: a physical approach," *IEEE Trans. PA&S*, vol. PAS-95, pp. 512–520, 1976.

[37] P.N. Mikropoulos, and T.E. Tsovilis, "Striking distance and interception probability," *IEEE Trans. Power Delivery*, vol. 23, no. 3, pp. 1571–1580, 2008.

[38] P.N. Mikropoulos, and T.E. Tsovilis, "Interception probability and shielding against lightning," *IEEE Trans. Power Delivery*, vol. 24, no. 2, pp. 863–873, 2009.

[39] P.N. Mikropoulos, and T.E. Tsovilis, "Interception probability and proximity effects: Implications in shielding design against lightning," *IEEE Trans. Power Delivery*, vol. 25, no. 3, pp. 1940–1951, 2010.

[40] T. Suzuki, K. Miyake, and T. Shindo, "Discharge path model in model test of lightning strokes to tall mast," *IEEE Trans. Power App. Syst.*, vol. PAS-100, no. 7, pp. 3553–3562, 1981.

[41] S. Grzybowski, and G. Gao, "Laboratory study of Franklin rod height impact on striking distance," *Proceedings of the 25th International Conference on Lightning Protection*, Rhodes, Greece, pp. 334–339, 2000.

[42] Y. Goto, Y. Sato, I. Ono, and T. Sato, "Model experiment of lightning discharge characteristics to power line," *Proceedings of the 25th International Conference on Lightning Protection*, Rhodes, Greece, pp. 311–317, 2000.

[43] B.F.J. Schonland, D.J. Malan, and H. Collens, "Progressive lightning II," *Proc. R. Soc.*, A152, pp. 595–625, 1935.

[44] K. Berger, "Novel observations of lightning discharges: results of research on Mount S. Salvatore 3," *J. Frankl. Inst.*, vol. 283, pp. 478–525, 1967.

[45] E.M. Bazelyan, B.N. Gorin, and V.I. Levitov, *Physical and Engineering Foundations of Lightning Protection*, Leningrad: Gidrometeoizdat, 1978, 223 pp.

[46] B.N. Gorin, V.I. Levitov, and A.V. Shkilev, "Some principles of leader discharge of air gaps with a strong non-uniform field," *IEE Conference Publication on Gas Discharges*, vol. 143, pp. 274–278, 1976.

[47] B.N. Gorin, "Mathematical modelling of the lightning return stroke," *Elektrichestvo*, vol. 4, pp. 10–16, 1985.

[48] V.A. Rakov, "Some inferences on th propagation mechanisms of dart leaders and return strokes," *J. Geophys. Res.*, vol. 103, pp. 1879–1987, 1998.

[49] S. Larigaldie, A. Roussaud, and B. Jecko, "Mechanisms of high current pulses in lightning and long spark," *J. App. Phys.*, vol. 72, no. 5, pp. 1729–1739, 1992.

[50] A. Bondiou-Clergie, G.L. Bacchiega, A. Castellani, P. Lalande, P. Laroche, and I. Gallimberti, "Experimental and theoretical study of the bi-leader process. Part I: experimental investigation," *Proceedings of the 10th International Conference on Atmospheric Electricity*, Osaka, Japan, 1996, pp. 244–247.

[51] L. Dellera, A. Pigini, and E. Garbagnati, "A lightning model based on the similarity between lightning phenomena and long laboratory sparks," *Proceedings of ICLP 16th International Conference on Lightning Protection*, Szeged (H), 1981.

[52] L. Dellera, and E. Garbagnati, "Lightning stroke simulation by means of the leader progression model. Part I: description of the model and evaluation of exposure of free-standing structures," *IEEE PWRD*, vol. 5, no. 4, pp. 2009–2022, 1990.

[53] R. Cortina, L. Dellera, E. Garbagnati, *et al.*, "Some aspects of the evaluation of the lightning performance of electric systems," *CIGRE 1980 Session Report 33-13*.

[54] L. Dellera, A. Pigini, and E. Garbagnati, "Lightning simulation by means of a leader progression model," *Proceedings of International Aerospace Conference on Lightning and Static Electricity*, Oxford, March 1982.

[55] L. Dellera, and E. Garbagnati, "Lightning stroke simulation by means of the leader progression model. Part II. Exposure and shielding failure evaluation of overhead lines with assessment of application graphs," *IEEE PWRD*, vol. 5, no. 4, pp. 2023–2029, 1990.

[56] M. Bernardi, L. Dellera, E. Garbagnati *et al.*, "Leader progression model of lightning: updating of the model on the basis of recent test results," *Proceedings of ICLP 23rd International conference on Lightning Protection*, Florence (I), 1996.

[57] A. Borghetti, F. Napolitano, C.A. Nucci, M. Paolone, and M. Bernardi, "Numerical solution of the leader progression model by means of the finite element method," *Proceedings of the 30th International Conference on Lightning Protection (ICLP)*, Cagliari, Italy, 2010.

[58] F. Rizk, "Switching impulse strength of air insulation: leader inception criterion," *IEEE Trans.*, vol. PWRD-4, no. 4, pp. 2187–2195, 1989.

[59] F. Rizk, "A model for switching impulse leader inception and breakdown of long air-gaps," *IEEE Trans.*, PWRD-4, no. 1, pp. 596–606, 1989.

[60] F.A.M. Rizk, "Modeling of transmission line exposure to direct lightning strokes," *IEEE Trans. Power Delivery*, vol. 5, pp. 1983–1997, 1990.

[61] F. Rizk, "Modelling of lightning incidence to tall structures. Part I: theory," *IEEE Trans.*, PWRD-9, no. 1, pp. 162–171, 1994.

[62] F. Rizk, "Modelling of lightning incidence to tall structures. Part II: application," *IEEE Trans.*, PWRD-9, no. 1, pp. 172–182, 1994.

[63] F. Rizk, "Critical switching impulse strength of long air gaps: modelling of air density effects," *IEEE Trans.*, PWRD-7, no. 3, pp. 1507–1515, 1992.

[64] M. Becerra, and V. Cooray, "A simplified physical model to determine the lightning upward connecting leader inception," *IEEE Trans. Power Delivery*, vol. 21, no. 2, pp. 897–908, 2006.

[65] M. Becerra, and V. Cooray, "Time dependent evaluation of the lightning upward connecting leader inception," *J. Phys. D: Appl. Phys.*, vol. 39, no. 21, pp. 4695–4702, 2006.

[66] M. Becerra, and V. Cooray, "A self-consistent upward leader propagation model," *J. Phys. D: Appl. Phys.*, vol. 39, no. 16, pp. 3708–3715, 2006.

[67] M. Becerra, and V. Cooray, "Dynamic modelling of the lightning upward connecting leader inception," *Proceedings of the 28th International Conference on Lightning Protection*, ICLP, Kanazawa, Japan, 2006, pp. 543–548.

[68] M. Becerra, and V. Cooray, "An improved upward leader propagation model," *Proceedings of the 28th International Conference on Lightning Protection*, ICLP, Kanazawa, Japan, 2006, pp. 581–586.

[69] V. Cooray, "A review of simulation procedures utilized to study the attachment of lightning flashes to grounded structures," document for CIGRE WG C4.405 "lightning interception," 2008.

[70] V. Cooray, and M. Becerra, "Attractive radius and the volume of protection of vertical and horizontal conductors evaluated using a self-consistent leader inception and propagation model – SLIM," *30th International Conference on Lightning Protection*, Cagliari, Italy, Paper No. 1062, 2010.

[71] M. Becerra, "Self-consistent leader inception and propagation model – SLIM: response to some criticism," *Proceedings of the 31st International Conference on Lightning Protection (ICLP)*, Wien, Austria, 2012.
[72] I. Gallimberti, "The mechanism of long spark formation," *J. Phys. Coll.*, Vol. 40, C7, Suppl. 7, pp. 193–250, 1972.
[73] N.I. Petrov, V.R. Avanskii, and N.V. Bombenkova, "Measurement of the electric field in the streamer zone and in the sheath of the channel of a leader discharge," *Tech. Phys.*, vol. 39, p. 546, 1994.
[74] A. Bondiou-Clergie, and I. Gallimberti, "Theoretical modelling of the development of the positive spark in long air gaps," *J. Phys. D: Appl. Phys.*, vol. 27, pp. 1252–1266, 1994.
[75] P. Lalande, "Study of the lightning stroke conditions on a grounded structure," Doctoral Thesis, Pub. ONERA, 1996.
[76] N. Goelian, P. Lalande, A. Bondiou-Clergie, G.L. Bacchiega, A. Gazzani, and I. Gallimberti, "A simplified model for the simulation of positive-spark developments in long air gaps," *J. Phys. D: Appl. Phys.*, vol. 30, pp. 2441–2452, 1997.
[77] W. Sima, Y. Li, V.A. Rakov, Q. Yang, T. Yuan, and M. Yang, "An analytical method for estimation of lightning performance of transmission lines based on a leader progression model," *IEEE Trans. Elec. Comp.*, vol. 56, no. 6, pp. 1530–1539, 2014.
[78] J. He, Y. Tu, R. Zeng, J.B. Lee, S.H. Chang, and Z. Guan, "Numeral analysis model for shielding failure of transmission line under lightning stroke," *IEEE Trans. Power Delivery*, vol. 20, no. 2, 2005.
[79] H. Singer, H. Steinbigler, and P. Weiss, "A charge simulation method for the calculation of high voltage fields," *IEEE Power Apparatus Syst.*, 1974, pp. 1660–1668.
[80] J.G. Anderson, Lightning performance of transmission lines," *Transmission Lines Reference Book*, Chapter 12, EPRI, 1982.
[81] Les Renardieres Group, "Research on long air gap discharges at Les Renardieres," *Electra*, no. 29, pp. 53–157, 1972.
[82] Les Renardieres Group, "Research on long air gap discharges at Les Renardieres – 1973 Results," *Electra*, no. 35, pp. 49–156, 1974.
[83] Les Renardieres Group, "Positive discharges in long air gap discharges at Les Renardieres – 1975 results and conclusions," *Electra*, no. 53, pp. 31–153, 1977.
[84] Les Renardieres Group, "Negative discharges in long air gap discharges at Les Renardieres," *Electra*, no. 74, pp. 67–216, 1981.
[85] A. Brambilla, E. Garbagnati, A. Pigini, *et al.*, "Switching impulse strength of very large air gaps," *Third International Symposium on High Voltage Engineering – ISH*, Milan, 1979.
[86] G.D. McCann, "The measurements of lightning currents in direct strokes," *AIEE Trans.*, vol. 63, pp. 1157–1164, 1944.
[87] E.T. Pierce, "Triggered lightning and same unsuspected lightning hazards," Standard Research Institute, *Scientific Note* 15, January 1972.

[88] K. Berger, and E. Vogelsanger, "Measurements and results of lightning records at Monte S. Salvatore from 1955 to 1963," *Bull. SEV*, vol. 56, no. 1, 1965 (in German).

[89] K. Berger, "Novel observations of lightning discharges: results of research on Mount S. Salvatore 3," *J. Frankl. Inst.*, vol. 283, pp. 478–525, 1967.

[90] K. Berger, "Method and results of research on lightning on Mount S. Salvatore 1963, 1971," *Bull. SEV* (ASE), vol. 63, no. 24, p. 1403, 1972 (in German).

[91] K. Berger, "The earth flash," Chapter 5, in *Lightning*, vol. 1, Academic Press, London, 1977.

[92] R.B. Anderson, and A.J. Eriksson, "Lightning parameters for engineering applications," *Electra*, no. 69, p. 65, 1980.

[93] A. Borghetti, F. Napolitano, C.A. Nucci, M. Paolone, and M. Bernardi, "Numerical solution of the leader progression model by means of the finite element method," *Proceedings of the 30th International Conference on Lightning Protection (ICLP)*, Cagliari, Italy, 2010.

[94] V. Cooray, V. Rakov, and N. Theethayi, "The relationship between the leader charge and the return stroke current – Berger's data revisited," *Proceedings of the 27th International Conference on Lightning Protection*, Avignon, France, 2004.

[95] C.B. Moore, G.D. Aulich, and W. Rison, "Measurements of lightning rod responses to nearby strikes," *Geo. Res. Lett.*, vol. 27, pp. 1487–1490.

[96] C.B. Moore, W. Rison, J. Mathis, and G.D. Aulich, "Lightning rod improvements," *J. App. Meteo.*, vol. 39, pp. 593–609.

[97] P. Lalande, A. Bondiou-Clergie, G. Bacchiega, and I. Gallimberti, "Observations and modelling of lightning leaders," *C.R. Phys.*, no. 3, pp. 1375–1392, 2002.

[98] G. Baldo, "Lightning protection and the physics of discharge," *Proceedings of ISH*, 1999.

[99] A. Borghetti, C.A. Nucci, M. Paolone, and M. Bernardi, "Effect of the lateral distance expression and of the presence of shielding wires on the evaluation of the number of lightning induced voltages," *Proceedings of the 25th International Conference on Lightning Protection (ICLP)*, Rhodes, Greece, 2000.

[100] CIGRE Document No. 118, "Lightning exposure of structures and interception efficiency of air terminals," October 1997.

[101] J. Tarchini, "The use of leader progression model to predict lightning incidence in power lines," *Proceedings of the 27th International Conference on Lightning Protection (ICLP)*, Avignon, France, 2004.

[102] Lightning and Insulator Subcommittee of the T&D Committee, "Parameters of lightning strokes: a review," *IEEE Trans. Power Delivery*, vol. 20, no. 1, pp. 346–358, 2005.

[103] CIGRE Working Group 33.01, "Guide to procedures for estimating the lightning performance of transmission lines," *Tech. Bull.*, vol. 63, 1991.

[104] Z.G. Datsios, P.N. Mikropoulos and T.E. Tsovilis, "Estimation of the minimum shielding failure flashover current for first and subsequent lightning strokes to overhead transmission lines," *Electr. Power Syst. Res.*, vol. 113 (SI), pp. 141–150, 2014.

[105] P.N. Mikropoulos, and T.E. Tsovilis, "Estimation of lightning incidence to overhead transmission lines," *IEEE Trans. Power Delivery*, vol. 25, no. 3, pp. 1855–1865, 2010.

[106] V.A. Rakov, A. Borghetti, C. Bouquegneau, *et al.*, "Lightning parameters for engineering applications," *Electra*, no. 269_7, 2013. Technical Brochure no. WG C4.407.

[107] S. Visacro, C.R. Mesquita, A. De Conti, and F.H. Silveira, "Updated statistics of lightning currents measured at Morro do Cachimbo station," *Atmos. Res.*, vol. 117, pp. 55–63, 2012.

[108] P.N. Mikropoulos, and T.E. Tsovilis, "Estimation of the shielding performance of overhead transmission lines: the effects of lightning attachment model and lightning crest current distribution," *IEEE Trans. Dielectr. Electr. Insul.*, vol. 19, no. 6, pp. 2155–2164, 2012.

Chapter 6

Field-to-transmission line coupling models

Vernon Cooray[1], *Carlo Alberto Nucci*[2], *Alexandre Piantini*[3], *Farhad Rachidi*[4] *and Marcos Rubinstein*[5]

This chapter presents the coupling of lightning electromagnetic fields to overhead and underground lines based on the transmission line approximation. The chapter starts with an introductory section in which the advantages and the drawbacks of three approaches, namely the quasi-static approach, the transmission line approach, and the antenna approach, are presented and discussed. The conditions of validity of the application of the transmission line approximation and its suitability for the case of lightning generated fields are also presented in that section.

Section 6.2 is devoted to the derivation of the coupling equations for the most widely used and extensively validated field-to-transmission line coupling models: the Taylor *et al.* model, the Agrawal *et al.* model, the Rachidi model, and the Rusck model.

The single-wire derivations are next extended to account for the finite ground conductivity case and then the case of multiconductor lines.

It is also demonstrated that the underlying equations of all of the models are equivalent and that they therefore all produce identical results if the line as well as the excitation fields are the same.

The coupling models presented in Section 6.2 were initially developed for the case of overhead lines, but they have been subsequently adapted to be also applicable to underground cables. The coupling to underground lines is dealt with in Section 6.3.

A brief discussion of a time domain representation of the field-to-transmission line coupling equations is provided in Section 6.4, while in Section 6.5 comparisons of measured and calculated lightning-induced voltages on overhead lines are provided which illustrate the validity of the coupling models presented in Section 6.2.

[1]Uppsala University, Department of Engineering Sciences, Uppsala, Sweden
[2]University of Bologna, Dept. of Electrical, Electronic and Information Engineering, Bologna, Italy
[3]University of São Paulo, Institute of Energy and Environment, São Paulo, Brazil
[4]Swiss Federal Institute of Technology (EPFL), Institute of Electrical Engineering, Lausanne, Switzerland
[5]University of Applied Sciences of Western Switzerland, Advanced Communication Systems Group, Yverdon-les-Bains, Switzerland

6.1 Introduction (TL approximation, QS approximation, and full-wave approach)

Electromagnetic fields from natural sources such as lightning, or from artificial sources such as intentional electromagnetic interference, can propagate over long distances and can create disturbances on transmission and distribution lines, frequently leading to outages and damages to line elements or electronics at consumer premises. An understanding of the mechanisms that govern the coupling of external electromagnetic fields to overhead and underground lines is of great importance as a tool in the development and testing of effective protection methods.

Depending on the required degree of precision, the available resources and the relation between the line dimensions and the frequencies of interest, approaches can be used for the calculation of the voltages and currents induced by external electromagnetic fields on overhead and underground lines: the quasi-static approach, the transmission line approach, and the antenna approach, also called antenna theory approach or full-wave approach. The remainder of this section is devoted to a brief discussion of the main advantages and disadvantages of each of these three approaches.

The simplest of the three approaches is based on the quasi-static approximation [1]. This approach is only applicable if the wavelengths involved are much longer than any of the relevant dimensions of the line or, in a more general sense, of the victim system. Under these conditions, propagation can be neglected and, therefore, the overhead or underground cables, as well as their terminations and any other elements of the system, can be modeled as lumped elements. The quasi-static approach is however not generally applicable to lightning fields that couple to transmission and distribution lines since the wavelength condition is rarely satisfied. Indeed, power networks have lengths typically measured in kilometers and the smallest wavelength associated with lightning is of the order of 100 m [2].

The antenna approach is based on the numerical resolution of Maxwell's equations [1]. This approach is the most accurate and versatile of the three approaches presented here since it can account for nonideal line characteristics such as a nonuniform cross section, realistic terrain elevation profiles, the presence of multiple modes, and so on. However, it is computationally intensive in terms of memory and the required number of operations when, as is often the case in practical applications, electrically long lines are involved.

For numerous practical applications, the third approach, based on the transmission line theory, represents an excellent trade-off between the inaccuracies of the simple quasi-static approximation method and the computational cost associated with the antenna approach. Indeed, the transmission line approach can yield results with excellent accuracy using reasonable computer resources.

The main assumptions on which the transmission line approximation is based can be described for an overhead line composed of horizontal wires parallel to the ground. The assumptions are (1) that the propagation of the coupled voltages and currents occurs along the line axis, (2) that the net sum of the line currents at any cross section of the line is zero, and (3) that the response of the line to the electromagnetic fields is quasi-transverse electromagnetic (quasi-TEM). The last assumption implies that the electric and magnetic field components from the induced electric charges and currents along the line are approximately in the transverse plane of the line, perpendicular to the line axis (only small components of the electric and magnetic fields can exist in the direction of propagation).

The assumption of propagation along the line is a good approximation for lines of electrically small cross-sectional dimensions. The condition of net-zero current at any position along the line is satisfied by lines whose currents can be obtained by the method of images. This is the case for lines with a ground made of a sufficiently high conductivity. Finally, the condition that the response of the line is quasi-TEM is only satisfied up to a threshold frequency, above which higher-order modes can also propagate [1]. For typical transmission lines and for the frequency content of lightning fields, this last condition is generally satisfied.

The transmission line approach is the most widely used technique due to its accuracy and flexibility and it is the main subject of the rest of the chapter.

6.2 Field-to-transmission line coupling models for overhead lines

The first complete model based on the transmission line approach for the coupling of external electromagnetic fields to transmission lines was presented by Taylor *et al.* [3] in 1965. Two other complete coupling models, also based on the transmission line approach, the model of Agrawal *et al.* [4] and the Rachidi Model [5], were introduced, respectively, in 1980 and in 1995. Note that 7 years before Taylor *et al.*'s original paper, Rusck presented a somewhat less general coupling model [6] that, in its original form, was applicable only to the case of straight and vertical sources and lossless lines.

We will now present the different coupling models starting with the analysis of an ideal, single-wire, horizontal line above a flat ground. Both, the wire and the ground will be initially considered to be perfect conductors. We will then build on the coupling equations derived for this canonical case to include the effect of losses due to a finitely conducting ground and wire, and to extend the models to the case of multiconductor lines.

The horizontal wire is assumed to be cylindrical in shape with a radius a and a height h above the ground, as illustrated in Figure 6.1. The line is assumed to be terminated on the left-hand side and the right-hand side, respectively, in impedances Z_A and Z_B.

Physically, the coupling of external fields to overhead lines can be understood as follows: a distant source (a lightning strike in this case) produces electromagnetic radiation that propagates over the ground, inducing currents in it that also radiate. The superposition of the fields from the lightning and the fields from the currents induced in the ground (in absence of the overhead line) forms the exciting fields that will eventually reach the line. These exciting electric and magnetic fields

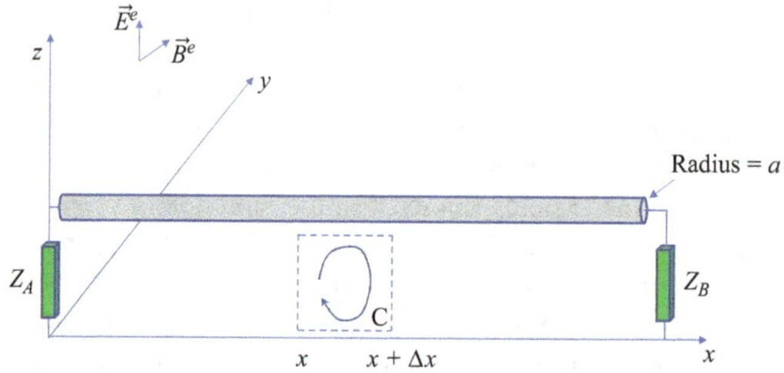

Figure 6.1 A single-conductor line terminated in impedances Z_A and Z_B. $\vec{E}^e$ and $\vec{B}^e$ are the exciting electric and magnetic fields. The rectangular contour C is used for the derivation of the first generalized Telegrapher's equation

are denoted, respectively, $\vec{E}^e$ and $\vec{B}^e$ and they are included in Figure 6.1. As the line becomes immersed in the exciting fields, currents are induced both, in the conducting overhead wire and in the terminations. These newly generated currents act themselves as sources of electromagnetic radiation that will induce new currents in the ground that will themselves reradiate until an equilibrium is reached. The fields generated by the currents and charges flowing in the line conductor and by the corresponding currents and charges induced in the ground are called scattered fields. These newly generated scattered fields add to the original impinging exciting fields by superposition to form the total fields as discussed further in the next paragraph.

Formally, the external exciting electric and magnetic fields $\vec{E}^e$ and $\vec{B}^e$ are defined as the fields from the source, including the effect of the ground but in absence of the overhead line conductor. The total fields $\vec{E}$ and $\vec{B}$, on the other hand, are given by the sum of the excitation fields and the scattered fields. The latter were defined in the previous paragraph and they are denoted as $\vec{E}^s$ and $\vec{B}^s$.

The four complete models mentioned earlier in this section, namely the model of Taylor *et al.* [3], the model of Agrawal *et al.* [4], the Rachidi model [5], and the modified Rusck model [7] will be presented in Subsections 6.2.1 through 6.2.5.

6.2.1 Derivation of the generalized Telegrapher's equations for the model of Taylor et al.

The propagation of voltages and currents in transmission lines with voltage or current sources at their extremities are governed by two coupled differential equations known as the Telegrapher's equations, derived in the 1880s by Olivier Heaviside. For lossless transmission lines, these equations are

$$\frac{dV(x)}{dx} + j\omega L' I(x) = 0 \tag{6.1}$$

$$\frac{dI(x)}{dx} + j\omega C' V(x) = 0 \tag{6.2}$$

This version of the Telegrapher's equations is not directly applicable in the presence of radiation. We will now derive the generalized Telegrapher's equations for the case of external exciting electromagnetic fields according to the formulation of Taylor *et al.* [3] for a single conductor line with reference to Figure 6.1.

6.2.1.1 Derivation of the first generalized Telegrapher's equation

We begin by writing Maxwell–Faraday's equation, $\nabla \times \vec{E} = -j\omega\vec{B}$, in integral form:

$$\oint_C \vec{E} \cdot \vec{dl} = -j\omega \iint_S \vec{B} \cdot \vec{e}_y \, dS \tag{6.3}$$

Let us apply (6.3) to the area enclosed by the dashed rectangle under the overhead line shown in Figure 6.1. The line integral on the left side of (6.3) will be carried out along the contour C and the surface integral on the right-hand side will be performed on the surface S defined by C. Note that the surface vector, according to the right-hand rule, points in the positive y-direction. The result is

$$\int_0^h [E_z(x,z) - E_z(x+\Delta x, z)]dz + \int_x^{x+\Delta x} [E_x(x,h) - E_x(x,0)]dx$$
$$= -j\omega \int_0^h \int_x^{x+\Delta x} B_y(x,z)dxdz \tag{6.4}$$

Note that, in (6.4), $y = 0$ (see Figure 6.1). From now on, we will only write the y dependence explicitly whenever y is not constant.

Dividing (6.4) by Δx and taking the limit as Δx approaches zero yields

$$-\frac{\partial}{\partial x}\int_0^h E_z(x,z)dz + E_x(x,h) - E_x(x,0) = -j\omega \int_0^h B_y(x,z)dz \tag{6.5}$$

Since the wire and the ground are assumed to be perfect conductors, the total tangential electric fields, $E_x(x,h)$ and $E_x(x,0)$, must equal zero due to the continuity of the tangential electric field and the fact that the electric field inside a perfect conductor must be identical to zero. Let us now define the total transverse voltage $V(x)$ in the quasistatic sense (since $h \ll \lambda$) as

$$V(x) = -\int_0^h E_z(x,z)dz \tag{6.6}$$

where the integral on the right-hand side is carried out along a straight vertical line from the ground to the line conductor. Using (6.6), (6.5) can be written as

$$\frac{dV(x)}{dx} = -j\omega \int_0^h B_y(x,z)dz = -j\omega \int_0^h B_y^e(x,z)dz - j\omega \int_0^h B_y^s(x,z)dz \tag{6.7}$$

where the two terms on the right-hand side stem from the decomposition of the B field into its excitation and scattered components, indicated by the superscripts e and s.

The last integral in (6.7) represents the magnetic flux between the conductor and the ground produced by the current $I(x)$ flowing in the conductor.

To derive the relation between the current and the magnetic flux, we will use Ampère–Maxwell's equation in integral form, which we will apply here to the scattered fields:

$$\oint \vec{B}^s \cdot d\vec{l} = \mu_o I + j\omega\mu_o \iint \vec{D}^s \cdot d\vec{s} \tag{6.8}$$

Let us define an integration path with contour C' in the transverse plane, defined by a constant x in such a manner that the overhead conductor passes through it, as shown in Figure 6.2.

Applying (6.8) to C', we can write

$$\oint \vec{B}^s_T(x,y,z) \cdot d\vec{l} = \mu_o I(x) + j\omega\mu_o \iint \vec{D}^s_x(x,y,z) \cdot d\vec{S} \tag{6.9}$$

where the subscript T is used to indicate that the field is in the transverse direction, and where we have explicitly included the dependence of the fields on the three Cartesian coordinates.

Assuming that the response of the wire is TEM as required by the transmission line approximation, the scattered electric flux density in the x direction, D^s_x, is zero and (6.9) can be written as

$$\oint_{C'} \vec{B}^s_T(x,y,z) \cdot d\vec{l} = \mu_o I(x) \tag{6.10}$$

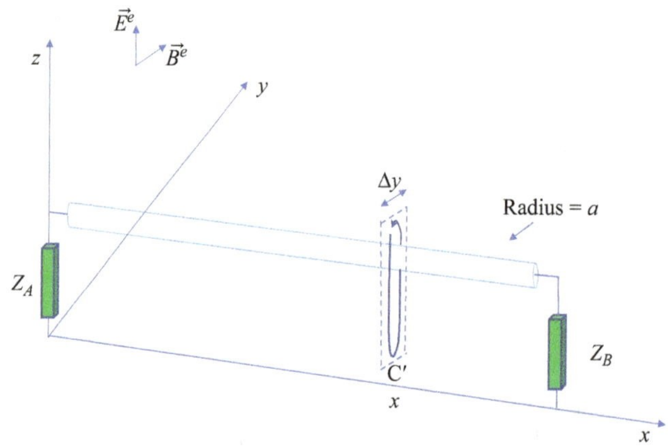

Figure 6.2 Single-conductor line terminated in impedances Z_A and Z_B. Rectangular contour C' used for the integration of Ampère–Maxwell's equation

As can be seen from (6.10), $I(x)$ is the only source of $\vec{B}_T^S(x,y,z)$. It is also clear from (6.10) that $\vec{B}_T^S(x,y,z)$ is directly proportional to $I(x)$. In addition, for a line with uniform cross-section, the proportionality factor between $I(x)$ and $\vec{B}_T^S(x,y,z)$ must be independent of x.

Let us now concentrate on the y-component of $\vec{B}_T^S(x,y,z)$. Using the facts we just established that $I(x)$ and $\vec{B}_T^S(x,y,z)$ are proportional and that the proportionality factor is independent of x, we can now write

$$B_y^s(x,y,z) = k(y,z)I(x) \tag{6.11}$$

where $k(y,z)$ is the proportionality constant that includes μ_0 as a factor.

The last integral in (6.7) represents the per unit length magnetic flux under the line, which was evaluated for $y = 0$, along a straight line under the line (see Figure 6.1). Particularizing (6.11) for $y = 0$ and substituting the result into the last integral in (6.7), we obtain

$$\int_0^h B_y^S(x,z)dz = \int_0^h k(y=0,z)I(x)dz \tag{6.12}$$

We can rewrite (6.12) as follows

$$\int_0^h B_y^s(x,z)dz = I(x)\int_0^h k(y=0,z)dz \tag{6.13}$$

Equation (6.13) implies that the per-unit-length scattered magnetic flux under the line at any point along it is proportional to the current at that point. The proportionality constant, which is the per-unit-length inductance of the line L', is given by

$$L' = \int_0^h k(y=0,z)dz. \tag{6.14}$$

Plugging (6.14) into (6.13), we obtain the well-known linear relationship between the magnetic flux and the line current, the proportionality constant being the line's per-unit-length inductance:

$$\int_0^h B_y^s(x,z)dz = L'I(x) \tag{6.15}$$

Assuming that the transverse dimension of the line is much greater than the radius of the wire, ($h \gg a$), the magnetic flux density can be calculated using Ampere's Law and the integral can be evaluated analytically [1].

Inserting (6.15) into (6.7), we obtain the first generalized Telegrapher's equation

$$\frac{dV(x)}{dx} + j\omega L'I(x) = -j\omega\int_0^h B_y^e(x,z)dz \tag{6.16}$$

Unlike the classical Telegrapher's equations, in which no external excitation is present (see the right-hand sides of (6.1) and (6.2)), the external excitation field results in a forcing function that is expressed in terms of the exciting magnetic flux. This forcing function can be viewed as a distributed voltage source along the line, as we will see in the equivalent circuit that will be given in the next Section 6.2.2.

Attention must be paid to the fact that, as mentioned when (6.6) was introduced above, the integration path for the calculation of the voltage $V(x)$ in (6.6) must be a straight and vertical line ending in the overhead conductor since the curl of the electric field in the integrand is generally not zero, which makes integral dependent on the integration path. A discussion of the choice of the integration path for the coupling calculations can be found in [7].

6.2.1.2 Derivation of the second generalized Telegrapher's equation

The second Telegrapher's equation can be derived starting from Ampère–Maxwell's equation in differential form

$$\nabla \times \vec{B} = \mu_o \vec{J} + j\omega\mu_o\varepsilon_o\vec{E} \tag{6.17}$$

Writing the z component of (6.17) in Cartesian coordinates, and after straightforward algebraic manipulations, one can write

$$j\omega E_z(x,z) = \frac{1}{\varepsilon_o\mu_o}\left[\frac{\partial B_y(x,z)}{\partial x} - \frac{\partial B_x(x,z)}{\partial y}\right] - \frac{J_z}{\varepsilon_o} \tag{6.18}$$

We will assume that the air conductivity is sufficiently low that the current density J_z in (6.18) can be neglected (this condition will be relaxed later on).

With this assumption and decomposing $B_x(x,z)$ and $B_y(x,z)$ on the right-hand side of (6.18) into their exciting and scattered components, we get

$$j\omega E_z(x,z) = \frac{1}{\varepsilon_o\mu_o}\left[\frac{\partial B_y^e(x,z)}{\partial x} - \frac{\partial B_x^e(x,z)}{\partial y}\right] + \frac{1}{\varepsilon_o\mu_o}\left[\frac{\partial B_y^s(x,z)}{\partial x} - \frac{\partial B_x^s(x,z)}{\partial y}\right] \tag{6.19}$$

Now, integrating both sides of (6.19) with respect to z from 0 to h, and making use of (6.6) on the left-hand side, we obtain

$$\begin{aligned} -j\omega V(x) = {} & \frac{1}{\varepsilon_o\mu_o}\int_0^h \left[\frac{\partial B_y^e(x,z)}{\partial x} - \frac{\partial B_x^e(x,z)}{\partial y}\right] dz \\ & + \frac{1}{\varepsilon_o\mu_o}\int_0^h \left[\frac{\partial B_y^s(x,z)}{\partial x} - \frac{\partial B_x^s(x,z)}{\partial y}\right] dz \end{aligned} \tag{6.20}$$

Since the excitation fields must also satisfy Ampère–Maxwell's equation, we can apply (6.18) to the excitation electromagnetic field components only to get

$$j\omega E_z^e(x,z) = \frac{1}{\varepsilon_o\mu_o}\left[\frac{\partial B_y^e(x,z)}{\partial x} - \frac{\partial B_x^e(x,z)}{\partial y}\right] \tag{6.21}$$

Integrating along z from 0 to h yields

$$j\omega \int_0^h E_z^e dz = \frac{1}{\varepsilon_o\mu_o}\int_0^h \left[\frac{\partial B_y^e}{\partial x} - \frac{\partial B_x^e}{\partial y}\right] dz \tag{6.22}$$

Plugging (6.22) into (6.20) and using the fact that the x component of the magnetic flux B_x^s is equal to zero since the scattered fields are assumed to be TEM, we can write

$$-j\omega V(x) = j\omega \int_0^h E_z^e dz + \frac{1}{\varepsilon_o \mu_o} \frac{\partial}{\partial x} \int_0^h B_y^s(x,z)dz \tag{6.23}$$

Note that, in the last term of (6.23), we have taken the partial derivative with respect to x out of the integral.

Now, plugging (6.15) into (6.23) yields,

$$-j\omega V(x) = j\omega \int_0^h E_z^e dz + \frac{1}{\varepsilon_o \mu_o} L' \frac{dI(x)}{dx} \tag{6.24}$$

In (6.24), the partial derivative of the current was written as a total derivative since $I(x)$ depends only on x.

Finally, rearranging terms and using the fact that the per-unit-length line capacitance C' is related to the per-unit-length inductance L' through $\varepsilon_0 \mu_0 = L'C'$, (6.24) can be rewritten as

$$\frac{dI(x)}{dx} + j\omega C' V(x) = -j\omega C' \int_0^h E_z^e(x,z)dz \tag{6.25}$$

Equation (6.25) is the second Telegrapher's equation for field-to-transmission line coupling in the model of Taylor *et al.*

For a line of finite length L, as in Figure 6.1, the boundary conditions at the extremities are given by

$$V(0) = -Z_A I(0) \tag{6.26}$$

$$V(L) = Z_B I(L) \tag{6.27}$$

6.2.2 Equivalent circuit

Equations (6.16) and (6.25) are the generalized Telegrapher's equations for the Taylor *et al.* coupling model. Their equivalent circuit representation is given in Figure 6.3. The distributed series voltage and parallel current sources that appear in the equivalent circuit correspond to the driving functions in the Telegrapher's equations (the right-side of (6.16) and (6.25)), which stem from the excitation fields.

6.2.3 The model of Agrawal, Price, and Gurbaxani

Agrawal, Price, and Gurbaxani [4] derived a set of generalized Telegrapher's equations in which the only forcing function along the line is the horizontal exciting electric field in the direction of the line. Although this appears to contradict the model presented by Taylor *et al.* 15 years earlier, in which the forcing sources are the vertical electric field and the horizontal magnetic flux

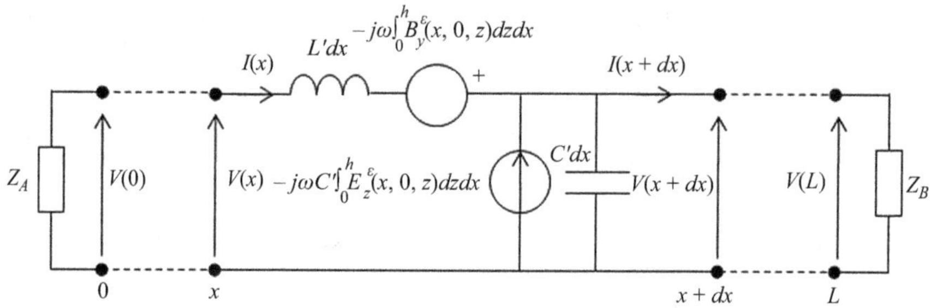

Figure 6.3 Equivalent circuit of the Taylor et al. model for the case of a lossless overhead line consisting of a single wire subject to an external electromagnetic field

density, both of the models, as well as the Rachidi model to be introduced in the next section, have been shown to be entirely equivalent, as further discussed in Section 6.2.4.

Agrawal *et al.* derived their Telegrapher's equations (as did Taylor *et al.*), based on Maxwell's equations and on the transmission line approximation. The two equations of the model of Agrawal *et al.* are [1][*]:

$$\frac{dV^s(x)}{dx} + j\omega L' I(x) = E_x^e(x,0,h) \tag{6.28}$$

$$\frac{dI(x)}{dx} + j\omega C' V^s(x) = 0 \tag{6.29}$$

in which, L' (introduced in (6.14) and (6.15)), and C' (already used in writing (6.25)) are the per-unit-length inductance and capacitance of the line, respectively. $I(x)$ is the induced current in the line and $V^s(x)$ is the so-called scattered voltage, defined as

$$V^S(x) = -\int_0^h E_z^S(x,z)dz \tag{6.30}$$

This definition is similar to the definition of the total voltage used in the derivation of the model of Taylor *et al.* in (6.6) but for the scattered electric field only. Since the scattered electric field is quasi-TEM, the path of integration in (6.3) does not need to be along a straight line, as long as it stays within a transverse plane.

The scattered voltage $V^s(x)$ is related to the total voltage $V(x)$ by way of

$$V(x) = V^s(x) + V^e(x) \tag{6.31}$$

[*]Note that in the original paper of Agrawal *et al.*, the considered configuration was a multiconductor line in free space. Here, the equations are written for the case of a conductor above a ground plane.

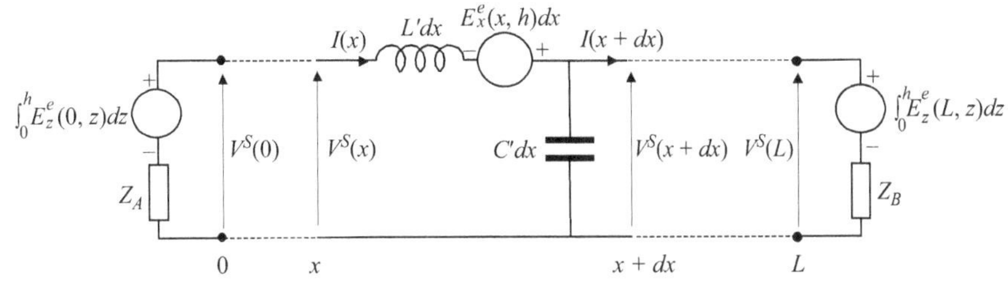

Figure 6.4 Equivalent circuit or the Agrawal et al. model for the case of a lossless overhead line consisting of a single wire subject to an external electromagnetic field in terms of the total current and the scattered voltage. The total voltage is calculated using (6.31)

where the exciting voltage $V^e(x)$ is defined as

$$V^e(x) = -\int_0^h E_z^e(x, z)dz \tag{6.32}$$

The integral in (6.32) must be carried out over a straight and vertical line since the excitation field is not conservative in the transverse plane of the line.

The forcing function on the right-hand side of (6.28), $E_z^e(x, h)$, is the horizontal component of the exciting electric field at line height.

The terminal conditions in terms of the scattered voltage and the total current used in (6.28) and (6.29) are given by

$$V^s(0) = -Z_A I(0) + \int_0^h E_z^e(0, z)dz \tag{6.33}$$

$$V^s(L) = Z_B I(L) + \int_0^h E_z^e(L, z)dz \tag{6.34}$$

Figure 6.4 presents the equivalent circuit representation of (6.28), (6.29), (6.33), and (6.34).

In the Agrawal *et al.* model, an exciting voltage source, whose per-unit-length value is equal to the exciting electric field $E_x^e(x, h)$ tangential to the line conductor acts as a distributed voltage source along the line. In this model, the terminal conditions (6.33) and (6.34) entail also the use of voltage sources equal to the additive inverse of the exciting voltage in series with the line terminations Z_A and Z_B.

6.2.4 The Rachidi model

Rachidi [5] presented in 1993 a new formulation of the generalized Telegrapher's equations for coupling calculations whose forcing functions are different from those in the Agrawal *et al.* and the Taylor *et al.* models, but that is nevertheless rigorously equivalent to them. In these coupling equations, the only forcing function is the exciting magnetic flux density. Rachidi pointed out in [5] that the set of coupling equations in his model can be seen as a dual of that of Agrawal *et al.*, in which only the electric field appears as a forcing function as we saw in the previous Section 6.2.3.

Rachidi's model formulation is given in (6.35) and (6.36):

$$\frac{dV(x)}{dx} + j\omega L' I^s(x) = 0 \tag{6.35}$$

$$\frac{dI^s(x)}{dx} + j\omega C' V(x) = \frac{1}{L'} \int_0^h \frac{\partial B_x^e(x,z)}{\partial y} dz \tag{6.36}$$

In these equations, $I^s(x)$ is the so-called scattered current, which is related to the total current by

$$I(x) = I^s(x) + I^e(x) \tag{6.37}$$

where the excitation current $I^e(x)$ is defined as

$$I^e(x) = -\frac{1}{L'} \int_0^h B_y^e(x, 0, z) dz \tag{6.38}$$

The boundary conditions at the ends of the line are

$$I^s(0) = -\frac{V(0)}{Z_A} + \frac{1}{L'} \int_0^h B_y^e(0, 0, z) dz \tag{6.39}$$

and

$$I^s(L) = \frac{V(L)}{Z_B} + \frac{1}{L'} \int_0^h B_y^e(L, 0, z) dz \tag{6.40}$$

As with the first two models presented in the previous two Sections 6.2.1 and 6.2.3, the Rachidi model's equations can be represented by an equivalent circuit, shown in Figure 6.5.

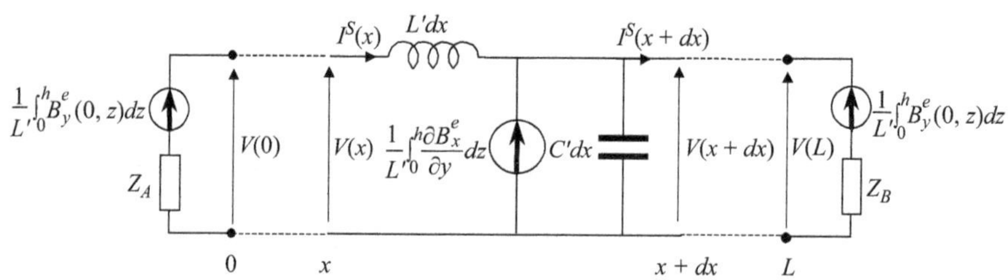

Figure 6.5 Equivalent circuit or the Rachidi model for the case of a lossless overhead line consisting of a single wire subject to an external electromagnetic field in terms of the scattered current and the total voltage. The total current is calculated using (6.37)

We have already mentioned that the coupling formulations of Taylor *et al.*, Agrawal *et al.* and Rachidi are rigorously equivalent, even though the forcing functions are different for each one of the approaches. One must therefore specify the coupling model being used when discussing the relative contributions of the different field components to the induced voltages and currents [9].

6.2.5 Rusck/modified Rusck model

In 1957, Rusck [10] introduced a coupling model applicable to the case of a perfectly conducting ground. Cooray [11] found that Rusck's original model is not complete since some of the forcing terms are missing. Recently, Cooray *et al.* [7] modified Rusck's model by adding the missing terms and extended the model to make it applicable to the case of a lossy ground. The equations of the model of Rusck and of the modified model are expressed in terms of the scalar and vector potentials.

The field-to-transmission line coupling equations for the modified Rusck's model, in the frequency domain, in terms of the scalar and vector potentials are summarized here for the considered case of a lossless conductor above a perfectly-conducting ground:

$$\frac{dV^q(x)}{dx} + j\omega L' I(x) = -j\omega A_x^i(x,h) \tag{6.41}$$

$$\frac{dI(x)}{dx} + j\omega C' V^q(x) = -j\omega C' \phi^i(x,h) \tag{6.42}$$

in which V^q is defined as

$$V^q(x) = -\int_0^h E_z^s(x,z) \cdot dz + \phi^i(x,h) \tag{6.43}$$

The total voltage can be obtained using

$$V(x) = V^q(x) + \int_0^h j\omega A_z^i dz \tag{6.44}$$

The boundary conditions at the ends of the transmission line at $x = 0$ and $x = L$ for resistive terminations are given by

$$V^q(0) = -Z_A \cdot I(0) - \int_0^h j\omega A_z^i(0,z)dz \tag{6.45}$$

$$V^q(L) = Z_B \cdot I(L) - \int_0^h j\omega A_z^i(L,z)dz \tag{6.46}$$

An equivalent circuit for the modified Rusck model is shown in Figure 6.6.

Note that the version of the modified Rusck's model given here is applicable for perfectly conducting lines. Equations taking into account line losses can be found in [7].

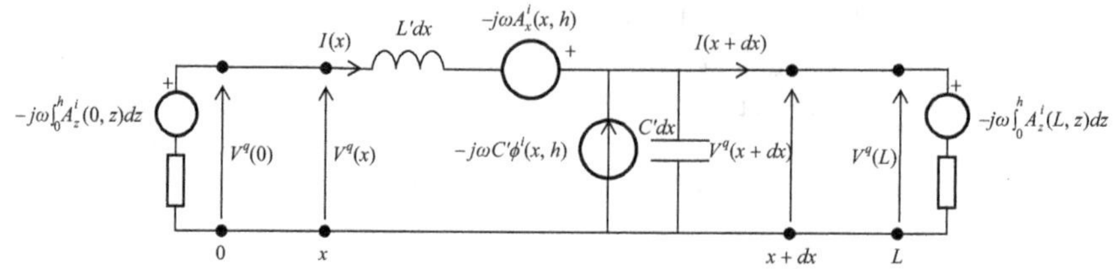

Figure 6.6 Equivalent circuit of the modified Rusck coupling model for the case of a lossless conductor above a perfectly conducting ground

An extension of the original Rusck's model to take into account the effect of the finite ground conductivity was also presented by Piantini [12], who demonstrated that lightning-induced voltages on an overhead line over a lossy ground can be obtained by adding to the voltage calculated for the case of perfectly conducting ground a term associated with the horizontal component of the electric field.

6.2.6 Finite ground and medium conductivity

Up to this point, we have assumed for the models of Taylor *et al.*, Agrawal *et al.*, Rachidi, and the modified Rusck model that both the ground and the line conductor have perfect conductivity and that the medium between the line and the ground is a perfect dielectric. However, waveforms of the induced overvoltages and currents can be significantly influenced by losses due to the finite conductivity of the ground and the wires [13] and, to a lesser extent in realistic cases, due to a nonzero conductivity of the dielectric between the conductors.

For the wire, let us denote the finite conductivity σ_w and the relative permittivity ε_{rw}. For the ground, let us assume a conductivity σ_g and a relative permittivity ε_{rg}.

To compare the first generalized Telegrapher's equation for the case of a perfectly conducting ground and the equation modified to account for the effect of a finite ground conductivity, we first rewrite here for convenience (6.5), used earlier in the chapter in the derivation of the first coupling equations of the model of Taylor *et al.*:

$$-\frac{\partial}{\partial x}\int_0^0 E_z(x,z)dz + E_x(x,h) - E_x(x,0) = -j\omega\int_0^0 B_y(x,z)dz \tag{6.47}$$

The second and third terms in (6.47), which represent the electric fields tangential to the overhead wire and the ground, respectively, were neglected in Section 6.2.1 since electric fields tangential to perfect conductors must equal zero by virtue of the fact that the tangential electric field is continuous across the interface between two media and that the fields are zero inside a perfect conductor. Since the conductivities are assumed to be finite in this section, those two terms can no longer be neglected. Instead, they are replaced by the product (in the frequency domain) of impedances times the current in the given conductor ($I(x)$ in the wire and $I_g(x)$ in the ground):

$$E_x(x,h) = Z_w^{'} I(x) \tag{6.48}$$

and

$$E_x(x,0) = Z_g^{'} I_g(x) \tag{6.49}$$

Since the quasi-TEM approximation implies that $I(x) = -I_{g(x)}$, the two terms that had been neglected for the perfectly conducting ground case can now be written as

$$E_x(x,h) - E_x(x,0) = (Z_w^{'} + Z_g^{'})I(x) \tag{6.50}$$

Equation (6.50) represents an extra term that must be added to the first Telegrapher's equation (to (6.16) in Section 6.2.1). With this addition made to (6.16), the first generalized Telegrapher's equation for Taylor *et al.* model that takes into account the finite conductivity of the ground is

$$\frac{dV(x)}{dx} + j\omega L'I(x) + (Z_w^{'} + Z_g^{'})I(x) = -j\omega \int_0^h B_y^e(x,z)dz \tag{6.51}$$

The inductive, ground, and wire impedances can be combined into the so-called longitudinal per-unit-length impedance

$$Z' = j\omega L' + Z_w^{'} + Z_g^{'} \tag{6.52}$$

and (6.51) can be rewritten as

$$\frac{dV(x)}{dx} + Z'I(x) = -j\omega \int_0^h B_y^e(x,z)dz \tag{6.53}$$

The second coupling equation needs to be modified to account for the nonzero conductivity of the medium between the line and the ground and for the imperfect ground conductivity.

The second equation for a perfectly conducting ground for the model of Taylor *et al.*, given in (6.25) and rewritten here for convenience is

$$\frac{dI(x)}{dx} + j\omega C'V(x) = -j\omega C' \int_0^h E_z^e(x,z)dz \tag{6.54}$$

The equation taking into account the losses is

$$\frac{dI(x)}{dx} + Y'V(x) = -Y' \int_0^h E_z^e(x,z)dz \tag{6.55}$$

Repeating the same exercise for the model of Agrawal *et al.*, we can write the coupling equations extended to the present case of a wire above an imperfectly conducting ground as (a step-by-step derivation for the Agrawal model can be found in [1]):

$$\frac{dV^s(x)}{dx} + Z'I(x) = E_x^e(x,0,h) \tag{6.56}$$

$$\frac{dI(x)}{dx} + Y'V^s(x) = 0 \tag{6.57}$$

In (6.53) through (57), Z' and Y' are the longitudinal and transverse per-unit-length impedance and admittance, respectively, given by [1,13]†:

$$Z' = j\omega L' + Z_w' + Z_g' \tag{6.58}$$

$$Y' = \frac{(G' + j\omega C')Y_g'}{G' + j\omega C' + Y_g'} \tag{6.59}$$

in which L', C', and G' are the per-unit-length longitudinal inductance, transverse capacitance, and transverse conductance, respectively, calculated for a lossless wire above a perfectly conducting ground:

$$L' = \frac{\mu_o}{2\pi}\cosh^{-1}\left(\frac{h}{a}\right) \cong \frac{\mu_o}{2\pi}\ln\left(\frac{2h}{a}\right) \quad \text{for } h \gg a \tag{6.60}$$

$$C' = \frac{2\pi\varepsilon_o}{\cosh^{-1}(h/a)} \cong \frac{2\pi\varepsilon_o}{\ln(2h/a)} \quad \text{for } h \gg a \tag{6.61}$$

$$G' = \frac{\sigma_{\text{air}}}{\varepsilon_o} C' \tag{6.62}$$

Z'_w is the per-unit-length internal impedance of the wire; assuming a round wire and an axial symmetry for the current, the following expression can be derived for the wire internal impedance (e.g., [14]):

$$Z_w' = \frac{\gamma_w I_o(\gamma_w a)}{2\pi a \sigma_w I_1(\gamma_w a)} \tag{6.63}$$

where $\gamma_w = \sqrt{j\omega\mu_o(\sigma_w + j\omega\varepsilon_o\varepsilon_{rw})}$ is the propagation constant in the wire and I_o and I_1 are the modified Bessel functions of zero and first order, respectively.

Z_g' is the per-unit-length ground impedance, which is defined as [2,15]:

$$Z_g' = \frac{j\omega\int_{-\infty}^{h} B_y^s(x,z)dx}{I} - j\omega L' \tag{6.64}$$

†In [1], the per unit length transverse conductance was disregarded.

where B_y^s is the y-component of the scattered magnetic induction field. and $Y_g^{'}$ is the ground admittance defined as [1]:

$$Y_g^{'} \cong \frac{\gamma_g^2}{Z_g^{'}} \tag{6.65}$$

For typical overhead power lines, the effect of ground admittance and the wire impedance is negligible and can be disregarded in the computation [2,16]. Note that for buried cables, the effect of the ground admittance is no longer negligible [17].

Several expressions for the ground impedance have been proposed in the literature (see for instance [2] for a survey). Sunde [18] derived a general expression for the ground impedance, which is given by

$$Z_g^{'} = \frac{j\omega\mu_o}{\pi} \int_0^{\infty} \frac{e^{-2hx}}{\sqrt{x^2+\gamma_g^2}+x} dx \tag{6.66}$$

where $\gamma_g = \sqrt{j\omega\mu_o(\sigma_g + j\omega\varepsilon_o\varepsilon_{rg})}$ is the propagation constant in the ground.

As noted in [19], Sunde's expression (6.66) is directly connected to the general expressions obtained from scattering theory. Indeed, it is shown in [1] that the general expression for the ground impedance derived using scattering theory reduces to the Sunde approximation when considering the transmission line approximation. Also, the results obtained using (6.66) have been shown to be accurate within the limits of the transmission line approximation [1].

The general expression (6.66) is not suitable for numerical evaluation since it involves an integral over an infinitely long interval. Several approximations for the expression of the ground impedance of a single-wire line have been proposed in the literature (see [2] for a survey). One of the simplest and most accurate was proposed by Sunde himself and it is given by the following logarithmic function:

$$Z_g^{'} \cong \frac{j\omega\mu_o}{2\pi} \ln\left(\frac{1+\gamma_g h}{\gamma_g h}\right) \tag{6.67}$$

It has been shown [2] that the above logarithmic expression represents an excellent approximation to the general expression (6.66) over the frequency range of interest.

6.2.7 *Multiconductor lines*

Although all of the presented models are equivalent and the choice of the coupling model for a given case is, in principle, arbitrary, the model of Agrawal *et al.* has been traditionally adopted for validation, perhaps due to the fact that the forcing function along the line is a single component of the electric field. We will present here the extension of that model to the case of multiconductor lines.

The geometrical parameters of an overhead multiconductor line include the number of wires, the wire diameters and their location in the transverse plane in terms of height and horizontal position. These parameters are defined in Figure 6.7.

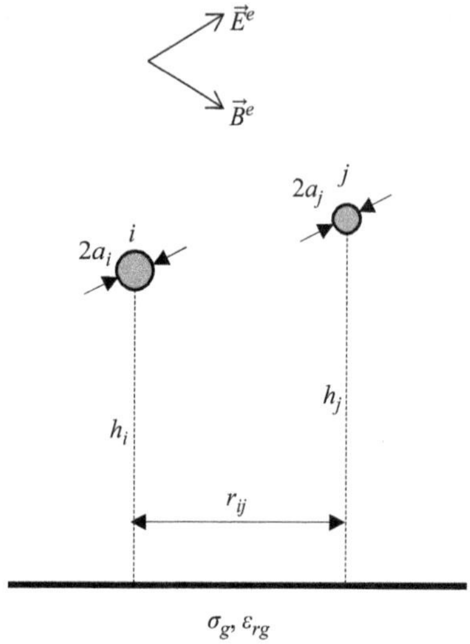

Figure 6.7 Cross-sectional geometry of a multiconductor line in presence of an external electromagnetic field

In the model of Agrawal *et al.* for multiconductor lines, the coupling equations are written in matrix form and they are given by [1,2,19]:

$$\frac{d}{dx}[V_i^s(x)] + j\omega[L'_{ij}][I_i(x)] + [Z'_{g_{ij}}][I_i(x)] = [E_x^e(x, 0, h_i)] \tag{6.68}$$

$$\frac{d}{dx}[I_i(x)] + [G'_{ij}][V_i^s(x)] + j\omega[C'_{ij}][V_i^s(x)] = [0] \tag{6.69}$$

In (6.68) and (6.69), $[V_i^s(x)]$ and $[I_i(x)]$ are frequency-domain vectors of the scattered voltage and the current along the line; the vector $[E_x^e(x, h_i)]$ contains the component of the excitation electric fields tangential to the line conductors; [0] is the zero-matrix; $[L'_{ij}]$ is the per-unit-length line inductance matrix. For the practical case in which the spacing between conductors is much greater than their radii, the mutual inductance between pairs of conductors i and j is given by [1]:

$$L'_{ij} = \frac{\mu_o}{2\pi} \ln\left(\frac{r_{ij}^2 + (h_i + h_j)^2}{r_{ij}^2 + (h_i - h_j)^2}\right) \tag{6.70}$$

The self-inductance for conductor i is given by

$$L'_{ii} = \frac{\mu_o}{2\pi} \ln\left(\frac{2h_i}{r_{ii}}\right) \tag{6.71}$$

The per-unit-length line capacitance matrix $[C'_{ij}]$ can be evaluated, for a homogeneous line, from the inductance matrix using the following expression [6]:

$$\left[C'_{ij}\right] = \varepsilon_o \mu_o \left[L'_{ij}\right]^{-1} \tag{6.72}$$

The per-unit-length transverse conductance matrix $[G'_{ij}]$ is obtained starting either from the capacitance matrix or the inductance matrix using the following relations

$$\left[G'_{ij}\right] = \frac{\sigma_{\text{air}}}{\varepsilon_o}\left[C'_{ij}\right] = \sigma_{\text{air}}\mu_o\left[L'_{ij}\right]^{-1} \tag{6.73}$$

In the majority of the cases in practice, the transverse conductance matrix elements can be neglected.

Finally, $[Z'_{g_{ij}}]$ is the ground impedance matrix, in which the mutual ground impedance between two conductors i and j, as derived by Sunde, is given by [18]:

$$Z'_{g_{ij}} = \frac{j\omega\mu_o}{\pi}\int_0^\infty \frac{e^{-(h_i+h_j)x}}{\sqrt{x^2+\gamma_g^2}+x}\cos(r_{ij}x)dx \tag{6.74}$$

In a similar way to that proposed by Sunde for the case of a single-wire line, an accurate logarithmic approximation is proposed by Rachidi *et al.* [19] which is given by

$$Z'_{gij} \cong \frac{j\omega\mu_o}{4\pi}\ln\left[\frac{\left(1+\gamma_g\left(\frac{h_i+h_j}{2}\right)\right)^2+\left(\gamma_g\frac{r_{ij}}{2}\right)^2}{\left(\gamma_g\frac{h_i+h_j}{2}\right)^2+\left(\gamma_g\frac{r_{ij}}{2}\right)^2}\right] \tag{6.75}$$

Note that in (6.68) and (6.69), the terms corresponding to the wire impedance and the so-called ground admittance have been neglected. This approximation is valid for typical overhead power lines [13].

The boundary conditions for the two-line terminations are given by

$$[V_i^s(0)] = -[Z_A][I_i(0)] + \left[\int_0^{h_i} E_z^e(0,0,z)dz\right] \tag{6.76}$$

$$[V_i^s(L)] = [Z_B][I_i(L)] + \left[\int_0^{h_i} E_z^e(L,0,z)dz\right] \tag{6.77}$$

in which $[Z_A]$ and $[Z_B]$ are the impedance matrices at the two-line terminations, respectively.

6.2.8 Equivalence of the coupling models

Using a numerical example, Nucci and Rachidi [9] illustrated the fact that, as predicted by the fact that the theoretical derivations of the models of Taylor *et al.*, Agrawal *et al.*, and Rachidi are based on identical assumptions applied to

Maxwell's equations, the induced voltage waveforms obtained using those three coupling models, presented in Sections 6.2.1, 6.2.3, and 6.2.4, are identical. Interestingly, since the forcing functions associated with each model are different, the contribution of a given component of the exciting electromagnetic field to the total induced voltage and current differs for each model. Indeed, the three coupling models (and the modified Rusck model presented in Section 6.2.5) are different but equivalent approaches that yield identical total voltages and total currents, in spite of the fact that they take into account the electromagnetic coupling in different ways. In other words, the presented models are different expressions of the same equations, cast in terms of different combinations of the various electromagnetic field components, which are related through Maxwell's equations.

6.2.9 Source terms in field-to-transmission line coupling models

The source terms in the different coupling models described above are expressed in terms of different components of the electromagnetic fields generated by a lightning discharge, the calculation of which is described in Chapter 3 of this volume. It is worth noting that, within the context of the Agrawal *et al.* coupling model, the vertical electric field radiated by the lightning channel is generally calculated assuming the ground as a perfectly conducting plane, since that field component is not significantly affected by the soil resistivity in the frequency and distance range of interest. For the calculation of the horizontal electric field component, a commonly used approach is the use of the Cooray–Rubinstein formula [20,21].

6.3 Field-to-transmission coupling models for buried cables

6.3.1 Preliminary remarks

Power and communication systems contain sensitive electronics and processors that are exposed to damages or logic upset caused by low levels of induced electromagnetic interference. The electronic circuits can be connected to underground cables and, for that reason, the evaluation of lightning-induced overvoltages on buried cables has been the subject of numerous studies (e.g., [17,22–30]). Typical examples are submarine fiberoptic cables and earthling buried telecommunication cables which include repeater power supply cables. Concerning submarine cables, buried sections as long as 20 km may run from the shore to the supply units, exposing them to the lightning threat.

This section presents calculation techniques that allow the estimation of lightning-induced disturbances on buried cables. In this chapter, we will consider only the case of a homogeneous ground exhibiting constant electric conductivity and permittivity. For information on the evaluation of the lightning fields in an inhomogeneous ground, refer to the relevant literature (e.g., [31–33]). Note that the frequency dependence of the soil electrical parameters (e.g., [34]) is not considered here.

We will present first in Section 6.4.2 the methods for the calculation of lightning electromagnetic fields penetrating into the soil. Sections 6.4.2 and 6.4.3 present the coupling of an electromagnetic fields on a buried cable. As with overhead lines, the theory is based on the transmission line assumption.

6.3.2 Calculation of the lightning electric field under the ground

As for the coupling of lightning fields to overhead lines, the calculation of the lightning-induced currents and voltages in buried cables requires the knowledge of the lightning return stroke electromagnetic fields in the vicinity of the buried cables. In the case of buried cables, this implies the calculation of the fields below the ground surface. We will use the usual representation of the lightning channel as a vertical antenna. The ground will be assumed here to be a uniform half-space with constant conductivity σ_g and relative permittivity ε_{rg}. The general expressions in the frequency domain for the vertical electric field dE_z and the horizontal electric field dE_r measured at a distance r from an elementary dipole located at height above the ground, and at a depth d under the ground, are given by the following equations (see Figure 6.8 for the relevant geometry) [35]:

$$dE_r(r,d,z') = \frac{j\omega\mu_0 I(z')dz'}{4\pi}\frac{\partial^2 V(r,d,z')}{\partial r \partial z} \tag{6.78}$$

$$dE_z(r,d,z') = \frac{j\omega\mu_0 I(z')dz'}{4\pi}\left\{\left(\frac{\partial^2}{\partial z^2} + k_g^2\right)V(r,d,z')\right\} \tag{6.79}$$

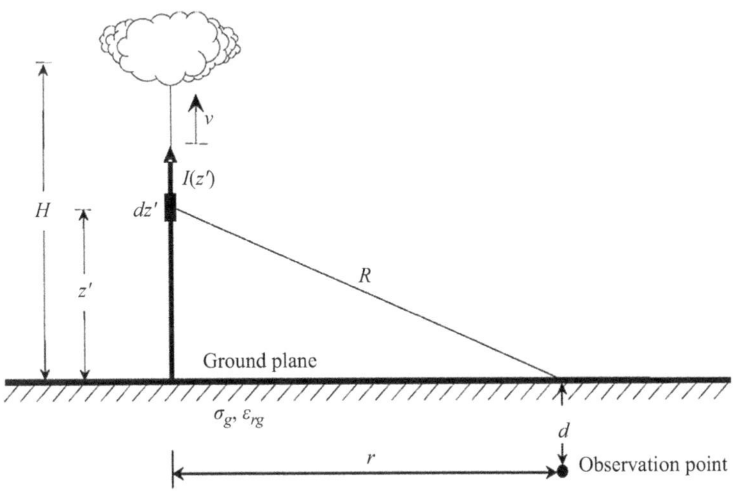

Figure 6.8 Geometry for the calculation of electromagnetic field below the ground surface radiated by a vertical lightning channel

in which

$$V(r,d,z') = \int_0^\infty \frac{\exp(\gamma_g d - \gamma_0 z') J_0(\lambda r) \lambda d\lambda}{k_g^2 \gamma_0 + k_0^2 \gamma_g} \tag{6.80}$$

where k_0 and k_g are the propagation constants in the air and in the ground given, respectively, by

$$k_0^2 = \omega^2 \mu_0 \varepsilon_0 \tag{6.81}$$

and

$$k_g^2 = \omega^2 \mu_0 \varepsilon_0 \varepsilon_{rg} + j\omega\mu_0\sigma_g, \tag{6.82}$$

And where

$$\gamma_0^2 = \lambda^2 - k_0^2,\ \gamma_g^2 = \lambda^2 - k_g^2 \tag{6.83}$$

Equation (6.80) is the so-called Sommerfeld integral which requires CPU and memory intensive algorithms (e.g., [36]) for its numerical evaluation.

To obtain the incident electric field in the time domain, the integration of (6.78) and (6.79) along the lightning channel followed by an inverse Fourier transformation is also required. Therefore, the direct use of (6.78) and (6.79) is extremely costly in terms of computation time, particularly in view of the fact that, for coupling calculations, the fields have to be determined at numerous points along the whole buried cable. For that reason, simplified expressions have been proposed for the determination of the electric fields below the ground surface produced by a vertical lightning channel. Cooray [37] proposed expressions as a function of the electric field at the air–soil interface. The expressions proposed by Cooray for the vertical and horizontal components of the electric fields in the frequency domain are given by

$$E_z(j\omega, r, d) = E_z(j\omega, r, 0) \frac{\varepsilon_0 \exp(-k_g d)}{\sigma_g + j\omega\varepsilon_0\varepsilon_{rg}} \tag{6.84}$$

$$E_r(j\omega, r, d) = E_r(j\omega, r, 0) \exp(-k_g d) \tag{6.85}$$

In the time-domain, (6.84) and (6.85) read, respectively

$$E_z(t, r, d) = \int_0^t E_z(t-\tau, r, 0) \Psi(\tau) d\tau \tag{6.86}$$

and

$$E_r(t, r, d) = \int_0^t E_r(t-\tau, r, 0) Y(\tau) d\tau \tag{6.87}$$

where $\Psi(t)$ and $Y(t)$ are the inverse Fourier transforms of $\frac{\varepsilon_0 \exp(-k_g d)}{\sigma_g + j\omega\varepsilon_0\varepsilon_{rg}}$ and $\exp(-k_g d)$, respectively, given by [4,17]

$$\Psi(t) = \int_0^t \Psi_1(t-\tau)Y(\tau)d\tau \tag{6.88}$$

$$\Psi_1(t) = \frac{1}{\varepsilon_{rg}}\exp(-at) \tag{6.89}$$

and

$$Y(t) = \frac{\exp(-at/2)at_z}{2\sqrt{t^2 - t_z^2}} I_1\left(\frac{a\sqrt{t^2 - t_z^2}}{2}\right)u(t - t_z) + \exp(-at_z/2)\delta(t - t_z) \tag{6.90}$$

in which

$$a = \frac{\sigma_g}{\varepsilon_0\varepsilon_{rg}} \quad \text{and} \quad t_z = d\sqrt{\mu_0\varepsilon_0\varepsilon_{rg}}.$$

At the air–soil interface, when $d = 0$, (6.90) reduces to $Y(t)|_{d=0} = \delta(t)$, ensuring the continuity of the horizontal electric field.

In (6.84) and (6.85), the vertical and horizontal electric field components at the ground surface $E_z(t, r, 0)$ and $E_r(t, r, 0)$ can be calculated with reasonable accuracy assuming a perfectly conducting ground for the vertical electric field component [16], and the Cooray–Rubinstein approximation for the horizontal component [20,38].

Petrache *et al.* [17] compared the horizontal electric field calculated using Cooray's simplified expressions and the nearly exact numerical solutions of (6.78) published by Zeddam in [39], for different depths and different values for the ground conductivity, their analysis confirmed that the simplified approach proposed by Cooray yields satisfactory results.

6.3.3 Coupling to buried cables

Theethayi *et al.* [29] used FDTD simulations as a reference to demonstrate that the transmission line approximation remains applicable when it is applied to buried cables for frequencies up to a few MHz (typical of lightning return strokes). The coupling equations applicable to the case of buried cables have therefore the same form as the generalized Telegrapher's equations derived earlier in this chapter for overhead wires.

Let us consider a buried horizontal cable of length L with an insulated jacket located along the x-axis at a depth d. The fields under the ground are characterized by a small vertical component compared to the corresponding vertical field above the ground. Assuming that the vertical component can be neglected under the ground [37], the induced voltages and currents along the cable due to a lightning return stroke can be calculated using the Agrawal *et al.* field-to-transmission line equations with per-unit-length parameter calculated for the case of buried cables [17]:

$$\frac{dV(x)}{dx} + Z'I(x) = E_x^e(x, z = -d) \tag{6.91}$$

$$\frac{dI(x)}{dx} + Y'V(x) = 0 \tag{6.92}$$

where the longitudinal impedance is

$$Z' = j\omega L' + Z_w' + Z_g' \tag{6.93}$$

and the transversal admittance is

$$Y' = \frac{(G + j\omega C') \cdot Y_g'}{(G + j\omega C') + Y_g'} \tag{6.94}$$

in which (see Figures 6.9 and 6.10 for the definitions and the geometrical parameters):

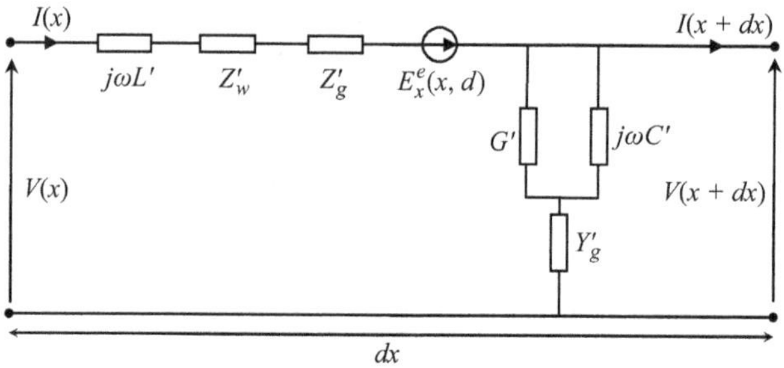

Figure 6.9 Differential equivalent coupling circuit for a buried cable illuminated by an external field. Adapted from [17]

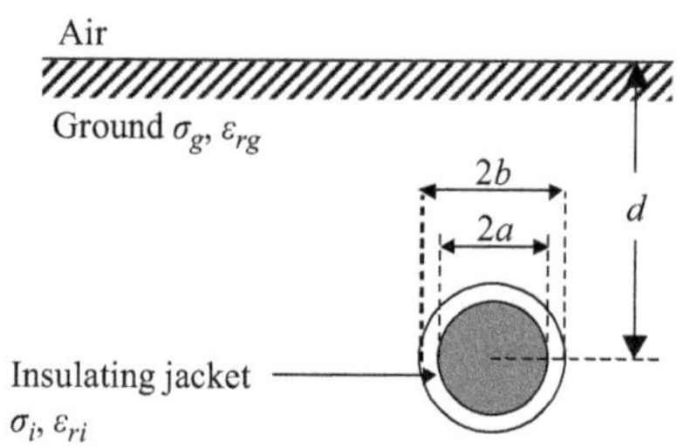

Figure 6.10 Geometry of the insulated buried cable. Adapted from [17]

L' and C' are, respectively, the per-unit-length longitudinal inductance and transverse capacitance of the cable, given by

$$L' = \frac{\mu_o}{2\pi}\ln\left(\frac{b}{a}\right) \tag{6.95}$$

$$C' = \frac{2\pi\varepsilon_0\varepsilon_{ri}}{\ln(b/a)} \tag{6.96}$$

G' is the per-unit-length transverse conductance of the cable:

$$G' = \frac{\sigma_i}{\varepsilon_0\varepsilon_{ri}}C' \tag{6.97}$$

Z'_w is the per-unit-length internal impedance of the wire. Assuming a round wire and an axial symmetry for the current, the following expression can be adopted [40]:

$$Z'_w = \frac{\gamma_w I_o(\gamma_w a)}{2\pi a\sigma_w I_1(\gamma_w a)} \tag{6.98}$$

where $\gamma_w = \sqrt{j\omega\mu_0(\sigma_w + j\omega\varepsilon_0\varepsilon_{rw})}$ is the propagation constant in the wire and ε_{rw} is the relative permittivity of the wire.

Z'_g and Y'_g are the per-unit-length ground impedance and ground admittance, respectively. These two quantities are related through the following expression [1]:

$$Y'_g \cong \frac{\gamma_g^2}{Z'_g} \tag{6.99}$$

where γ_g is the propagation constant inside the ground which can be expressed as $\gamma_g = \sqrt{j\omega\mu_0(\sigma_g + j\omega\varepsilon_0\varepsilon_{rg})}$.

Note that the coupling equations (6.91) and (6.92) are expressed in terms of the total voltage, because, as mentioned earlier, the vertical electric field inside the ground has been assumed to be negligible, so that the scattered voltage is equal to the total voltage.

A number of expressions have been proposed in the literature for the ground impedance (see [17] for a review).

Petrache *et al.* [17] proposed the following logarithmic approximation for the ground impedance

$$Z'_g = \frac{j\omega\mu_0}{2\pi}\ln\left(\frac{1+\gamma_g b}{\gamma_g b}\right) \tag{6.100}$$

which has been shown to be in excellent agreement with the general Sunde's expression. Furthermore, its implementation is very simple and it does not require any numerical treatment. Finally, unlike most of the considered approximations, it has an asymptotic behavior at high frequencies. Hence, the corresponding transient ground resistance in the time-domain has no singularity at $t = 0$ [41].

Petrache *et al.* [17] showed that, within the frequency range of interest, the wire impedance can be neglected due to its small contribution to the overall longitudinal impedance of the line. The ground admittance, however, can play an important role at high frequencies (1 MHz or so) especially in the case of poor ground conductivity, and it needs to be taken into account in the calculation of lightning-induced currents and voltages on buried cables. This is in contrast with the case of overhead lines, in which its contribution is generally negligible even in the MHz range.

Neglecting the transverse conductance, the transverse admittance per-unit-length can be written in the following form

$$Y' = j\omega C' + Y'_{add}, \quad \text{where } Y'_{add} = -\frac{(j\omega C')^2}{j\omega C' + Y'_g} \tag{6.101}$$

6.4 Coupling equations in time domain

A time-domain representation of the field-to-transmission line coupling equations can be useful when dealing with nonlinear components or phenomena, as well as in the case of a time-variable line topology [2]. On the other hand, frequency-dependent parameters, such as the ground impedance, need to be represented using convolution integrals.

The field-to-transmission line coupling equations for either overhead or underground lines can be straightforwardly converted into the time domain (see., e.g., [42,43] for overhead lines and [17] for underground cables). For the case of lossy lines, however, the frequency dependence of line parameters needs to be taken into account using convolution integrals. Specific expressions for the line parameters in time domain can be found in [41,44] for overhead lines, and in [17] for buried cables.

6.5 Experimental validation

The field-to-transmission line coupling models have been extensively validated using experimental data obtained using different sources as the exciting field, namely reduced-scale models (e.g., [12,45–50]), nuclear electromagnetic pulse simulators (e.g., [51]), and natural or artificially initiated lightning (e.g., [12,22,49,52–58]). In this section, some examples are presented which illustrate the validity of the models discussed in Section 6.2. As demonstrated there, the equations of all of the models are equivalent and produce identical results if the line and the excitation fields are the same.

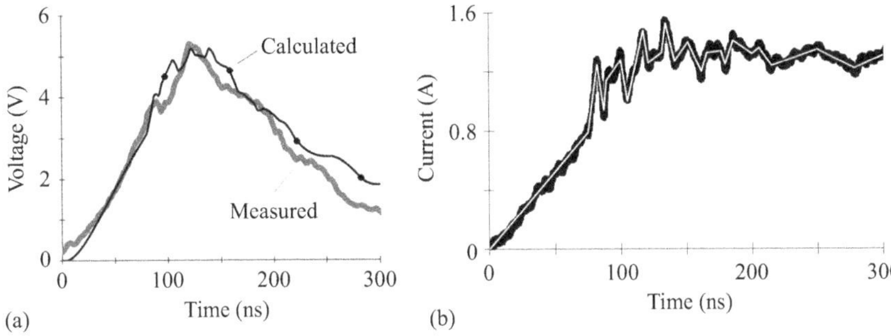

Figure 6.11 Measured and calculated (using the extended Rusck model [12]) phase-to-ground-induced voltages: (a) induced voltages; (b) measured current and straight-line approximation superimposed. Adapted from [49]

Figure 6.11 shows a comparison involving measured and calculated lightning-induced voltages. The experiment was performed under controlled conditions, on a 1:50 scale model, and the induced voltage at the point of the line closest to the stroke location was computed using the extended Rusck model (ERM) [12]. The single-phase straight line was 28 m long (corresponding to a length of 1.4 km in the full-scale system), matched at both ends, and 1.4 m (equivalent to 70 m) from the model of the stroke channel, which was equidistant from the line terminations. The copper conductor was supported by polyvinyl chloride (PVC) poles spaced every 60 cm (corresponding to 30 m), and its diameter was 0.4 mm. The conductor height was 20 cm (equivalent to 10 m) and its measured surge impedance was 455 Ω. The ground was simulated by means of interconnected aluminum plates and could therefore be reasonably assumed as perfectly conducting. The calculation was performed with the recorded current waveform approximated by straight lines, which in Figure 6.11(b) are shown superimposed to the measured current. Further details of the scaled system can be found in [48,50].

Though neither attenuation nor distortion of the current along the structure that simulated the stroke channel was considered in the calculations, a good agreement is observed between measured and calculated voltages. Many other comparisons which demonstrate the validity of the ERM considering the presence of a multi-grounded neutral or shield wire and equipment such as transformers and surge arresters, as well as the incidence of lightning flashes to nearby elevated objects or a line over a lossy ground can be found in [12,49]. Some of these comparisons are presented in Chapter 5 of Volume 2.

Experiments carried out on the same scale model were used by Piantini *et al.* [48] to validate the LIOV-EMTP code [16,51,59], which allows for the computation of voltages induced by lightning return strokes on homogeneous, multi-conductor, lossy overhead lines using the Agrawal *et al.* coupling model [4].

The LIOV-EMTP is based on two already existing computer programs, the LIOV—developed in the framework of an international collaboration involving the University of Bologna, the Swiss Federal Institute of Technology of Lausanne, and the University of Rome—and the EMTP [60]. Many comparisons were made between measured and calculated induced voltages, involving power distribution networks with different topologies. One of the test configurations, representative of an urban distribution network, is depicted in Figure 6.12, where the dimensions are referred to the full-scale system. The line was three-phase and the heights of the phase and neutral conductors were 10 m and 8 m, respectively. The distance between adjacent phases was 0.75 m and the main feeder was matched at both ends. Each transformer was represented by capacitors connected between each phase and the neutral. The neutral was grounded at every transformer and also at the middle of the laterals through ground resistances of 50 Ω except at the measuring point, where the ground resistance was zero. The stroke current could be reasonably represented by a triangular waveform with peak value of 70 kA, front time of 2 μs and time to half-value of 85 μs.

Figure 6.13 shows a good agreement between measured and calculated induced voltages. Besides other comparisons involving reduced models considering different network configurations, the LIOV-EMTP code has also been validated using data obtained from triggered lightning experiments and nuclear electromagnetic pulse simulators. For an overview of the experimental validation, the readers are invited to refer to [2].

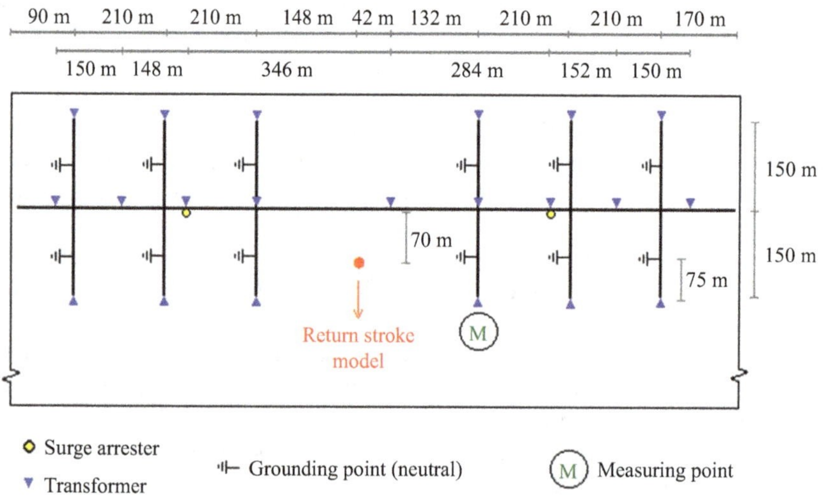

Figure 6.12 Power distribution network configuration corresponding to the comparison shown in Figure 6.13. Dimensions referred to the full-scale system. Adapted from [48]

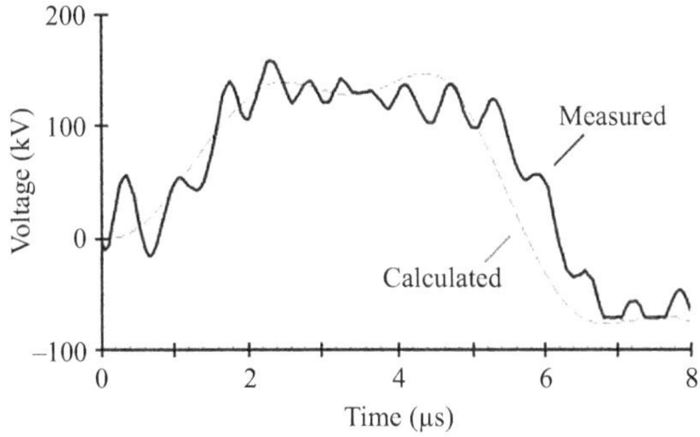

Figure 6.13 Measured and calculated (using the Agrawal et al. coupling model) phase-to-ground-induced voltages at point M of Figure 6.12, at the phase closest to the return stroke model. Voltage and time scales referred to the full-scale system. Adapted from [48]

References

[1] F. M. Tesche, M. Ianoz, and T. Karlsson, *EMC Analysis Methods and Computational Models*. New York: Wiley Interscience, 1997.

[2] C. A. Nucci and F. Rachidi, "Interaction of electromagnetic fields generated by lightning with overhead electrical networks," in V. Cooray (Ed.), *The Lightning Flash*, 2nd ed., London, IET, 2014, pp. 559–610.

[3] C. D. Taylor, R. S. Satterwhite, and C. W. Harrison, "The response of a terminated two-wire transmission line excited by a nonuniform electromagnetic field," *IEEE Trans. Antennas Propag.*, vol. 13, no. 6, pp. 987–989, 1965.

[4] A. K. Agrawal, H. J. Price, and S. H. Gurbaxani, "Transient response of multiconductor transmission lines excited by a nonuniform electromagnetic field," *IEEE Trans. Electromagn. Compat.*, vol. 22, no. 2, pp. 119–129, 1980.

[5] F. Rachidi, "Formulation of the field-to-transmission line coupling equations in terms of magnetic excitation field," *IEEE Trans. Electromagn. Compat.*, vol. 35, no. 3, pp. 404–407, 1993.

[6] S. Rusck, "Induced lightning over-voltages on power-transmission lines with special reference to the over-voltage protection of low voltage networks," *Trans R. Intitute Technol. Stockh.*, 1958.

[7] V. Cooray, F. Rachidi, and M. Rubinstein, "Formulation of the field-to-transmission line coupling equations in terms of scalar and vector potentials," *IEEE Trans. Electromagn. Compat.*, vol. PP, no. 99, pp. 1–6, 2017.

[8] V. Cooray, J. Guo, F. Rachidi, and M. Rubinstien, "On the definition of total line voltage in the interaction of electromagnetic fields with overhead electrical networks," *IEEE Trans. Electromagn. Compat.*, Submitted, 2019.

[9] C. A. Nucci and F. Rachidi, "On the contribution of the electromagnetic field components in field-to-transmission lines interaction," *IEEE Trans. Electromagn. Compat.*, vol. 37, no. 4, pp. 505–508, 1995.
[10] S. Rusck, "Induced lightning over-voltages on power-transmission lines with special reference to the overvoltage protection of low voltage networks," Ph.D. dissertation, Stockholm, Royal Institute of Technology, 1957.
[11] V. Cooray, "Calculating lightning-induced overvoltages in power lines: a comparison of two coupling models," *IEEE Trans. Electromagn. Compat.*, vol. 36, no. 3, pp. 179–182, 1994.
[12] A. Piantini, "Extension of the Rusck model for calculating lightning-induced voltages on overhead lines considering the soil electrical parameters," *IEEE Trans Electromagn. Compat.*, 2017.
[13] F. Rachidi, C. A. Nucci, M. Ianoz, and C. Mazzetti, "Influence of a lossy ground on lightning induced voltages on overhead lines," *IEEE Trans. Electromagn. Compat.*, vol. 38, no. 3, pp. 250–264, 1996.
[14] S. Ramo, J. R. Whinnery, and T. Van Duzer, *Fields and Waves in Communication Electronics*, 3rd ed., New York, NY, John Wiley & Sons, 1994.
[15] F. M. Tesche, "Comparison of the transmission line and scattering models for computing the HEMP response of overhead cables," *IEEE Trans. Electromagn. Compat.*, vol. 34, no. 2, pp. 93–99, 1992.
[16] F. Rachidi, C. A. Nucci, I. M., and C. Mazzetti, "Influence of a lossy ground on lightning-induced voltages on overhead lines," *IEEE Trans. Electromagn. Compat.*, vol. 38, no. 3, pp. 250–263, 1996.
[17] E. Petrache, F. Rachidi, M. Paolone, C. Nucci, V. A. Rakov, and M. A. Uman, "Lightning-induced voltages on buried cables. Part I: theory," *IEEE Trans. Electromagn. Compat.*, vol. 47, no. 3, 2005.
[18] E. D. Sunde, *Earth Conduction Effects in Transmission Systems*. New York: D. Van Nostrand Company, 1949.
[19] F. Rachidi, C. A. Nucci, and M. Ianoz, "Transient analysis of multiconductor lines above a lossy ground," *IEEE Trans. Power Deliv.*, vol. 14, no. 1, pp. 294–302, 1999.
[20] M. Rubinstein, "An approximate formula for the calculation of the horizontal electric field from lightning at close, intermediate, and long range," *IEEE Trans. Electromagn. Compat.*, vol. 38, no. 3, pp. 531–5, 1996.
[21] V. Cooray, "Some considerations on the 'Cooray-Rubinstein' approximation used in deriving the horizontal electric field over finitely conducting ground," presented at the 24th International Conference on Lightning Protection, 1998, pp. 282–286.
[22] M. Paolone, E. Petrache, F. Rachidi, *et al.*, "Lightning-induced voltages on buried cables. Part II: experiment and model validation," *IEEE Trans. Electromagn. Compat.*, vol. 47, no. 3, 2005.
[23] N. Theethayi, R. Thottappillil, M. Paolone, C. A. Nucci, and F. Rachidi, "External impedance and admittance of buried horizontal wires for transient

studies using transmission line analysis," *IEEE Trans Dielectr. Electr. Insul.*, vol. 14, no. 3, pp. 751–761, 2007.

[24] A. Andreotti, D. Assante, F. Mottola, and L. Verolino, "Fast and accurate evaluation of the underground lightning electromagnetic field," in *2008 International Symposium on Electromagnetic Compatibility - EMC Europe*, 2008, pp. 1–6.

[25] L. Diaz, M. Martinez, J. Ramírez, and M. Rubinstein, "Lightning induced voltages in simple configurations of underground cables," in *2012 International Conference on Lightning Protection (ICLP)*, 2012, pp. 1–8.

[26] B. Yang, B. H. Zhou, B. Chen, J. B. Wang, and X. Meng, "Numerical study of lightning-induced currents on buried cables and shield wire protection method," *IEEE Trans. Electromagn. Compat.*, vol. 54, no. 2, pp. 323–331, 2012.

[27] J. Paknahad, K. Sheshyekani, F. Rachidi, M. Paolone, and A. Mimouni, "Evaluation of lightning-induced currents on cables buried in a lossy dispersive ground," *IEEE Trans. Electromagn. Compat.*, vol. 56, no. 6, pp. 1522–1529, 2014.

[28] B. Zhang, J. Zou, X. Du, J. Lee, and M. N. Ju, "Ground admittance of an underground insulated conductor and its characteristic in lightning induced disturbance problems," *IEEE Trans. Electromagn. Compat.*, vol. 59, no. 3, pp. 894–901, 2017.

[29] N. Theethayi, Y. Baba, F. Rachidi, and R. Thottappillil, "On the choice between transmission line equations and full-wave Maxwell's equations for transient analysis of buried wires," *IEEE Trans. Electromagn. Compat.*, vol. 50, no. 2, pp. 347–357, 2008.

[30] H. Tanaka, Y. Baba, and C. F. Barbosa, "Effect of shield wires on the lightning-induced currents on buried cables," *IEEE Trans. Electromagn. Compat.*, vol. 58, no. 3, pp. 738–746, 2016.

[31] J. O. S. Paulino, C. F. Barbosa, and W. d C. Boaventura, "Lightning-induced current in a cable buried in the first layer of a two-layer ground," *IEEE Trans. Electromagn. Compat.*, vol. 56, no. 4, pp. 956–963, 2014.

[32] J. Paknahad, K. Sheshyekani, and F. Rachidi, "Lightning electromagnetic fields and their induced currents on buried cables. Part I: the effect of an ocean–land mixed propagation path," *IEEE Trans. Electromagn. Compat.*, vol. 56, no. 5, pp. 1137–1145, 2014.

[33] J. Paknahad, K. Sheshyekani, F. Rachidi, and M. Paolone, "Lightning electromagnetic fields and their induced currents on buried cables. Part II: the effect of a horizontally stratified ground," *IEEE Trans. Electromagn. Compat.*, vol. 56, no. 5, pp. 1146–1154, 2014.

[34] D. Cavka, N. Mora, and F. Rachidi, "A comparison of frequency-dependent soil models: application to the analysis of grounding systems," *IEEE Trans. Electromagn. Compat.*, vol. 56, no. 1, pp. 177–187, 2014.

[35] A. Baños, *Dipole Radiation in the Presence of a Conducting Half-Space*. Oxford, 1966.

[36] P. Parhami, Y. Rahmat-Samii, and R. Mittra, "An efficient approach for evaluating Sommerfeld integrals encountered in the problem of a current element radiating over lossy ground," *IEEE Trans. Antennas Propag.*, vol. 28, no. 1, pp. 100–104, 1980.

[37] V. Cooray, "Underground electromagnetic fields generated by the return strokes of lightning flashes," *IEEE Trans. Electromagn. Compat.*, vol. 43, no. 1, pp. 75–84, 2001.

[38] V. Cooray, "Horizontal fields generated by return strokes," *Radio Sci.*, vol. 27, no. 4, pp. 529–37, 1992.

[39] A. Zeddam, "Couplage d'une onde électromagnétique rayonnée par une décharge orageuse à un câble de télécommunications," Lille University of Sciences and Technologies, Lille, France, 1988.

[40] S. Ramo, J. R. Whinnery, and T. van Duzer, *Fields and Waves in Communication Electronics*, 3rd ed., New York, NY, Wiley, 1994.

[41] F. Rachidi, S. Loyka, C. A. Nucci, and M. Ianoz, "A new expression for the ground transient resistance matrix elements of multiconductor overhead transmission lines," *Electr. Power Syst. Res. J.*, vol. 65, pp. 41–46, 2003.

[42] C. A. Nucci, F. Rachidi, and A. Rubinstein, "Derivation of Telegrapher's equations and field-to-transmission line interaction," in *Electromagnetic Field Interaction with Transmission Lines: From Classical Theory to HF Radiation Effects*, F. Rachidi and S. Tkachenko (Eds.), Southampton: WIT Press, 2008.

[43] F. Rachidi, C. A. Nucci, and M. Ianoz, "Transient analysis of multiconductor lines above a lossy ground," *IEEE Trans. PWDR*, vol. 14, no. 1, pp. 294–302, 1999.

[44] R. Araneo and S. Cellozi, "Direct time domain analysis of transmission lines above a lossy ground," *IEE Proc Sci. Meas. Technol.*, vol. 148, no. 2, pp. 73–79, 2001.

[45] M. Ishii, K. Michishita, Y. Hongo, and S. Ogume, "Lightning-induced voltage on an overhead wire dependent on ground conductivity," *IEEE Trans. Power Deliv.*, vol. 9, no. 1, pp. 109–118, 1994.

[46] A. Piantini and J. M. Janiszewski, "An Experimental study of lightning induced voltages by means of a scale model," presented at the 21st International Conference on Lightning Protection (ICLP), 1992, pp. 195–199.

[47] C. A. Nucci, A. Borghetti, A. Piantini, and W. Janischewskyj, "Lightning-induced voltages on distribution overhead lines: comparison between experimental results from a reduced-scale model and most recent approaches," presented at the 24th International Conference on Lightning Protection, ICLP, 1998, pp. 314–320.

[48] A. Piantini, J. M. Janiszewski, A. Borghetti, C. A. Nucci, and A. Paolone, "A scale model for the study of the LEMP response of complex power distribution networks," *IEEE Trans. Power Deliv.*, vol. 22, no. 1, pp. 710–720, 2007.

[49] A. Piantini and J. M. Janiszewski, "Lightning-induced voltages on overhead lines: application of the extended Rusck model," *IEEE Trans. Electromagn. Compat.*, vol. 51, no. 3, pp. 548–558, 2009.

[50] A. Piantini and J. M. Janiszewski, "Scale models and their application to the study of lightning transients in power systems," in *Lightning Electromagnetics*, London: IET, Chapter 19.

[51] A. Borghetti, A. Gutierrez, C. A. Nucci, M. Paolone, E. Petrache, and F. Rachidi, "Lightning-induced voltages on complex distribution systems: models, advanced software tools and experimental validation," *J. Electrost.*, vol. 60, pp. 163–174, 2004.

[52] E. Petrache, M. Paolone, F. Rachidi, *et al.*, "Measurement of lightning-induced currents in an experimental coaxial buried cable," presented at the IEEE Power Engineering Society Summer Meeting, 2003.

[53] S. Yokoyama, H. Mitani, K. Miyake, and N. Yamazaki, "Corroboration of induced lightning voltage suppressing effect of overhead ground wire using natural lightning," *Electr. Eng. Jpn.*, vol. 105, no. 2, pp. 75–85, 1985.

[54] M. J. Master, M. A. Uman, W. Beasley, and M. Darveniza, "Lightning induced voltages on power lines: experiment," *IEEE Trans. Power Appar. Syst.*, no. 9, pp. 2519–29, 1984.

[55] M. Rubinstein, A. Y. Tzeng, M. A. Uman, P. J. Medelius, and E. M. Thomson, "An experimental test of a theory of lightning-induced voltages on an overhead wire," *IEEE Trans. Electromagn. Compat.*, vol. 31, no. 4, pp. 376–83, 1989.

[56] F. De la Rosa, R. Valdivia, H. Pérez, and J. Loza, "Discussion about the inducing effects of lightning in an experimental power distribution line in Mexico," *IEEE Trans. Power Deliv.*, vol. 3, no. 3, 1988.

[57] P. P. Barker, T. A. Short, A. Eybert-Berard, and J. B. Berlandis, "Induced voltage measurements on an experimental distribution line during nearby rocket triggered lightning flashes," *IEEE Trans. Power Deliv.*, vol. 11, pp. 980–995, 1996.

[58] M. Paolone, F. Rachidi, A. Borghetti, *et al.*, "Lightning electromagnetic field coupling to overhead lines: theory, numerical simulations and experimental validation," *IEEE Trans. Electromagn. Compat. Press*, vol. 51, no. 3, pp. 532–547, 2009.

[59] C. A. Nucci, F. Rachidi, M. Ianoz, and C. Mazzetti, "Lightning-induced voltages on overhead power lines," *IEEE Trans. Electromagn. Compat.*, vol. 35, no. 1, pp. 75–86, 1993.

[60] H. W. Dommel, Electromagnetic Transient Program Reference Manual (EMTP Theory Book), Bonneville Power Administration, Portland, OR, 1986.

Chapter 7
Lightning response of grounding electrodes

Silverio Visacro[1]

The response of electric power systems subject to lightning-related voltages and currents can be strongly influenced by the behavior of their ground terminations.

In order to assess how these terminations affect the systems' response, one must first understand the lightning response of grounding electrodes, which is quite different from that observed when electrodes are subjected to low-frequency currents. This chapter addresses this issue.

Initially, a simplified approach is used for explaining basic concepts related to grounding electrodes, beginning with their representation by equivalent circuits and their response in frequency domain. This comprises a qualitative discussion of the involved physical quantities and characterization of related parameters, notably the low-frequency grounding resistance and harmonic grounding impedance.

Following, the impulse response of electrodes is analyzed. Based on experimental evidences and theoretical considerations, parameters that translate this response are considered, such as the impulse grounding impedance. The main factors that influence this response are discussed and, using an elaborate electromagnetic model, sensitivity analysis are developed to reveal their relevance.

Then, the particular impulse response of electrodes subjected to lightning currents of first and subsequent return strokes is discussed and fundamental influent parameters, such as the effective length of electrodes, soil ionization, and frequency dependence of electrical parameters of soil, are considered.

Finally, the representation of grounding electrodes in lightning studies and protection applications is addressed. The accuracy of the results obtained from concise representations of electrodes, notably from the impulse impedances of first and subsequent return stroke, is discussed, along with the efficiency of using them. Ways for determining a concise representation of grounding systems for prompt application are presented.

Most of content of this chapter is based on the concepts, results, and discussions presented by the author in a limited number of works published in the last decade.

[1]Lightning Research Center, Federal University of Minas Gerais (UFMG), Belo Horizonte, Brazil

7.1 Basic concepts

The interaction between lightning and power systems is frequently responsible for the flow of intense currents through their ground terminations, resulting either from direct strikes to the system components or from voltages induced by nearby strikes [1,2].

7.1.1 Characterizing grounding systems

Irrespective of their physical arrangement, all these ground terminations comprise three components that compose a grounding system: (i) the metallic conductors that drive the current to the electrodes, (ii) the metallic electrodes buried in the soil, and (iii) the earth surrounding the electrodes [3]. The last one is the most relevant component [4]. Figure 7.1 illustrates such components.

A grounding electrode consists of any buried metallic body responsible for dispersing current to the soil. The specific application has significant influence on the electrodes' arrangement and dimension. The arrangement can vary from a simple rod, connected to the transformer of a distribution line, to horizontal electrodes buried about 0.5 m deep in the soil composing the counterpoise wires of transmission-line tower footings, or the typical grounding grids of power system substations. The area covered by electrodes may vary widely, from a few to thousands of square meters, depending on the application and on the soil resistivity value [4,5].

The characteristic of the soil surrounding the electrodes, notably its resistivity, has great influence on their response when they are subjected to impressed currents.

Although the specific application can demand a grounding system to perform particular functions, in most cases, these functions aim ensuring safety or an expected performance of the electrical system when it is subjected to electrical unbalances, short circuits or transients [4]. In particular, in the conditions imposed by lightning stresses, these functions pursue providing a low impedance path for the flow of lightning-related currents to the ground and ensuring safety conditions, in terms of distribution of the electric potentials over the ground in the region where electrodes disperse lightning currents into the soil [4,5].

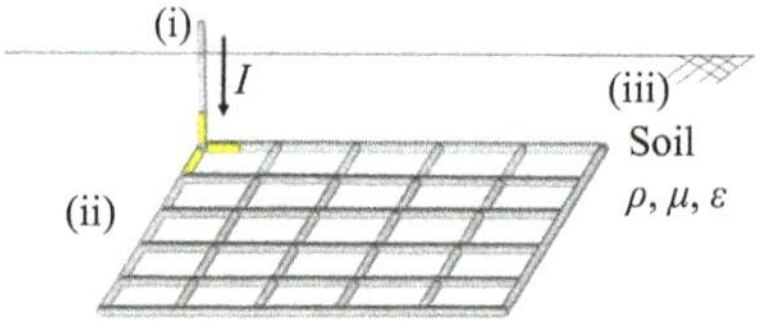

Figure 7.1 Illustration of grounding system components

7.1.2 Simplified representation of grounding system by equivalent circuits

The primary response of grounding electrodes subjected to impressed currents consists of the grounding potential rise (GPR) in relation to a distant region, where the electrical potential is null. In practical cases, this region, known as remote earth, can be considered at distances from electrodes exhibiting very reduced potentials, around a few hundredth of the GPR, typically between ten and twenty times longer than the linear dimension of the area covered by the grounding system [2].

Any termination to earth presents resistive, capacitive, and inductive effects. Figure 7.2 considers this aspect, exhibiting a set of "grounding elements," consisting of a short length of electrode and the portion of surrounding earth, taken from the grounding system.

The current of each grounding element comprises two components: the leakage transversal current I_T and the longitudinal current I_L transferred to the remaining electrode length. The leakage current flowing into the soil causes a potential rise of the electrode and surrounding soil in relation to remote earth. The flow of the longitudinal current generates a voltage drop along the electrode. The equivalent circuit of each element indicated in Figure 7.2(b) has to represent these effects [4].

The upper bar in the circuit represents the electrode where the longitudinal impedances of the elements are distributed. The inferior bar corresponds to the remote earth where the electric potential has a null value [2].

The current that enters the electrode is partially dispersed into soil, flowing across the transversal parameters G and C toward the remote earth, represented in this circuit by the inferior bar. The remaining current is transmitted to other grounding elements, where it is ultimately dispersed to soil. This leakage current, associated with the electric field in the soil, has conductive and capacitive

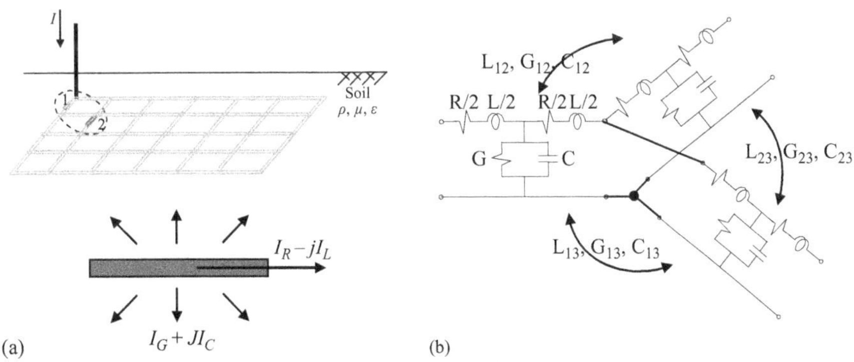

Figure 7.2 (a) Elements of a grounding system and (b) their representation by an equivalent circuit. Adapted from [2,5]

components. The density of the conductive component J is directly related to the soil resistivity and the E-field intensity ($J = 1/\rho \cdot E$, where ρ is the electrical resistivity), whereas the density of the capacitive current is related to the variation of the electric field ($\partial \epsilon E/\partial t$, where ϵ is the electrical permittivity of soil). The branch with the parallel conductance and capacitance (G, C) represents the effects associated to the leakage current. An interesting feature of the leakage current is that the ratio between conductive and capacitive currents in the soil does not depend on the electrode geometry but only on the relation "$\sigma/\omega\epsilon$" ($\sigma = 1/\rho$ and $\omega = 2\pi f$ is the angular frequency). The current that flows along the electrode yields a magnetic field inside and around the electrode and causes internal losses along it. The RL parameters in Figure 7.2 (R: resistance, L: inductance) are responsible for these effects and for the voltage drop along the electrode during the flow of current [6].

The equivalent circuit has also to include the mutual effects corresponding to the electromagnetic couplings among the grounding elements. Capacitive, inductive, and resistive couplings between each pair of elements are taken into account in the circuit, as represented in Figure 7.2(b).

Thus, the evaluation of the response of the grounding system to an impressed current requires the solution of a series of circuits similar to the presented one, connected according to the topology of electrodes and taking into account mutual effects among all the grounding elements. This solution can be expressed by the impedance seen from the point where current is impressed on, which can be represented as the ratio between the ground potential rise (potential developed at the electrode V_T in relation to a remote earth) and the impressed current I_T. The presence of reactive circuit elements makes the solution of this circuit frequency dependent, leading to the definition of an harmonic impedance $Z(\omega)$ [6].

7.2 The frequency response of grounding systems: a qualitative approach

7.2.1 Introduction

In spite of its simplicity, the equivalent circuit of Figure 7.2 allows developing qualitative analyses of the transient behavior of electrodes based on their response in frequency-domain. Addressing this frequency response allows understanding some basic concepts, which are still valid when considering the lightning response of electrodes. Furthermore, it is feasible calculating the time-domain response of electrodes from the frequency components of the impressed current and corresponding GPR, by using Fourier transform.

As the equivalent circuit comprises reactive elements, the circuit response varies with frequency. Thus, the relevance of resistive, capacitive, and inductive components of this response depends very much on the frequency content of the electrical current impressed on electrodes and on the characteristics of soil. In particular, the inductive component is also significantly affected by the arrangement and length of electrodes.

7.2.2 The harmonic impedance

The parameter that express the frequency response of electrodes is the so-called harmonic impedance, $Z(\omega)$, given by the ratio between the potential rise V_T of the point of the electrode where the current is impressed (in relation to a remote earth) and the impressed current I_T (7.1) [6]:

$$Z(\omega) = \frac{V_T(\omega)}{I_T(\omega)} \tag{7.1}$$

Figure 7.3 illustrates this response for a horizontal electrode buried in soils of low (250 Ωm) and high (2,500 Ωm) resistivity values. The results were calculated using an electromagnetic model [7] but their analysis was developed based on the equivalent circuit.

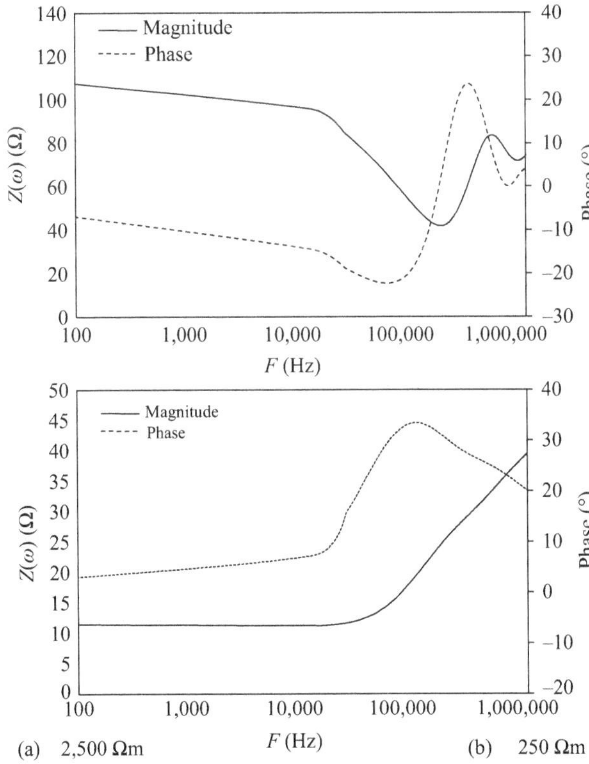

Figure 7.3 Harmonic grounding impedance within the frequency range of interest in power-system applications. 40-m long horizontal electrode buried 0.5 m deep in (a) 2,500 Ωm and (b) 250 Ωm soil. Electrode radius: 0.5 cm. Variation of the electric resistivity and permittivity of soil given by the Visacro–Alipio expressions [8]. Results simulated using the HEM model [7]

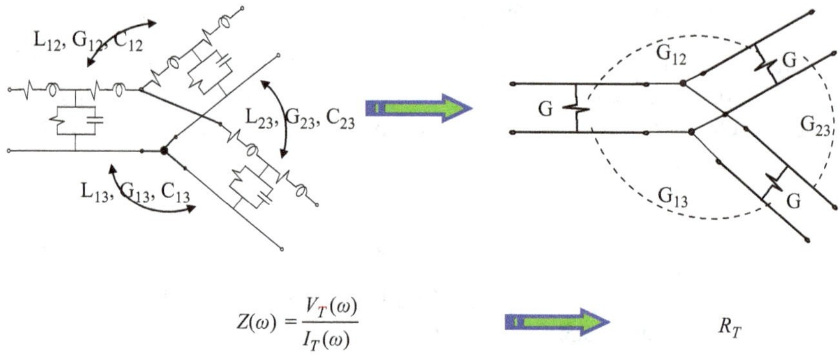

Figure 7.4 Representing the low-frequency grounding resistance as a particular condition of the harmonic grounding impedance. Adapted from [2,5]

The harmonic impedance is almost constant and equivalent to a resistance (phase angle approximately null) at the low-frequency range. In this range, reactive and skin effects are practically negligible: the inductance and electrode's internal resistance yields no voltage drop along the electrode and, practically, no capacitive currents flow into the soil. In addition, the resistance per unit length of electrodes is too low due to their typical conductivity. As a result, the corresponding reactances and resistances can be dropped from the equivalent circuit, which becomes simply a set of conductances, including mutual conductive effects among them [2,5], as depicted in Figure 7.4.

7.2.3 The low-frequency resistance

The solution of this low-frequency equivalent circuit consists of a conductance, whose inverse is the low-frequency grounding resistance R_T, a real number given by the ratio between the potential developed by electrodes in relation to remote earth and the impressed current (7.2). Thus, in fact, the so-called grounding resistance R_T is simply the harmonic impedance $Z(\omega)$ in the specific low-frequency condition [6]:

$$R_T = \frac{V_T}{I_T} \tag{7.2}$$

It worth mentioning that in this low-frequency condition, the electrical potential remains constant at a same value all along the electrodes, since there is no voltage drop along them. Thus, the electrode is an equipotential volume in this condition.

Considering the curves of Figure 7.3(a), as frequency rises, above a few kilohertz, capacitive current becomes relevant (see the negative impedance phase, whose magnitude rises with increasing frequency). As inductive effect is not important in this frequency range, the effect of the parallel G–C circuit prevails, and the capacitive current contributes to reduce the impedance in relation to that observed at the low frequency range (at that range only conductive current flows into the soil). Increasing further the frequency, the inductive effect becomes relevant, as shown by the decrease of the negative phase in Figure 7.3. Initially, the interaction of inductive and capacitive effects yields additional decrease of the harmonic impedance. This occurs until the harmonic impedance reaches the lowest value, around the frequency in which the phase crosses the null line. After that the inductive effect begins to prevail as denoted by the positive impedance phase and the impedance increases continuously until the frequency reaches a few Megahertz. All through this range, the voltage drop along the electrodes becomes significant and the equipotential assumption is no longer valid for electrodes [2,6].

As soil resistivity diminishes, two effects are observed. First, the decrease of impedance in the inferior frequency-spectrum range is less pronounced and the minimum impedance value, as well. Furthermore, the increase of impedance after reaching a minimum value due to inductive effect occurs ate earlier frequencies. Figure 7.3(b), which refers to low-resistivity soil, illustrates both aspects [2,6].

Important facts arise from the previous analysis. Above a few kilohertz, two main assumptions ordinarily valid when electrodes are subjected to slow varying currents are no longer valid. It is not possible to represent the grounding system by means of a simple resistance R_T nor to consider the electrodes to be equipotential volumes. In this frequency range, reactive and propagation effects become relevant and the electric potential can pronouncedly vary along the electrodes.

7.2.4 Propagation effects

Within determined limits, one can get a piece of information about propagation effects in the soil from the equivalent circuit.

The configuration of electrodes (thin wires surrounded by a dispersive medium, the soil) allows representing them as a sequence of connected circuit elements similar to those of Figure 7.2, in which R, L, G, and C correspond to per-unit-length parameters of a transmission line embedded in the soil. According to this representation, R is associated to internal losses of the electrode and G to losses in the soil [6].

Expressions (7.3) and (7.4) represent the harmonic current and voltage at position z along the electrode, resulting from the impression of current I_0 on the electrode at position $z = 0$ m. Expressions (7.5) and (7.6) give the propagation constant γ and the characteristic impedance Z of the line at frequency ω [9]:

$$I(\omega) = I_0 e^{-\alpha z} \cos(\omega t - \beta \cdot z) \tag{7.3}$$

$$V(\omega) = Z \cdot I_0 e^{-\alpha z} \cos(\omega t - \beta \cdot z - \theta_Z) \tag{7.4}$$

$$\gamma = \alpha + j\beta = \sqrt{(R + j\omega L)(G + j\omega C)} \tag{7.5}$$

$$Z = \sqrt{\frac{R + j\omega L}{G + j\omega C}} \tag{7.6}$$

All aspects of the frequency response of the electrodes of Figure 7.3, commented in the paragraphs following (7.2), can be qualitatively explained in (7.3)–(7.6).

The most important propagation effect in grounding electrodes consists of the strong attenuation the current and voltages waves are subjected to, while propagating along the electrode. The analysis of the attenuation constant α allows explaining this effect. If the electrode's internal losses (associated with R) and losses in the soil (associated with G) are negligible, no attenuation would be observed in the current and voltage waves propagating along the electrode (null value for α).

At the low-frequency range, propagation effects are negligible. The propagation constant is insignificant, as the values of R and ωL are too low and can be disregarded in (7.5). At this frequency range, as shown in Figure 7.4, the equivalent circuit representing the grounding system turns equal to a conductance and the corresponding impedance becomes equal to R_T, the low-frequency grounding resistance [2,5,6].

However, at the high frequency range the value of α becomes significant, resulting in pronounced attenuation. Considering representative conditions of soil, one may prospect on the parameter(s) responsible for such attenuation. In this frequency range, R has frequently a negligible value (compared with that of the inductive reactance ωL) and the conductance G has always very significant value compared with that of the capacitive susceptance ωC. Thus, from the modified (7.5), $\gamma = (j\omega L \cdot G)^{0.5}$, it becomes evident that, in these conditions, G is effectively the parameter responsible for attenuation along the electrode and that frequency ω has the same weight on the propagation [2,5,10].

As the conductance G is basically proportional to the apparent soil conductivity σ, it is clear that attenuation increases with decreasing soil resistivity. It is also clear that attenuation increases with increasing frequency. References [2,6] discuss in detail the propagation of current and associated voltage waves along grounding electrodes in frequency domain.

7.2.5 *The frequency dependence of soil resistivity and permittivity*

An important property of soils, which affect significantly the frequency response of electrodes is the strong frequency dependence of soil resistivity ρ and permittivity ε [11–17], though the magnetic permeability has a constant value equal to the permeability of air. This effect used to be ignored in engineering applications until recently, as there was no reliable general formulation for expressing it. In applications, the soil resistivity used to be assumed constant and equal to the value measured by using low-frequency measuring instruments and the relative permittivity used to be assumed to vary from 4 to 81, according to the soil humidity [6].

Recently, Visacro and collaborators proved the relevance of this effect in the response of grounding electrodes [8,16] and developed a new experimental methodology to determine this frequency dependence in the range between 100 Hz and

4 MHz, based on measurements performed in field conditions. This methodology was validated by comparing the experimental response of electrodes subjected to impulsive currents with a large spectrum of frequency components with that simulated using parameters determined for the soil where the electrodes were buried by using the new methodology. The results show that the response, in terms of the GPR resulting from the impression of different impulsive currents, are practically identical and quite different from that obtained under the assumption of constant parameters (see Figure 7.6). Reference [16] explains all details of this methodology. The curves of Figure 7.5 illustrates the result of the application of this experimental methodology in a soil of low-frequency resistivity of 2,420 Ωm.

Typically, soil resistivity decreases with increasing frequency, as shown in Figure 7.5(a) and the decrease becomes more pronounced in soils of higher resistivity. As discussed in [8,16], this increase is due to increasing losses occurring due to polarization processes, as the frequency of electric-field alternation is increased. Such losses are responsible for increasing soil conductivity.

However, the relative permittivity of soil decreases with increasing frequency, from a value about several hundreds at low-frequency range to several tens around a few Megahertz. The permittivity tends to be larger for low resistivity soils. As discussed in [8,16], this effect is due to the inability of certain polarization processes of higher inertia to follow the electric-field alternation, as the frequency increases. Above a threshold frequency, these processes no longer contribute to the total polarization and the drop of their contribution is responsible for decreasing the electric permittivity of soil.

After the consistency of this new methodology was demonstrated, it was extensively applied to large number of soils in natural conditions, with resistivity values ranging from a few tens of ohms-meters to about ten thousand of ohms-meters. Based on the experimental curves determined for soil resistivity and permittivity variation in frequency spectrum curving fitting techniques were applied,

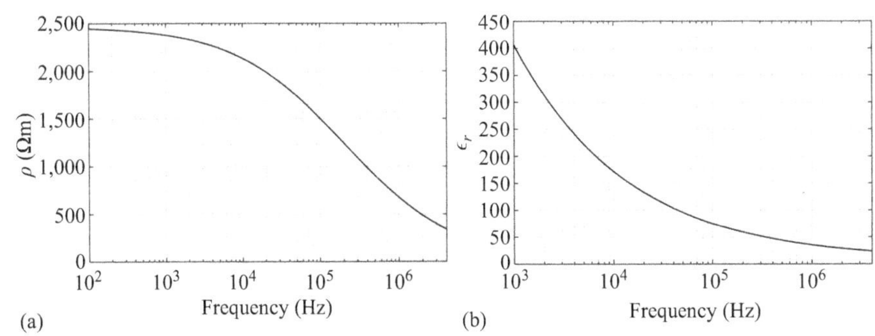

Figure 7.5 Experimental result of the frequency dependence of electrical parameters of a 2,420-Ωm soil in the frequency range of 100 Hz to 4 MHz. Variation of measured (a) resistivity ρ and (b) relative permittivity ε_r obtained from the application of the new methodology

following a conservative approach, to determine expressions able to predict the frequency dependence of both parameters, the so-called Visacro–Alipio expressions (7.7) and (7.8), as described in [8]:

$$\rho = \rho_0\{1 + [1.2 \cdot 10^{-6} \cdot {\rho_0}^{0.73}] \cdot [(f - 100)^{0.65}]\}^{-1} \tag{7.7}$$

$$\varepsilon_r = 7.6 \cdot 10^3 f^{-0.4} + 1.3 \tag{7.8}$$

In the expressions above, ρ and ε_r are the soil resistivity and relative permittivity at frequency f in Hz and ρ_0 is the soil resistivity at 100 Hz. Expressions (7.7) and (7.8) are applicable to the 100-Hz to 4-MHz range. The soil resistivity can be considered constant below 100 Hz.

The general impact of this frequency dependence is decreasing the grounding impedance in the high frequency range in relation to that obtained under the assumption of constant soil parameters. This decrease becomes more significant with increasing low-frequency soil resistivity.

7.3 The impulse response of grounding electrodes

As discussed in [2,6], for very fast time-varying currents, the primary response of electrodes, consisting of the GPR, is quite different from that resulting from the impression of low-frequency currents. In addition to conductive effects in the soil, inductive and capacitive effects are also important, along with propagation and frequency dependence effects. Grounding electrodes subjected to impulsive currents are not equipotential volumes, unlike in the slow-current conditions. Furthermore, the representation of the grounding system by a pure resistance is no longer valid.

This section addresses the behavior of electrodes subject to pulses of currents, specifically those with front times in the range of microseconds. The particular case of lightning-return-stroke currents is addressed in Section 7.4.

7.3.1 Fundamental aspects of the impulse response of electrodes and impulse grounding impedance

To provide elements for discussing the impulse response of grounding electrodes, Figure 7.6 presents the measured GPR of a horizontal electrode buried in a 1,400-Ωm soil, subjected to impulsive currents. It also presents the GPR simulated by using an electromagnetic model (HEM [7]), under the assumptions of constant and frequency dependent soil resistivity and permittivity (values determined by applying the experimental methodology referred to in Section 7.2.5).

The accuracy of the results provided by this electromagnetic model is evident from the comparison of the simulated and experimental results. Note that the GPR simulated, using the frequency-dependent soil parameters measured for that specific soil using the methodology mentioned in Section 7.2.5, practically matches the experimental GPR. Results exhibiting the same quality were obtained by applying this model to determine the GPR of general arrangements of electrodes, for instance of grounding grids and counterpoise wires [18–21]. However, the GPR

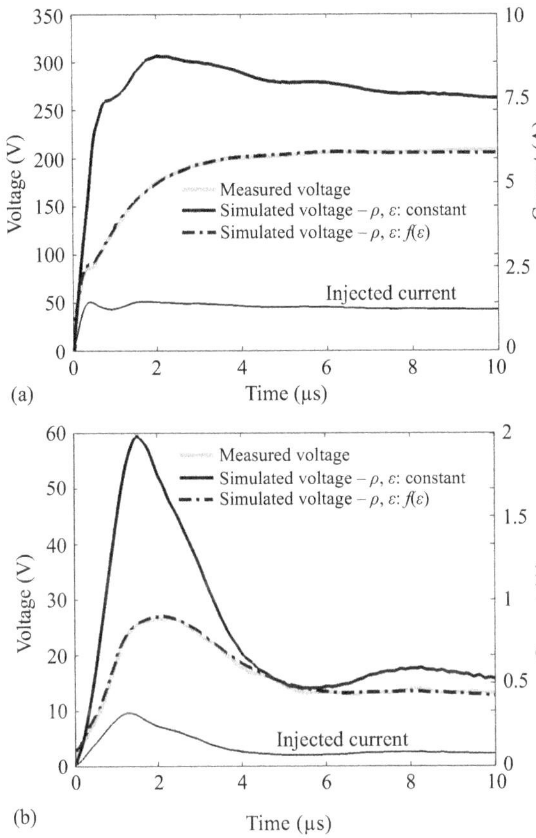

Figure 7.6 Measured and simulated grounding potential rise of 9.6-m-long horizontal electrode buried 0.5 m deep in a 1,400-Ωm soil and subjected to the impression of impulsive currents with front times of 0.4 μs and 1.4 μs (radius of 0.5 cm). Adapted from [16]

simulated under the assumption of constant soil parameters is quite different from the measured one.

The relevance of the factors that mostly affect the impulse response of grounding electrodes, namely the apparent soil resistivity and the electrode length, in addition to the front time of the impulsive current are demonstrated by applying the HEM model. Figure 7.7 shows the simulated GPR of a horizontal electrode, subjected to impulsive currents, under the variation of these influent factors.

The results presented in Figure 7.7 reveal important aspects related to the impulse response of electrodes. The current waveforms and corresponding GPRs are relatively similar, though their peaks are not simultaneous due to reactive and propagation effects in the soil, notably for the current waves with the shortest front time [Figure 7.7(c) and (d)].

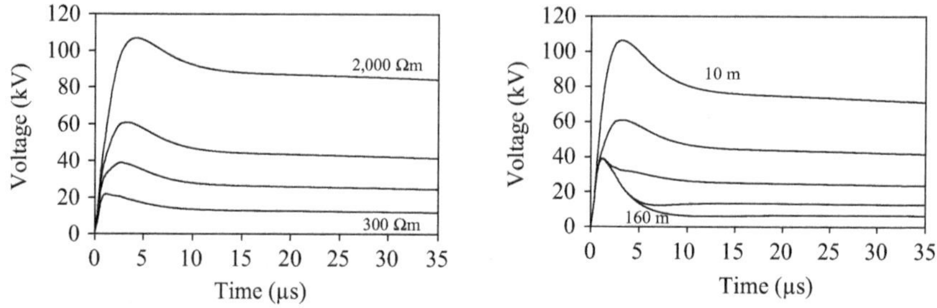

Length L (m)	20
Soil resistivity (Ωm)	2k, 1k, 600, 300
Waveform: $T_{d30}/T_{half\text{-}wave}$	2/50 μs

(a)

Length L (m)	10, 20, 40, 80, 160
Soil resistivity (Ωm)	1,000
Waveform: $T_{d30}/T_{half\text{-}wave}$	2/50 μs

(b)

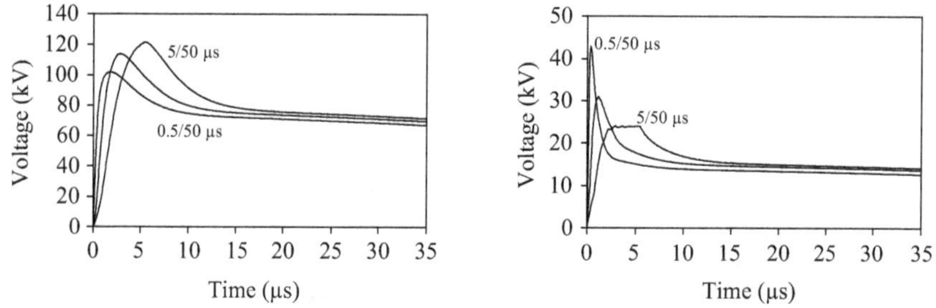

Length L (m)	5
Soil resistivity (Ωm)	600
Waveform: $T_{d30}/T_{half\text{-}wave}$: x/50	X: 0.5, 2, 5 μs

(c)

Length L (m)	30
Soil resistivity (Ωm)	600
Waveform: $T_{d30}/T_{half\text{-}wave}$: x/50	X: 0.5, 2, 5 μs

(d)

Figure 7.7 Impulse response of a horizontal electrode of length L (radius of 0.5 cm) buried 0.5 m deep, assuming frequency dependent soil parameters according to Visacro–Alipio expression: (a) varying soil resistivity; (b) varying electrode length; (c) varying the front-time for a short electrode: 5 m; and (d) varying the front-time for a long electrode: 30 m. Peak current: 1 kA

As shown in Figure 7.7(a), the variation of soil resistivity has the most relevant effect on the GPR. For a given impressed current, the amplitude of the resulting GPR is almost proportional to the low-frequency resistivity value.

Increasing the electrode length is very effective for diminishing the GPR, as shown in Figure 7.7(b). Nevertheless, the effectiveness decreases as the length becomes too long. Beyond a threshold length, the decrease of the GPR peak tends to saturate, though the decrease remains significant at the wave tail.

The results of the analyses developed above for the horizontal electrodes are all consistent with those developed in [22] for counterpoise wires of transmission lines.

In order to improve the interpretation of the results of Figure 7.7, it is useful to resort to the concept of the so-called impulse impedance Z_P, defined as the ratio

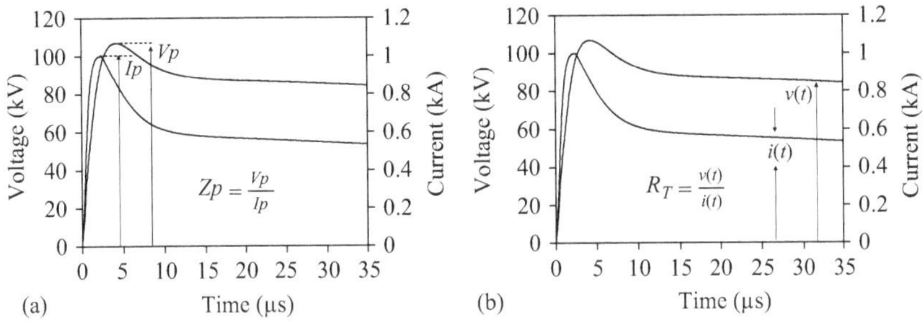

Figure 7.8 GPR of a 10 m long horizontal electrode buried 0.5 m deep in 1,000-Ωm soil. Front time of current wave: 2 μs; $Z_P = 108\ \Omega$; $R_{LF} = 160\ \Omega$

between the peak values of the GPR V_P and impressed current I_P (7.9):

$$Z_P = V_P/I_P \tag{7.9}$$

Figure 7.8 illustrates the concept of this impedance, referring to the curves of Figure 7.7(a). Although, in most cases, the peak values of GPR and current are not simultaneous, using the impulse impedance calculated as their ratio allows promptly determining the maximum GPR value for any given peak current. Figure 7.7(b) indicates that the low-frequency grounding resistance R_{LF} can be calculated from the impulsive GPR and current waves, as the ratio of their respective instantaneous value at the wave tail.

Returning to Figure 7.7(c) and (d), using as reference a 600-Ωm soil, they depict the very relevant effect of the front-time on the impulse response of short and long electrodes.

Considering first the short electrode, the decrease of the front time results in lower GPR amplitude and corresponding impulse impedance Z_P. This occurs because waves with short front time have higher frequency components. Both the capacitive currents in the soil and the soil conductivity are larger at such higher frequencies. As inductive effects are not relevant due to the short electrode length, the impulse impedance (and, therefore, the GPR) becomes lower. Note the decrease of Z_P from 125 to 116 and to 111 Ω, as the front-time is diminished from 5 to 2 and to 0.5 μs. Also, note that, for short electrodes, in all cases, the Z_P value is lower than that of R_{LF} (~153 Ω): $Z_P < R_T$.

Different from the condition considered above, for long electrodes, the higher frequency components of waves with short front time increase the inductive current and this effect can prevail over their effect of increasing the capacitive currents and decreasing soil resistivity. For long electrodes, the impulse impedance and respective GPR increases with decreasing front times, as shown in Figure 7.7(d). Note the increase of Z_P from 25 to 30 and to 42 Ω, as the front-time is decreased from 5 to 2 and to 0.5 μs. Also, note that, for long electrodes, Z_P has a value higher than that of R_{LF} (~36 Ω) only for the wave with the lowest front time (0.5 μs).

A relevant conclusion of the results commented above is that the impulse impedance Z_P is strongly dependent of the current-wave front time.

A question naturally arises from the above analysis: "what is a short and long electrode?" To answer this question, Section 7.3.2 discusses attenuation effects in the soil.

7.3.2 Attenuation of impulsive currents propagating along electrodes and effective length

Propagation effects are very pronounced in the soil when waves of short front times are concerned. Basically, these effects are expressed by the attenuation and distortion of the current and associate voltage waves propagating along electrodes [2,6].

Attenuation is related to the losses occurring during propagation, mainly those associated to the current flowing in the soil, as the losses along the electrodes are very low due to their high conductivity. As soil conductivity rises, losses in the soil increase. The distortion, observed notably at the wave front, is related to the different level of attenuation the different frequency components of the current are subjected to. As mentioned in Section 7.2.4, frequency rise results in increase of losses, subjecting high frequency components to higher attenuation. Thus, current waves with short front time are subjected to more pronounced attenuation [2,6]. Figure 7.9 exhibits the waveform of a fast pulse of current propagating along a 40-m long electrode at three different positions, including the point of current impression.

Thus, both attenuation and distortion become more pronounced, as soil resistivity and wave front time decrease.

The parameter that quantifies the current-wave attenuation in time-domain is the effective length of electrode L_{EF}. As discussed in [6,23,24], L_{EF} corresponds to a limiting length. Increasing the electrode length further does not yield reduction of the impulse grounding impedance. This occurs because for longer electrodes the high-frequency components of current associated with the wavefront are strongly attenuated and present negligible amplitude.

In order to clarify the effective length concept, Figure 7.10 exhibits the variation of the impulse grounding impedance as a function of the length of a horizontal electrode and indicates how to determine L_{EF}.

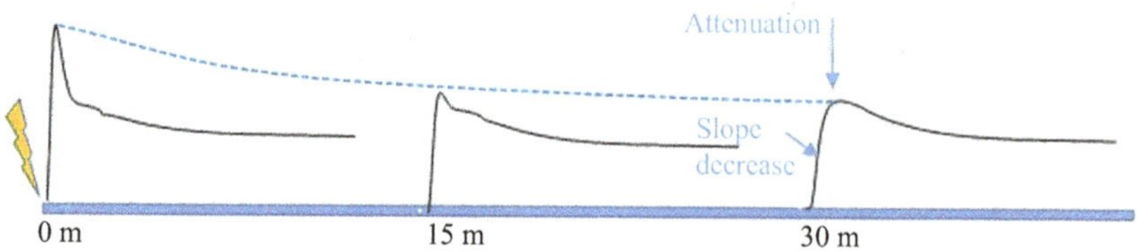

Figure 7.9 Attenuation and distortion of the impulsive current wave propagating along the electrodes attributed to losses in the soil. Adapted from [6]

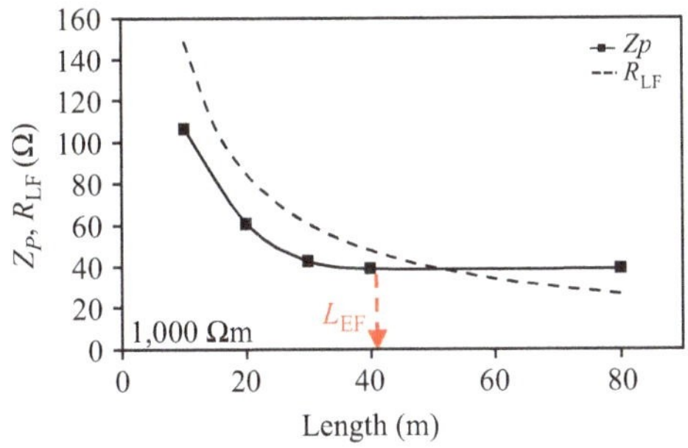

Figure 7.10 Identifying the effective length of electrode L_{EF} at the point where the decrease of the impulse impedance with increasing electrode length begins to saturate. Horizontal electrode (radius = 0.5 cm) buried 0.5 deep in a 1,000-Ωm soil, subjected to a 2/50-μs current wave impressed on one extremity

Note that, initially, as the electrode becomes longer the impulse impedance is decreased. However, after reaching a determined length, the rate of Z_P decrease is diminished due to attenuation and, soon, the decrease is saturated. Beyond that length, the impulse impedance remains constant. As indicated by the arrow in the figure, the effective length corresponds to the length in which the decrease of Z_P becomes saturated.

As soil resistivity decrease and frequency rise increase attenuation, L_{EF} is shorter in low resistivity soils and it is also shorter for current pulses of short front times, as they present higher frequency components [2,6].

The concept of effective length can be applied to other arrangement of electrodes, for instance to substation grounding grids. In this case, the electrode length has to be considered from the point where current is impressed on. Also, with respect to grids, it is usual to refer to an effective area, as an extension of the concept of the effective length of electrodes. This specific issue is discussed in [21,23].

7.3.3 The impulse coefficient

One practical issue of engineering interest consists of the way the impulse response of electrodes is related to their low-frequency response. The impulse coefficient I_C, given by the ratio of the impulse impedance Z_P and the low-frequency resistance R_T was defined for expressing this relationship:

$$I_C = Z_P/R_T \tag{7.10}$$

This parameter is very useful, as it allows promptly determining the impulse impedance from the low-frequency resistance, which, in most cases, is the only available parameter, as it can be easily obtained by measurement or calculation. Figure 7.11 depicts the typical profile of I_C as a function of the electrode length.

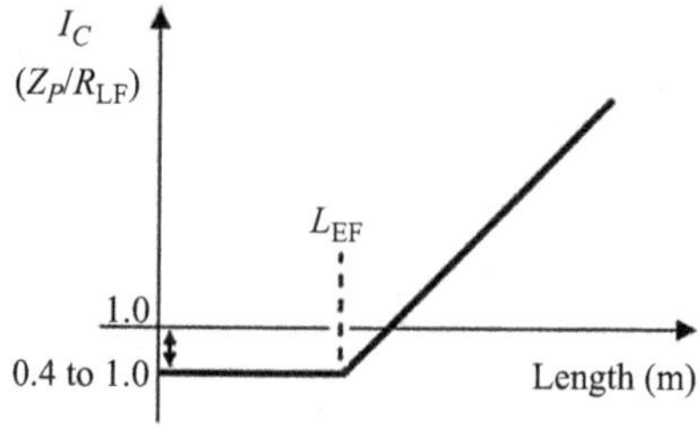

Figure 7.11 Typical profile of the impulse coefficient I_C as a function of electrode length for current pulses with front-times in the microsecond range. Adapted from [28]

The curve of Figure 7.11 denotes that this coefficient varies from a value below one for short electrodes to values higher than it for electrodes longer than L_{EF} [16,18]. The values below one obtained for electrodes shorter than L_{EF} are attributed to the superimposed effects of capacitive currents in the soil and of soil-resistivity decrease due to the frequency dependence of soil, both of which contribute toward making Z_P lower than R_T [20,25]. The lowest values of this ratio (from about 0.4 to 0.7) are found for short electrodes buried in high resistivity soils. Referring to Figure 7.10, one can see that, as the electrode length increases, the impulse impedance and the low-frequency resistance tends to decrease. However, for electrodes longer than L_{EF}, while the low frequency resistance continuously decreases by increasing electrode length, the value of Z_P remains constant, leading to increasing values of I_C, as discussed in [21,25–27]. Thus, long electrodes buried in low resistivity soils tend to present values of I_C much higher than one, when subjected to current waves with very short front time.

It is worth mentioning that the curve of Figure 7.11 presents the general qualitative profile of I_C. The numerical values of this coefficient are defined according to the specific front-time and waveform of the impressed current, in addition to the specific characteristics of the soil (notably, the soil resistivity).

This figure allows responding the final question in Section 7.3.1: what is a short and long electrode? The effective length can be used to answer this question, as it separates two domains, in which the impedance is first lower and, then, higher than the low-frequency resistance, while the electrode length is increased. Thus, short electrodes can be considered those shorter than L_{EF} and long electrodes are considered those longer than L_{EF}, in the specific soil into consideration. Note the relativity of the "short/long electrode" concept, as it depends on the soil resistivity, like the value of L_{EF}.

The analysis above is consistent with the results obtained for the 30-m long electrode buried in the 600-Ωm soil of Figure 7.7(d). Those results exhibit a Z_P value lower than that of R_T for the current waves with front times of 5 and 2 μs but higher than that of R_T for a wave with 0.5-μs front-time (values of Z_P, respectively of 25 to 32 and 42 Ω against $R_T = 36$ Ω). This happens because the effective length

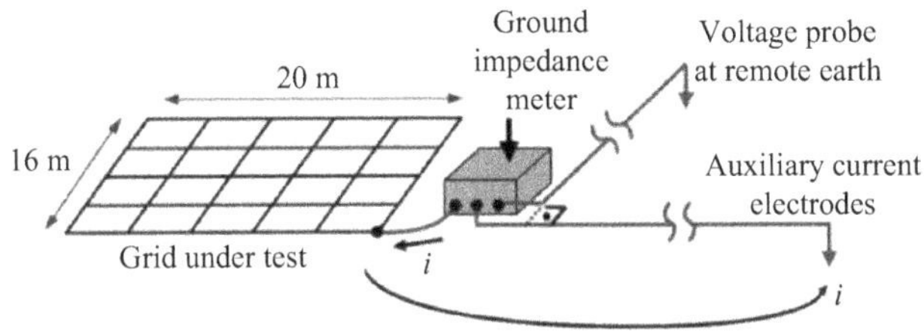

Soil (Ωm)	160	2,000
Measured Z_P (Ω)	11	22
Measured R_{LF} (Ω)	4.3	54
Ratio Z_P/R_{LF}	2.6	0.41

Figure 7.12 Experimental response of a same grounding grid buried 0.5 m deep in soils of 160 and 2,000 Ωm. The grid consists of 20 4 × 4-m² meshes of 0.5-cm-radius electrode. Current injected at the grid corner. Adapted from [22,28]

in that soil is longer than 30 m for the longer front times (slightly longer for the 2-μs front time) but shorter for the 0.5-μs front time (L_{EF} shorter than 16 m).

Also, an experimental result confirms the above analysis, denoting that Z_P can be either higher or lower than R_{LF}, depending on the electrode length and its effective length in the specific soil in which the electrodes are buried. Figure 7.12 exhibits the setup used for measuring the impulse impedance of a same grid, buried in 160- and 2,000-Ωm soils, subjected to impulsive currents of about 3.8/200 μs. It also shows the parameters determined from the measurements .

Z_P values of 11 and 22 Ω and R_T values of 4.3 and 54 Ω were measured for the low and high resistivity soils, respectively. Thus, Z_P exhibits values significantly higher and significantly lower than that of R_{LF}, respectively for the low- and high-resistivity soils. As in the 160-Ωm soil, the length of the grid's electrodes largely exceeds L_{EF} ($L_{EF} < 15$ m for a 4/200-μs wave), Z_P is larger than R_T (about 2.6 times larger). In the 2,000-Ωm soil, the electrodes' length is much shorter than L_{EF} ($L_{EF} > 80$ m), leading to a Z_P value lower than that of R_T (about 2.4 times smaller).

7.3.4 Soil ionization effect

Soil ionization occurs when very high currents are impressed on short electrodes, leading to very high electric fields at the electrode surface and disruptive process in the surrounding soil. The electric discharges resulting in the soil around the electrodes promotes an effect equivalent to increasing the current-dissipation area, reducing the electrode impedance while the process occurs.

Figure 7.13 illustrates the general effect of soil-ionization occurrence, which consists of the reduction of the impulse impedance Z_P [6,29,30] and consequent decrease of the electrode GPR.

In Figure 7.13(a) the impression of a low peak current on a short vertical electrode results in an impulse impedance of about 25 Ω. As the peak current increases very significantly [about 200 times in Figure 7.13(b)], preserving the current waveform, the impulse impedance is decreased to about the half value. The ionization process in the soil increased the electrode dissipation area, decreasing the resulting GPR in about 200-times and, so, the impulse impedance. Thus,

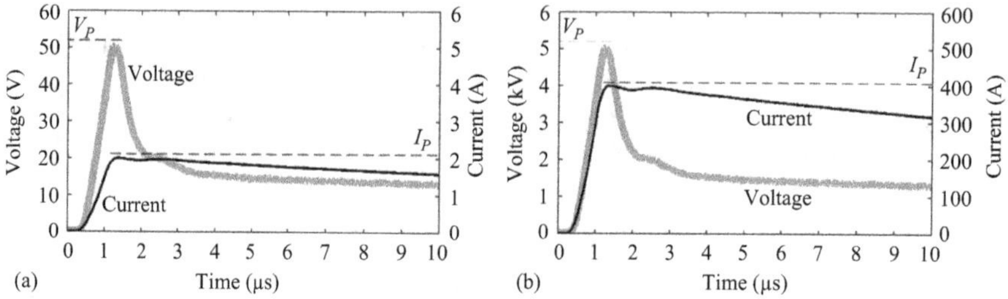

Figure 7.13 The effect of soil ionization

ionization effect can produce a relevant reduction of the electrode impedance, in conditions of very high density of currents at the electrode surface. This happens when extremely intense currents are impressed on short electrodes.

The dynamics of soil ionization has already been modeled by different approaches [31–35] and a large number of works have addressed the impact of this effect on the electrode performance (e.g., [32–34,36,37]). However, the main complexities involved in determining the impact of this effect, namely defining the critical electric-field value (corresponding to the onset of the ionization process) and the nonlinear distribution of the effect along the electrode due to propagation effects, have not been properly addressed yet. Experimental results demonstrate that the critical electric field can vary in a wide range from 200 to 1,700 kV/m, according to the soil type and moisture content [35,38,39].

Due to the mentioned uncertainties, the simplified approached given by the so-called Weck's formula (7.11) and (7.12) [34] is still used frequently to determine the soil ionization impact, in terms of the reduction it promotes on the grounding resistance. Equation (7.12) indicates the correction (reduction) of the low-frequency resistance value (R_T) to R_I, due to the increase of the peak current. The ratio "I/I_{PI}" gives the number of times the impressed peak current I is larger than I_{PI}, the threshold value of current corresponding to the onset of ionization in the soil of low-frequency resistivity of ρ. In this equation, the critical electric field E_0 is frequently assumed as 400 kV/m:

$$R_I = \frac{R_T}{\sqrt{1 + I/I_{PI}}} \tag{7.11}$$

$$I_{PI} = \frac{E_0 \rho}{2\pi {R_T}^2} \tag{7.12}$$

A simple but consistent improvement of the Weck's formula can be easily achieved replacing R_T by Z_P in (7.11):

$$Z_{PI} = \frac{Z_P}{\sqrt{1 + I/I_{PI}}} \tag{7.13}$$

To conclude this topic, it is worth weighting that soil ionization has a relevant effect in the response of short electrodes subjected to intense currents. In practical conditions, it is required to verify whether the density of current along the electrode (and, therefore, the electric field intensity at the electrode surface) is high enough to cause significant ionization or not. The result depends on both current intensity and electrode length. According to the result, ionization process can have significant or negligible impact, in terms of the grounding impedance reduction.

7.4 Response of grounding electrodes subjected to lightning currents

7.4.1 Introduction

To discuss the lightning response of electrodes, first this section objectively considers the relevant features of lightning currents that affect this response.

Then, influent parameters related to the impulse response of electrodes are discussed, based on experimental results and results simulated by using an elaborate electromagnetic model.

Finally, the relevance of specific behaviors of soil, namely the frequency dependence of soil parameters and soil ionization, is addressed, in terms of their impact on the response of power-systems components subjected to lightning-related currents.

7.4.2 Characteristics of return stroke currents

The lightning current is the source of the hazardous effects produced by this phenomenon. As mentioned in Chapter 2 of this volume, this current has different characteristics according to the type of lightning event. Concerning the influence of grounding systems in the interaction between lightning and power systems, the pulses of current, corresponding to first and subsequent return strokes are, in fact, what matters most.

Literature suggests different waveforms for application in lightning-related simulations, for instance double-exponential, triangular and Heidler-function-derived waveforms of current.

In the analysis of this chapter, the waveforms of current in Figure 7.14, which exhibit the main features of first and subsequent return strokes, were used. In particular, the curves of Figure 7.14(a) reproduce the initial concavity, the abrupt rise about the half peak that lasts until the first peak and a second peak larger than the first, typical of first strokes [40–42]. The curves also exhibit all median time and peak parameters of first- and subsequent-return-stroke currents of negative cloud-to-ground lightning, measured at Monte San Salvatore (MSS) [43,44] and at Morro do Cachimbo Station (MCS) [45,46].

As shown in Figure 7.14, the median peak current of first return strokes is about three times higher than that of subsequent ones, although, in a few cases, the peak current of one of the subsequent strokes in individual flashes are in fact higher

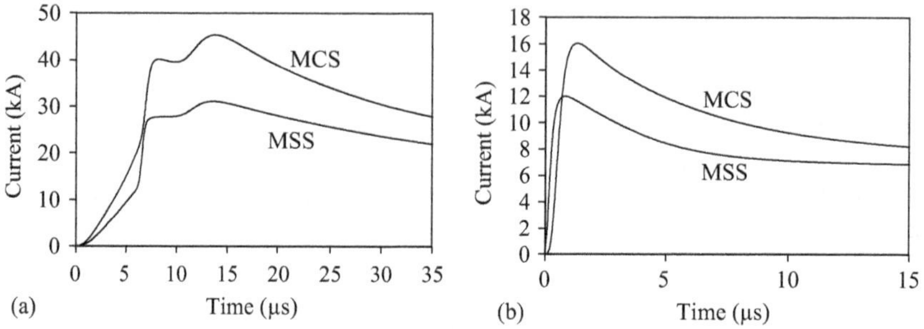

Database	Event	I_{P1} (kA)	I_{P2} (kA)	T_{10} (μs)	T_{30} (μs)	T_{50} (μs)	dI/dt_{MAX} (kA/μs)
MCS	First	40.3	45.3	5.6	2.9	53.5	19.4
	Subs.	16.3	–	0.7	0.4	16.4	29.9
MSS	First	27.7	31.1	5.6	3.8	75.0	24.3
	Subs.	11.8	–	0.75	0.67	32.0	39.9

I_{P1} and I_{P2} refer to the first and second return-stroke current peak value, T_{10} and T_{30} correspond respectively to the time between 0.1 I_{P1} and 0.9 I_{P1} and between 0.3 I_{P1} and 0.9 I_{P1}, T_{50} is the time interval required for the current to decay to 0.5 I_{P2}, and dI/dt_{MAX} is the maximum time derivative.

(c)

Figure 7.14 Representative waveforms of lightning currents, which reproduces all median parameters of time and amplitude of return-stroke currents: (a) first, (b) subsequent return strokes, and (c) median parameters [43–46]. Waveforms reproduced by using Heidler functions, according to the procedures described in [47] and [48] for first- and subsequent-return strokes

than that of the first stroke. However, this happens only with first strokes of low peak current. The maximum peak of currents measured at instrumented towers is of 62 kA for subsequent strokes, against 153 kA for first strokes [46]. However, the median first strokes' front time is about five times longer, ranging from about 1.5 to 10 μs, against 0.2 to 1.4 μs for subsequent strokes [44,46]. Thus, the frequency content of subsequent strokes is significantly wider and includes superior components of 4 MHz and so. The comments above are valid for the data of both MSS and MCS, though the median peak-current values of MCS data are significantly higher, about 50% and 30% higher, respectively for currents of first and subsequent return strokes [44–46].

7.4.3 Lightning response of grounding electrodes

Taking the representative waveforms of first- and subsequent-return-stroke currents of MSS (Figure 7.14) as reference, the GPR resulting from their impression on a typical arrangement of transmission-line grounding electrodes was simulated by applying the HEM Model and assuming frequency-dependent soil parameters given by the Visacro–Alipio expressions. The results depicted in Figures 7.15 and 7.16 reveal some interesting aspects.

Figure 7.15 shows the developed GPR of tower-footing electrodes, consisting of counterpoise wires of a 138-kV line [Figure 7.15(c)] as a function of the low-frequency soil resistivity, considering two counterpoise-wire lengths, relatively short (10 m) and long (30 m).

As shown in Figure 7.15, the variation of soil resistivity has a relevant effect on the GPR, whose peak is almost proportional to the low-frequency resistivity value, for the short electrodes. For long electrodes, this holds true only in first-return-stroke case. For the subsequent stroke, the relationship becomes less than proportional due

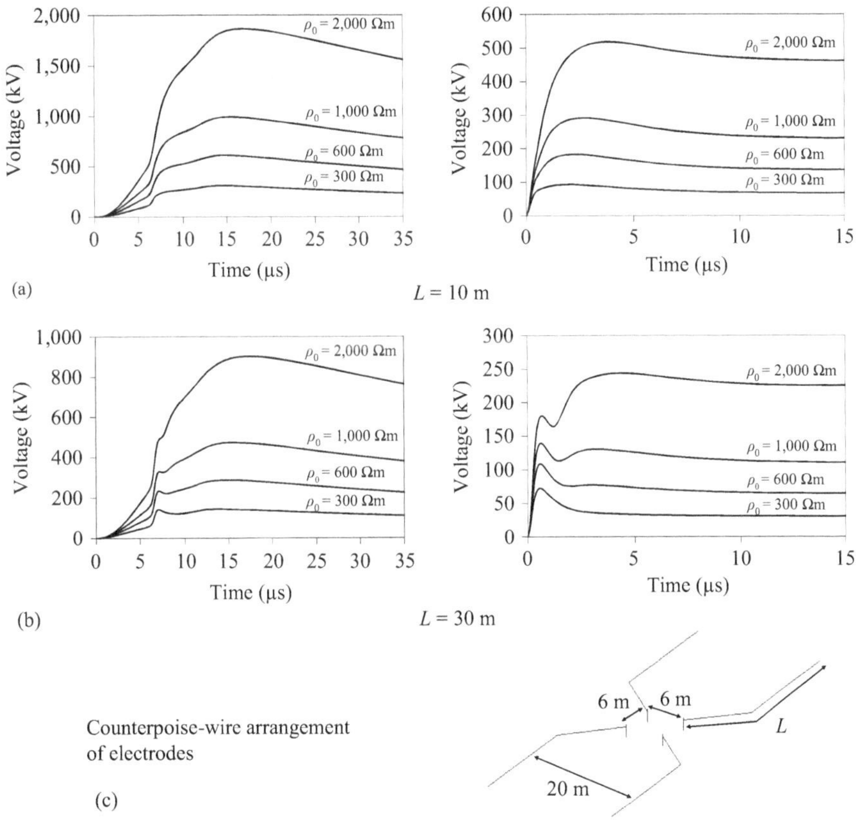

Figure 7.15 Simulated tower-footing potential rise as a function of the low-frequency soil resistivity for counterpoise wires (a) 10 and (b) 30 m long. Results for first- and subsequent-return-stroke currents on the left and right, respectively. Arrangement of electrode (radius: 0.5 cm; deep: 0.5 m), including four vertical 2.5-m long electrodes of 5-cm radius that represent the metallic parts of the tower penetrating the soil (c). Adapted from [22,49]

to the higher attenuation effect at low soil-resistivity values (for instance of 600 and 300 Ωm), as the high frequency components of currents are dissipated to the soil only along a length of the electrode shorter than L_{EF}.

Note that, for the condition of high resistivity soil and short electrode, the impulse impedance is lower for subsequent strokes. For instance, in the 2,000-Ωm soil, it is about 44 Ω against 59 Ω, for first return stroke. In the condition of low resistivity soil and long electrode, this trend is inverted: the impulse impedance becomes higher for subsequent strokes, for instance, about 5.9 Ω against 4.8 Ω, in the 300-Ωm soil. These results are consistent with the discussion in Section 7.3.3, which explains the existence of impulse impedances lower and higher than the low frequency resistance, respectively for electrode length shorter and longer than the effective length.

Other aspect concerns the difference on the absolute response of electrodes subjected to representative first and subsequent return stroke currents. Note the larger amplitude of the first-stroke GPR in all cases, about three times larger, like the ratio between first- and subsequent-return-stroke peak currents.

Figure 7.16 shows the developed GPR of the same electrode arrangement [Figure 7.15(c)] as a function of the counterpoise-wire length (L), considering it in conditions of low- (300 Ωm) and high-resistivity soils (2,000 Ωm).

As the electrode length increases, the GPR peak significantly diminishes. However, the ratio of decrease tends to diminish as the length becomes too long,

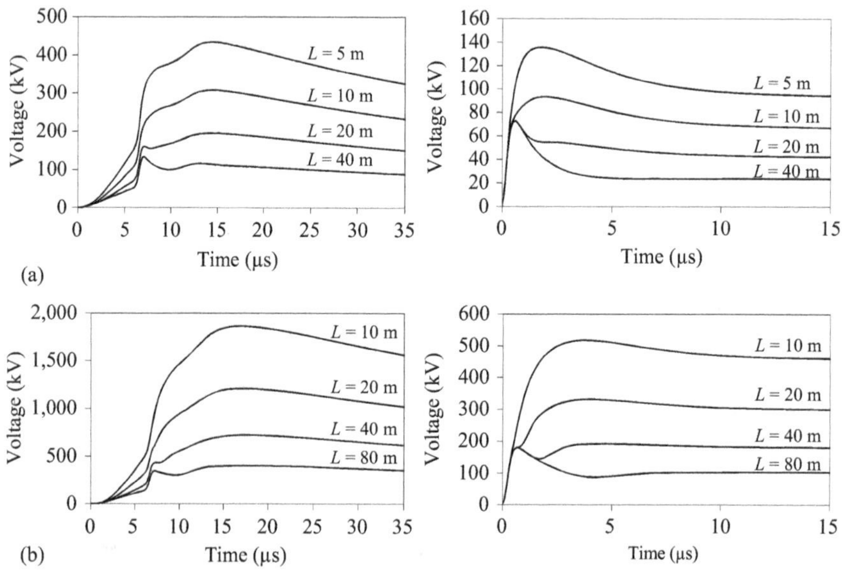

Figure 7.16 Simulated tower-footing GPR as a function of electrode length, considering (a) a low- (300 Ωm) and (b) a high- (2,000 Ωm) resistivity soils. Response for representative currents of first and subsequent return strokes on the left and right, respectively. Adapted from [22,49]

approaching the L_{EF} value, though the decrease remains significant at the wave tail. This trend is stronger for subsequent strokes, as their effective length is shorter. Note that for first strokes, the ratio of GPR decrease remains significant for both high and low resistivity soils, as their effective length is longer than the considered lengths.

Although the quantitative results change for different arrangements of electrodes, for instance grids, vertical and horizontal electrodes, the qualitative analyses presented above holds true for all of them. To quantify the response of electrodes subject to lightning current in each case, one may apply sophisticate tools, such as the HEM model. However, in most cases, it is still possible to use simplified approaches for doing that, as those presented in Section 7.5.

7.4.4 Effective length of electrodes for lightning currents

As discussed before, the attenuation of impulsive current and voltage waves propagating along electrodes is dependent on their frequency content, which is affect by the wave front time. Considering the very significant differences on the representative front times of first and subsequent stroke currents, about 0.7 μs against 4 μs median values, significant differences are expected on their attenuation and, therefore, on their effective length.

Using the same procedure of Section 7.3.2, the effective length of electrodes subject to representative currents of first (L_{EF1st}) and subsequent (L_{EFSub}) return strokes was estimated in each case from curves of Z_P versus the electrode length, like in Figure 7.10. The results are presented in Table 7.1, corresponding

Table 7.1 Approximate effective length of electrodes of horizontal electrodes and counterpoise wires

Apparent ρ_0 (Ωm)	**L_{EF}* (m)**	
	L_{EF1st}	**L_{EFSub}**
Horizontal electrode buried 0.5 m deep		
100	12	7
300	22	11
600	34	16
1,000	45	21
2,000	77	34
4,000	126	57
Counterpoise wires as in Figure 7.15(c)		
100	20	10
300	30	15
600	40	22
1,000	60	30
2,000	90	45
4,000	>150	75

*L_{EF} determined at the inflection point in I_C curve. Adapted from [28] and [22].

to horizontal electrodes and counterpoise wires, with the arrangement of Figure 7.15(c).

As discussed in [22], L_{EF1st} and L_{EFSub} values estimated for a horizontal electrode are still valid for vertical electrodes and grids, as long as the current is impressed on a single point. Note that the values of L_{EF} are significantly longer for counterpoise wires of transmission lines, as discussed in [28]. It is worth mentioning that, in real transmission-lines, the length of counterpoise wires is defined according to the soil-resistivity value. Long counterpoise wires are used only for towers installed over high-resistivity soils, aiming to achieve low values of tower-footing impedance, comparable to those of towers installed over low-resistivity soils. In practical cases, the length of counterpoise wires is always expected to be shorter than the first-return-stroke effective length. The values in Table 7.1 can be used as reference for approximately delimiting domains of constant and increasing impulse coefficient for most electrode arrangements.

7.4.5 *Remarks on the frequency dependence and soil ionization effects*

The frequency content of lightning currents makes the effect of the frequency dependence of soil resistivity and permittivity very relevant, contributing to significantly diminishing the grounding impedance, except for in low resistivity soils ($\rho_0 < 500$ Ωm). The impedance decrease is about 20% to 50% and this reduction is more pronounced for high resistivity soils [8,16,17,50]. It is also more pronounced for subsequent strokes, due to their higher frequency content.

Different from soil ionization, this effect occurs independently of the lightning current amplitude, what leads to the recommendation of always considering it for electrodes buried in soils of 500 Ωm and above [16,50].

However, it is impressive that, in terms of the impact of this effect in power-system lightning performance, the effect is much more pronounced for first return strokes, as discussed in [50]. In fact, this occurs because the influence of ground terminations on the severity of lightning stresses on power system components is less significant for subsequent strokes. When a power-system component is subjected to lightning current or overvoltage, the time required for the ground-termination response influencing the overvoltage amplitude (time required for the current wave to propagate from the stressing point to the closest ground termination and to return) is usually longer than or comparable to the subsequent-stroke front time. Thus, usually this response is not able promote significant reduction of lightning overvoltage. In most cases, the condition is different for first strokes, whose front times is typically about five times longer than that of subsequent strokes. This aspect is discussed in [51].

As mentioned before, ionization effect can produce a relevant reduction of the electrode impedance, in conditions of very high density of currents at the electrode surface. This happens when extremely intense currents are impressed on short electrodes.

In this respect, the probability of soil ionization is lower for subsequent return strokes, as they have median peak currents about three times lower.

Considering the conditions of power-system lightning protection, short electrodes are typically used in low resistivity soils, in which low impedances values are easily obtained. Reducing further the low impedance value due to soil ionization tends to have irrelevant impact on the lightning performance of the grounded system and, in most cases, taking this effect into account does not change the picture.

However, in high resistivity soils, long electrodes are always used to ensure low grounding impedance values, typically required in lightning protection applications. Thus, their current density tends to be low, reducing the probability of soil ionization occurrence or the intensity of this effect. In applications involving long electrodes, such as the common counterpoise wires of transmission lines, neglecting the favorable effect of soil ionization is considered prudent [6], notably when taking the favorable effect of the frequency dependent parameters of soil into account.

7.5 Representation of grounding systems in lightning protection studies

7.5.1 Introduction

As the lightning response of grounding electrodes has relevant influence on the performance of electrical systems, in the last decades significant research efforts have been dedicated to developing numerical models for predicting the response of electrodes subjected to lightning currents [7,52–54]. In particular, models based on electromagnetic approaches are able to determine the accurate response of electrodes subject to impulsive currents, provided that correct constitutive parameters are assumed for the soil. This has been demonstrated in experimental works, for instance those by Visacro and collaborators, in which the measured GPR of electrodes subjected to impulsive currents with lightning-patterned waveforms practically matches the GPR simulated by electromagnetic models [8,16,20]. Thus, presently, such models are available for accurately determining the lightning response of general conditions of soil and of arrangement of electrodes.

However, in spite of the accuracy of results provided by these models, in lightning protection applications, concise representation of grounding electrodes is frequently required for expediting the estimates of the lightning response of the whole grounded electrical system, for instance in the assessment of the lightning performance of transmission lines.

Despite the known fact that grounding resistance is not able to represent the lightning response of electrodes [2,6], except in very specific conditions, it is still the parameter used in most lightning-protection studies for representing grounding systems, as it can be easily obtained, by using low-frequency measuring instruments or by calculation [6].

In the last years, several researches have been developed to determine other concise representations of electrodes. For instance, curve-fitting techniques have been applied for determining equivalent circuits or rational functions that are able to reproduce the lightning response of electrodes, taking as reference their grounding

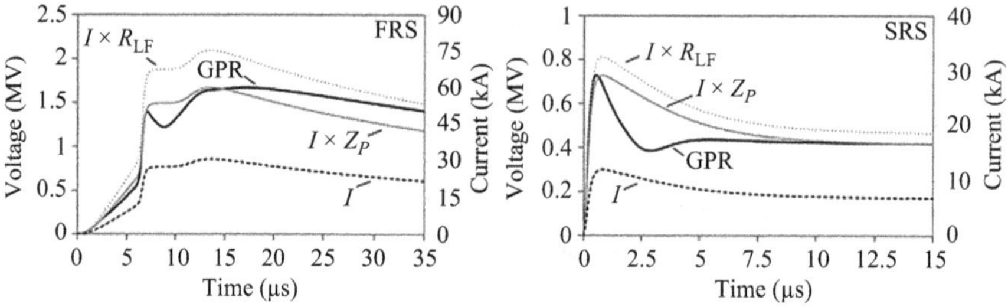

Figure 7.17 Response of a 60-m long horizontal electrode (radius of 0.5 cm) buried 0.5 m deep in a 2,000-Ωm soil subjected to representative currents of first and subsequent return strokes, under different representations: physical representation, low-frequency resistance representation ($R_{LF} = 69\ \Omega$), and impulse impedance representation ($Z_{P1st} = 54\ \Omega$ and $Z_{PSub} = 61\ \Omega$). Adapted from [28]

potential rise simulated using elaborate electromagnetic models (e.g., [55]). Nevertheless, developing this approach still requires using electromagnetic models.

A recent work has shown that, in most lightning-protection engineering applications, using the impulse impedance as concise representation of electrodes is worth. In most cases, the errors resulting from using this representation are irrelevant, typically lower than 2% in reference to results of GPR and insulation failures obtained with rigorous electromagnetic models that consider the entire electrode arrangement. This section addresses this representation.

Figure 7.17 compares the response of a horizontal electrode subject to representative currents of first and subsequent return strokes, in terms of the GPR curves, obtained under the physical representation of electrodes with those obtained under the concise representations Z_P and R_T that is also represented by R_{LF}.

Note that, though different curves of GPR are obtained under the different representations, the curves obtained under the physical and Z_P representations are closer and present practically the same peak value. This result, illustrated for a horizontal electrode buried in a 2,000-Ωm soil, is also found along the whole range of soil resistivity and for general arrangements of electrodes.

7.5.2 Using Z_P as a concise representation of grounding electrodes subject to lightning currents

In spite of the time shift between the current and GPR peaks, some features of the impulse grounding impedance make worth using it to represent grounding electrodes in lightning protection studies, as considered next. The following topics summarizes the discussion in [28].

In this respect, the first aspect to examine about Z_P is that it varies with the current waveforms, as their frequency content changes. For instance, the impulse impedance calculated from the curves of Figure 7.17(a) and (b) are 54 Ω and 61 Ω, respectively. Although the variation of Z_P with the waveform would be an apparent limitation to applying this representation, it is possible to overcome this limitation simply by using specific impulse impedances Z_{P1st} and Z_{PSub} to address the response of electrodes subject to representative currents of first and subsequent return strokes.

Exploring this aspect, Figure 7.18 exhibits the variation of both parameters, along with that of the low frequency resistance, as a function of the length of a horizontal electrode buried in a low- and high-resistivity soil.

The curves of Figure 7.18 allows observing two properties of the impulse impedance of first and subsequent strokes, concerning their relative values:

- Electrodes shorter than L_{EFSub} always exhibit values of Z_{PSub} lower than that of Z_{P1st}. The larger capacitive current in the soil (while $L < L_{EF}$) and greater frequency dependence of soil parameters for subsequent strokes are responsible for this behavior.
- However, the minimum value of Z_{P1st} value (found for $L = L_{EF1st}$) is always lower than the minimum value of Z_{PSub} (corresponding to $L = L_{EFSub}$).

For reference, Table 7.1 provided the effective length of electrodes subjected to currents of first and subsequent return strokes.

Simulations developed by using the HEM Model demonstrate that these features are general. They are observed for other electrode arrangements, such as grounding grids and counterpoise wires, within the whole usual range of soil resistivity values.

Another very attractive feature of Z_P is that, though the values of Z_{P1st} and Z_{PSub} of any arrangement of electrodes buried in any soil are significantly different, in most cases, for real lightning, the value of each one of them varies only slightly within their respective front-time ranges, notably the value of Z_{P1st}. The data in Table 7.2 exhibit the impact of the front-time variation in the value of Z_P within realistic front-time ranges for currents of first and subsequent return strokes. The results, obtained specifically for a horizontal electrode buried in soils with apparent

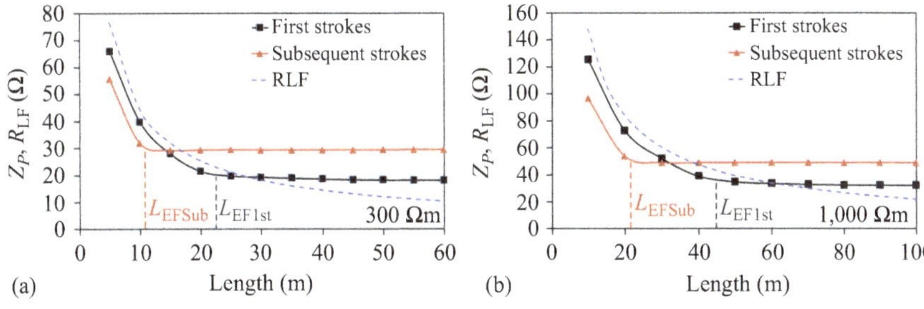

Figure 7.18 Impulse impedance of horizontal electrode (radius of 0.5 cm) buried 0.5 m deep in a (a) 300- and (b) 1,000-Ωm soil subjected to representative currents of first (Z_{P1st}) and subsequent return strokes (Z_{PSub}) as a function of electrode length

resistivity of 100 to 4,000 Ωm, are presented for the median front-time (x) and for the inferior and superior front-time limits.

The very important result depicted in Table 7.2 is that Z_{P1st} remains practically constant within its front-time range, with a variation lower than 1% in most cases (about 4% only for 4,000 Ωm-soil). Similar results are observed for other electrode arrangements (e.g., grid and counterpoise wires) and lengths. Thus, a relevant and general feature of Z_{P1st} is that it is practically nonsensitive to the variation of the front time along the whole range this parameter varies in real lightning.

The variation of Z_{PSub} within the respective front-time range is not so slight, as the frequency content of subsequent return stroke currents is much wider, reaching the order of 10%.

Another relevant property related to the features of the impulse impedance is the invariability of the impulse coefficient Z_P/R_{LF} in any given soil, valid for electrode lengths shorter than L_{EF}. In this domain, I_C does not depend on the electrode arrangement but only on the soil parameters. Table 7.3 demonstrates this property, considering the impulse impedance of three different electrode arrangements buried in low- and high-resistivity soils, determined by using the HEM Model.

Although, in a same soil, changing the electrode arrangement makes Z_P and R_{LF} to vary, their relative variation is practically the same, preserving the impulse-coefficient value. In Table 7.3, the corresponding variation of I_C value is only about 1%–2% for first strokes. The variation is not so slight for subsequent strokes, reaching the order of 10%–20%, as all electrodes in the table are longer than L_{EFSub}. It would be slight as well, if electrodes were shorter than this effective length.

If one considers that electrodes shorter than L_{EF} practically do not exhibit variation of the impulse coefficient and, however, shows a constant value (Section 7.3.3), a useful general property can be derived: in a given soil, since $L < L_{EF}$, I_C has a defined value, regardless the electrode arrangement and length L.

Table 7.2 Impulse impedance of horizontal electrodes buried 0.5 m deep in the soil as a function of the current front time. Adapted from [28]

ρ_0 (Ωm)	Z_{P1st} (Ω)				Z_{PSub} (Ω)			
	L (m)	**First stroke front time (μs)**			**L (m)**	**Subs. stroke front time (μs)**		
		1.5	**x = 3.8**	**10**		**0.4**	**x = 0.67**	**1.2**
100	5	22.2	22.4	22.4	5	18.7	19.7	20.6
300	15	28.7	28.9	29.1	10	32.9	32.8	35.0
600	25	37.2	37.5	37.8	15	44.8	44.5	48.2
1,000	40	40.8	40.9	41.3	20	53.8	55.0	60.5
2,000	50	63.0	63.3	64.0	30	64.9	69.4	78.1
4,000	60	96.6	96.4	99.9	40	90.6	99.3	105.1

Notes: ρ_0: low-frequency soil resistivity; x: median front-time value.

Table 7.3 Ratio Z_P/R_{LF} of different arrangement of electrodes buried 0.5 m in the soil (length shorter than L_{EF1st}). Adapted from [28]

ρ_0 (Ωm)	Arrangement	L (m)	First stroke			Subsequent stroke		
			Z_P (Ω)	R_{LF} (Ω)	I_C	Z_P (Ω)	R_{LF} (Ω)	I_C
600	Horizontal	30	32.3	36.4	0.89	40.1	36.4	1.10
	Counterpoise*	30	10.8	11.8	0.91	13.6	11.8	1.15
	Grid**	15 × 15	15.7	18.0	0.88	17.1	18.0	0.95
2,000	Horizontal	50	63.3	79.2	0.80	60.6	79.2	0.77
	Counterpoise*	50	20.8	26.2	0.79	16.3	26.2	0.62
	Grid**	25 × 25	29.8	37.4	0.80	25.4	37.4	0.68

*Anchors of counterpoise wires: 6 m × 6 m.
**Meshes of 5 m × 5 m.

7.5.3 *When using Z_P to represent the grounding system: applications*

In the lightning protection of power systems, the parameters of major interest are the maximum GPR developed when lightning currents are impressed on their electrodes (as a result of direct strikes or inducing effects of nearby strikes), or the potential of the corresponding overvoltage developed across the systems' insulators to cause insulation failures. This potential can be determined from the peak overvoltage (using the so-called $V \times T$ curves) or using a time-integral flashover criterion, such as the disruptive effect model (DE) [56–59].

What makes using the impulse impedance for representing the grounding system worthy is that the maximum GPR of first and subsequent return stroke currents and the corresponding potential of insulator overvoltages to produce failures calculated under this representation are practically the same calculated under the physical representation of electrodes. In addition, Z_P can be easily determined, as shown in Section 7.5.4.

This relevant aspect is illustrated for a specific application, consisting of the assessment of the lightning performance of transmission lines, considering the results presented in [22]. Such results show that outage rates of transmission lines calculated under the Z_{P1st} representation of tower-footing electrodes is practically the same obtained under the physical representation of electrodes. Figure 7.19 exhibits these result for two specific soils of low and high resistivity value.

This conclusion is validated in a wide range of soil-resistivity values and electrode length [16]. The errors resulting from this representation are lower than 2%, in terms of maximum GPR and insulation failures, in relation to results obtained with rigorous electromagnetic models that consider the entire electrode arrangement, whereas the errors resulting from the R_{LF} representation can exceed 20%–30% [16].

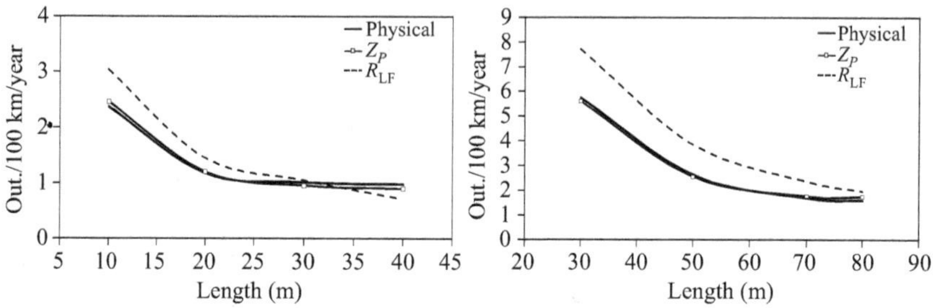

Figure 7.19 Outage rates of a 138-kV transmission line obtained under Z_{P1st}, R_{LF}, and physical representations of electrodes, using an electromagnetic model and the DE method (flashover criterion). Soils of 300 and 1,000 Ωm. Electrodes: counterpoise wires of length L. Towers: 30 m high and single shielding wire. N_g of 2.4 flashes/km^2/year. Calculated as described in [22]. Adapted from [22]

7.5.4 How to determine the impulse impedance

After showing the worth of using Z_P in lightning protection studies, this section addresses alternatives to determine it.

First, this impedance can be measured by using grounding impedance meters, such as that shown in Figure 7.20 [60].

A few portable instruments, provided with an internal generator to impress impulsive currents on electrodes, able to record current and GPR waves and to provide estimates of their Z_P and R_{LF} values are presently available. Though their application in field measurements is feasible, it is still somewhat complex and requires following specific procedures to prevent coupling effects among the electrodes and the leads used in the measurement, which can corrupt the measured results (see recommendations in [18]). These effects can be especially important when long electrodes are involved.

Another alternative consists of using analytical expressions provided in literature to determine Z_P. Reliable expressions are found in [22] and [21] to determine $Z_{P1\mathrm{st}}$ and $Z_{P\mathrm{Sub}}$ of largely used arrangements of horizontal and vertical electrodes. Different from other expressions, those presented in these references take the frequency dependence of soil parameters into account and were developed specifically from the results obtained for representative currents of first and subsequent return strokes. In particular, considering the large number of applications involving the lightning performance of transmission lines, it is worth recommending the simple expressions (7.14) and (7.15) from [22] to determine $Z_{P1\mathrm{st}}$ of counterpoise wires, of length L shorter than L_{EF1st}, which is a condition always observed in real transmission lines:

$$Z_{P1\mathrm{st}} = 0.16 \cdot \rho_0 \cdot L^{-0.687} \quad (\text{for } 100 \leq \rho_0 \leq 600\,\Omega\mathrm{m}) \tag{7.14}$$

$$Z_{P1\mathrm{st}} = 0.4 \cdot {\rho_0}^{0.89} \cdot L^{-0.75} \quad (\text{for } 600 \leq \rho_0 \leq 400\,\Omega\mathrm{m}) \tag{7.15}$$

Figure 7.20 View of a grounding impedance meter. This DSP-based measuring instrument stores both impressed current (with typical front times of first and subsequent return strokes' currents) and GPR waves and displays the calculated Z_{P1st}, Z_{Psub}, and R_{LF} values on the LCD screen. Adapted from [60]

Nevertheless, the most attractive alternative to determine Z_P consists of obtaining it directly from its low-frequency grounding resistance (measured or calculated R_{LF} [61]), simply by multiplying R_{LF} by I_C.

As the value of the impulse coefficient does not depend on electrode arrangement but on the soil parameters only (since $L < L_{\mathrm{EF}}$, Section 7.4.3), one may use the I_C value, determined for any electrode arrangement buried in that soil, calculated by using simple expressions, to obtain Z_P of any other arrangement of electrodes buried in that same soil. In particular, expressions (7.16) and (7.17) obtained for counterpoise wires [22] and horizontal electrodes [21], respectively, are recommended for determining I_C of first- and subsequent-return-stroke currents:

$$I_{C1\mathrm{st}} = 0.89 - 5 \cdot 10^{-5} \cdot \rho_0 \tag{7.16}$$

$$I_{C\mathrm{Sub}} = 1.24 - 0.095 \times \rho_0^{0.25} \tag{7.17}$$

As the value of Z_P becomes saturated with increasing electrode length when the effective length is reached, the same procedure described above can be used to determine Z_P of electrodes longer than L_{EF}, by simply using the resistance R_{LF} determined for an electrode length equal to L_{EF}.

This procedure can be used for electrodes connected to several down-conductors (e.g., LPS or multigrounded grids) but only if the electrode is shorter than or equal to L_{EF}.

The concept of the impulse grounding impedance is not applicable to arrangements comprising very long electrodes ($L > \mathrm{L_{EF}}$) and multiple down conductors, as the potential rise of the down conductors are not necessarily equal. In this case, one has to define a specific point above the ground, where the total current is impressed to determine the potential rise, for instance the top of a transmission-line tower. Electromagnetic models can be used for determining this potential rise in relation to remote earth and, then, to calculate the impulse impedance seen by the impressed current. However, this impedance, is not the impulse grounding impedance.

Finally, it is worth to remind that a correction can be applied in the calculated Z_P to roughly consider soil ionization by means of the modified Weck's approximation, expression (7.13), as mentioned in Section 7.3.4.

References

[1] A. P. Meliopoulos, *Power System Grounding and Transients*, New York, NY, Marsel Dekker Inc., 1988.

[2] S. Visacro, Transients on grounding systems, in A. Ametani (ed.), *Numerical Analysis of Power System Transients and Dynamics*, 1st edn, vol. 1, London, IET, 2014, pp. 481–507.

[3] R. Rudenberg, "Fundamental considerations on grounding currents," *Electrical Engineering*, vol. 64, no. 1, 1945.

[4] S. Visacro, Grounding and Earthing: Basic Concepts, Measurements and Instrumentation, Grounding Strategies, 2nd edn, São Paulo, Brazil, ArtLiber Edit., 2002, pp. 1–159 (in Portuguese).

[5] S. Visacro, Low-frequency grounding resistance and lightning protection, in V. Cooray (ed.), *Lightning Protection*, 1st edn, London, IET, 2009, vol. 1, pp. 475–502 (Chapter 9).

[6] S. Visacro, "A comprehensive approach to the grounding response to lightning currents," *IEEE Transactions on Power Delivery*, vol. 22, pp. 381–386, 2007.

[7] S. Visacro, and A. Soares Jr., "HEM: a model for simulation of lightning-related engineering problems," *IEEE Transactions on Power Delivery*, vol. 20, no. 2, pp. 1026–1208, 2005.

[8] S. Visacro, and R. Alipio, "Frequency dependence of soil parameters: experimental results, predicting formula and influence on the lightning response of grounding electrodes," *IEEE Transactions on Power Delivery*, vol. 27, no. 2, pp. 927–935, 2012.

[9] S. Ramo, J. R. Whinnery, and T. V. Duzer, *Fields and Waves in Communication Electronics*, New York, NY, John Wiley & Sons, 1965.

[10] S. Visacro, B. Hermoso, M. T. Almeida, *et al.*, "The response of grounding electrodes to lightning currents: summary report CIGRE WG C4.406," *Electra*, vol. 246, pp. 18–21, 2009.

[11] R. L. Smith-Rose, "The electrical properties of soils for alternating currents at radiofrequencies," *Proceedings of Royal Society*, vol. 140, no. 841 A, pp. 359–377, 1933.

[12] J. H. Scott, Electrical and magnetic properties of rock and soil, U.S. Geological Survey, U.S. Department of the Interior, Washington, D.C., 1966.

[13] C. L. Longmire, and K. S. Smith, "A universal impedance for soils," *Topical Report*, July 1–September 30, Defense Nuclear Agency, Santa Barbara, California, 1975.

[14] S. Visacro, and C. M. Portela, "Soil permittivity and conductivity behavior on frequency range of transient phenomena in electric power systems," in *Proceedings of the 1987 International Symposium on High Voltage Engineering*, vol. 93, no. 6, pp. 1–4, Germany.

[15] C. M. Portela, "Measurement and modeling of soil electromagnetic behavior," in *Proceedings of the IEEE International Symposium on Electromagnetic Compatibility*, Seattle, WA, 1999, pp. 1004–1009.

[16] S. Visacro, R. Alipio, M. H. Murta Vale, and C. Pereira, "The response of grounding electrodes to lightning currents: the effect of frequency-dependent soil resistivity and permittivity," *IEEE Transactions on Electromagnetic Compatibility*, vol. 53, no. 2, pp. 401–406, 2011.

[17] R. Alipio, and S. Visacro, "Modeling the frequency dependence of electrical parameters of soil," *IEEE Transactions on Electromagnetic Compatibility*, doi:10.1109/TEMC.2014.2313977, 2014.

[18] S. Visacro, and G. Rosado, "Response of grounding electrodes to impulsive currents: an experimental evaluation," *IEEE Transactions on Electromagnetic Compatibility*, vol. 51, pp. 161–164, 2009.

[19] S. Visacro, M. Guimarães, and L. Araujo, "Experimental impulse response of grounding grids," *Electric Power Systems Research*, vol. 94, pp. 92–98, doi.org/10.1016/j.epsr.2012.04.011, 2013.

[20] S. Visacro, R. Alipio, C. Pereira, M. Guimarães, and M. A. Schroeder, "Lightning response of grounding grids: simulated and experimental results," *IEEE Transactions on Electromagnetic Compatibility*, vol. 57, no. 1, pp. 121–127, 2015, doi:10.1109/TEMC.2014.2362091.

[21] R. Alípio, and S. Visacro, "Impulse efficiency of grounding electrodes: effect of frequency-dependent soil parameters," *IEEE Transactions on Power Delivery*, vol. 29, pp. 716–723, 2014.

[22] S. Visacro, and F. H. Silveira, "Lightning performance of transmission lines: requirements of tower-footing electrodes consisting of long counterpoise wires," *IEEE Transactions on Power Delivery*, 2016, doi:10.1109/TPWRD.2015.2494520.

[23] B. R. Gupta, and B. Thapar, "Impulsive impedance of grounding grids," *IEEE PAS-99*, no. 6, 1980.

[24] L. V. Bewley, "Theory and test of the counterpoise," *Electrical Engineering*, pp. 1163–1172, 1934.

[25] R. Alípio, and S. Visacro, Frequency dependence of soil parameters: effect on the lightning response of grounding electrodes. *IEEE Transactions on Electromagnetic Compatibility*, vol. 55, pp. 132–139, 2013.

[26] L. Grcev, "Impulse efficiency of ground electrodes," *IEEE Transactions on Power Delivery*, vol. 24, no. 1, pp. 441–451, 2009.

[27] L. Grcev, "Lightning surge efficiency of grounding grids," *IEEE Transactions on Power Delivery*, vol. 26, no. 3, pp. 223–237, 2011.

[28] S. Visacro, "The use of the Impulse Impedance as a concise representation of grounding electrodes in lightning protection applications," *IEEE Transactions on Electromagnetic Compatibility*, doi:10.1109/TEMC.2017.2788565, 2018.

[29] S. Sekioka, T. Sonoda, and A. Ametani, "Experimental study of current-dependent grounding resistance of rod electrode," *IEEE Transactions on Power Delivery*, vol. 20, no. 2, pp. 1579–1576, 2005

[30] A. Geri, "Behavior of grounding systems excited by high impulse currents: the model and its validation," *IEEE Transactions on Power Delivery*, vol. 14, no. 3, pp. 1008–1017, 1999.

[31] A. C. Liew, and M. Darveniza, "Dynamic model of impulse characteristic of concentrated earths," *IEE Proceedings*, vol. 121, no. 2, pp. 123–135, 1974.

[32] S. Visacro, and A. Soares Jr., "Sensitivity analysis for the effect of lightning current intensity on the behavior of earthing systems," in *Proceedings of the 1994 International Conference on Lightning Protection*, Hungary, pp. R3a-01(1–5).

[33] M. E. Almeida, and M. T. Correia de Bassos, "Accurate modelling of rod driven tower footing," *IEEE Transactions on Power Delivery*, vol. 11, no. 3, 1996.

[34] *CIGRE Guide to Procedures for Estimating the Lightning Performance of Transmission Lines, Technical Brochure 63, WG 01 (Lightning)*, Study Committee 33, October 1991.

[35] J. L. C. Lima, and S. Visacro, "Experimental developments on soil ionization: new findings," in *Proceedings of the 2008 International Conference on Grounding and Earthing (GROUND 2008)*, Florianópolis, Brazil, pp. 174–179, 2008.

[36] R. Zeng, X. Gong, J. He, B. Zhang, and Y. Gao, "Lightning impulse performances of grounding grids for substations considering soil ionization," *IEEE Transactions on Power Delivery*, vol. 23, no. 2, pp. 667–675, 2008.

[37] J. He, B. Zhang, R. Zeng, and B. Zhang, "Experimental studies of impulse breakdown delay characteristics of soil," *IEEE Transactions on Power Delivery*, vol. 26, no. 3, pp. 1600, 1607, 2011.

[38] E. E. Oettle, "A new general estimation curve for predicting the impulse impedance of concentrated earth electrodes," *IEEE Transactions on Power Delivery*, vol. 3, no. 4, pp. 2020–2029, 1988.

[39] A. M. Mousa, "The soil ionization gradient associated with discharge of high currents into concentrated electrodes," *IEEE Transactions on Power Delivery*, vol. 9, no. 3, pp. 1669–1677, 1994.

[40] S. Visacro, M. H. Vale, G. Correa, and A. Teixeira, "Early phase of lightning currents measured in a short tower associated with direct and nearby lightning strikes," *Journal of Geophysical Research*, vol. 115, no. D16104, 1-11, doi:10.1029/2010JD014097, 2010.

[41] V. Rakov, and M. A. Uman, *Lightning: Physics and Effects*, Cambridge University Press, 2003.

[42] S. Visacro, *Lightning: An Engineering Approach*, ArtLiber Edit., São Paulo, Brazil, 2005, pp. 1–276 (in Portuguese).

[43] K. Berger, R. B. Anderson, and H. Kröninger, “Parameters of lightning flashes,” *Electra*, vol. 41, pp. 23–37, 1975.

[44] R. B. Anderson, and A. J. Eriksson, “Lightning parameters for engineering application,” *Electra*, vol. 69, pp. 65–102, 1980.

[45] S. Visacro, M. A. O. Schroeder, A. J. Soares, L. C. L. Cherchiglia, and V. J. Sousa, “Statistical analysis of lightning current parameters: measurements at Morro do Cachimbo Station,” *Journal on Geophysical Research*, vol. 109, no. D01105, pp. 1–11, doi:10.1029/2003JD003662, 2004.

[46] S. Visacro, C. R. Mesquita, A. R. De Conti, and F. H. Silveira, “Updated statistics of lightning currents measured at Morro do Cachimbo Station,” *Atmospheric Research*, vol. 117, pp. 55–63, 2012.

[47] S. Visacro, “A representative curve for lightning current waveshape of first negative stroke,” *Geophysical Research Letters*, vol. 31, p. L07112, 2004.

[48] A. R. Conti, and S. Visacro, “Analytical representation of single- and double-peaked lightning current waveforms,” *IEEE Transactions on Electromagnetic Compatibility*, vol. 49, pp. 448–451, 2007.

[49] S. Visacro, and F. H. Silveira, Lightning performance of transmission lines: methodology to design grounding electrodes to ensure an expected outage rate. *IEEE Transactions on Power Delivery*, doi:10.1109/TPWRD.2014.2332457, 2015.

[50] S. Visacro, and F. H. Silveira, “The impact of the frequency dependence of soil parameters on the lightning performance of transmission lines,” *IEEE Transactions on Electromagnetic Compatibility*, 2015, pp. 434–441, doi:10.1109/TEMC.2014.2384029.

[51] F. H. Silveira, S. Visacro, A. R. Conti, and C. R. Mesquita, “Backflashovers of transmission lines due to subsequent lightning strokes,” *IEEE Transactions on Electromagnetic Compatibility*, vol. 54, pp. 316–322, 2012.

[52] G. J. Burke, and A. J. Poggio, Numerical electromagnetics code (NEC): method of moment, Lawrence Livermore National Laboratory, Livermore, CA, January 1981.

[53] L. Grcev, and F. Dawalibi, “An electromagnetic model for transients in grounding systems,” *IEEE Transactions on Power Delivery*, vol. 5, no. 4, pp. 1773–1781, 1990.

[54] M. Tsumura, Y. Baba, N. Nagaoka, and A. Ametani, “FDTD simulation of a horizontal grounding electrode and modeling of its equivalent circuit,” *IEEE Transactions on Electromagnetic Compatibility*, vol. 48, no. 4, pp. 817–825, 2006.

[55] S. Sheshyekani, M. Akbari, B. Tabei, and R. Kazemi, “Wideband modeling of large grounding systems to interface with electromagnetic transient,” *IEEE Transactions on Power Delivery*, vol. 29, no. 4, pp. 1868–1876, 2014.

[56] J. H. Hagenguth, “Volt–time areas of impulse spark-over,” *AIEE Transactions*, vol. 60, pp. 803–810, 1941.

[57] A. H. Hileman, *Insulation Coordination for Power Systems*, Boca Raton, FL, CRC, 1999, pp. 627–640.

[58] M. Darveniza, and A. E. Vlastos, "The generalized integration method for predicting impulse volt-time characteristics for non-standard wave shapes: a theoretical basis," *IEEE Transactions on Dielectrics and Electrical Insulation*, vol. 23, no. 3, pp. 373–381, 1988.

[59] W. A. Chisholm, "New challenges in lightning impulse flashover modeling of air gaps and insulators," *IEEE Electrical Insulation Magazine*, vol. 26, no. 2, 2010.

[60] B. D. Rodrigues, and S. Visacro, "Portable grounding impedance meter based on DSP," *IEEE Transactions on Instrumentation and Measurement*, doi:10.1109/TIM.2014.2303532, 2014.

[61] R. J. Heppe, "Computation of potential at surface face above an energized grid or other electrode, allowing for non-uniform current distributions," *IEEE Transactions on Power Delivery*, vol. 13, pp. 762–767, 1998.

Chapter 8

Surge-protective devices

Georgij V. Podporkin[1], *Martin Wetter*[2] *and Holger Heckler*[2]

This chapter is divided into two parts. The first deals with surge-protective devices (SPDs) used in transmission and distribution overhead lines (OHLs), while the second is related to low-voltage (LV) distribution systems.

Metal oxide arresters (MOAs), with and without external gaps, and multi-chamber arresters (MCAs) are used for protecting high-voltage (HV) and medium-voltage (MV) lines against lightning overvoltages. In the first part of this chapter, the operating principle and main designs of MOA and MCA are presented.

The main element of MCA is a multi-chamber system (MCS). MCS contains a large number of electrodes integrated into silicon rubber profile. There are two types of MCS developed: the first ensures quenching of power-follow current when the grid voltage crosses zero and the second provides impulse arc quenching without follow current of the grid. MCS can be installed on various supporting bases, e.g. rods or rings (both the insulation and metal), and could be obtained optimal arrester designs for any rated MV.

By installing MCA around insulation shed of an insulator, a new apparatus is developed: multi-chamber insulator arrester (MCIA). It combines the properties of insulator and arrester. Cup-and-pin MCIA can be arranged in strings and using them, it can be provided lightning protection of overhead power lines of any rated voltage.

For lightning protection of LV power systems, so-called SPDs are used. During normal operation conditions, SPDs are high-ohmic. In the very moment when there is a transient overvoltage, SPDs become immediately low-ohmic. The protection effect of SPDs is based on the following characteristics:

- safe diversion of lightning currents and surge current;
- limitation of the voltage differences between conductors – to a level which is no longer harmful for equipment to be protected.

More and more sensitive electrical equipment is used in LV power systems. Sensitive equipment is threatened by natural and by man-made surges. The main causes for surges are direct or indirect lightning strikes and man-made switching overvoltages. For an efficient protection of LV power systems against voltage or

[1]Streamer Electric Company, St. Petersburg, Russia
[2]Phoenix Contact, Blomberg, Germany

current surges, so-called LV surge protective devices are frequently used. The second part of this chapter gives an overview about surge protection for LV power systems.

To be able to design, produce, select, install and to maintain SPDs for LV power systems properly, many different factors have to be considered. Some relevant topics are addressed in this chapter: terms and definitions; function principle of SPDs; testing of SPDs; approvals from certified bodies; surge-protective components; selection of SPDs; locations for the installation of SPDs; effective protective distance; installation of SPDs; and inspection and testing of SPDs.

8.1 Common definitions and general function principle of SPDs used in HV, MV and LV systems

8.1.1 Common definitions

AC	alternating current (current that has sinusoidal shape, usually with frequency 50 or 60 Hz)
DLS	direct lightning strikes (lightning strikes that hit elements of overhead power lines such as towers, poles, shielding wires and conductors)
HV	high voltage (rated voltage of electrical equipment from 69 to 220 kV)
IEC	International Electric Committee (international organization that prepares standards and recommendations for electrical equipment and applications)
LV	low voltage (rated voltage of electrical equipment up to 1 kV)
MCA	multi-chamber arrester (surge arrester at which MCS is used)
MCIA	multi-chamber insulator arrester (it is insulator at perimeter of its shed MCS is installed. MCIA combines properties of insulator and arrester)
MCS	multi-chamber system (silicon rubber length at which consecutively connected many small arc-quenching chambers are installed)
MO	metal oxide (material with very strong non-linear voltage – current characteristic, which is used for MOA)
MOA	metal oxide arrester (arrester which is made with the use of MOV)
MOV	metal oxide varistors (non-linear resistors which are made in a form of MO discs)
MV	medium voltage (rated voltage of electrical equipment from 3 to 69 kV)
SA	surge arrester (a device that is intended to limit transient overvoltages and divert currents in MV and HV grids)
SPD	surge-protective device (a device that is intended to limit transient overvoltages and divert currents in LV grids)
SVU	series varistor unit (non-linear MO resistor part, contained in a housing)

8.1.2 General function principle

There are different principles of operation and different names for the devices that can be used to protect installations and equipment in LV power systems [1]. It has to be differentiated between two, in principle, different technologies to limit overvoltages: a spark gap (SG), which has a switching function, and a non-linear resistor, which has a voltage-limiting (clamping) function. Due to the different technologies and fields of applications, different wordings are used.

In [2], the characteristics, operating mechanisms and classifications of SPDs are discussed in general. In [3], the SPDs typically used in MV systems and their application and co-ordination are given a more technical description. However, it appears to be useful to give here for clarification the most common definitions and typical applications.

An SPD is a device that is intended to limit transient overvoltages and divert currents. It contains at least one non-linear component. This is the general definition in the IEC 61643 series. An SPD can be a spark gap, a non-linear resistor (which is the same as a variable resistor – i.e. varistor) or a combination of both. The term SPD is mainly used in the LV installation community. These SPDs are usually mounted on DIN rails.

A voltage limiting type SPD has high impedance when no surge is present, but this impedance is continuously reduced with increased surge current and voltage. Common examples of components used as non-linear devices are varistors and suppressor diodes. These SPDs are sometimes called 'voltage-limiting type' to differentiate them from the voltage-switching types.

A surge protective device (SPD) is a device intended to limit transient overvoltages to a specific limit; see EN 50526-1 [4]. Surge arrester or, shorter, 'arrester' is a more general term for metal oxide arrester (MOA). Surge protective device(s) contain one or more non-linear MO resistors. The wording surge arrester is mainly used in the HV and MV communities and describes different designs of MOAs. In the LV field, surge protective device(s) are used in traction systems. On LV power lines, generally overhead lines, surge protective device(s) are sometimes called secondary arresters or LV arresters. Such surge protective device(s) are typically metal-oxide varistors (MOVs).

Finally, a MOA without gaps is an arrester having non-linear MO resistors connected in series and/or in parallel without any integrated series or parallel spark gaps; see IEC 60099-4 [5].

Surge arresters are voltage-limiting type SPDs and are mainly made of MOVs (same technology as for the MV and HV networks) [6,7]. The active part (non-linear resistor), made usually of a single MOV, is encapsulated inside porcelain or polymeric housings, with or without sheds to take care of pollution and rain effect on the insulation, and has two terminals. Figure 8.1 shows the voltage–current characteristic of a typical MOV. The MOV has a non-linear behaviour and operates as soon as a surge occurs, whatever is the magnitude of the surge. The voltage protection level U_{pl} is determined for a current value I_{n} stated by the manufacturer.

In the lower part 1 of Figure 8.1, at the maximum continuous operating voltage U_{c}, the current i_{c} is in the range of 1 mA and below. At the knee point 2, at the reference voltage U_{ref}, the arrester starts to conduct, and the ohmic content of the current increases rapidly with a slight voltage increase. At U_{ref}, the current i_{ref} has a dominant ohmic component. Temporary overvoltages have to be considered in the voltage region between points 2 and 3.

In the low-current region up to point 2 (range of continuous operation), power frequency currents and voltages have to be considered. In the region above point 2, the protective characteristic of the MOA is of importance. That is why in this range the voltage–current characteristic is defined by impulse currents of different wave

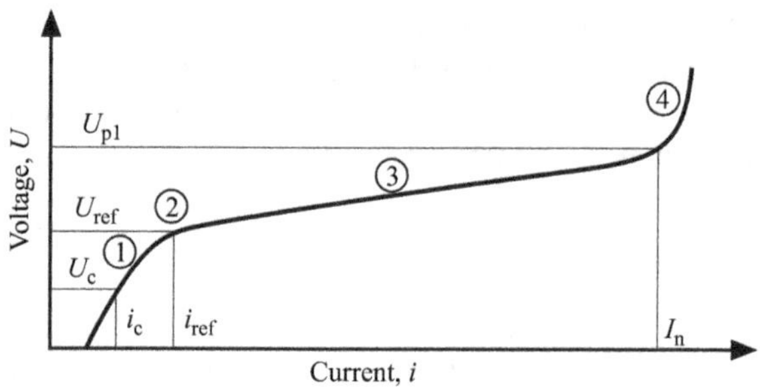

Figure 8.1 Voltage–current (U–I) characteristic of a MOA: 1, lower part (capacitive); 2, knee point; 3, strongly non-linear part; 4, upper part ('turn up' area); U_c, continuous operating voltage; U_{ref}, reference voltage; U_{pl}, protective level (at nominal discharge current I_n)

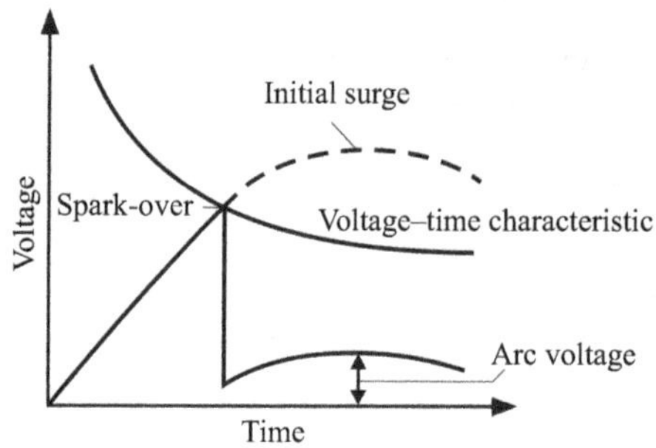

Figure 8.2 Protection performance of a typical gap

shapes and current magnitudes. The most important parameter is the lightning impulse protective level U_{pl}. This is the maximum permissible peak voltage on the terminals of an arrester subjected to the nominal discharge current I_n [8].

For simulations of the performance of MOAs, a voltage–current characteristic is normally used that starts at some 10 A in region 3 and goes up to maximum of 40 kA in region 4. Usually in HV systems, an impulse current wave shape of 8/20 μs is considered.

SPDs based on an spark gap are voltage switching devices. The spark gaps are typically encapsulated in an insulating housing. Triggered spark gaps provide a better (in this case lower) and more precise protection level. In LV systems, a lot of SPDs are build-up of a combination of several components and may include a disconnecting device and auxiliary equipment for state indication purpose.

Spark gaps behave in a different way than an MO resistor: until the spark-over voltage is reached, the gap behaves as an open circuit. When the spark-over voltage is reached (the level depends on voltage magnitude and the rate of rise), the gap behaves almost as a short circuit (the voltage value is given by the arc voltage). Figure 8.2 shows the typical performance of an spark gap, Figure 8.2 shows that, at a constant steepness, the spark overvoltage may still vary (i.e. there is a spread in the spark overvoltage values) and this is inherent to the (untriggered) spark technology.

In case an SPD needs to be disconnected automatically from the network, e.g. because of increasing continuous current or increasing temperature in the SPD, a thermal disconnector is integrated in its housing.

8.2 SPDs used in transmission and distribution (HV and MV) overhead lines

8.2.1 Metal oxide arresters

8.2.1.1 Construction and mode of operation

Major parts of the construction of MOA are those shown on Figure 8.3, non-linear elements (varistors), insulation housing, and upper and bottom flanges.

Under normal operating conditions when operating phase to neutral voltage influences MOA, the conduction current flows through varistors – its value makes fractions of milliamperes. If overvoltage occurs, the current rises sharply in varistors and the energy of overvoltage is diffused by way of heat in MOA and its ground circuit. Hence, to ensure the faultless operation of MOA, the heat removal through its lateral surface (Figure 8.3) needs to be more intensive than the process of heat removal in varistors.

Successive time intervals between lightning and/or switching overvoltages are, as a rule, sufficient to let MOA cool down and go to the initial state when the temperature of varistors is equal or slightly exceeds the ambient temperature. To the contrary, transient overvoltages may result in multiple actuations of MOA within a moment that is insufficient for varistors cooling down. That is why they say that MOAs are designed to protect the insulation against lightning and switching overvoltages, but they are not designed to protect them against transient overvoltages.

While the porcelain housing was used as external insulation for manufacturing valve-type arresters (VAs), MOAs came to the energy industry together with

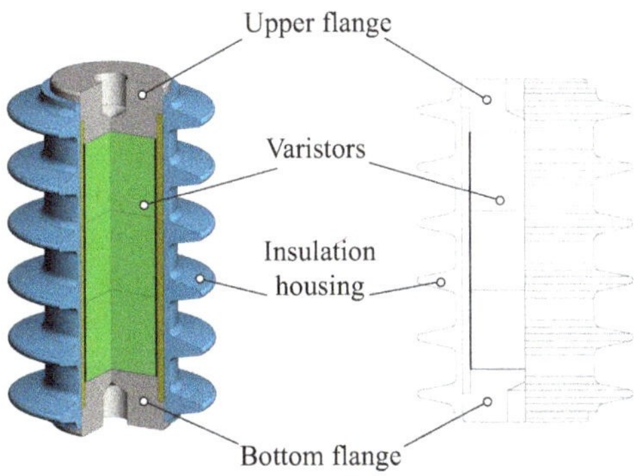

Figure 8.3 Major parts of MOA construction

polymeric insulation. Advantages of polymeric insulation have been proved more than once during its service, which are the following: high hydrophobic properties (and as a consequence a low probability of insulation flashover in wet condition), explosion proofness, vandal resistivity, light weight, capability to operate under conditions of natural and industrial pollutions, wide range of service temperatures, etc.

Due to the absence of SGs in the construction of MOA and the use of polymeric insulation, it became possible to simplify at the most its construction and to make it cheaper and explosion resistant, and this explains the general industrial application of MOAs.

Large sizes and weights made impossible the large-scale installation of VAs on OHLs while the light and compact design of modern MOAs with polymeric insulation made them attractive for the protection of OHL insulation. The following, alas erroneous, assumptions served as a special push to make 'popular' the engineering solution of MOA installation on OHLs [9]:

- to ensure a reliable protection of OHL insulation it is sufficient to install a certain number of MOAs along the OHL route, it means that the OHL protection zone makes many hundreds of meters;
- to ensure a reliable operation of MOAs installed on OHL it is sufficient to use MOAs rated for lower currents as compared to MOAs used in switchgears, it means that MOAs designed for OHL may be considerably lighter and cheaper than those ones designed for switchgears.

Actually, in order to ensure the protection of OHL a required number of MOAs may be comparable to the number of towers and the current and energy characteristics of suspended MOA must be no worse than those ones of substation equipment. Yet, MOAs are largely applied and will find further application on OHLs.

8.2.1.2 MOAs for OHLs

Application of line arresters is quite common and well established mainly in Japan, still in the 'trial phase' in many other countries around the world [10]. Line arresters are installed in the line, directly in parallel to the insulators. They help avoiding insulator flashovers in case of missing shield wires or in case of high back flashover risk due to high footing impedances, e.g. in rocky ground areas. Line arresters may be of non-gapped (NGLA) or externally gapped (EGLA) design (Figure 8.4) While in IEC standardization NGLAs are considered as standard (gapless MO) arresters for special applications, EGLAs are covered by a separate IEC standard [11], as their technology and application differs much from that of conventional gapless MOAs.

Comparative characteristics of the considered types of arresters are given in Table 8.1 [12]. Even without resorting to a detailed techno-economic analysis, it can be concluded that arresters with external gap are preferable when protecting a long line sections, when it is necessary to install a large number of such arresters. To the fore in this case are such factors as lower cost, less risk of damage to the arrester from the point of view of implications for the work of the line, ease of installation and maintenance.

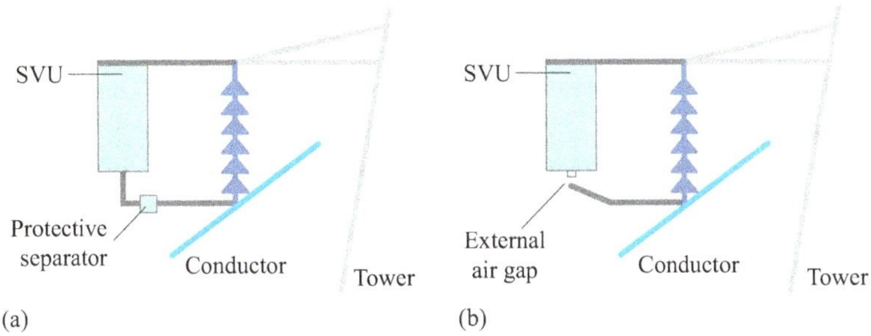

Figure 8.4 Configurations of arresters with insulator: (a) NGLA; (b) EGLA

Table 8.1 EGLA and NGLA comparison

EGLA	NGLA
Protection against lightning overvoltages (only)	Protection against lightning and switching overvoltages
SVU is not connected directly to conductor. Therefore, less number of varistor units can be used for the nominal voltage and SVU cost is reduced	SVU is connected directly to conductor. Therefore, more varistor units should be used for the nominal voltage and SVU cost is increased
Detection of failed arrester is complicated – only by damage of housing	The failed arrester can be detected visually due to disconnecting device
Pollution has no influence on SVU	Pollution of housing may lead to SVU overheating
Breakdown of external gap depends on atmospheric conditions	Atmospheric conditions have no influence on the arrester operation
Lightning current is distributed only between two arresters of the same span of the line	Lightning current is distributed among several arresters of adjacent spans of the line

NGLAs can be more preferable for the protection of local sections of OHLs, when the cost is not significant due to the small number of arresters, but require a higher reliability from the viewpoint of stability of the protective characteristics, or combining the functions of protection against lightning and switching surges.

Detailed information about the design of MOAs is presented, e.g. in [13], and application of arresters is discussed in Chapters 3 and 4 in Volume 2.

8.2.2 Multi-chamber arresters

8.2.2.1 Multi-chamber system

The base of MCAs, including MCIA (see Section 8.2.3), is the MCS shown in Figure 8.5. It comprises a large number of electrodes mounted in a length of silicon rubber. Holes drilled between the electrodes and going through the length act as miniature gas-discharge chambers. When a lightning overvoltage impulse is applied to the arrester, it breaks down gaps between electrodes. Discharges between electrodes take place inside chambers of a very small volume; the resulting high pressure drives spark discharge channels between electrodes to the surface of the insulating body and hence outside, into the air around the arrester. A blowout action and an elongation of inter-electrode channels lead to an increase in total resistance of all channels, i.e. that of the arrester, which limits the current.

Studies have shown [14] (see Section 8.2.4) that spark discharge quenching at MCS can take place in two instances: (1) when 50 Hz follow current crosses zero (this type of discharge quenching is further referred to as 'zero quenching'); (2) when the instantaneous value of lightning overvoltage impulse drops to a level equal to or larger than the instantaneous value of power frequency voltage; that is, lightning overvoltage current gets quenched with no follow current in the grid (this type of discharge quenching is further referred to as 'impulse quenching').

8.2.2.2 'Zero quenching' arresters

24 kV MCA

The principal components of a 24 kV MCA (see Figure 8.6) is an MCS, a fibreglass bearing rod and an assembly for securing arresters to insulator pins. Arresters are mounted on insulator pins with air gaps of 3–6 cm between top ends of arresters and the conductor (or conductor clamp). A lightning overvoltage first breaks down the air SG and next the arrester's MCS, which assures extinction of follow current.

Shown in Figure 8.6 is an arrester with 40 gas-discharge chambers intended for protection of 24 kV OHLs against induced overvoltages. One piece of this model is installed on each phase-interlacing pole (see Figure 8.7). In this case, the path of AC follow currents, which are associated with lightning overvoltage-induced multi-phase, includes the tower-grounding resistance circuits. Thanks to an extra resistance of the pole grounding circuit, follow currents are made lower, which raises the quenching efficiency of the arrester. Arc-quenching tests are presented in Section 8.2.4.

Main technical parameters of MCA-24 are listed in Table 8.2.

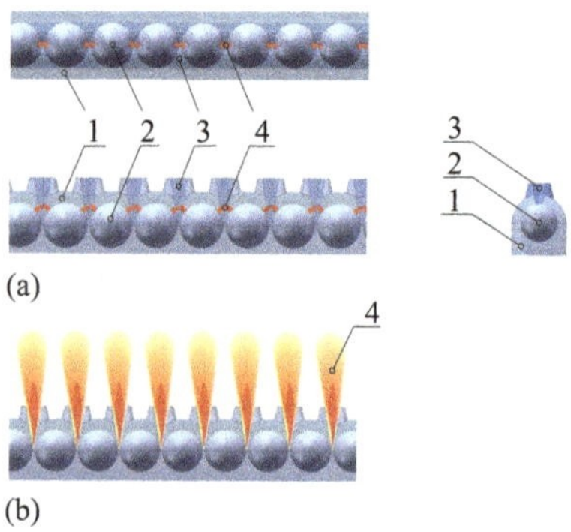

Figure 8.5 MCS: (a) diagram showing the discharge onset instant; (b) diagram showing the discharge end instant; 1 – silicon rubber length; 2 – electrodes; 3 – arc-quenching chamber; 4 – discharge channel

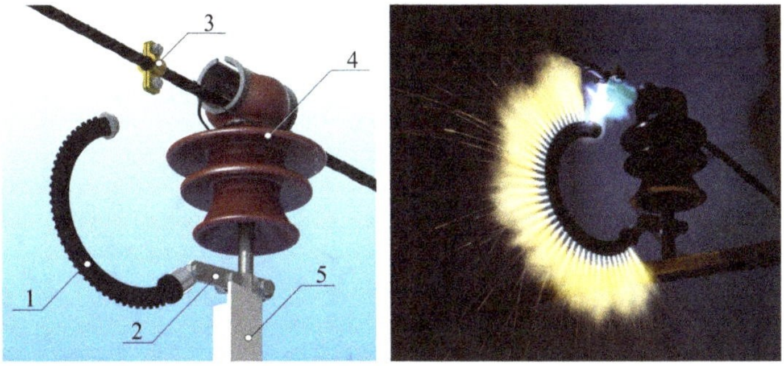

Figure 8.6 *A 24 kV multi-chamber arrester, MCA-24, for protection against induced overvoltages: 1 – MCS; 2 – rod clamp; 3 – conductor clamp; 4 – insulator; 5 – metallic pole*

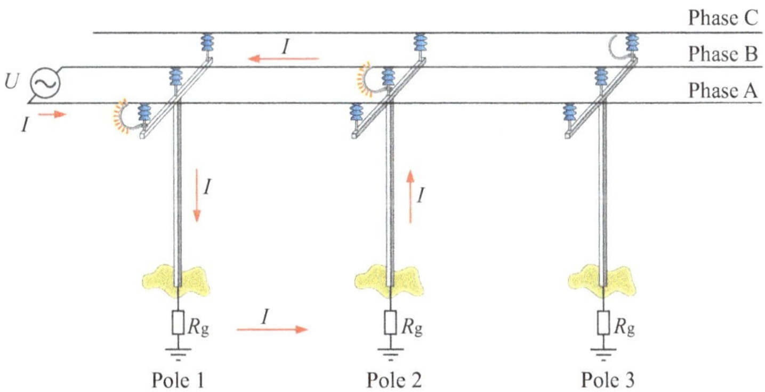

Figure 8.7 *Schematic of MCA installation on a distribution line*

Ring Type MCA35 kV

Studies have shown that while meeting specific technical conditions a steel toroidal rod (ring) can be used as arrester housing base [18]. Along with technology simplification, a steel carrier ring exhibits an extra- and long-term strength.

Figure 8.8 presents a fragment of steel-ring based MCS. It contains a large number of electrodes and steel carrier ring integrated into silicon rubber profile. Cylindrical quenching chambers of *ad hoc* configuration with SGs generated inside have been moulded between MCS adjacent electrodes.

The arrester consists of two rings electrically installed in parallel to the protected insulation on upper and lower fittings (Figure 8.9).

The first MCS end of lower ring is connected via the fittings of insulator to line conductor, while the second one is provided with an upward rod lead.

The first end of MCS of upper ring is provided with a rod lead directed downwards, while the second end of MCS is connected via insulator fixture with

Table 8.2 Main technical parameters of MCA-24 kV

Characteristic	Value
Highest voltage for equipment*, kV	24
Maximum prospective fault current, kA	1.5
External air gap, mm	70 ± 10
50% flashover voltage, kV	120
Power frequency withstand voltage**, kV	50
Lightning discharge capability (200 μs)***, C	2.4
High current impulse (4/10 μs), kA	65
Maximum fault quenching lightning current (1/50 μs), kA	3
Minimum withstand amount of operations	10
Additional power losses on the line, %	0
Average life time expectancy, years	20
UV resistance****, h	1,000
Weight	0.45
Maintenance	1 visual verification per year

*According to IEC 60038.
**According to IEC 60071-1.
***According to IEC 60099-8.
****According to ISO 4892-2, method A, IEC 62217.

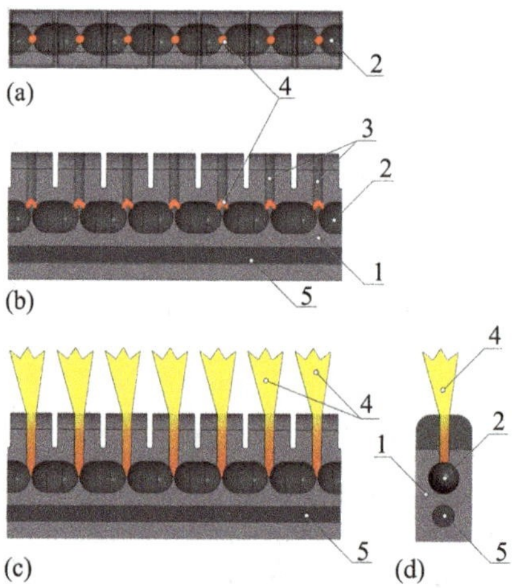

Figure 8.8 Fragment of steel-ring-based MCS: 1 – silicon rubber profile; 2 – intermediate electrodes; 3 – arc-quenching chambers; 4 – discharge channels; 5 – steel ring

OHL cross-arm. Required length of SG between rod leads of upper and lower rings can be set up by rotation of one ring around insulator's end fitting.

In case of surge impact on OHL, for example, under direct lightning strike (DLS) on conductor, SG between rod leads of lower and upper rings is flashed over

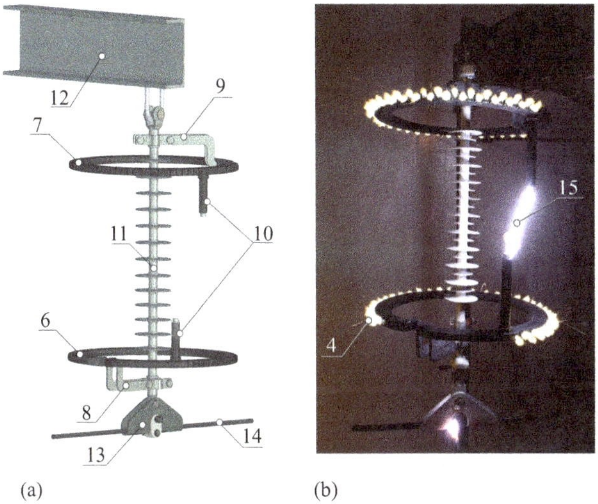

Figure 8.9 A 35 kV ring-type arrester: (a) general view; (b) test photo: 4 – discharge channels; 6, 7 – upper and lower rings with MCS; 8, 9 – upper and lower lead-in; 10 – rod leads; 11 – insulator; 12 – cross-arm; 13 – supporting fittings; 14 – conductor; 15 – SG

thus connecting their MCS sequentially. The current of lightning overvoltage flows from conductor through lower ring MCS, along SG between rod leads, through upper ring MCS and then leaks into earth via tower (Figure 8.9(b)).

It should be noted that the capability of arc-quenching chambers to direct the arc outside MCS makes it possible to disperse the bulk of flowing energy in ambient space. It makes the MCS design resistant to multiple electrodynamic and thermal actions of impulse (DLSs included) and follow currents. Power follow-current arc quenching is described in Section 8.2.4.

A steel carrier ring of lower ring connected via fittings to conductor gets a high potential while that one of upper ring connected via fittings to tower has a zero potential. High electric strength of silicon rubber layer between intermediate electrodes and metal ring is essential for successful arrestor operation. So, it becomes the most topical quality issue (see Figure 8.8).

Arrester electrical characteristics should ensure a reliable coordination of its operation with all types of protected insulators and contribute to repeated quenching of the follow-current arc under various lightning-generated surges, including those produced by DLS to line conductor. The arrester should withstand any transient internal overvoltages without being flashed over. A steel carrier ring and MCS external materials should be resistant to mechanical wind and ice loads as well as to environmental stresses such as dust, salts and solar radiation.

Ring-type arrester can be applied to string of cap and pin insulators and to composite insulators as well. To provide reliable operation of composite insulators, it is appropriate to minimize electric field along fibreglass rod. Arrester's steel rings function as grading screens ensuring uniform distribution of electric field along fibreglass rod. Results of calculation of composite insulator axial electric field are depicted in Figure 8.10.

As it can be seen from Figure 8.10, installation of arrester (composed of two rings) results in drop of maximal electric field from 4.2 to 1.5 kV/cm, i.e. almost three times lower increasing its reliability under applied power-frequency voltage.

Main characteristics of ring-type arrester are given in Table 8.3.

8.2.2.3 MCS: 'impulse quenching'

To increase the follow-current quenching efficiency of an MCS, it is offered to have a 4–20-fold longer elementary gap of a discharge chamber, compared to the MCS shown in Figure 8.10. A low discharge voltage of such an advanced MCS can be attained through use of creeping discharge and cascading operation of MCS circuit chambers.

Creeping discharge flashover voltage is known to depend little on the electrode spacing; that is, a fairly large gap can be flashed over even at a relatively LV (see, e.g., [14]).

Cascading is caused by effect of an additional electrode set-up along the entire MCS (Figure 8.11). It is connected to the last electrode of the last chamber and isolated from all the other electrodes.

The additional electrode is connected to the ground and thus has a zero potential. As the MCS gets actuated, the high potential U is applied to the first

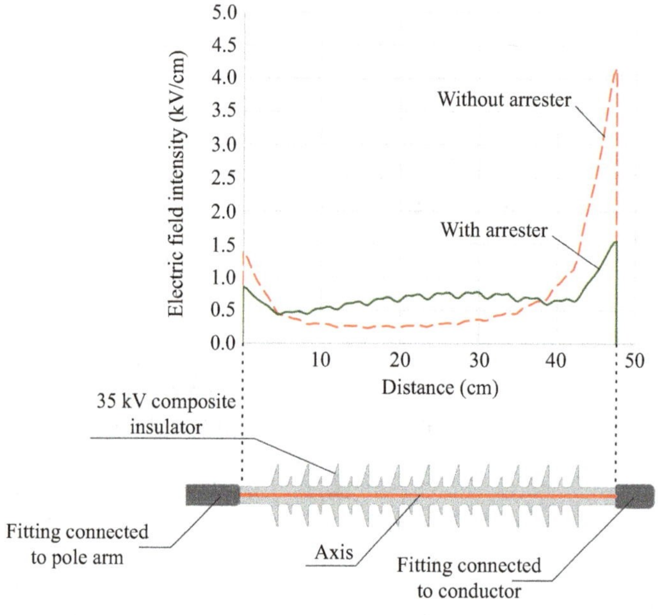

Figure 8.10 Electric field along fibreglass rod of composite insulator 35 kV

Table 8.3 Main characteristics of 35 kV ring-type arrester

Characteristic	Value
Highest voltage for equipment*, kV	40.5
Maximum prospective fault current, kA	5
External air gap, mm	150 ± 30
50% flashover voltage, kV	230
Power frequency withstand voltage**, kV	80
Lightning discharge capability (200 μs)***, C	2.4
High current impulse (4/10 μs), kA	65
Maximum fault quenching lightning current (8/50 μs), kA	30
Minimum withstand amount of operations	10
Additional power losses on the line, %	0
Average life time expectancy, years	20
UV resistance****, h	1,000
Weight	5.3
Maintenance	1 visual verification per year

*According to IEC 60038.
**According to IEC 60071-1.
***According to IEC 60099-8.
****According to ISO 4892-2, method A, IEC 62217.

electrode. The voltage gets distributed among chambers' SGs most unevenly, as follows from the circuit diagram (Figure 8.11(b)).

Let us estimate voltage that is applied between the electrodes in the first chamber. The entire capacitance circuit in Figure 8.11(b) can be visualized as a string of two capacitances (see Figure 8.11(c)), viz. that between the first and second electrodes C_1 and the equivalent ground capacitance C_{eq} of the remaining capacitance string excepting C_1. It is to be noted that C_{eq} is determined basically by the chamber electrode-additional electrode capacitance C_0, i.e. $C_{eq} \approx C_0$. Capacitances C_1 and C_0 are series connected (see Figure 8.11(c)). Their voltages are distributed in inverse proportion to their values; thus, the voltage across electrodes of the first chamber is $U_1 \approx U/(1 + C_1/C_0)$.

Thanks to a relatively large area of the chamber electrode's surface that faces the additional electrode, as well as because permittivity of a solid dielectric ε is much higher than that of air ε_0 (generally, $\varepsilon/\varepsilon_0 \approx 2$–3), capacitance of the intermediate electrode to the additional electrode (i.e. capacitance of this intermediate electrode to ground) is substantially larger than its capacitance to the adjacent intermediate electrode, i.e. $C_0 > C_1$ and respectively $C_1/C_0 < 1$. With the ratio C_1/C_0 ranging from 0.1 to 0.9, voltage U_1 remains within $U_1 = (0.53\text{–}0.91)U$. Therefore, as the MCS gets exposed to voltage U, a larger part of the voltage drop (at least more than half) occurs in the first SG between the first and second electrodes. Under effect of this voltage the first gap gets sparked over, the potential of the second electrode rises to that of the first HV electrode, while the potential of the next intermediate electrode becomes U_0. With the spark-over pattern repeating again and again, gaps between electrodes get sparked over in series. The cascade

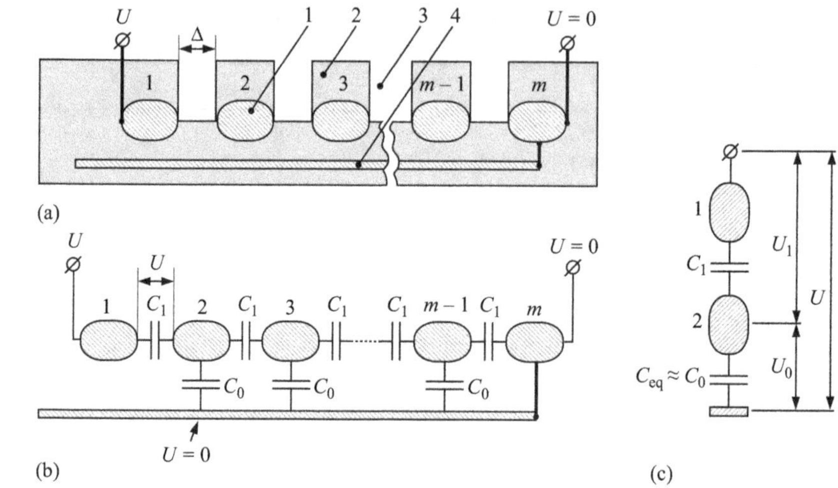

Figure 8.11 Illustration of cascading operation of MCS: (a) sketch of MCS; (b) circuit diagram of MCS; (c) circuit diagram of a chamber: 1 – MCS electrode; 2 – silicon rubber shape; 3 – gas-discharge chambers; 4 – additional electrode; Δ – length of chamber's discharge gap; C_0 – capacitance of chamber's electrode to additional electrode; C_1 – capacitance of chamber's electrode to another electrode of same chamber; U – overall MCS voltage; U_1 – voltage between chamber's electrodes; U_0 – voltage between the chamber's second electrode and the additional electrode

operation of discharge gaps assures needed low flashover voltages for actuation of an MCS as a whole.

Impulse quenching lasting only a few microseconds – or a few dozens of microseconds at most – involves next to no erosion even after numerous firings of MCS (see Section 8.2.5). As far as there is no power follow current, MCS with impulse quenching can be used in arresters for grids with high values of short-circuit currents [15].

Shown in Figure 8.12 is an MCA design with electrodes as pieces of stainless steel tube and additional electrode passing through these electrodes. Length of breakdown gaps is additionally increased by using diagonal discharge splits. Due to such design, MCA becomes more compact. Besides capacitance between tube electrodes and additional electrode of a discharge chamber, C_0 is much higher than between adjacent electrodes of the chamber C_1. This insures very non-uniform distribution of voltage among discharge chambers and consequently decreases discharge voltage.

Figure 8.13 presents MCA for protection 12 kV line against induced over-voltages (MCA12-I) [16]. MCS of the arrester consists of 10 chambers made in accordance with Figure 8.12. For avoiding the connection between plasma clouds outgoing from discharge chambers at their operation, the chambers alternately

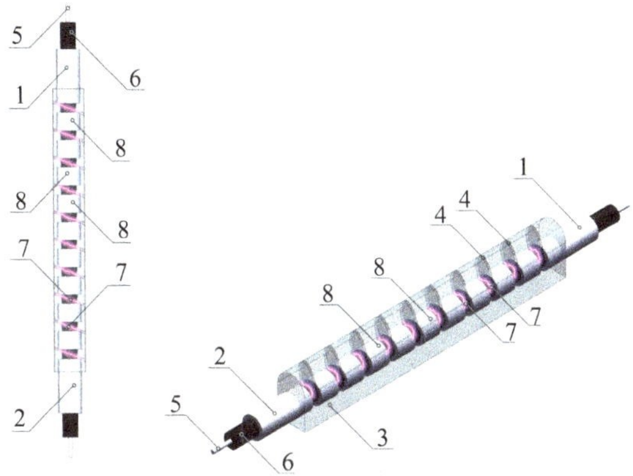

Figure 8.12 *MCA with additional electrode passing via metallic tube electrodes: 1 – main high potential electrode; 2 – main low potential electrode; 3 – silicone rubber; 4 – discharge split; 5 – additional electrode (cable conductor); 6 – cable insulation; 7 – discharge channel; 8 – intermediate electrodes*

directed in opposite sides: five odd chambers directed in one side and five even in opposite side.

MCA12-I is intended for protection of overhead distribution 12 kV lines and quenches impulse arc without power follow current. Therefore, it can be used in grids with solidly grounded neutral and four conductors OHLs (three conductors for three phases plus one conductor for grounded neutral).

Test results

Most important and complicated are arc-quenching tests. The tests procedure is presented in Section 8.2.4. Table 8.4 presents arc-quenching test results of MCA12-I as well as other electrical and test results.

MCA12-I should be installed at OHLs using one arrester per pole with phase interlacing (see Figure 8.7, Section 8.2.2).

MCA12-I quenches discharge impulse arc of induced overvoltages without power follow current. Conductor erosion caused by impulse current with amplitude of about 1 kA and duration 4–5 μs is insignificant. This enables to use the arrester without additional clamps on conductors (bared and covered as well).

In short, main advantages of MCA12-I are the following:

1. small dimensions and weight;
2. ease of installation, including those live line;
3. no need for conductor clamp;
4. no change of spark-over gap due to conductor slip.

Shown in Figure 8.14 is a sketch of a gas-discharge chamber intended for use in arresters protecting OHLs against DLSs. The discharge chamber is strengthened

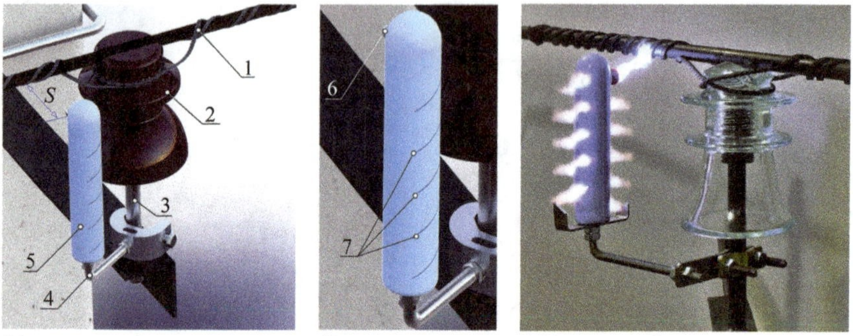

(a) (b) (c)

Figure 8.13 MCA for protection 12 kV line against induced overvoltages (MCA12-I): (a) general view; (b) close-up view; (c) test photo: 1 – conductor; 2 – insulator; 3 – rod; 4 – clamp; 5 – silicon rubber body; 6 – discharge splits; 7 – electrode; S – spark-over gap

Table 8.4 Test results of MCA12-I

No.	Tests	Units	Results
1	MCA maximum external gap installed in parallel to 12 kV insulator under application of lightning impulse 300 kV, 1.2/50 μs	mm	50
2	Check-up of internal insulation by application impulse with high rate of voltage increase of 2 MV/μs	–	OK
3	MCA minimum external gap under application power frequency voltage	mm	30
4	Withstand power frequency voltage at dry/rain conditions	kV	42/28
5	50 Hz grid voltage at which arc is quenched at application of induced impulse current (2.5 kA, 1/4 μs)	kV	6
6	Electrodynamic stability at DLS (impulse current 20 kA, 8/50 μs)	Number of shots	3

mechanically by a fibreglass plastic sleeve. Each electrode consists of two nested tubes – inner and outer tubes. While the outer tube is copper, the material of inner tube is tungsten. Under the lightning, overvoltage electrical breakdown occurs. Due to increase in current discharge, channel diameter increases rapidly, which leads to intensive pressure build-up, plasma outflow and eventually to arc extinction. In numerous performed experiments, certain discharge chamber constructions demonstrated enhanced quenching capability. For example, it was discovered that additional cavities made inside electrodes (electrode volume) increases chamber performance in general. Fast heating of gas in discharge gap during the current rise stage (first 8 μs) causes gas to flow out of the chamber with high speed. Later after lightning impulse is almost decayed, the flow reverses its direction and relatively cold gas from electrode volume goes to discharge gap causing the decrease in temperature, which accelerates arc extinction.

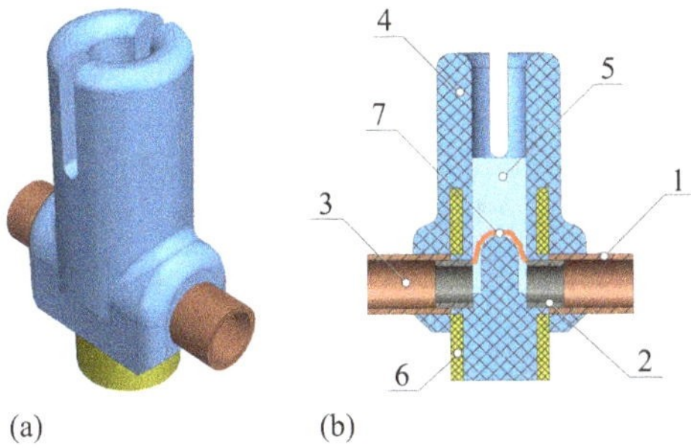

Figure 8.14 *Discharge chamber of MCA: (a) general view; (b) cross-section: 1 – outer tube; 2 – inner tube; 3 – cavity; 4 – silicone rubber; 5 – discharge slot; 6 – fibreglass plastic sleeve; 7 – discharge channel*

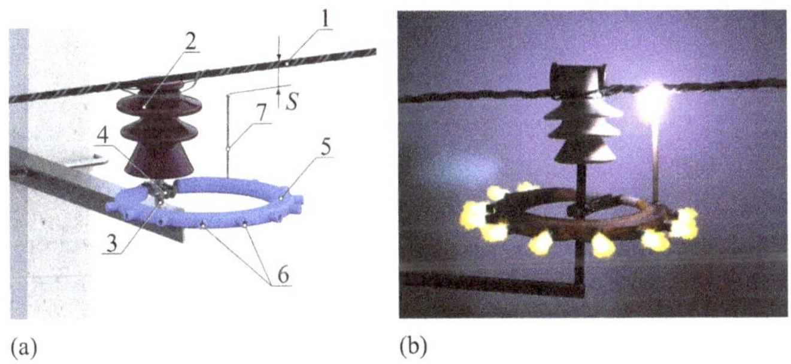

Figure 8.15 *MCA for protection 12 kV line against DLSs (MCA12-D): (a) general view; (b) test photo. 1 – conductor; 2 – insulator; 3 – rod; 4 – ring of steel rod; 5 – silicon rubber body; 6 – discharge splits; 7 – electrode; S – spark-over gap*

In Figure 8.15, a prototype of MCA for protection 12 kV line against DLSs (MCA12-D) is presented. MCS of the arrester consists of ten chambers made in accordance with Figure 8.14. Due to lightning overvoltage at the line conductor, spark-over gap *S* between the conductor and arrester electrode breaks down and the MCS operates.

8.2.3 Multi-chamber insulator arresters

MCIA combines properties and functions of arrester and insulator. Application of MCIAs makes it possible to ensure lightning protection of OHLs of any voltage ratings: the higher the line voltage, the larger the number of units in a string and thus the higher the rated voltage and the arc-quenching capacity of a string of insulators arresters.

Various designs of insulators with arrester properties are possible. An MCIA is generally a production glass, porcelain or composite insulator fitted with an MCS. Installation of an MCS confers arrester properties to the insulator without any deterioration of its insulating capacity. For this reason, application of MCIA on OHLs makes shield wire redundant, while the height, weight and cost of poles or towers goes down. At a lower overall cost and with a better lightning performance, such a line features a noticeably reduced number of lightning failures, cuts damage from undersupply of energy and lowers maintenance costs.

Figure 8.16 shows an MCIA based on SDI-37 pin insulator. The MCS is mounted over three quarters of the circumference of an insulator shed. The left and right ends of the MCS are approached by the upper and lower feed electrodes, respectively, which are installed on the upper and lower terminals; there are spark air gaps between the feed electrodes and the ends of the MCS.

When the MCIA is stressed by an overvoltage, the air gaps get sparked over first, the MCS coming next. The lightning overvoltage current flows from the conductor and its feed electrode via the spark channel of the upper SG to the MCS and on to the insulator rod via the discharge channel of the lower SG and the lower feed electrode. Note that there are no intervening electrodes between the upper and lower feed electrodes on the MCS-bearing silicon rubber shed; thus, the discharge develops over the MCS taking some three quarters of the shed's circumference, rather than between the feed electrodes.

The MCIA shown in Figure 8.20 is intended for protection of 6–20 kV lines against induced overvoltages.

Strings of MCIAs are intended for protecting 35–220 kV and above OHLs against DLSs. The MCS of an insulator-arrester comprises 14 chambers described in Section 8.2.2.3 (Figure 8.14) [15]. As a line conductor gets exposed to lightning overvoltage, air gaps between electrodes and respective taps, as well as gaps between taps of adjacent insulators, are sparked over actuating the MCS as a whole (Figures 8.17 and 8.18).

Discharges between electrodes take place inside chambers of a small volume; the resulting high pressure drives spark discharge channels between electrodes to the surface of the insulating body and hence outside, into the air around the MCS.

A blowout action and an elongation of inter-electrode channels lead to an increase in total resistance of all channels, i.e. that of the MCS, which limits the lightning overvoltage impulse current and quenches impulse arc. Results of arc-quenching tests are presented in Section 8.2.4.

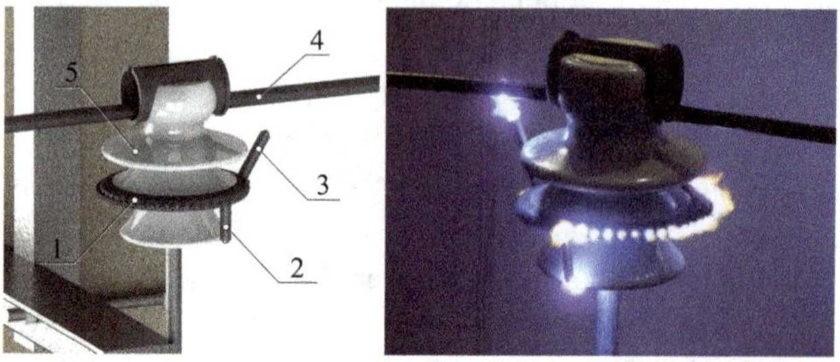

Figure 8.16 A 20 kV MCIA based on SDI-37 insulator: 1 – MCS; 2, 3 – taps; 4 – conductor; 5 – insulator

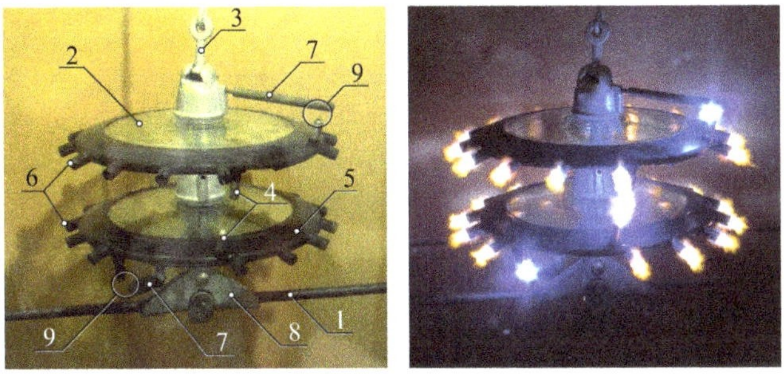

(a) (b)

Figure 8.17 String of two MCIAs based on U120AD insulator: (a) general view; (b) during tests: 1 – conductor; 2 – U120AD insulator; 3 – ball eyes; 4 – taps; 5 – arrester's body; 6 – discharge chambers; 7 – electrodes; 8 – suspension clamp; 9 – upper and lower coordination SGs

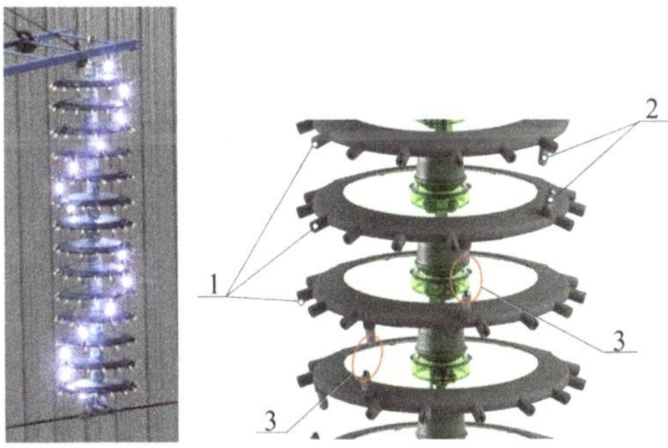

Figure 8.18 A 220 kV string of 14 MCIAs: 1 – MCS, 2 – taps, 3 – coordination SGs

As shown in [14–17], 220 kV MCIA strings have successfully passed the specified package of electrical, mechanical and climatic tests and studies listed in Table 8.5.

To obtain the accurate information about thunderstorm activity along the line and to record the discharge parameters of DLSs into the line, a complex lightning monitoring system was developed. Besides all MCIA strings were equipped with indicators (Figure 8.19(a)) which can be broken after operation of the string during lightning flashover (Figure 8.19(b)).

The efficiency of the MCIAs was proved in 3-year pilot operation on certain operating OHL220 kV without shielding wire protection (Figure 8.20) located in the region with high lightning activity [16].

Table 8.5 List of tests and studies for 220 kV MCIA string

No.	Tests
1	Flashover characteristics at lightning and switching impulses
2	Flashover characteristics at operational voltage at clean, wet and polluted conditions
3	Flashover characteristics at operational voltage and ice
4	Power follow-current quenching*
5	Lightning charge transfer capability
6	Radio interference
7	Climatic tests
8	Mechanical tests

*The results are presented in Section 8.2.5.

(a) (b)

Figure 8.19 MCIA strings with indicators: (a) indicator is not broken, if the string was not flashed over; (b) indicator is broken after flashover of the string

Figure 8.20 Experimental overhead power line 220 kV with MCIAs

8.2.4 Arc-quenching tests

8.2.4.1 Test procedure

The circuit diagram of the tests is shown in Figure 8.21. Follow-current quenching tests were carried out according to the procedure described in [14] for three modes:

1. *Induced overvoltages* (surge capacitance of voltage and current impulse generator $C_g = 0.02$ μF; impulse current $I_{max} \approx 2.5$ kA; 1/4 μs);
2. *Back flashover overvoltages* ($C_g = 0.5$ μF; $I_{max} \approx 3$ kA; 1.2/50 μs);
3. *DLS overvoltages* ($C_g = 6.5$ μF; $I_{max} \approx 30$ kA; 8/50 μs).

Negative polarity lightnings account for some 90% of the total number. For this reason, impulses simulating the lightning overvoltage impulse were taken to be negative.

A lightning can strike at any instantaneous value of the grid voltage. The worst possible case is a negative DLS on a line conductor at negative instantaneous grid voltage. Here the total current across the arrester, made up by the overvoltage impulse current and the follow current of the grid, tends to reach the fault current level of the grid without crossing zero. That is why most of the tests concentrated on this particular ratio of overvoltage impulse and grid polarities (−/−). However in some cases, (−/+) ratio was used also.

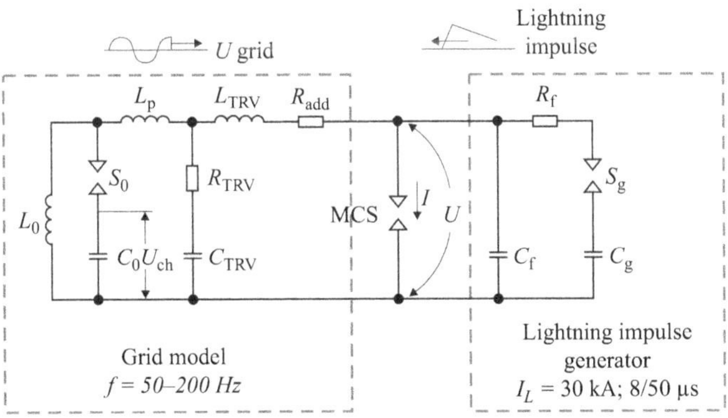

Grid simulator:
C_0 = 350 μF – capacitance of oscillatory circuit;
L_0 = 22 mH – inductance of oscillatory circuit;
L_p = 11 mH – inductance;
L_{TRV} = 1 mH – transient return voltage (TRV)-governing inductance;
R_{TRV} = 50 ohm – TRV-governing resistance;
C_{TRV} = 0.125–3.375 μF – TRV-governing capacitance;
R_{add} = 0–10 ohm – additional resistance;
S_0 – spark discharge gap of grid simulator;

Lightning voltage and current impulse generator:
C_g = 0.02–6.5 μF – capacitance of impulse generator;
R_f = 5–100 ohm – front resistance;
C_f = 4,500 pF – front capacitance;
S_g – total spark discharge gap of impulse generator;
MCS – test object (e.g. MCS).

Figure 8.21 Circuit diagram of test set-up

The test procedure was as follows: first, the capacitor bank C_0 and the impulse generator was charged; operation of the impulse generator led to breakdown of the test object (MCS) and the auxiliary arrester S_0. Thus, both a lightning overvoltage impulse and the AC voltage were applied to the test MCS simultaneously. As the lightning overvoltage impulse ends, only power frequency voltage remains applied to the arrester.

8.2.4.2 Principal test results

Voltage and current oscillograms were recorded during the tests (see Figure 8.22). Figure 8.22(b) also presents additional computer oscilloscope patterns of arc dynamic resistance R_{dyn} obtained by dividing the digital oscilloscope pattern of voltage U by the oscilloscope pattern of current I.

Studies have shown that spark discharge quenching can take place in two instances: (1) when the instantaneous value of lightning overvoltage impulse drops to a level equal to or larger than the instantaneous value of power frequency voltage; that is, lightning overvoltage current gets extinguished with no follow current in the grid (this type of discharge quenching is further referred to as 'impulse quenching'; see Figure 8.22(a)); (2) when 50 Hz follow current crosses zero (this

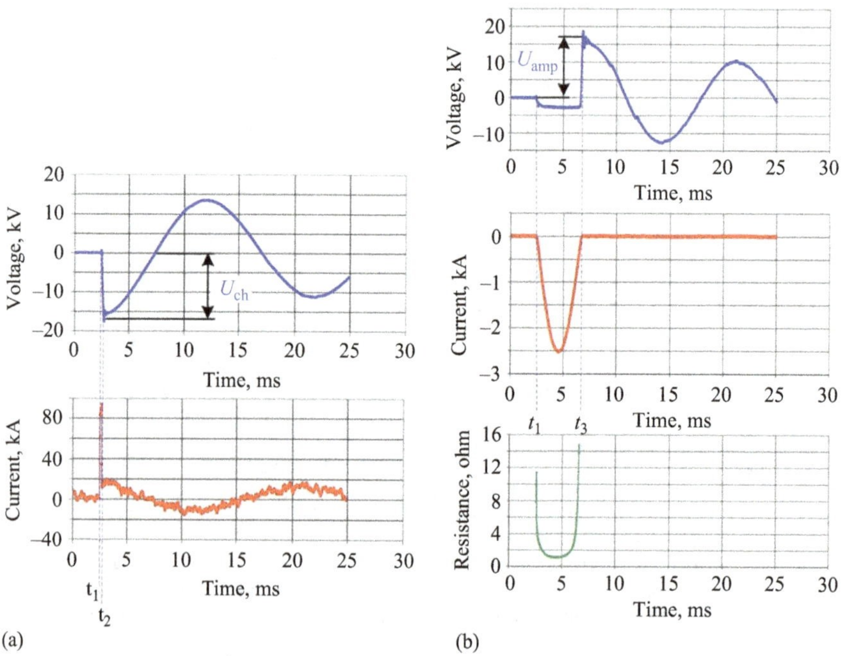

Figure 8.22 Typical voltage, current and resistance oscillograms in power follow current quenching tests: (a) impulse quenching; (b) zero quenching; t_1 – application of AC and lightning impulse; t_2 – quenching of lightning impulse; t_3 – quenching of power follow current

type of discharge quenching is further referred to as 'zero quenching'; see Figure 8.26(b)).

8.2.4.3 MCA zero quenching (see Section 8.2.2)

It was shown by the tests that quenching occurs 'in impulse' at low values of U_{ch} but 'in zero' as U_{ch} increases. Of interest is the fact that both at impulse quenching (Figure 8.22(a)) and at zero quenching (Figure 8.22(b)), voltage does not get chopped to zero, as it happens in standard gaps (e.g. rod–plane and rod–rod), and a considerable residual voltage exists. Shown in Figure 8.23 are oscilloscope patterns obtained for various numbers of MCA chambers.

Figure 8.24 shows experimental values of grid voltage at which follow current is quenched versus the number of MCS chambers. The data of Figure 8.24 make it possible to estimate the needed number of MCS chambers for arresters of different rated voltages.

8.2.4.4 MCA and MCIA impulse quenching (see Sections 8.2.2 and 8.2.3)

Table 8.6 lists the successful test results for arresters intended for protection 12 kV lines against induced overvoltages (MCA12-I, see Figure 8.13), against DLS (MCA12-D, see Figure 8.15) and a string of two insulator-arresters (see Figure 8.17).

MCA12-I was tested with current impulse of induced overvoltage with maximum value $I_{max} = 2.5$ kA and impulse shape 1/4 μs.

MCA12-D and string of two MCIA were tested with current impulse of DLS with maximum value $I_{max} = 30$ kA and impulse shape 8/50 μs.

MCIAs were tested for 220 kV OHL (see Figures 8.17–8.19) with fault current 30 kA. The grid simulator had to imitate the operating environment at crest value of fault current $I_f^* = 30 \cdot V2 = 42.4$ kA (ampl.). Earlier studies demonstrated that the issue of impulse or zero follow-current quenching pattern is settled close to the instant $t = 100$ μs. With a sine form of 50 Hz current, its instantaneous value at this

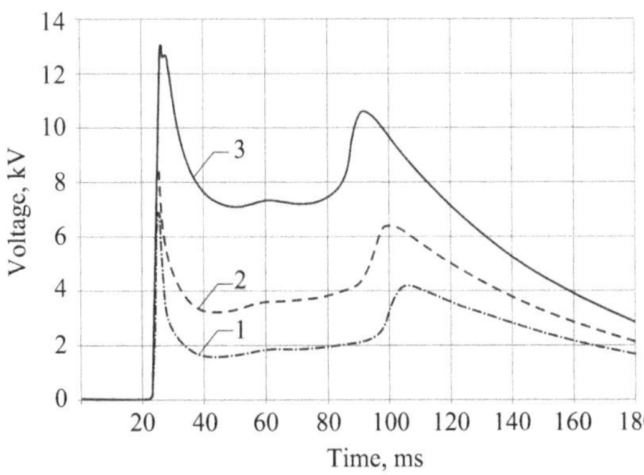

Figure 8.23 Oscilloscope patterns for MCS with different chamber numbers m: 1 – 50; 2 – 100; 3 – 200

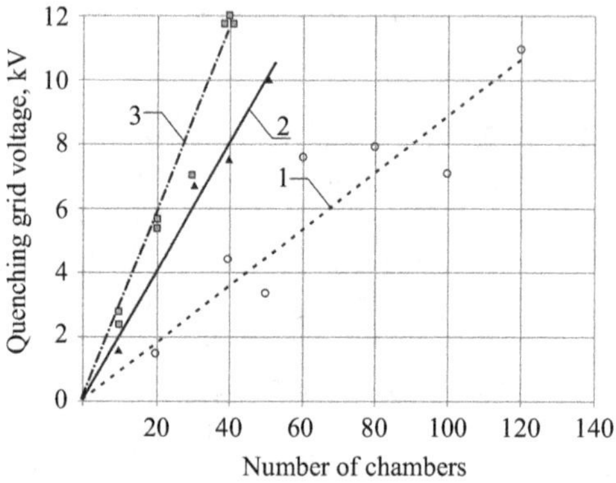

Figure 8.24 Follow-current-quenching grid voltage vs. number of MCS chambers: 1 – impulse quenching (instantaneous value) ○; 2 – zero quenching at $R_g = 0$ (effective value) ▲; 3 – zero quenching at $R_g = 10$ ohm (effective value) □

Table 8.6 Impulse quenching test results

Test object/figure no.	**Current impulse**		**Overvoltage**	U_{ch}/U_{quench} **(kV)**
	I_{max} **(kA)**	**Time (μs)**		
MCA12-I/Figure 8.13	2.5	1/4	Induced	12/8.5
MCA12-D/Figure 8.15	30	8/50	Direct strike	12/8.5
Two MCIA/Figure 8.17	30	8/50	Direct strike	30/21

Notes: U_{ch} – charging voltage of grid unit capacitors; $U_{quench} = U_{ch}/\sqrt{2}$ – respective effective phase voltage of grid.

moment equals $I = I^*_{f.} \sin(\omega t) = 42.4 \sin(314 \cdot 100 \cdot 10^{-6}) \approx 1.3$ kA. That was why the grid simulation was set up so as to assure a near-linear build-up of current from zero to 1.3 kA in 100 μs. This condition is met with frequency of the grid's oscillatory circuit being $f = 200$ Hz.

Table 8.6 shows principal test findings for MCIA string comprising two insulators-arresters (see Figure 8.17). The crest voltage capacity of the grid simulator is within $U_{ch} = 30$ kV. This determined the number of MCIA in a string during the tests.

The maximum permissible phase voltage U_{max} of a 220 kV line is 146 kV. As seen from Table 8.6, a two-insulator MCIA assures arc quenching at U_{2MCIA} of 21 kV. At 14 units per string, an MCIA can be believed to assure quenching at U_{14MCIA} of $7 \cdot 21 = 147$ kV. Thus a 14-unit insulator MCIA can assure quenching of lightning overvoltage impulse arc without generating follow current.

8.3 SPDs for LV power systems

8.3.1 Terms and definitions

1.2/50 μs (surge) voltage impulse is a transient voltage impulse with a virtual front time of 1.2 μs and a time to half-value of 50 μs (see Figure 8.25). This impulse represents

- lightning-induced voltages,
- man-made switching overvoltages and
- electrostatic discharges with relatively low energy content.

The internal resistance of commercially available 1.2/50 μs surge-voltage generators is usually in the range between 40 and 500 ohm.

8/20 μs (surge) current impulse is a transient current impulse with a virtual front time of 8 μs and a time to half-value of 20 μs (see Figure 8.26). This impulse represents

- man-made switching surges and
- lightning-induced surge currents with medium-high energy content.

1.2/50 and 8/20 μs combination wave (hybrid impulse) is a combination of a voltage impulse (1.2/50 μs) and a current impulse (8/20 μs). For this impulse, a

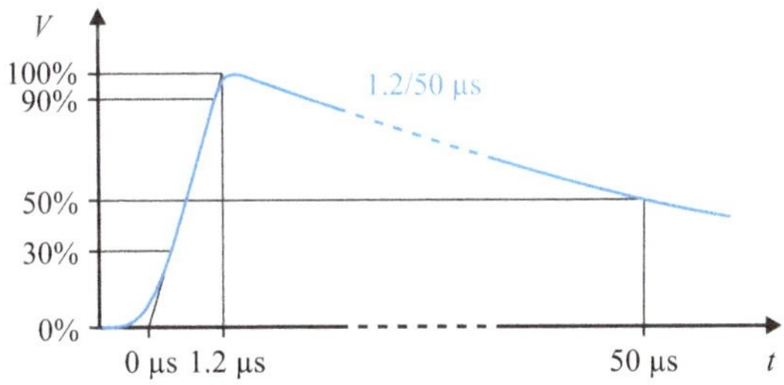

Figure 8.25 Surge-voltage impulse (1.2/50 μs)

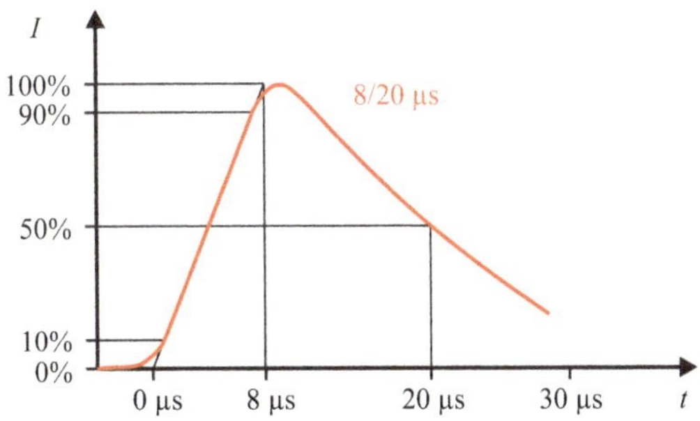

Figure 8.26 Surge-current impulse (8/20 μs)

combination wave generator is used. In open-circuit mode, the generator produces a surge-voltage impulse with a virtual front time of 1.2 μs and a time to half-value of 50 μs. In closed-circuit mode (when the device under test becomes conductive), the generator produces a surge-current impulse with a virtual front time of 8 μs and a time to half-value of 20 μs. The charge voltage of the combination wave generator is called open-circuit (charge) voltage U_{OC}. The internal resistance of the combination wave generator is usually 2 ohm. Combination wave generators with an internal resistance of 12 ohm are as well used for testing of electrical equipment in accordance to relevant electromagnetic compatibility (EMC) standards.

10/350 μs (lightning) current impulse is a transient current impulse with a virtual front time of 10 μs and a time to half-value of 350 μs (see Figure 8.27). This impulse represents lightning-induced surge currents with high energy content.

Nominal discharge current I_n refers to a surge-current impulse of the wave shape 8/20 μs. Type 1 and Type 2 SPD get tested multiple times with the nominal discharge current I_n. The nominal discharge current represents

- man-made switching surges and
- lightning-induced surge currents with medium-high energy content.

Maximum discharge current I_{max} refers to a surge-current impulse of the wave shape 8/20 μs. The peak value I_{max} has to be equal or greater than the I_n value. According to actual IEC surge-protection standards, it's not mandatory to state an I_{max} value. If an I_{max} is stated by the manufacturer of the SPD, then it's mandatory to test the respective SPD – in accordance to IEC surge-protection standards – with the maximum discharge current I_{max}. The maximum discharge current I_{max} represents

- man-made switching surges and
- lightning-induced surge currents with medium-high energy content.

Impulse discharge current I_{imp} refers to a surge-current (lightning current) impulse of the wave shape 10/350 μs (see Figure 8.27). Type 1 SPDs and some SPDs for measurement and control circuits and data lines get tested with I_{imp}. The I_{imp} represents lightning-induced surge currents with high energy content.

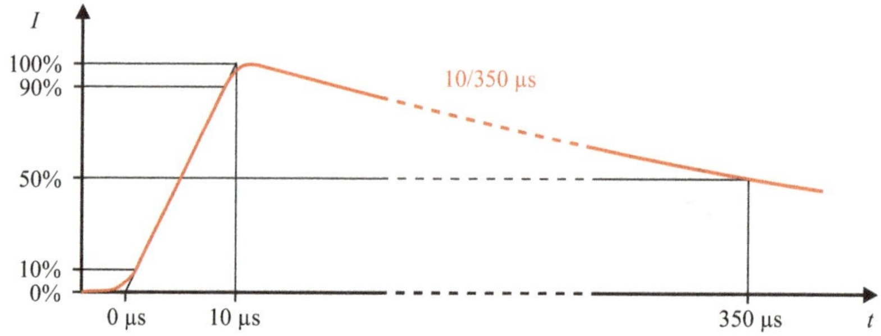

Figure 8.27 Lightning current impulse (10/350 μs)

Maximum continuous operating voltage (MCOV) U_C is the maximum AC or DC voltage, which may be continuously applied to an SPD or a certain mode of protection of an SPD. In AC circuits, the U_C value refers to the RMS value of the voltage.

Voltage protection level U_p is a technical data given by the manufacturer of the SPD. The U_p refers to the maximum voltage level at the terminals of an SPD which shall not be exceeded during the testing with defined test impulses used for the testing of SPDs in accordance to the respective test standard. The 'given' voltage protection level U_p can exceed the measured limiting voltage values recorded during the testing of SPDs. The comparison of the voltage protection levels U_p of different SPDs can be misleading, because the procedure to determine the voltage protection level U_p depends on the respective Type/Class of an SPD and on the given nominal discharge current I_n. A direct comparison of the voltage protection levels U_p of two SPDs is only possible if the U_p values refer to the same wave shape and amplitude of the respective given nominal discharge current I_n.

Measured limiting voltage refers to the peak voltage level at the terminals of an SPD during the testing with a test impulse of specific wave shape and amplitude.

Residual voltage U_{res} refers to the peak voltage level at the terminals of an SPD during the passage of a discharge current.

Open-circuit voltage U_{OC} refers to the open-circuit charge voltage of a combination wave generator. See topic '1.2/50 & 8/20 μs combination wave (hybrid impulse)'.

Short-circuit current I_{CW} of a combination wave generator refers to the prospective short-circuit current of a combination wave generator during the testing of a device under test (DUT) with a negligibly low impedance. Example: A combination wave generator with an internal resistance of 2 ohm and an open-circuit (charge) voltage U_{OC} of 6 kV is capable to produce a short-circuit current I_{CW} of 3 kA (8/20 μs) – when testing a DUT with negligible low impedance.

Prospective short-circuit current I_p of a power supply system refers to the prospective short-circuit current of a power supply system at a given location – if the impedance of a short circuit is negligibly low.

Short-circuit current rating I_{SCCR} of an SPD refers to the maximum permissible prospective short-circuit current of a power supply system at the (intended) installation location of an SPD – in conjunction with the disconnector (e.g. fuse, circuit breaker and internal disconnector) specified by the manufacturer of an SPD.

Follow current interrupt rating I_{fi} of an SPD refers to the prospective short-circuit current which an SPD is capable to interrupt (on its own) without the operation of a disconnector (e.g. fuse, circuit breaker and internal disconnector).

Status indicator is a device or mechanism which indicates the operational status of an SPD or the operational status of a part of an SPD. If a status indicator indicates the status 'no longer functioning', an SPD or a part of an SPD has to be replaced. In most cases the status indicator is linked to a (thermal) disconnector and indicates the status of the (thermal) disconnector of an SPD. A comprehensive electrical evaluation of the status of an SPD is only possible by carrying out electrical testing with a suitable high-voltage test device. Such testing has to be carried out in accordance with the manufacturer's guidelines and equipment provided by the manufacturer of the SPD (see IEC 62305-3).

8.3.2 Standards

Depending on the type of power supply system and the legal situation in a certain country or territory, different lightning protection standards are used:

- Countries with IEC-style power systems usually follow IEC surge protection and lightning protection standards
- Countries with US-style power systems usually follow UL surge protection and lightning protection standards

On a world-wide perspective, the most comprehensive lightning protection standards are the IEC 62305 lightning protection standards. This series of standards consists of four parts and covers all relevant aspects of lightning physics, risk management, external lightning protection and internal lightning protection (surge protection) for LV systems:

- IEC 62305-1 (2010): Protection against lightning – Part 1: General principles
- IEC 62305-2 (2010): Protection against lightning – Part 2: Risk management
- IEC 62305-3 (2010): Protection against lightning – Part 3: Physical damage to structure and life hazard *(with the focus on external lightning protection)*
- IEC 62305-4 (2010): Protection against lightning – Part 4: Electrical and electronic systems within structures *(with the focus on internal lightning protection)*

Product-specific IEC standard for the testing of SPDs for the protection of LV power systems:

- IEC 61643-11 (2011): Low-voltage surge protective devices – Part 11: Surge protective devices connected to low-voltage power distribution systems – Requirements and testing methods

Application-specific standards for SPDs for the protection of low-voltage power systems:

- IEC 61643-12 (2008): Low-voltage surge protective devices – Part 12: Surge protective connected to low-voltage power distribution systems – Selection and application principles
- IEC 60364-4-44 (2007 + AMD1:2015): Low-voltage electrical installations – Part 4-44: Protection for safety – Protection against voltage disturbances and electromagnetic disturbances
- IEC 60364-5-53 (2001 + AMD1:2002 + AMD2:2015): Electrical installations of buildings – Part 5-53: Selection and erection of electrical equipment – Isolation, switching and control – Section 534 Devices for protection against transient overvoltages

General information on lightning physics and lightning protection:

- IEC TR 62066 (2002): Surge overvoltages and surge protection in low-voltage a.c. power systems – General basic information
- IEEE C 62.41.2 (2002): Practice on characterization of surges in low-voltage (1,000 V and less) ac power circuits

General information on power systems:

- IEC 60364-1 (2005): Low-voltage electrical installations – Part 1: Fundamental principles, assessment of general characteristics, definitions

General information on selection and erection of controlgear and on over-current selectivity

- HD 60364-5-53 (2015): Low-voltage electrical installations – Part 5-53: Selection and erection of electrical equipment – Switchgear and controlgear (see section 536 Co-ordination of electrical equipment for protection, isolation, switching and control)

Insulation coordination and (required) rated impulse voltage (1.2/50 μs) for equipment:

- IEC 60664-1 (2007): Insulation coordination for equipment within low-voltage systems – Part 1: Principles, requirements and tests

Product-specific UL standard for the testing of SPDs for the protection of low-voltage power systems:

- UL 1449 (2014): SPDs

8.3.3 Introduction to surge protection for LV power systems

Nowadays more and more sensitive electrical equipment is used. This is especially true for electronics. Electronic devices get smaller and smaller. The smaller electronic devices get, the more challenging it is to protect them against man-made surges (e.g. switching overvoltages) and against lightning-induced surges. Surges can affect the proper functioning of electrical equipment. Powerful surges can damage or destroy electrical equipment. If electrical equipment gets damaged by less-powerful surges, it's sometimes difficult to detect this kind of damage with the naked eye. Powerful surges usually cause visible damage. Surges can cause as well so-called 'dangerous sparking'. Dangerous sparking can, e.g., cause subsequent short-circuit currents in electrical systems. In many cases damages, caused by subsequent short-circuit currents, are more severe than damages caused by the initial surge or by dangerous sparking.

To protect LV power systems against surges and against dangerous sparking, SPDs are used. SPDs are *not* capable to protect electrical equipment against temporary overvoltages (TOVs) and long-duration overvoltages – e.g. caused by insufficient regulation of the supply voltage. Nevertheless SPDs are capable to protect electrical equipment against short-duration man-made surges and against short-duration lightning-induced surges. This kind of short-duration surges usually lasts less than a millisecond. Short-duration surges are sometimes called 'transients'. SPDs are a kind of 'transient protection devices', capable to protect against transient surge-voltage impulses and against transient surge-current impulses.

During normal operation SPDs have a resistance which in the range of several megaohm. In the very moment, when there is a sufficiently high transient overvoltage, SPDs immediately become low-ohmic. During the conduction phase, the resistance of SPDs is significantly lower than one ohm. When SPDs become conductive, they basically create a low-ohmic short circuit. As soon as the transient overvoltage is gone and transient currents got discharged, an SPD immediately becomes high-ohmic again. The main tasks of SPDs are:

- Lightning currents and surge currents shall be 'redirected' in a way that no conductors or conductive parts get overheated and that there is no unwanted sparking or arcing.
- Surge-voltage impulses and lightning-induced voltages shall be limited in a way that downstream electrical equipment does not get damaged or destroyed.
- The peak voltage of an SPD shall be suitable (sufficiently low) for the protection of the downstream electrical equipment (see IEC 62305-4, IEC 61643-12).
- During the conduction phase of an SPD the course of the let-through voltage shall be sufficiently low – for an efficient protection during longer-duration lightning impulses.

Surge-protective components can have a voltage-switching or a voltage-limiting characteristic. The following surge-protective components are used inside SPDs:

- Spark gaps
- Metal-oxide varistors (MOVs)
- Gas-discharge tubes (GDTs)
- Suppressor diodes (transient-voltage suppression diodes, TVS diodes).

Voltage-switching components: Spark gaps and GDTs are voltage-switching components. They are in the high-ohmic state when no surge is present, but they can have a sudden change to the low-ohmic state in case of a surge. The operation behaviour of voltage-switching components, connected to energized conductors, depends very much on the let-through voltage of the respective voltage-switching component.

N/PE spark gaps and GDTs have a let-through voltage which is usually significantly lower than the max. continuous operating voltage (MCOV, U_C) of the respective surge-protective component. The ability of such voltage-switching components to quench line follow currents is usually very limited. Therefore GDTs and N/PE spark gaps are usually *not* installed between two energized conductors and they are usually *not* installed between an energized conductor and a grounded conductor.

L/N spark gaps got specially designed for the installation between two energized conductors, or between an energized conductor and a grounded conductor. These voltage-switching components have increased let-through voltages. The increased let-through voltage can be, e.g., in the range of the maximum continuous operating voltage – which is applicable for a specific mode of protection (e.g. between L and N). Because of the increased let-through voltage, these voltage-switching components have an increased ability to limit or to avoid line follow currents. If the let-through voltage is high enough, then no line follow current will flow. State-of-the-art L/N spark gaps are free of line follow currents. L/N spark gaps with a sufficiently high let-through voltage are suitable for the installation in direct proximity to powerful transformers – e.g. at installation locations with prospective short-circuit currents of up to 100 kA (RMS).

Voltage-limiting components: MOVs and suppressor diodes (TVS diodes) are voltage-limiting components. They are in the high-ohmic state when no surge is present, but they reduce the resistance continuously in response to the voltage applied to the terminals of the respective voltage-limiting components. The let-through voltage of a voltage-limiting component is always higher than the max. continuous operating voltage (MCOV, U_C) of the respective surge-protective component. Therefore SPDs with voltage-limiting components are usually free of line follow currents.

Combinations of surge-protective components with voltage-switching and voltage-limiting characteristics can have a variety of operation characteristics – depending on the respective combination of surge-protective components. The interaction of different SPDs, installed in a power distribution system, can be quite complex. Suitable simulation software can be very helpful to understand the operation characteristics of SPDs during the discharge of surge-voltage and surge-current impulses.

For the simulation of the operation characteristics of electronic components, nowadays simulation software is used on a regular basis. A popular general-purpose open-source simulation program, used for such kind of simulations, is called SPICE (Simulation Program with Integrated Circuit Emphasis). SPICE can be as well a part of a more complex software package (e.g. LTspice from Linear Technology). Vendors of electric and electronic components provide electronic models which can be used in SPICE-based simulation software. Such electronic models are available for voltage-limiting surge-protective components (e.g. varistors and suppressor diodes). The accuracy of the simulation of voltage-limiting components is quite high. Unfortunately the simulation of voltage-switching components (e.g. spark gaps and GDTs) is much more challenging – in comparison to the simulation of voltage-limiting components.

For all conductors, which carry surge currents, the stray inductance L has to be considered. A straight solid round conductor has a stray inductance of about 1 μH/m. The stray inductance of a round conductor is nearly independent from the cross-section of a conductor. If a fast-changing current flows through a conductor, then there will be an inductive voltage drop along the conductor. See Faraday's law of induction: $V = -L*di/dt$.

Surge-current impulses usually have a very short rise time of just a few microseconds. Because of the stray inductance of conductors and because of the inductive voltage drop along conductors, fast-changing surge-current impulses can have a huge impact on the protection effect of SPDs and surge protection schemes. Internal lightning protection systems with SPDs should be designed in a way that such inductive voltage drops are as low as possible. To be able to achieve a good protection effect, the total inductance of an SPD (including connecting leads) should be as low as possible. Therefore leads and other conductors, which have to carry fast-changing surge currents, should be as short as possible and they should be routed as straight as possible.

8.3.4 Multi-stage surge protection schemes

IEC lightning protection standards stipulate to install SPDs, for the protection of LV power systems, at certain preferred installation locations. Depending on the respective installation location and on the required discharge capacity, different types of SPDs are needed (see IEC 62305-4, IEC 61643-11, IEC 61643-12):

- Type 1 [T1] SPDs *or*
- Type 2 [T2] SPDs *or*
- Type 3 [T3] SPDs.

A single SPD, installed at the origin of the installation, is usually not capable of protecting a bigger electrical system. Additional SPDs are needed downstream – to limit and to divert

- surges which are left over by the first or subsequent stages of protection,
- surges which are induced due to magnetic or electric fields,
- surges which are coupled in galvanically via other grounded services or other conductors,
- surges caused by man-made switching action and to
- attenuate oscillations which can happen in electrical systems.

According to IEC lightning protection standards, the most powerful surge currents have to be expected at installation locations where power lines enter or leave a building (see Figure 8.28). Therefore, an SPD with a high discharge

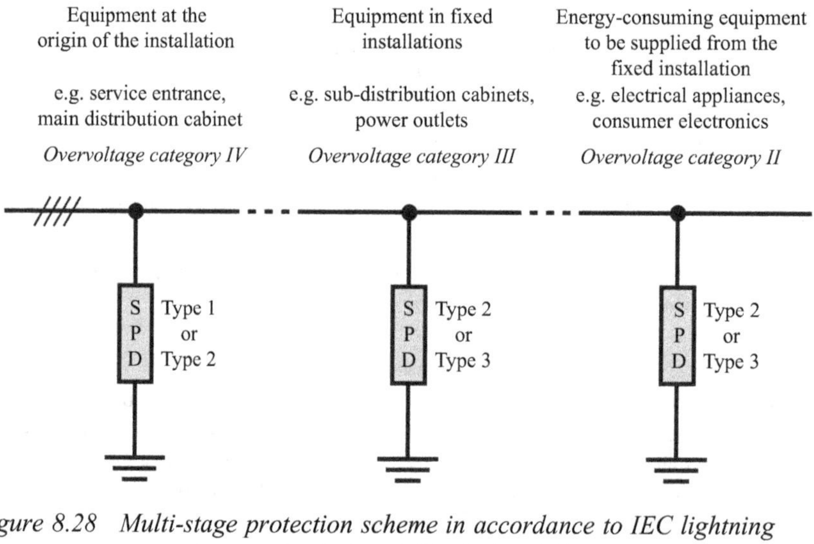

Figure 8.28 Multi-stage protection scheme in accordance to IEC lightning protection standards (see IEC 60364-5-53)

capacity shall be placed, as a first stage of protection, at the origin of the installation (service entrance, main distribution cabinet). Type 1 SPDs are capable to 'redirect' high-energy long-duration lightning currents to ground and as well to other directly or indirectly grounded conductors. Therefore, Type 1 SPDs shall be placed at all installation locations where high-energy long-duration lightning currents can be expected. At installation locations where only medium- or low-energy surges are expected, Type 2 or Type 3 SPDs are usually sufficient. Depending on the overvoltage category (OVC) of electrical equipment, installed at different installation locations, SPDs have to be chosen in a way that they meet the requirements of the respective OVC at a certain installation location (see Table 8.7).

8.3.5 Lightning protection zones

An electrical system of a facility is divided into different lightning protection zones (LPZs). A suitable SPD shall get installed at the boundary from one LPZ to the next LPZ. The following LPZs are defined in IEC standards (see Table 8.8):

- **LPZ 0_A:** outdoors; not protected against direct lightning flashes
- **LPZ 0_B:** outdoors; protected against direct lightning flashes
- **LPZ 1:** indoors (inside LPZ 0); protected against direct lightning flashes; reduced surge currents due to current sharing and/or isolating interfaces and/or SPDs at the boundary between LPZ 0 and LPZ 1

Table 8.7 Rated impulse voltages (1.2/50 μs) for equipment energized directly from the LV mains (see IEC 60664-1 Annex F)

Nominal voltage of the supply system based on IEC 60038		Voltage line to neutral derived from nominal voltages AC or DC up to and including (V)	Rated impulse voltage (1.2/50 μs) OVC			
Three phase (V)	Single phase (V)		I (V)	II (V)	III (V)	IV (V)
		50	330	500	800	1,500
		100	500	800	1,500	2,500
	120–240	150	800	1,500	2,500	4,000
230/400, 277/480		300	1,500	2,500	4,000	6,000
400/690		600	2,500	4,000	6,000	8,000
1,000		1,000	4,000	6,000	8,000	12,000

OVC IV: Equipment for the use at the origin of the installation.
OVC III: Equipment in fixed installations and for cases where there are special requirements regarding reliability and availability.
OVC II: Energy consuming equipment which is supplied from the fixed installation.
OVC I: Equipment in circuits where transient overvoltages already got limited to an appropriately low level.

Table 8.8 SPD types installed at the boundaries of lightning protection zones (LPZs) (see IEC 60364-5-53)

Boundaries of LPZs	Types of SPD
LPZ 0_A → LPZ 1	Type 1 [T1], Class I
LPZ 0_B → LPZ 1	Type 2 [T2], Class II
LPZ 1 → LPZ 2	Type 2 [T2], Class II
LPZ 2 → LPZ 3	Type 3 [T3], Class III

- **LPZ 2:** indoors (inside LPZ 1); protected against direct lightning flashes; further reduced surge currents due to current sharing and/or isolating interfaces and/or SPDs at the boundary between LPZ 1 and LPZ 2
- **LPZ 3:** indoors (inside LPZ 2); protected against direct lightning flashes; further reduced surge currents due to current sharing and/or isolating interfaces and/or SPDs at the boundary between LPZ 2 and LPZ 3

8.3.6 Types of SPDs

Type 1 [T1]/Class I SPDs (see Figures 8.29 and 8.30) are installed at the boundary between LPZ 0_A and LPZ 1. Facilities with external lightning protection system *or* facilities supplied by overhead lines have to be equipped with Type 1 SPDs. The typical installation location for this kind of SPD is the service entrance or the main distribution cabinet. An installation in close proximity to the main earthing busbar is recommended. Type 1 SPDs have a high discharge capacity (lightning current impulse I_{imp} (10/350 μs), surge-current impulse I_n (8/20 μs)). Typical surge-protective components used in Type 1 SPDs are spark gaps, big varistors and big GDTs.

Type 2 [T2]/Class 2 SPDs (see Figure 8.31) are installed at the boundary between LPZ 0_B and LPZ 1 *or* at the boundary between LPZ 1 and LPZ 2. The typical installation locations for this kind of SPDs are the main distribution cabinet (of facilities supplied by buried cables *and* without external lightning protection system) or sub-distribution cabinets. Type 2 SPDs have a medium-high discharge capacity (surge-current impulse I_n (8/20 μs)). Typical surge-protective components used in Type 2 SPDs are medium-sized varistors and medium-sized GDTs.

Type 3 [T3]/Class 3 SPDs (see Figure 8.32) are installed at the boundary between LPZ 2 and LPZ 3. The typical installation location for this kind of SPD is a location close to sensitive equipment. Type 3 SPDs have a relatively low discharge capacity (combination wave impulse, 1.2/50 μs, 8/20 μs, 2 ohm, U_{OC}). Typical surge-protective components used in Type 3 SPDs are small varistors, small GDTs and sometimes suppressor diodes.

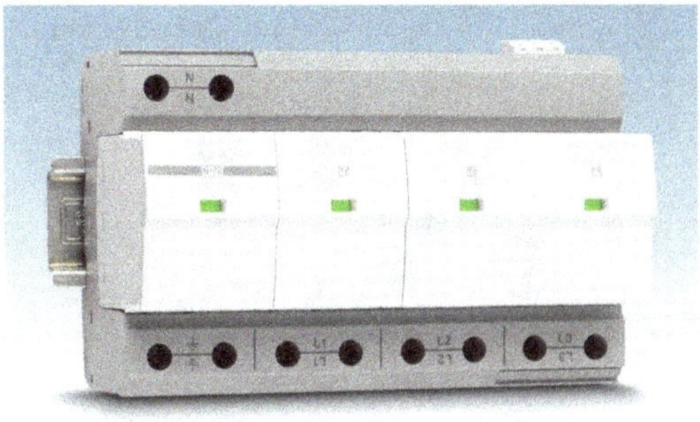

Figure 8.29 *Pluggable Type 1 SPD with three L/N spark gaps and one N/PE spark gap; for the diversion of lighting currents (10/350 μs); four modes of protection; suitable for the protection of three-phase five-wire systems*

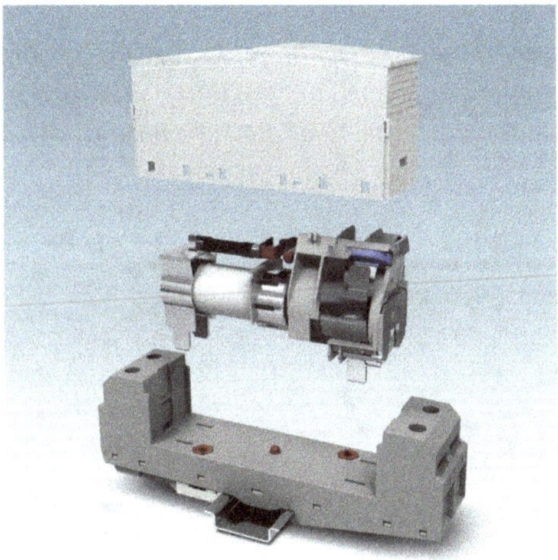

Figure 8.30 *Pluggable Type 1 SPD with one spark gap; for the diversion of lighting currents (10/350 μs); one mode of protection; with integrated back-up fuse*

8.3.7 Surge-protective components

8.3.7.1 Spark gaps

Sparks gaps are voltage-switching components. Older designs of spark gaps had been untriggered. Many newer designs of spark gaps are triggered. Due to their robust design, spark gaps are capable to divert high-energy long-duration lightning currents of the wave shape 10/350 μs. Because of their high discharge capacity, they are rated as Type 1 SPDs.

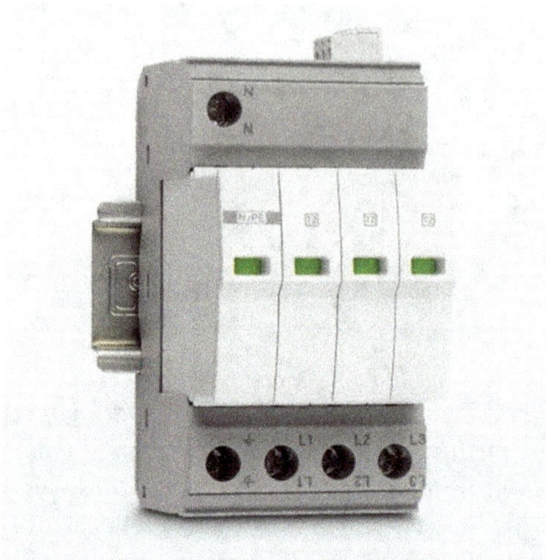

Figure 8.31 Pluggable Type 2 SPD with three varistors and one GDT; for the diversion of surge currents (8/20 μs); four modes of protection; suitable for the protection of three-phase five-wire systems

Figure 8.32 Pluggable Type 3 SPDs; left: suitable for single-phase three-wire systems; right: suitable for the protection of three-phase five-wire systems

As soon as a spark gap has become fully conductive, the level of the let-through voltage (residual voltage) is so low that there is no longer any electrical stress for downstream equipment. Because of the low level of the let-through voltage, during the conduction phase, Type 1 spark gaps are the best choice for the first stage of protection. The protection effect of spark gaps is usually significantly better than the protection effect of SPDs with voltage-limiting components.

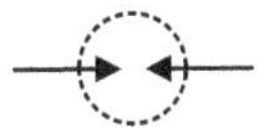

Figure 8.33 Single-stage spark gaps

Figure 8.34 Multi-stage spark gaps

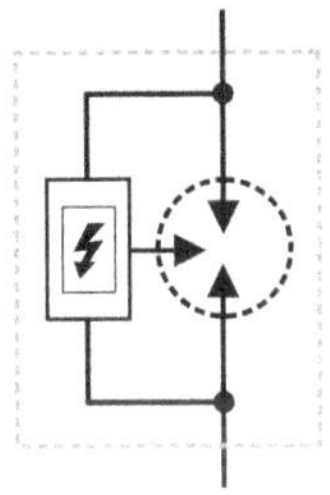

Figure 8.35 Triggered single-stage spark gaps

Single-stage spark gaps have two main electrodes. To achieve a low protection level, state-of-the-art single-stage spark gaps are equipped with a triggering electrode and a triggering circuit (see Figures 8.35 and 8.36). Modern single-stage spark gaps, for the use between L and N conductors (L/N spark gaps), have an arc-burning voltage which is high enough to interrupt or to avoid line follow currents. Single-stage spark gaps, for the use between N and PE conductors (N/PE spark gaps) or between other non-energized conductors, have a relatively low arc-burning voltage. Their ability to interrupt or to avoid line follow currents is relatively low.

Multi-stage spark gaps usually consist of an arrangement of several carbon discs which are arranged in a way that the space between the individual discs acts as a series connection of individual spark gaps (see Figure 8.34). To achieve a low protection level, modern multi-stage carbon SGs are equipped with a triggering circuit which is connected to the individual carbon discs (see Figures 8.37 and 8.38). The arc-burning voltage of the individual carbon SGs is relatively low. Due to the series arrangement of carbon spark gaps, it's possible to increase the total arc-burning voltage of a multi-stage spark gaps to a level which is high enough to interrupt or to avoid line follow currents.

The gap between the electrodes of spark gaps is usually filled with ambient air. During normal operation the space between the main electrodes is non-conductive, and the isolation resistance between the main electrodes is in the megaohm range. When a spark gaps becomes conductive, then there is an electric arc between the main electrodes. The voltage drop between electrodes is called arc-burning voltage (arc drop voltage). The course or the arc-burning voltage – during the initial stage

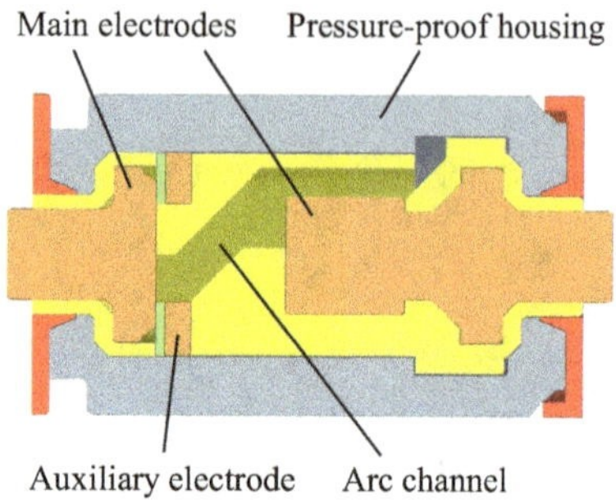

Figure 8.36 Encapsulated and triggered single-stage Type 1 L/N spark gap; barrel-shaped housing; free of line follow current

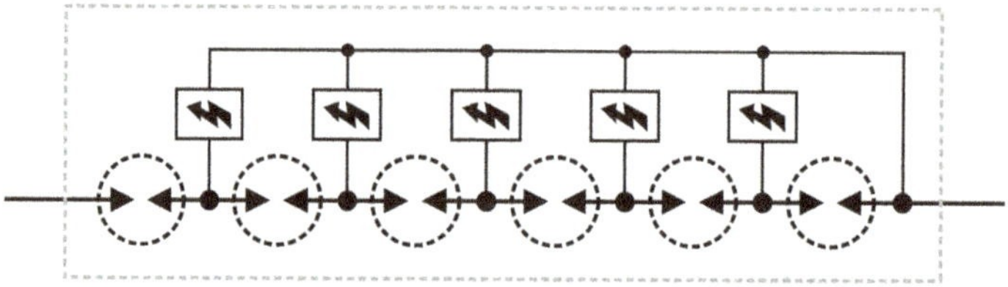

Figure 8.37 Encapsulated and triggered multi-stage spark gap; six stages

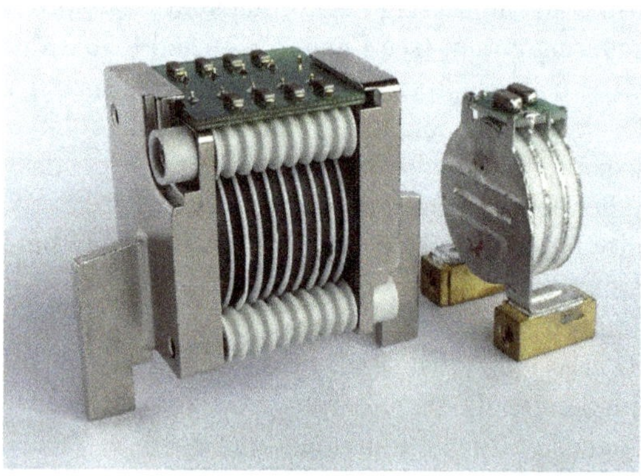

Figure 8.38 Encapsulated and triggered multi-stage Type 1 L/N spark gap; capacitive triggering; left: nine stages; right: two stages

of the discharge of surge currents and lightning currents and during the conduction phase of a spark gaps – depends very much on the specific electro-mechanical design of a spark gaps. For some spark gaps designs (L/N spark gaps), it's favourable to design them in a way that the arc-burning voltage stays at a relatively high voltage level. Depending on the specific design of a spark gap, one or more of the following physical principles and design features can be used to increase the arc-burning voltage:

- lengthening of the electric arc,
- splitting of the one electric arc into several electric arcs,
- arc cooling,
- quenching gas,
- increase in pressure in the gap between the electrodes (currently a matter of contention in the relevant literature).

To be able to control the arc-burning voltage – during the initial stage of the discharge of surge currents and lightning currents – modern spark gaps are equipped with a trigger circuit and one or more triggering electrodes. Because of the trigger circuit, nowadays spark gaps for 230/400 V AC power systems are capable to limit the peak level of the let-through voltage during the conduction phase to less than 1.5 kV. Spark gaps for 400/690 V AC systems are nowadays capable to limit the peak level of the voltage during the conduction phase to less than 2.5 kV.

Each voltage-switching type SPD has a certain ability to quench short-circuit currents (line follow currents) without the help of an additional overcurrent protective device. The technical parameter which provides information about the ability of a spark gaps to interrupt (prospective) short-circuit currents by itself is called 'follow current interrupt rating I_{fi}'. In IEC 61643-12, it's stipulated that the power follow current interrupting rating I_{fi} of an SPD shall exceed the prospective short-circuit current at the installation location of an SPD.

Older designs of L/N spark gaps have (during the conduction phase) an arc-burning voltage which is lower than the peak voltage of the supplying power system. Such spark gaps have a limited ability to interrupt follow currents on their own. Because of their relatively low follow current interrupt capacity, this kind of spark gaps are not suitable for installation locations with higher prospective short-circuit currents.

Newer designs of L/N spark gaps have (during the conduction phase) an arc-burning voltage which is about as high or slightly higher than the peak voltage of the supplying power system. Such spark gaps have an increased ability to interrupt follow currents on their own. Nowadays L/N spark gaps with a very high follow current interrupt rating I_{fi} are commercially available (e.g. 100 kA RMS). Typical L/N spark gaps are, e.g., able to discharge high-energy long-duration lightning currents (10/350 μs) with amplitudes of, e.g., 25...50 kA (per mode of protection).

State-of-the-art L/N spark gaps are encapsulated, triggered, fast-acting and entirely free of follow current (see Figure 8.39). Their voltage protection level U_p and the course of their residual voltage – during the conduction phase – is low enough for the protection of sensitive electronic equipment.

Some spark gaps are specially designed for the use between N and PE conductors. These N/PE spark gaps can get installed between grounded conductors in solidly grounded power systems – at installation locations where the prospective short-circuit current is negligibly low. During the conduction phase of N/PE spark gaps, the residual voltage is at a very low voltage level. Therefore, the follow current

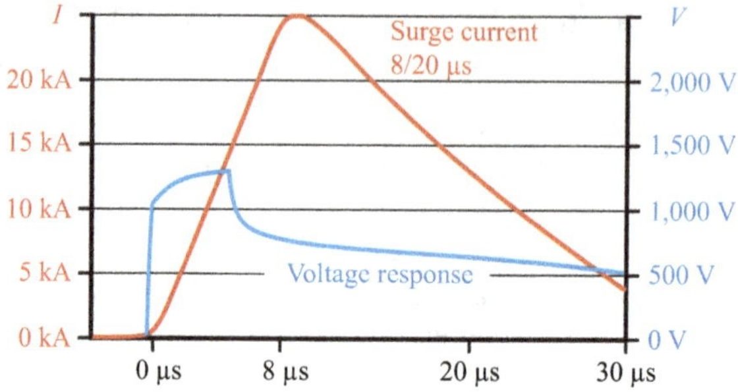

Figure 8.39 Voltage response of a Type 1 SPD (for $U_N = 230$ V AC) with an encapsulated and triggered L/N spark gap and a high arc-burning voltage; during the discharge of a 25 kA (8/20 μs) surge-current impulse

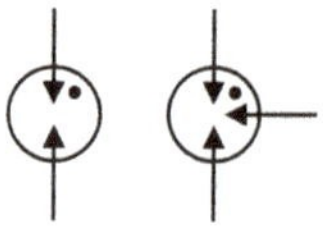

Figure 8.40 Two- and three-electrode gas-discharge tubes (GDTs)

interrupt rating I_{fi} of N/PE spark gaps is relatively low, but sufficiently high for the installation between N and PE conductors in solidly grounded power systems. Due to the very low level of the residual voltage, commercially available N/PE spark gaps are able to discharge high-energy long-duration lightning currents with amplitudes of e.g. 100 kA (10/350 μs).

8.3.7.2 Gas-discharge tubes

Gas-discharge tubes (GDTs) are fully encapsulated, non-triggered miniature spark gaps with two or three electrodes (see Figures 8.40 and 8.41). The gas-tight housing of GDTs is filled with inert gas. GDTs are used for Type 1, Type 2 and Type 3 SPDs. GDTs are voltage-switching components with a high discharge capacity. During the conduction phase of a GDT, the level of the residual voltage is very low. Because of the very low level of the residual voltage during the conduction phase, GDTs cannot interrupt line follow currents efficiently. Therefore, it's not permissible in power systems to install a GDT between two energized conductors or between an energized and a grounded conductor. GDTs usually get installed between N and PE conductors of solidly grounded power systems.

The spark-over voltage of GDTs depends on the rate of rise of the voltage. During the discharge of fast-rising voltage impulses, the spark-over voltage of a GDT is increased – in comparison to the discharge of slow-rising voltage impulses where the spark-over voltage of a GDT is at a lower voltage level.

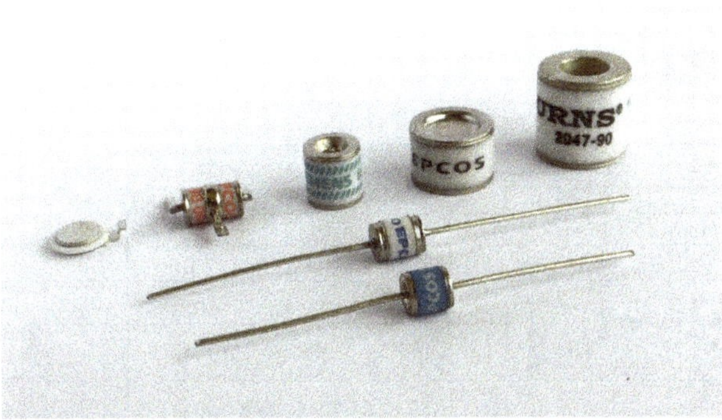

Figure 8.41 Gas-discharge tubes (GDTs)

Figure 8.42 Metal-oxide varistor (MOV)

8.3.7.3 Metal-oxide varistors

Varistors (variable resistors) are voltage-dependent resistors. Due to the fact that varistors consist of polycrystalline metal oxide ceramics, they are as well called metal-oxide varistors (MOVs) (see Figure 8.42). The resistance of varistors decreases with increasing voltage levels. This means that they are voltage-limiting components. During the conduction state, the resistance of a varistor is significantly lower than 1 ohm. Varistors are the most frequently used surge-protective components for the protection of LV power systems. They are the best choice to divert man-made switching overvoltages.

Commercially available varistors often have an electrical tolerance band of $\pm 10\%$. For the testing of varistors, it's common practice to use a current of 1 mA. A current source, which is capable to produce a constant current of 1 mA, is used for the testing of varistors. The voltage drop at a current of 1 mA is used as a reference value for varistors and it's used to evaluate if a varistor is still in the tolerance band specified by the varistor manufacturer (see Figure 8.43).

The *V/I* curve diagrams in data sheets of varistor manufacturers usually do not show the complete tolerance band of a varistor (see Figure 8.44). In many cases, *V/I* curve diagrams of varistors depict

- the negative tolerance – for currents less than 1 mA and
- the positive tolerance – for currents greater than 1 mA.

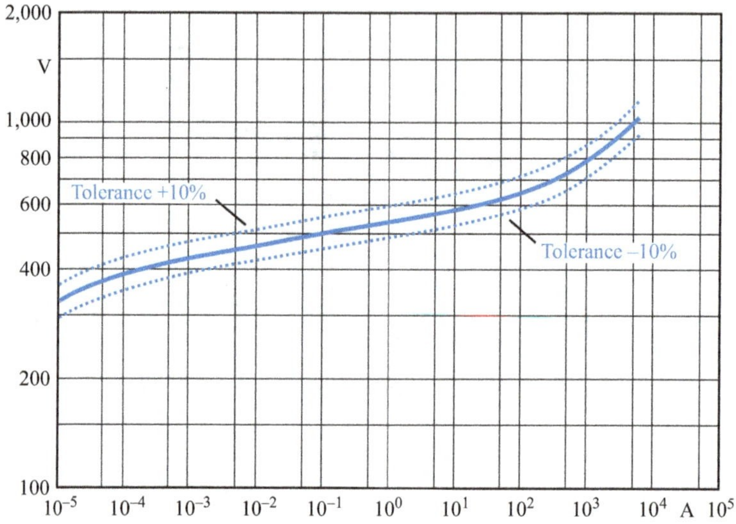

Figure 8.43 V/I curve of an MOV with a tolerance band of ±10% (MCOV = 275 V AC, I_{max} = 8 kA (8/20 μs))

The *V/I* curve of a varistor does not have a discontinuity point at 1 mA. The *V/I* curve shows a combination of two different curves in one figure (max. positive tolerance and max. negative tolerance of a varistor):

- *Currents less than 1 mA: V/I* curve of a varistor with max. negative tolerance (in this case: −10%). The left section of the characteristic curve (less than 1 mA) depicts the high-impedance state of the varistor. For high-impedance states, developers of electronic circuits are particularly interested in the highest possible leakage current at a given operating voltage. The lower limit of the tolerance band shows the maximum possible leakage current, flowing through a varistor, at any given voltage level.
- *Currents greater than 1 mA: V/I* curve of a varistor with max. positive tolerance (in this case: +10%). The right section of the characteristic curve (greater than 1 mA) depicts the low-impedance state of the varistor. For low-impedance states, developers of electronic circuits are interested in the worst-case voltage drop on the varistor. The upper limit of the tolerance band shows the maximum possible voltage drop on a varistor at any given surge-current amplitude.

Varistors are made of sintered metal oxide ceramics. Most varistors for low-voltage applications in power systems are disc-type varistors or block-type varistors. Varistors can be used in Type 1, Type 2 and Type 3 SPDs. The discharge capacity of varistors mainly depends on the footprint (size) of the respective varistor disc. If a high discharge capacity is needed, in most cases varistors with a bigger footprint are chosen. To achieve an increased discharge capacity, it's possible to connect varistors in parallel. Connecting varistors in parallel is common

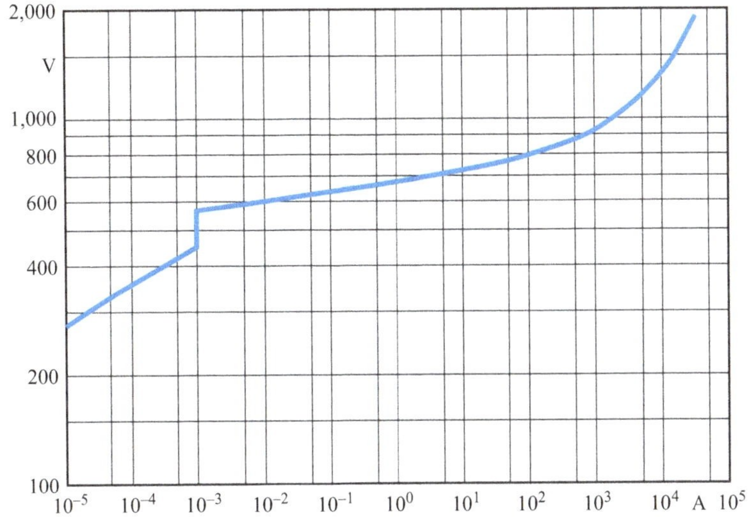

Figure 8.44 V/I curve of an MOV with a tolerance band of ±10% (MCOV = 320 V AC, I_{max} = 40 kA (8/20 μs))

practice for varistor-based Type 1 SPD and for SPDs used in countries with US-style power systems. The max. continuous operating voltage (MCOV, U_C) of a varistor depends on the thickness of the respective varistor disc (see Figures 8.45 and 8.46). The thicker the disc is, the higher is the MCOV.

Most varistors are designed for the discharge of 8/20 μs surge currents. Some varistors are capable of discharging high-energy long-duration 10/350 μs lightning currents. Varistors are voltage-limiting components, and their operation characteristic differs significantly from the operation characteristic of voltage-switching components like sparks gaps. During the whole conduction phase of a varistor, the course of the let-through voltage (residual voltage) is always significantly higher than the supply voltage of the power system. Therefore, varistors are free of line follow currents.

During the discharge of short-duration surge impulses (e.g. waveform 8/20 μs), varistors are capable to provide a sufficient protection effect for sensitive equipment (see Figure 8.47). Due to the relatively high level of the residual voltage, during the discharge of long-duration lightning currents (e.g. waveform 10/350 μs), varistors and other voltage-limiting components can cause undesirable electrical stress for downstream equipment. Therefore, varistors are not the best choice at installation locations where long-duration lightning currents (e.g. waveform 10/350 μs) can be expected. At installation locations with long-duration lightning currents, spark gaps are the better choice.

Varistors are prone to ageing. Ageing and degradation of varistors can be caused, e.g., by surge currents, lightning currents, man-made switching overvoltages, temporary overvoltages and repetitive voltage spikes (e.g. caused by power

Figure 8.45 Different MOVs for Type 1 and Type 2 SPDs

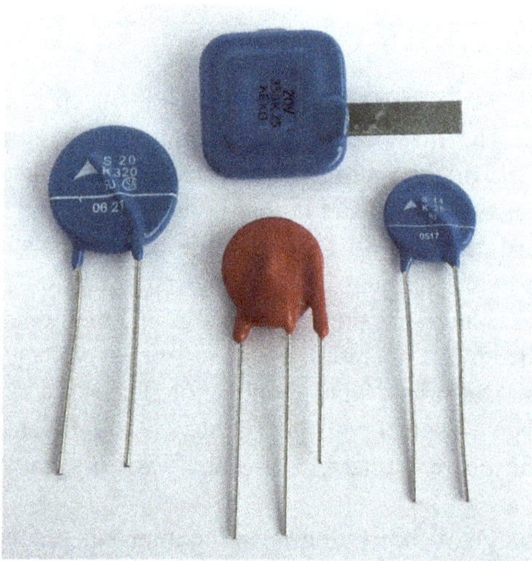

Figure 8.46 Different MOVs for Type 2 and Type 3 SPDs

electronics). Due to ageing, a leakage current can flow through a varistor and can heat up a varistor. Therefore, varistors, for the installation at such energized conductors where a build-up of heat may happen, are always equipped with a thermal disconnect device.

For small-sized varistors, thermal disconnect devices are commercially available. For bigger disc-type varistors, the manufacturers of SPDs design application-specific thermal disconnect devices. Most SPDs, which are equipped with varistors, have a local status indicator. If a thermal disconnect device disconnects an SPD from the

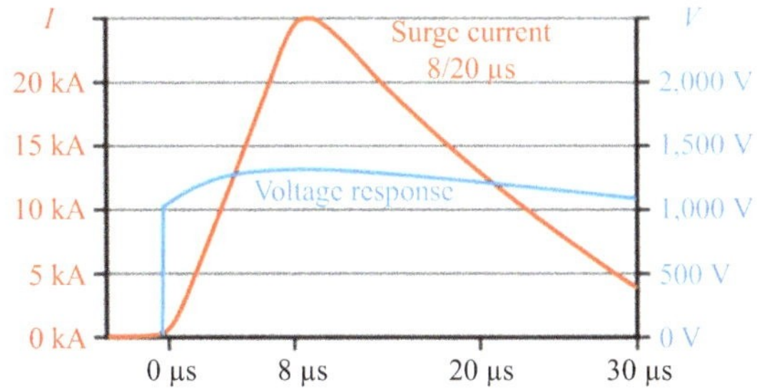

Figure 8.47 Voltage response of a Type 2 SPD with MOV (MCOV = 350 V AC, I_{max} = 40 kA (8/20 µs)); during the discharge of a 25 kA (8/20 µs) surge-current impulse

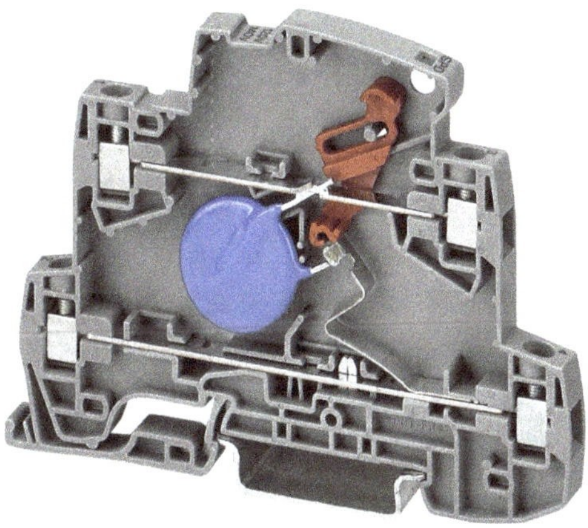

Figure 8.48 Single-stage SPD with varistor, thermal disconnect device and local status indicator; width: 6.2 mm

supply voltage, then this is indicated with the help of a local status indicator. It has to be considered that a status indicator is not able to provide any information about the ageing status of a varistor or if a varistor already got stressed by surge currents. To obtain reliable information about the 'health' of a varistor and other surge-protective components, suitable electrical testing is required.

The operation characteristic of an 'ideal varistor' only depends on the voltage level. Real-world varistors can have a quite complex operation characteristic. If a varistor is exposed to higher-frequency voltage changes (e.g. exceeding 20... 30 kHz), the stray capacitance of a varistor has to be taken into consideration. In addition to the stray capacitance, the stray inductance, the resistances of intergranular boundaries and the bulk resistance of the respective varistor have to be considered. Therefore the real-world equivalent circuit diagram of a varistor is more complex in comparison to a simplified equivalent circuit diagram of an 'ideal varistor' (see Figure 8.49).

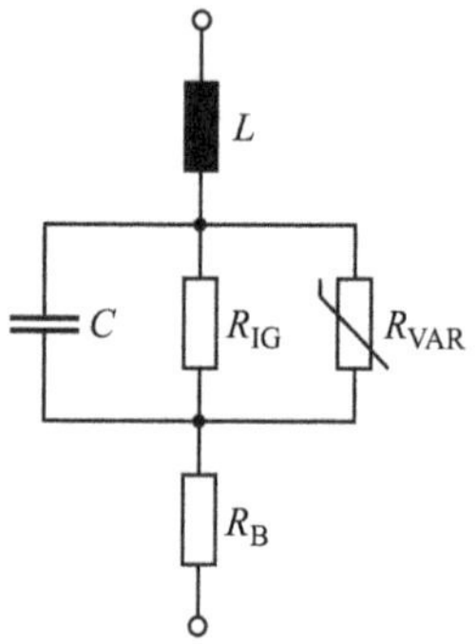

Key:

- L Inductance of leads ($\approx$ 1 μH/m)
- C Capacitance (depends on the specific varistor design)
- R_{IG} Resistance of intergranular boundaries ($\rho \approx 10^{12}$ to 10^{13} ohm cm)
- R_{VAR} Ideal varistor (few milliohm to ∞ ohm)
- R_B Bulk resistance of varistors ($\rho \approx$ 1 to 10 ohm cm)

Figure 8.49 Equivalent circuit of a varistor

Figure 8.50 Suppressor diode

The capacitance of a varistor depends on the hardware design of the specific type of varistor. Relevant factors, which have an influence on the capacitance of varistors, are, e.g., the size (footprint), the thickness and the MCOV of a varistor. Vendors of varistors provide information about the capacitance. In most cases, the capacitance of varistors, stated in data sheets, refers to a frequency of 1 kHz. It has to be considered that – in real-world applications – the oscillation frequencies in power systems can significantly exceed 1 kHz, e.g. due to power harmonics, due to the switching of controlled semiconductors or due to voltage and/or current surges. Due to the capacitance of varistors, varistor-based SPDs have a low-pass characteristic. A high capacitance can help to smooth steep surge-voltage edges (at least for low-energy surges).

8.3.7.4 Suppressor diodes

In comparison to varistors, suppressor diodes are less powerful. They are fast-acting voltage-limiting surge-protective components (see Figures 8.50 and 8.51). Their operation characteristic is comparable with the operation characteristic of varistors. The course of the let-through voltage (residual voltage) of a suppressor diode is usually lower than the let-through voltage of a comparable varistor with the same discharge capacity and the same max. continuous operating voltage. Suppressor diodes are less prone to ageing – in comparison to varistors. SPDs with

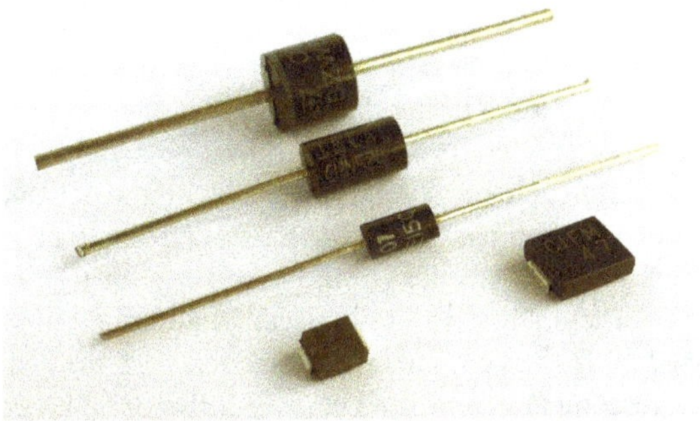

Figure 8.51 Suppressor diodes

Figure 8.52 Series connection of an varistor (MOV) and a GDT

suppressor diodes, for the protection of power systems, may be equipped with internal back-up fuses.

In countries with IEC-style power systems, suppressor diodes are mainly used in Type 3 SPDs for circuits with extra low voltage (ELV). The main reason for this is the relatively low discharge capacity of suppressor diodes. In countries with US-style power systems, some vendors offer SPDs with suppressor diodes connected in parallel and in series. Such SPDs are mainly used for the protection of power systems with system voltages exceeding the extra low voltage range.

8.3.8 Series and parallel connection of surge-protective components

8.3.8.1 Series connection of a varistor and a GDT

SPDs with a series connection of a varistor and a GDT are frequently used as Type 1, Type 2 and Type 3 SPDs (see Figure 8.52). They are suitable for general-purpose applications. Because of the high resistance of a GDT during normal operation, this series connection of surge-protective components is free of leakage currents. Therefore, this arrangement of components is frequently used for applications where leakage currents are not desirable or not permissible. This kind of SPDs is as well suitable for LV power systems where repetitive voltage spikes can be expected.

Repetitive voltage spikes can be, e.g., caused by electronics with insufficient filtering circuits.

Nowadays, SPDs with a series connection of a varistor and a GDT are as well used as first stage of protection. During the conduction phase of a series connection of a varistor and a GDT, the course of the let-through voltage (residual voltage) stays at a relatively high voltage level. This level of the let-through voltage is usually at a lower level in comparison to the let-through voltage of varistor-based SPDs. A high voltage level during longer-duration surge or lightning currents can cause undesirable electrical stress for downstream equipment.

Spark-gap-based SPDs have a much lower course of the let-through voltage (residual voltage) – in comparison to varistor-only SPDs or SPDs with a series connection of a varistor and a GDT. Due to the lower course of the let-through voltage during the conduction phase, spark-gap-based SPDs have a better protection effect as the first stage of protection – in comparison to other kind of SPDs with voltage-limiting components.

8.3.8.2 Parallel connection of varistors

For most applications, in IEC-style power systems, just one varistor per 'mode of protection' is used, but there are some applications where it's reasonable to connect varistors in parallel (see Figure 8.53).

The main reasons for the parallel connection of varistors are

- increased discharge capacity and
- increased availability and redundancy.

In SPDs in accordance to IEC lightning protection standards, varistors are connected in parallel to increase the nominal discharge current I_n (8/20 μs) and the maximum discharge current I_{max} (8/20 μs) and/or to increase the impulse discharge current I_{imp} (10/350 μs). During approval tests in accordance to IEC 61643-11 – carried out by independent and accredited laboratories – IEC-style SPDs are tested with all the 8/20 and 10/350 μs discharge currents stated by the manufacturer of the respective SPD.

SPDs for use in US-style power systems, which are designed and tested in accordance to the relevant UL standard for SPDs (UL 1449), often look different in comparison to IEC-style SPDs. There are significant technical differences in the

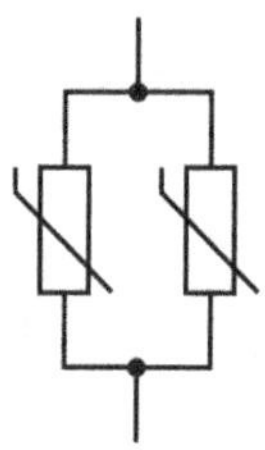

Figure 8.53 Parallel connection of metal-oxide varistors (MOVs)

design of IEC-style power systems and US-style power systems. These technical differences affect as well the design of SPDs. Manufacturers of US-style SPDs follow a 'design philosophy' which focuses on availability and redundancy.

In SPDs for US-style power systems, it's common practice to connect multiple surge-protective components in parallel. Some manufacturers put multiple varistors or suppressor diodes in parallel. Other manufacturers put multiple 'series connections of varistors and GDTs in parallel.

These arrangements of US-style protective components are usually installed in bigger-sized housings, and it's common practice to use internal back-up fuses – for the protection of the individual protective components against overheating and for the protection of the individual modes of protection against short-circuit currents. US-style SPDs are equipped with local status indicators which indicate a fault if one or more of the protective components are disconnected by (internal) back-up fuses. Due to the multiple protective components connected in parallel, there is a high level of availability and redundancy – even if one or more of the protective components get overloaded by surge currents or lightning currents.

In UL 1449 mainly the 8/20 μs wave shape is used for the testing of SPDs. The lightning current test wave shape 10/350 μs – well known from many IEC standards and some UL standards – is *not* used for the testing of US-style SPDs in accordance to UL 1449.

Most manufacturers of US-style SPDs (acc. to UL 1449) add up the individual discharge capacities of the protective components connected in parallel, and they state this 'calculated discharge capacity' in their data sheets – without having carried out any testing of the respective US-style SPDs with the 'calculated discharge capacity'.

Example:

- Let's assume that there are, in a US-style SPD (in acc. to UL 1449), eight varistors connected in parallel – per mode of protection. Each of the varistors is capable to discharge 40 kA (8/20 μs). The calculated – but not the tested – discharge capacity is therefore 320 kA (8/20 μs). The manufacturer states this calculated discharge capacity at the SPD and in the product literature, without the need to test the SPD with the calculated discharge capacity.

During approval testing (in accordance to UL 1449), SPDs are tested by UL (Underwriter Laboratories) with surge currents of up to 20 kA (8/20 μs). If a manufacturer states a 'calculated discharge capacity' which is higher than 20 kA (8/20 μs), UL does *not* test the respective SPDs with 8/20 μs surge currents exceeding 20 kA.

8.3.8.3 Parallel connection of a spark gap and a varistor

Sometimes it's favourable to connect different kinds of surge-protective components directly in parallel. For this purpose, some manufacturers of SPDs offer triggered spark gaps and varistors, which are specially designed to be connected in parallel and which work in a coordinated way. With this combination of different surge-protective components, it's possible to achieve an optimized protection effect.

Triggered Type 1 spark gaps are the best choice to divert high-energy long-duration lightning currents, and Type 2 varistors are the best choice to divert man-made short-duration switching overvoltages. A triggered Type 1 spark gap and a Type 2 varistor, connected in parallel right at the service entrance, are a very good solution for an efficient diversion and mitigation of short-duration and of long-duration surges and for the mitigation of man-made switching overvoltages (see Figure 8.54).

8.3.9 Connection types of SPDs

All over the world different power distribution systems, with different conductor arrangements and earthing systems, are in use. The most commonly used power systems in countries with IEC-style power systems are solidly grounded three-phase five-wire TN-S or TN-C-S systems (L1, L2, L3, N and PE) (see Figure 8.55). More general information about IEC-style power systems can be found in IEC 60364-1.

For five-wire systems, there are many possible *modes of protection* which can be equipped with surge-protective components. Especially in countries with US-style power systems, it's common practice to use as many modes of protection as possible – e.g. up to seven or even up to ten modes of protection for a three-phase five-wire system.

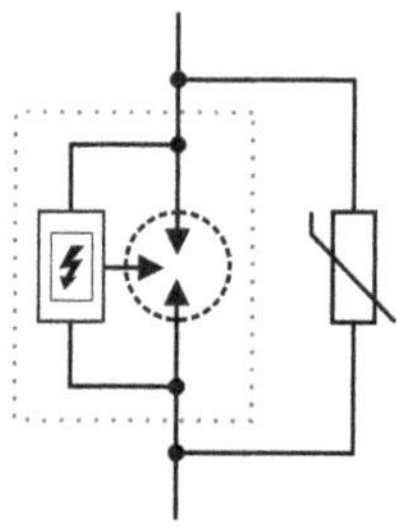

Figure 8.54 Parallel connection of an encapsulated and triggered Type 1 spark gap and a Type 2 varistor

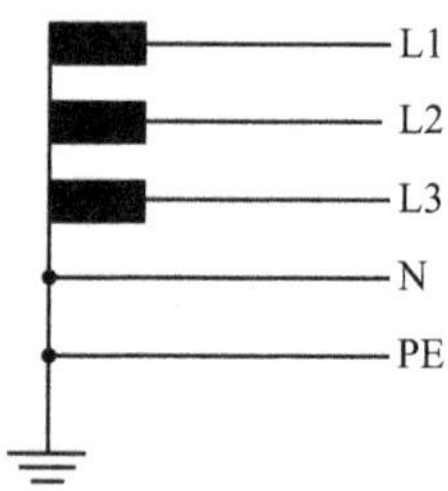

Figure 8.55 Solidly grounded three-phase five-wire TN-S system

Decade-long real-world experience with IEC-style power systems, installed in lighting-prone areas, has proved that three-phase five-wire power systems can be protected efficiently with just four modes of protection – even if the respective facility is equipped with an external lightning protection system which gets hit by a direct lightning strike. Therefore, in IEC 60364-5-53, the relevant IEC standard for the selection and installation of SPDs, two preferable connection types of SPDs are defined.

Connection Type 1 (CT 1) (see Figure 8.56): An SPD that has one mode of protection between each active conductor (phase conductor and neutral conductor, if applicable) and the PE conductor. This connection type is often designated as a '$x + 0$' connection type, where x represents the number of active conductors.

In the past, Connection Type 1 (CT 1) has been used frequently for the protection of power systems with separate neutral and PE conductor (see Figure 8.56).

Connection Type 2 (CT 2) (see Figure 8.57): An SPD that has one mode of protection between each phase conductor and the neutral conductor and an additional mode of protection between the neutral conductor and the PE conductor. This connection type is often designated as a '$x + 1$' connection type, where x represents the number of phase conductors.

Nowadays Connection Type 2 (CT 2) is frequently used for the protection of power systems with separate neutral and PE conductor – especially if sensitive single-phase devices are installed between L and N.

When choosing a connection type for SPDs, the dielectric strength of equipment has to be considered. For most single-phase devices, the dielectric strength between the individual conductors (L, N and PE) is usually as follows:

- Between L and N: low dielectric strength
- Between L and PE: high dielectric strength
- Between N and PE: high dielectric strength.

For an efficient protection of sensitive single-phase devices, a connection type should be selected which allows to achieve a low voltage protection level U_p between L and N conductor.

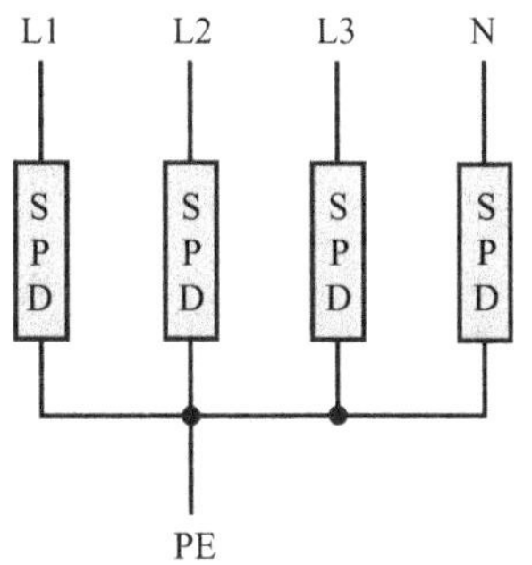

Figure 8.56 SPD with Connection Type 1 (CT 1) for the protection of three-phase five-wire systems (4 + 0 connection)

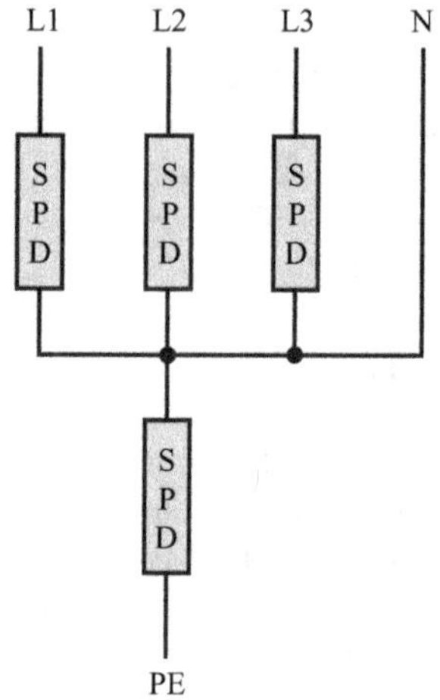

Figure 8.57 SPD with Connection Type 2 (CT 2) for the protection of three-phase five-wire systems (3 + 1 connection)

With Connection Type 1 (CT 1, $x + 0$ connection), there is a relatively high protection level between L and N conductor, because with Connection Type 1 there are always two surge-protective components (connected in series) between L and N. Therefore, Connection Type 1 and the 4 + 0 and 2 + 0 connection types are not the most suitable connection types for the protection of sensitive single-phase devices.

With Connection Type 2 (CT 2, $x + 1$ connection), there is a low protection level between L and N conductor, because there is only one surge-protective component between L and N. Therefore, the Connection Type 2 and the 3 + 1 and 1 + 1 connection types are nowadays used frequently for the protection of three-phase and single-phase power systems with separate N and PE conductor.

During normal operation, the voltage difference between N and PE conductor is usually negligibly low. Therefore, it's permissible to install, between the N and the PE conductor, a surge-protective component with a relatively low follow current interrupt rating I_{fi}. This kind of component, installed between N and PE is often called 'N/PE spark gap'.

If the Connection Type 2 (CT 2, $x + 1$ connection) is used, then the mode of protection between the N and the PE conductor has to be able to carry the whole impulse current the complete SPD is rated for (summation current, total discharge current I_{total}). Therefore, the discharge capacity of an N/PE spark gaps is usually higher than the discharge capacity of surge-protective components installed between L and N conductors. Modern N/PE spark gaps, for the installation at the origin of the installation or close to the origin of the installation, are suitable to discharge high-energy long-duration lightning currents – e.g. 100 kA of the wave shape 10/350 μs.

Table 8.9 Maximum period between inspections of a lightning protection system (see IEC 62305-3:2010 Annex E.7.1)

Lightning protection level LPL	Visual inspection (years)	Complete inspection (years)	Critical systems complete inspection (years)
I and II	1	2	1
III and IV	2	4	1

8.3.10 Inspection and field-testing of SPDs

To achieve high system availability, system operators must regularly inspect and maintain lightning protection systems. This is stipulated by legislators, supervisory authorities or professional associations. According to the lightning protection standard IEC 62305-3 (Annex E.7), regular testing and maintenance of lightning protection systems (external and internal lightning protection) is required (see Table 8.9). The maximum period of time between inspections depends on the lightning protection level LPL and the specific kind of installation.

The international standard IEC 62304-3 also demands that maintenance of lightning protection systems and SPDs is properly documented. The following points are particularly important to note:

- 'Critical systems complete inspection' relates to structures containing sensitive internal systems, office blocks, commercial buildings or places where a high number of people may be present. The electrical test of the installations should be carried out at least once a year.
- Lightning protection systems, in applications with a risk caused by explosive material, should undergo a visual check at least every 6 months.
- For systems with strict requirements in terms of required safety (e.g. petrochemical facilities and nuclear power plants), national authorities or institutions can prescribe a comprehensive check. This can be necessary if there has been a lightning strike within a certain radius of the respective system.
- Specialist knowledge is required in order to carry out professional inspection and testing of lightning protection systems. For this reason, this inspection and testing must be carried out by a lightning protection specialist. Inspecting the SPDs is also part of this.

During a visual inspection, it's not possible to assess the status of an SPD properly. A proper assessment is only possible by carrying out electrical testing in accordance with the manufacturer's guidelines and equipment provided by the manufacturer. For the comprehensive testing of SPDs, a suitable high-voltage test device is needed (see Figure 8.58). When an electrical test is carried out on SPDs, the test voltage is selected such that the SPD becomes partially conductive. The measurement results are then compared to reference values and they are evaluated. To be

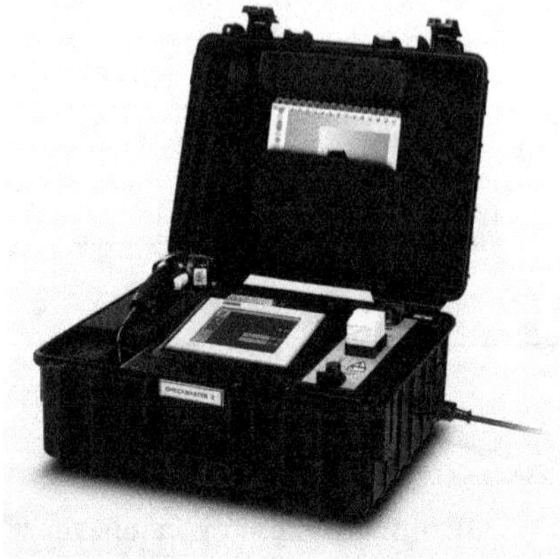

Figure 8.58 Fully automatic high-voltage test device for the touch-safe electrical testing of pluggable SPDs

able to test SPDs comfortably and in a safe way, it's recommended to use pluggable SPDs.

8.3.11 Test generators and test facilities for type testing

For the type testing of low-voltage SPDs, in accordance to IEC 61643-11, special high-voltage/high-current test equipment and special test facilities are required. SPDs with low discharge capacity can get tested with the help of desktop surge generators. For the type testing of Type 1 SPDs, suitable for the discharge of high-energy long-duration lightning currents, powerful lightning current generators are required. Typical Type 1 SPDs, for the protection of three-phase five-wire power systems, are tested with lightning impulse discharge currents I_{imp} of, e.g., up to 100 kA (10/350 μs); (see Figure 8.59).

Sometimes SPDs get installed at installation locations with prospective short-circuit currents of, e.g., 50 kA (RMS). An SPD – which is rated for the use at installation locations with high prospective short-circuit currents of, e.g., 50 kA (RMS) – has to be tested with multiple surge-current impulses, when it's hooked-up to a powerful transformer which is capable to produce such high-energy short-circuit currents (see Figure 8.60).

8.3.12 Approvals from certified bodies

The comprehensive type testing of powerful low-voltage SPDs, in accordance to IEC 61643-11, is time-consuming and complicate. Special purpose-built high-voltage and high-current test equipment is needed for testing. The test parameters and the test requirements for the full-scope testing of powerful SPDs go far beyond the test requirements for ordinary electrical equipment. Therefore, the type testing of powerful low-voltage SPDs is complicate and expensive. Only a few accredited

Figure 8.59 Lightning current generator for the testing of low-voltage SPDs with surge currents up to 100 kA (10/350 μs) or up to 200 kA (8/20 μs)

Figure 8.60 Adjustable custom-built three-phase transformer for short-circuit current testing (8.4 MVA, 50 kA RMS, 100...1,000 V AC)

lightning current test laboratories in the world are capable to carry out full-scope testing of powerful low-voltage SPDs.

To be able to design, test and manufacture high-quality SPDs, which really meet the requirements of all relevant IEC standards, a manufacturer of SPDs needs to have access to a sophisticated surge and lightning current test laboratory.

Approvals from independent and certified laboratories are an important indicator for the performance of an SPD. Therefore, vendors of high-quality SPDs – for the protection of low-voltage power systems – get their SPDs tested and approved by independent and accredited test laboratories (e.g. by ÖVE, KEMA, DEKRA and UL), and they make the respective test certificates publicly available.

References

[1] CIGRE WG C4.408, 'Lightning Protection of Low-Voltage Networks', CIGRE Technical Brochure, no. 550, Aug. 2013.

[2] CIGRE-CIRED JWG C4.4.02, 'Protection of MV and LV Networks against Lightning. Part 1: Common Topics', CIGRE Technical Brochure, vol. 287, p. 53, 2006.

[3] CIGRE-CIRED Joint WG C4.4.02, 'Protection of Medium Voltage and Low Voltage Networks against Lightning. Part 2: Lightning Protection of Medium Voltage Networks', CIGRE Technical Brochure, vol. 441, p. 39, 2010.

[4] EN 50526-1: Railway applications – Fixed installations – D.C. surge arresters and voltage limiting devices – Part 1: Surge arresters, 2012.

[5] IEC 60099-4, 'Surge arresters – Part 4: Metal-oxide surge arresters without gaps for a.c. systems', 2009.

[6] A. Rousseau and R. Gumley, 'Surge Protection', Webster Encyclopedia of Electrical and Electronics Engineering, vol. 21, pp. 153–165, 1999.

[7] IEC/TR 62066, 'Surge overvoltages and surge protection in low-voltage a.c. power systems – General basic information', 2002.

[8] Application Guidelines, 'Overvoltage protection: Metal oxide surge arresters in medium voltage systems', 5th revised edition May 2011. ABB Switzerland Ltd, Division surge arresters, Wettingen, Switzerland.

[9] M. V. Dmitriev, 'Use of Surge Protection Devices (SPDs) for the Protection of 6–750 kV OHL Insulation'. St. Petersburg: Publishing House of the Polytechnic University, 2009. 92 pp., (in Russian).

[10] V. Hinrichsen, 'Overview of IEC Standards' recommendations for lightning protection of electrical high-voltage power systems using surge arresters', International Conference on Lightning Protection (ICLP), 11–18 October 2014.

[11] IEC 60099-8, Edition 1.0, 2011-01: Surge arresters-Part 8: Metal-oxide surge arresters with external series gap (EGLA) for overhead transmission and distribution lines of a.c. systems above I kV.

[12] A. Gayvoronsky, 'Line Arresters – Radical Mean for Lightning Protection of Overhead Power Lines', News of Electrical Engineering, vol. 2, no. 38, pp. 113–116, 2006 (in Russian).

[13] V. Hinrichsen, 'Metal-oxide surge arresters in high-voltage power systems', www.siemens.com/energy/arrester.

[14] G. V. Podporkin, E. Yu. Enkin, E. S. Kalakutsky, V. E. Pilshikov and A. D. Sivaev, 'Overhead Lines Lightning Protection by Multi-Chamber Arresters and Insulator-Arresters', IEEE Transactions on Power Delivery, vol. 26, no. 1, pp. 214–221, 2011.

[15] A. R. Koryavin, O. V. Volkova and E. A. Milkin, 'Discharge Charcteristics of 220 kV Insulators-Arresters Strings', Elekrichestvo, vol. 8, pp. 32–36, 2014 (in Russian).

[16] G. V. Podporkin, E.Y. Enkin, V. E. Pilshikov and V.V Zhitenev, 'Lightning Protection of Overhead Distribution and Transmission Lines by Multi-

Chamber Arresters and Insulators-Arresters of a Novel Design', International Colloquium on Lightning and Power systems, Lyon, France, CIGRE SC C4 2014.
[17] G. V. Podporkin, E. Y. Enkin, E. S. Kalakutsky, V. E. Pilshikov and A. D. Sivaev, 'Development of Multi-Chamber Insulator-Arresters for Lightning Protection of 220 kV Overhead Transmission Lines', XI International Symposium on Lightning Protection (SIPDA), Fortoleza, Brazil, Rep. 7-2, 3–7 October 2011.
[18] G. V. Podporkin, E. Yu. Enkin, V. Ye. Pilschikov and D. O. Belko, 'Development of Ring-type Metal Based Lightning Multi-Chamber Arrester for 35 kV Overhead Line', The 10th Asia-Pacific International Conference on Lightning, Krabi, Thailand, May 16–19, 2017.

Chapter 9
Modelling of power transmission line components

Alberto De Conti[1] *and Fernando H. Silveira*[1]

The evaluation of the lightning performance of high-voltage transmission lines is generally focused on direct lightning strikes to the line, more specifically to towers, shield wires, and phase conductors. Such events can lead to severe overvoltages across insulator strings and, depending on the balance between the resulting overvoltage and the capability of line insulators to withstand it, flashovers can occur, in most cases leading to line outages. Since the critical flashover overvoltage (CFO) of high-voltage transmission lines is typically larger than 600 kV, voltages induced by nearby lightning are not prone to cause line outages and are typically disregarded in this kind of study.

Overvoltages across line insulators can lead to flashovers (associated with lightning incidence on phase conductors of unshielded transmission lines or to shielding failure of shielded transmission lines) or backflashovers (due to lightning incidence on shield wires or towers). Both phenomena can result in a line outage. In this context, the calculation of lightning overvoltages across line insulators is the key point to assessing the lightning performance of a transmission line. However, this calculation depends on several aspects, such as the lightning return-stroke current waveform, the characteristics of transmission line elements (e.g. tower geometry, number and location of shield wires and phase conductors, and type and number of insulator strings), the environment in which the line is installed (soil resistivity, relief, etc.), and the effects resulting from the interaction of the lightning current with the system elements (soil ionization, corona, non-linear response of surge arresters, etc.). The quality of the estimation will depend on the quality of the models assumed to represent each of the aforementioned aspects.

In addition, the assessment of the lightning performance of transmission lines is strongly affected by the considered calculation method, which is generally classified as follows: (i) analytical, (ii) circuit based, or (iii) electromagnetic-field

[1]Lightning Research Center, Department of Electrical Engineering, Federal University of Minas Gerais (UFMG), Belo Horizonte, Brazil

based. Each of these methods has specificities to represent transmission line elements that may affect the calculated lightning overvoltage.

Among the three cited methods, the analytical approach is the one that allows calculating lightning overvoltages in a simple and direct way without requiring advanced processing. The overvoltage calculation is based on a simplified representation of the system components (line, tower, and grounding) [1–3]. Such approach is adopted in procedures proposed by the IEEE and CIGRÉ to estimate the lightning performance of transmission lines and consists in using analytical formulations to describe the transient behaviour of the line when subjected to a direct lightning strike. In spite of its widespread use, some model approximations assumed in the analytical approach to speed up the calculation process are questionable and, in some cases, can lead to inaccurate results [4].

The increasing availability of electromagnetic transient programs in which improved models are available for describing both frequency-dependent and non-linear phenomena in time-domain simulations has allowed a more accurate calculation of lightning overvoltages in transmission lines than the analytical approach discussed above (e.g. [5,6]). Since these tools rely on lumped- and distributed-circuit theory, they are generically referred to as circuit-based approach.

Also, there is a trend to the application of models that follow an electromagnetic field approach for calculating lightning overvoltages across insulator strings (e.g. [7–11]). In spite of requiring longer processing time, the application of such models leads to results that are more accurate and general than those obtained using analytical and circuit-based approaches. The physical system is represented directly from the geometry of the problem and from the constitutive parameters of the media. The solution automatically takes into account the electromagnetic coupling between all system elements as well as propagation effects. An important application of electromagnetic models consists in its use to validate models based on analytical and circuit-based approaches.

The aim of this chapter is to present and discuss models of power transmission line components for the calculation of lightning overvoltages on transmission lines due to direct lightning strikes. The models presented in this chapter are based on lumped- and distributed-circuit theory. However, whenever necessary, comments are made to stress the limitations of the discussed models by taking as reference results obtained with the application of the electromagnetic field approach.

The structure of this chapter is as follows. First, some definitions related to transmission line theory are presented as a benchmark for the following sections, since most power transmission line components can be represented using such approach. The modelling of the basic elements of high-voltage transmission lines, such as phase conductors, shield wires, towers, grounding systems, insulator strings, and surge arresters, is described. For each model, parameters of interest and details of model applications related to the calculation of the lightning performance of transmission lines are presented.

9.1 Transmission lines

For an accurate assessment of the lightning performance of transmission lines, it is important to understand the propagation of voltage and current waves along shield wires and phase conductors. To discuss this subject, this section presents a brief overview of transmission line theory.

9.1.1 Transmission line equations

Voltages and currents associated with an overhead transmission line with n conductors oriented along the x-axis and parallel to ground are related in frequency domain by Telegrapher's equations:

$$\begin{aligned} \frac{d\boldsymbol{V}(x)}{dx} &= -\boldsymbol{ZI}(x) = -(\boldsymbol{Z}_i + \boldsymbol{Z}_e + \boldsymbol{Z}_g)\boldsymbol{I}(x) \\ \frac{d\boldsymbol{I}(x)}{dx} &= -\boldsymbol{YV}(x) = -[\boldsymbol{Y}_e^{-1} + \boldsymbol{Y}_g^{-1}]^{-1}\boldsymbol{V}(x) \end{aligned} \tag{9.1}$$

where $\boldsymbol{V}(x)$ and $\boldsymbol{I}(x)$ are $n \times 1$ vectors containing conductor voltages and currents, $\boldsymbol{Z}$ and $\boldsymbol{Y}$ are $n \times n$ matrices containing the per-unit-length impedance and admittance of the line, $\boldsymbol{Z}_i$ contains the internal impedance of the conductors, $\boldsymbol{Z}_e$ is the external impedance, $\boldsymbol{Z}_g$ is the ground-return impedance, $\boldsymbol{Y}_e$ is the external admittance, and $\boldsymbol{Y}_g$ is the admittance correction due to a finitely conducting ground.

The set of equations in (9.1) are strictly valid as long as (i) all conductors are infinite and parallel to the ground surface, (ii) the cross sections of the conductors are uniform along the line axis, (iii) the maximum separation between each pair of conductors, and between each conductor and the ground, is electrically short, (iv) the axial electric field on the surface of each conductor is negligible compared to the electric field components in the y- and z-axes, and (v) the sum of the conductor currents, including the ground-return current, is zero. These assumptions assure the existence of a transverse electromagnetic (TEM) field structure [12]. As a consequence, voltage can be uniquely defined as the integral of the electric field in the y–z plane transversal to the line axis regardless of the integration path, and, in addition, the per-unit-length capacitance and inductance matrices can be determined based on the laws of statics. Conditions (i)–(v) are violated, for example, at points near the line ends, in the analysis of the propagation of very fast voltage and current waveforms, or in the study of non-uniform transmission lines. Nevertheless, in most cases of practical interest, it is possible to accurately calculate $\boldsymbol{V}(x)$ and $\boldsymbol{I}(x)$ from (9.1) in frequencies up to some MHz [13], which is sufficient for studying lightning overvoltages.

9.1.2 Calculation of per-unit-length parameters

9.1.2.1 General formulation

For calculating $\boldsymbol{V}(x)$ and $\boldsymbol{I}(x)$ using (9.1), it is first necessary to determine the per-unit-length parameters of the line. Rigorous expressions for calculating the internal parameters of solid, tubular, and stranded conductors are based on Bessel's functions [14,15], while the calculation of the ground-return impedance and admittance depends on the solution of infinite integrals with complex integrands [16,17]. In practice, the elements of $\boldsymbol{Z}$ and $\boldsymbol{Y}$ can be determined using the approximate expressions listed in Table 9.1.

In Table 9.1, R_{ii}, μ_c, σ_c, and r_i are the dc resistance (Ω/m), permeability (H/m), conductivity (S/m), and external radius of the *i*th conductor (m), respectively; h_i and h_j are the average heights of the *i*th and *j*th conductors above the ground (m), obtained by subtracting 2/3 of the line sag from the height at the tower [14]; d_{ij} is the horizontal separation between conductors *i* and *j* (m); μ_0 and ε_0 are the permeability (H/m) and permittivity (F/m) of the vacuum, respectively; γ_g and γ_0 are the intrinsic propagation constants of ground and air (m^{-1}); σ_g and ε_{rg} are the conductivity (S/m) and relative permittivity of the ground, respectively; ω is the angular frequency; and $j = \sqrt{-1}$. The expressions for $\boldsymbol{Z_g}$ in Table 9.1 approximate the infinite integrals proposed by Nakagawa [18] for the particular case of homogeneous ground. As discussed in [17], if $(\varepsilon_{rg} - 1)$ is replaced by ε_{rg} in $\dot{p}$, then $\boldsymbol{Z_g}$ approximates Sunde's integral expressions [19,20]. Finally, if $\sigma_g \gg \omega(\varepsilon_{rg} - 1)\varepsilon_0$ in $\dot{p}$, that is, if the ground is a good conductor, or $\varepsilon_{rg} = 1$, then $\boldsymbol{Z_g}$ approximates Carson's integrals [21,22], which are used in popular electromagnetic transient simulators (e.g. [14,23]).

The logarithmic approximation for $\boldsymbol{Y}$ in Table 9.1 reproduces with good accuracy the correction term proposed by Wise [24] to account for the effect of an imperfectly conducting ground in the calculation of $\boldsymbol{Y}$ [16]. However, for lightning studies, it is usually possible to neglect $\boldsymbol{Y_g}$ without significant errors [25]. In this case, $\boldsymbol{Y} \approx j\omega 2\pi\varepsilon_0 \boldsymbol{M}^{-1}$, which is usually assumed in electromagnetic transient simulators (e.g. [14,23]). Although both σ_g and ε_{rg} vary with frequency [26–29], they can be assumed constant for the calculation of $\boldsymbol{Z_g}$ and $\boldsymbol{Y_g}$ for values of ground resistivity below 1,000 Ω m [25].

9.1.2.2 Reducing conductor bundles to a single equivalent conductor

In many cases, each phase of a transmission system consists of a bundle of conductors. To reduce the complexity of the problem, it is often convenient to represent this bundle as a single equivalent conductor. There are different techniques that can be used for that purpose. One possibility consists in assuming that the charge is equally divided between all subconductors in the bundle, and that they all share the same voltage. In this case, the bundle can be replaced by a single equivalent conductor located at the centre of the original bundle [1]. Another possibility, which is more general, considers that the equivalent conductor will carry the sum of the currents in the subconductors. By assuming equal subconductor voltages, $\boldsymbol{Z}$ and $\boldsymbol{Y}$ can be manipulated so that a reduced-order system is obtained [14]. In theory, the same approach can be used to combine shield wires in a single conductor for the study of a lightning strike to a tower. This is made by assuming that each conductor will carry the same current. However, although useful for hand calculations, this procedure should be avoided in a more rigorous analysis of lightning strikes to towers.

Table 9.1 Expressions for calculating the per-unit-length parameters of a transmission line

	Diagonal terms	Off-diagonal terms
$\boldsymbol{Z_i}$ [15]	$Z_{i,ii} = \sqrt{R_{ii}^2 + \frac{j\omega\mu_c}{4\sigma_c\pi^2 r_i^2}}$	$Z_{i,ij} = 0$
$\boldsymbol{Z_e}$ [14]	$Z_{e,ii} = \frac{j\omega\mu_0}{2\pi} M_{ii}$ $M_{ii} = ln\, \frac{2h_i}{r_i}$	$Z_{e,ij} = \frac{j\omega\mu_0}{2\pi} M_{ij}$ $M_{ij} = \ln \frac{\sqrt{(h_i + h_j)^2 + d_{ij}{}^2}}{\sqrt{(h_i - h_j)^2 + d_{ij}{}^2}}$
$\boldsymbol{Z_g}$ [17]	$Z_{g,ii} = \frac{j\omega\mu_0}{2\pi} \ln \frac{h_i + \dot{p}}{h_i}$	$Z_{g,ij} = \frac{j\omega\mu_0}{2\pi} \ln \frac{\sqrt{(h_i + h_j + 2\dot{p})^2 + d_{ij}{}^2}}{\sqrt{(h_i + h_j)^2 + d_{ij}{}^2}}$ $\dot{p} = \left\{j\omega\mu_0\left[\sigma_g + j\omega(\varepsilon_{rg} - 1)\varepsilon_0\right]\right\}^{-1/2}$
$\boldsymbol{Y} = j\omega 2\pi\varepsilon_0(\boldsymbol{M} + \boldsymbol{Q})^{-1}$ [16]	$Q_{ii} = Q_{ij}\big\vert_{h_i = h_j, d_{ij} = r_i}$	$Q_{ij} = \frac{2}{\upsilon^2 + 1} \ln\left(1 + \frac{\upsilon^2 + 1}{\eta\sqrt{(h_i + h_j)^2 + d_{ij}{}^2}}\right)$ $\upsilon = \frac{\gamma_g}{\gamma_0} = \frac{\sqrt{j\omega\mu_0(\sigma_g + j\omega\varepsilon_{rg}\varepsilon_0)}}{j\omega\sqrt{\mu_0\varepsilon_0}}$ $\eta = \sqrt{\gamma_g^2 - \gamma_0^2}$

9.1.3 Frequency-domain solution of the transmission line equations

The frequency-domain solution of (9.1) can be written as [15]:

$$\begin{aligned} \boldsymbol{V}(x) &= e^{-\sqrt{\boldsymbol{ZY}}x}\boldsymbol{V}_F + e^{\sqrt{\boldsymbol{ZY}}x}\boldsymbol{V}_B \\ \boldsymbol{I}(x) &= \boldsymbol{Y}_C\left(e^{-\sqrt{\boldsymbol{ZY}}x}\boldsymbol{V}_F - e^{\sqrt{\boldsymbol{ZY}}x}\boldsymbol{V}_B\right) \end{aligned} \tag{9.2}$$

where $\boldsymbol{V}_F$ and $\boldsymbol{V}_B$ are forward- and backward-propagating voltages, whose values will depend on the boundary conditions at the line ends, and

$$\boldsymbol{Y}_C = \boldsymbol{Z}^{-1}\sqrt{\boldsymbol{ZY}} \tag{9.3}$$

is the characteristic admittance, in S.

The matrix exponentials in (9.2) are responsible for attenuating $\boldsymbol{V}_F$ and $\boldsymbol{V}_B$ and modifying their phase angles along the x-axis. For a transmission line extending from $x = 0$ to $x = \ell$, $\boldsymbol{V}_F$ and $\boldsymbol{V}_B$ can be eliminated from (9.2), resulting in

$$\begin{bmatrix} \boldsymbol{V}(\ell) \\ \boldsymbol{I}(\ell) \end{bmatrix} = \begin{bmatrix} \cosh\left(\sqrt{\boldsymbol{ZY}}\ell\right) & -\mathrm{senh}\left(\sqrt{\boldsymbol{ZY}}\ell\right)\boldsymbol{Z}_C \\ -\boldsymbol{Y}_C\,\mathrm{senh}\left(\sqrt{\boldsymbol{ZY}}\ell\right) & \boldsymbol{Y}_C\cosh\left(\sqrt{\boldsymbol{ZY}}\ell\right)\boldsymbol{Z}_C \end{bmatrix} \begin{bmatrix} \boldsymbol{V}(0) \\ \boldsymbol{I}(0) \end{bmatrix} \tag{9.4}$$

which relates voltages and currents at the line ends. In this expression, $\boldsymbol{Z}_C = \boldsymbol{Y}_C^{-1}$ is the characteristic impedance of the line. From the load and source characteristics at $x = 0$ and $x = \ell$, it is possible to determine voltages and currents at these points. For example, if an ideal voltage source $\boldsymbol{V}_S$ is connected to the line at $x = 0$ and a resistive load $\boldsymbol{R}_L$ is connected at $x = \ell$, one can write $\boldsymbol{V}(\ell) = \boldsymbol{R}_L\boldsymbol{I}(\ell)$ and $\boldsymbol{V}(0) = \boldsymbol{V}_S$, and then solve (9.4) for $\boldsymbol{I}(0)$ and $\boldsymbol{I}(\ell)$.

9.1.4 Time-domain solution of the transmission line equations

Equation (9.2) can be written in time domain as

$$\begin{aligned} \boldsymbol{v}(x,t) &= \boldsymbol{a}_F(x,t) * \boldsymbol{v}_F(0,t) + \boldsymbol{a}_B(x,t) * \boldsymbol{v}_B(0,t) \\ \boldsymbol{i}(x,t) &= \boldsymbol{y}_C(t) * \left[\boldsymbol{a}_F(x,t) * \boldsymbol{v}_F(0,t) - \boldsymbol{a}_B(x,t) * \boldsymbol{v}_B(0,t)\right] \end{aligned} \tag{9.5}$$

where $\boldsymbol{v}(x,t)$, $\boldsymbol{i}(x,t)$, $\boldsymbol{v}_F(0,t)$, $\boldsymbol{v}_B(0,t)$, and $\boldsymbol{y}_C(t)$ are the time-domain equivalents of their counterparts in (9.2); $\boldsymbol{a}_F(x,t)$ and $\boldsymbol{a}_B(x,t)$ are the time-domain equivalents of $e^{-\sqrt{\boldsymbol{ZY}}x}$ and $e^{\sqrt{\boldsymbol{ZY}}x}$, respectively; and the symbol * corresponds to the convolution integral. An operation of the type $\boldsymbol{a}_F(x,t) * \boldsymbol{v}_F(0,t)$ results in the attenuation, distortion, and time delay of the voltage wave represented by $\boldsymbol{v}_F(0,t)$.

Numerical models based on the method of characteristics [12] can be used for solving (9.5) directly in time domain. Some of these models are available in popular electromagnetic transient simulators and can be readily used for studying direct lightning strikes on transmission lines [30,31]. However, it is often possible to neglect losses in this type of study, especially if qualitative analyses or hand calculations are to be performed. If this is the case, then $\boldsymbol{Z}_i = \boldsymbol{Z}_g = 0$ and $\boldsymbol{Y}_g \to \infty$

in (9.1). As a result, $\boldsymbol{Z}_C$ and $\boldsymbol{Y}_C$ reduce to the real matrices $\boldsymbol{Z}_0 = 60\boldsymbol{M}$ and $\boldsymbol{Y}_0 = (60\boldsymbol{M})^{-1}$, while $\boldsymbol{a}_F(x,t)$ and $\boldsymbol{a}_B(x,t)$ become diagonal matrices containing elements $\delta(t - x/v)$ and $\delta(t + x/v)$, respectively. In these expressions, δ is the Dirac delta function, v is the propagation speed, and $\boldsymbol{M}$ is given in Table 9.1. For a lossless overhead transmission line with bare cables, v is equal to the speed of light, c, and (9.5) can be rewritten as

$$\begin{aligned} \boldsymbol{v}(x,t) &= \boldsymbol{v}_F(0, t - x/c) + \boldsymbol{v}_B(0, t + x/c) \\ \boldsymbol{i}(x,t) &= \boldsymbol{Y}_0[\boldsymbol{v}_F(0, t - x/c) - \boldsymbol{v}_B(0, t + x/c)] \end{aligned} \tag{9.6}$$

Equation (9.6) shows that the solution of Telegrapher's equations (9.1) for a lossless multiconductor transmission line results in voltage and current waves that travel in the $+x$ and $-x$ directions without suffering attenuation or distortion. The existence of $\boldsymbol{v}_F$ or $\boldsymbol{v}_B$ will depend on the presence of sources, loads, and discontinuities along the line. For example, assuming a multiconductor transmission line extending from $x = 0$ to $x \to \infty$ excited at $x = 0$ by a voltage source $\boldsymbol{v}_S(t)$ with internal resistance $\boldsymbol{R}_S$, the forward propagating voltage will correspond to $\boldsymbol{v}_F(0,t) = (\boldsymbol{1} + \boldsymbol{R}_S\boldsymbol{Y}_0)^{-1}\boldsymbol{v}_S(t)$, where $\boldsymbol{1}$ is the identity matrix, while $\boldsymbol{v}_B(0,t) = 0$. However, if the transmission line is finite and terminated at $x = \ell$ with a resistive load $\boldsymbol{R}_L$, then part of the energy associated with the forward-moving waves will be transferred to the load, while part of it will be reflected back to the source. When the reflected waves reach the source, part of the associated energy will be dissipated in $\boldsymbol{R}_S$ and another part will be reflected back to the load and so forth. As a consequence of the successive reflections taking place at the line ends, the resulting voltages at $x = 0$ and $x = \ell$ will consist of the sum of delayed and attenuated versions of $\boldsymbol{v}_F(0,t)$, corresponding to $\boldsymbol{v}(0,t) = \boldsymbol{v}_F(0,t) + (\boldsymbol{1} + \boldsymbol{\rho}_S)[\boldsymbol{\rho}_L\boldsymbol{v}_F(0, t - 2\tau) + \boldsymbol{\rho}_L\boldsymbol{\rho}_S\boldsymbol{\rho}_L\boldsymbol{v}_F(0, t - 4\tau) + \cdots]$ and $\boldsymbol{v}(\ell,t) = (\boldsymbol{1} + \boldsymbol{\rho}_L)[\boldsymbol{v}_F(0, t - \tau) + \boldsymbol{\rho}_S\boldsymbol{\rho}_L\boldsymbol{v}_F(0, t - 3\tau) + \boldsymbol{\rho}_S\boldsymbol{\rho}_L\boldsymbol{\rho}_S\boldsymbol{\rho}_L\boldsymbol{v}_F(0, t - 5\tau) + \cdots]$. In these expressions, $\boldsymbol{\rho}_L = (\boldsymbol{R}_L - \boldsymbol{Z}_0)(\boldsymbol{R}_L + \boldsymbol{Z}_0)^{-1}$ and $\boldsymbol{\rho}_S = (\boldsymbol{R}_S - \boldsymbol{Z}_0)(\boldsymbol{R}_S + \boldsymbol{Z}_0)^{-1}$ are the voltage reflection coefficients at the line ends [12], written in matrix form.

As another example, a lightning strike intercepting the single shield wire of a lossless three-phase transmission line at the midspan will create two identical voltage waves, one travelling in the $+x$ direction and another one travelling in the $-x$ direction. Assuming that the lightning channel can be represented as a lumped current source $i_S(t)$ with internal impedance Z_S, neglecting corona, and assuming that the midspan is located at $x = 0$, the forward-propagating wave can be calculated as $\boldsymbol{v}_F(0,t) = (2\boldsymbol{Y}_0 + \boldsymbol{Y}_S)^{-1}\boldsymbol{i}_S(t)$, where $\boldsymbol{Y}_0$ is a 4×4 matrix representing the characteristic admittance of the line, $\boldsymbol{Y}_S$ is a matrix of zeros except at element (1, 1), which refers to the shield wire and corresponds to Z_S^{-1}, and $\boldsymbol{i}_S(t)$ is a 4×1 vector of zeros, except at element (1, 1), which corresponds to $i_S(t)$. If $\boldsymbol{Y}_S = 0$, then $\boldsymbol{v}_F(0,t) = 0.5\boldsymbol{Z}_0\boldsymbol{i}_S(t)$. The resulting voltage at $x = 0$ will then be affected by reflected voltage waves coming from the adjacent towers, as well as by the voltage at power frequency, which can be included as a constant term [2]. The difference is that now the voltage calculation will depend on the characterization of additional elements, such as towers, tower-footing impedance, surge arresters (if any), and insulators. The next sections are devoted to discussing the modelling of these elements.

9.2 Transmission towers

9.2.1 Overview

Towers are an important element of transmission lines for providing mechanical support to overhead cables, insulators, and surge arresters. For the study of the lightning performance of transmission lines, they can be viewed as a waveguide that provides a path to ground for impinging lightning currents.

One possibility to describe transient phenomena on a transmission tower is to use transmission line theory, for which two parameters are necessary: the characteristic impedance of the tower, also called tower surge impedance, and the propagation speed. Different approaches have been used for determining both parameters. They can be separated in three categories as follows: (i) numerical electromagnetic field analysis [7,8,32–34], (ii) experiments with real- and reduced-scale towers [35–41], and (iii) the analytical solution of Maxwell's equations considering equivalent tower geometries [42–50]. Approach (i) is the most rigorous and general among the listed methods. Approach (ii) is valuable for providing physical interpretation and data for the validation of theoretical models, but generalization of results to different tower geometries is difficult. Approach (iii) is the less rigorous among the listed methods, because it often overlooks certain aspects of the electromagnetic field distribution around the tower. In the next subsections, a discussion is presented on the modelling of transmission towers using transmission line theory.

9.2.2 Travelling wave analysis of a lightning strike to a tower

If the transmission tower illustrated in Figure 9.1(a) is represented as a lossless uniform transmission line with surge impedance Z_T, propagation speed υ, and length h, where h corresponds to the tower height, the voltage $v_{top}(t)$ at point T due to a direct lightning strike can be calculated using the equivalent circuit shown in Figure 9.1(b). In this figure, R_G is the tower footing resistance, $i_S(t)$ is the injected lightning current, Z_S is the internal impedance of the source representing the lightning return-stroke channel, and Z_{SW} is the characteristic impedance of the shield wire. If losses, corona, and the influence of adjacent towers are neglected, $v_{top}(t)$ is given by

$$v_{top}(t) = v_i(t) + (1 + \rho_t) \sum_{m=1}^{\infty} \rho_b^m \rho_t^{m-1} v_i(t - 2m\tau) \tag{9.7}$$

where $\rho_t = \left(Z_{eq} - Z_T\right)\left(Z_{eq} + Z_T\right)^{-1}$ is the voltage reflection coefficient at the tower top, $\rho_b = (R_G - Z_T)(R_G + Z_T)^{-1}$ is the voltage reflection coefficient at the tower base, $\tau = h/\upsilon$ is the travel time associated with the tower, $Z_{eq} = \left(2Z_{SW}^{-1} + Z_S^{-1}\right)^{-1}$ is the equivalent impedance seen by an upward moving voltage wave arriving at the tower top, and $v_i(t) = \left(2Z_{SW}^{-1} + Z_S^{-1} + Z_T^{-1}\right)^{-1} i_S(t)$ is the initial voltage at point T, given by the product of the injected current by the equivalent impedance seen by the source.

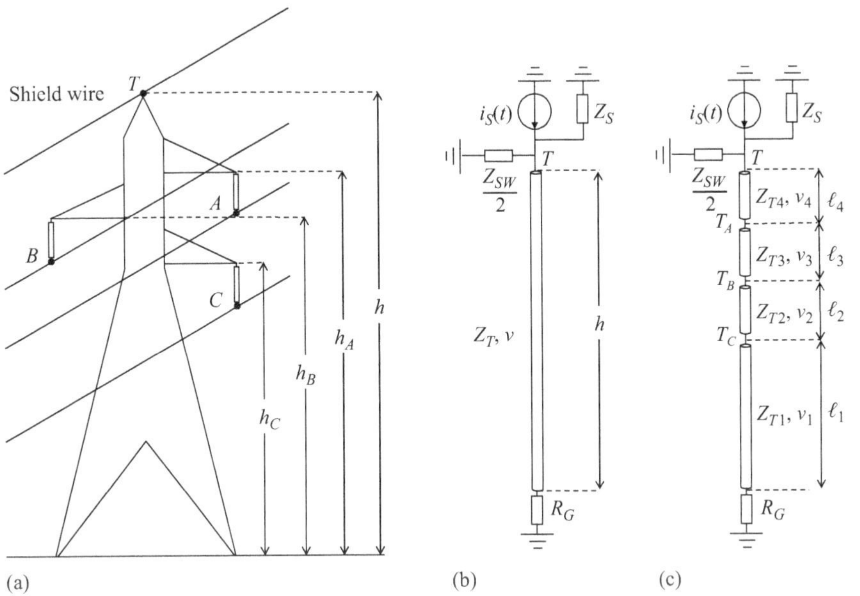

Figure 9.1 Tower modelling as a transmission line

Equation (9.7) indicates that the tower top voltage presents an oscillatory nature due to multiple reflections taking place at the tower ends. A similar analysis can be made in case of a lightning strike to the shield wire at the midspan, except that now $v_i(t)$ will be determined from the voltage waves that propagate on the transmission line and arrive at the tower. For a lightning strike to a phase conductor, the tower top voltage after a flashover can also be calculated using travelling wave theory, except that now the incoming current wave will be injected into the tower from one of its crossarms.

In practice, the study of direct lightning strikes to transmission towers is focused on assessing the probability of flashover occurrence, for which the voltage $\Delta\boldsymbol{v}(t)$ across the insulator strings is necessary. For the tower illustrated in Figure 9.1(a), it corresponds to the 3×1 vector $\Delta\boldsymbol{v}(t) = \boldsymbol{v}(t) + \boldsymbol{u}(t) - \boldsymbol{v}_T(t)$, where $\boldsymbol{v}(t)$ is the lightning-induced voltage on the phase conductors due to the electromagnetic coupling with the shield wires; $\boldsymbol{u}(t)$ is the voltage at power frequency at phases A, B, and C; and $\boldsymbol{v}_T(t)$ is the tower voltage at the crossarms. The lightning-induced voltage term can be expressed as $\boldsymbol{v}(t) = \boldsymbol{k}v_{top}(t)$, where

$$\boldsymbol{k} = \left[Z_{0,21}/Z_{0,11} \quad Z_{0,31}/Z_{0,11} \quad Z_{0,41}/Z_{0,11}\right]^t \tag{9.8}$$

is the TEM coupling factor between the shield wire and the phase conductors. In (9.8), $Z_{0,ij}$ corresponds to element i, j of the characteristic impedance matrix of the line for the lossless case, and superscript t denotes the transpose of the matrix.

To determine $\boldsymbol{v}_T(t)$, the tower can be modelled as a cascade of transmission lines with different lengths as illustrated in Figure 9.1(c). This will give access to points T_A, T_B, and T_C, where the elements of $\boldsymbol{v}_T(t)$ can be calculated. If all transmission line sections have identical parameters, it is straightforward to determine $\boldsymbol{v}_T(t)$ from travelling wave theory. However, if the parameters associated with each transmission line section are calculated to accommodate the tower characteristics at different heights, the determination of $\boldsymbol{v}_T(t)$ is more easily performed in electromagnetic transient simulators. This is also recommended if short open-ended transmission lines are connected to points T_A, T_B, and T_C to represent the effect of crossarms, as suggested in [45,49]. In this case, $\boldsymbol{v}_T(t)$ will contain the voltages calculated at the open end of these short transmission lines, whose characteristic impedance is usually assumed to correspond to that of an infinite horizontal conductor positioned at the same height [49].

9.2.3 Tower models

For calculating the voltage across the insulator strings of a tower struck by lightning, it is necessary to determine the tower voltages, for which the tower surge impedance and the associated propagation speed are necessary. The next subsections discuss different possibilities for the determination of both parameters.

9.2.3.1 Geometric tower models

Table 9.2 lists some of the equations proposed for calculating Z_T for the particular case of a vertical lightning strike to a tower, also referred to as vertical incidence. These equations were derived assuming the tower to be represented as an equivalent geometric figure with height h and radius r.

Four different equations are listed in Table 9.2 for calculating the surge impedance of a vertical cylinder. The observed differences are justified by different modelling assumptions. Jordan's equation [42], for example, was derived from the partial inductance of a vertical cylinder assuming quasi-statics. An error was later found in its derivation and a revised equation, also included in Table 9.2, was proposed by De Conti *et al.* [50]. Wagner and Hileman's equation was derived from the magnetic vector potential associated with the current distribution due to a step current injected at the top of a vertical cylinder [43]. Its time-varying nature indicates that the surge response of a vertical cylinder depends on the injected current waveform. Their equation was later extended by Sargent and Darveniza [44] to determine the maximum voltage at the top of a vertical cylinder at time $2h/c$ due to a ramp-type current waveform. The same authors proposed the time-independent impedance equation associated with a lightning strike to a vertical cone listed in Table 9.2 [44]. Gutiérrez *et al.* [49] proposed the representation of the tower as an inverted cone with vertex at the ground plane. If the cone radius is small compared to its height, the resulting equation approaches the one used for calculating the characteristic impedance of a horizontal line with infinite length.

Table 9.3 lists equations proposed for calculating Z_T in case of a lightning strike to the shield wire in the midspan, also referred to as horizontal incidence. The equation for the inverted cone with vertex at the ground plane proposed by

Table 9.2 Surge impedance of geometric tower models for vertical incidence

Geometry	Authors	Equation
Vertical cylinder	Jordan [42]	$Z_T = 60 \ln \frac{h}{r} + 90 \frac{r}{h} - 60$
	Wagner and Hileman [43]	$Z_T = 60 \ln\left(\sqrt{2}\frac{ct}{r}\right)$
	Sargent and Darveniza [44]	$Z_T = 60 \ln\left(\sqrt{2}\frac{2h}{r}\right) - 60$
	De Conti *et al.* [50]	$Z_T = 60 \ln\left(\frac{4h}{r}\right) - 60$
Vertical cone	Sargent and Darveniza [44]	$Z_T = 60 \ln\left(\sqrt{2}\frac{\sqrt{h^2+r^2}}{r}\right)$
Inverted cone	Gutiérrez *et al.* [49]	$Z_T = 60 \ln\left(\frac{\sqrt{h^2+r^2}+h}{r}\right)$

Table 9.3 Surge impedance of geometric tower models for horizontal incidence

Geometry	Authors	Equation
Vertical cylinder	Chisholm *et al.* [46]	$Z_T = 60 \ln\left\{\cot\left[0.5 \tan^{-1}\left(\frac{r}{h}\right)\right]\right\} - 60$
	Hara and Yamamoto [38]	$Z_T = 60 \ln\left(\sqrt{2}\frac{2h}{r}\right) - 120$
Inverted vertical cone	Chisholm *et al.* [45,46]	$Z_T = 60 \ln\left\{\cot\left[0.5 \tan^{-1}\left(\frac{r}{h}\right)\right]\right\}$

Chisholm *et al.* [45,46] can be shown to be equivalent to the one recommended by Gutiérrez *et al.* [49] for simulating vertical incidence to a tower. The equation for the vertical cylinder listed in Table 9.3 was obtained by averaging out the inverted cone equation assuming a time-varying height $h(t) = h - ct$ [45,46]. The equation proposed by Hara and Yamamoto [38] was derived from experimental tests with vertical cylinders of different heights fed by a horizontal current lead.

The use of different equations for determining Z_T depending on the direction of current incidence comes from the different electromagnetic phenomena experienced in each case. For example, a direct lightning strike to a tower creates electromagnetic waves that propagate spherically. After being reflected at the ground plane, these waves will modify the tower top voltage only after a time delay corresponding to 2τ. On the other hand, in case of a lightning strike to the shield wire at the midspan, the incoming electromagnetic waves will illuminate the whole tower simultaneously. In this case, induced tower currents will reduce the impedance seen from the tower top even before 2τ [39]. It must be noted that none of the equations listed in Tables 9.2 and 9.3 effectively includes the presence of vertical or horizontal current leads, except perhaps for the equation proposed by Hara and Yamamoto [38], which is based on experiments. Equations for calculating the transient impedance of a vertical wire including the influence of vertical or horizontal current leads are found in [48].

Determination of the radius r for use in the equations listed in Tables 9.2 and 9.3 depends on the actual tower geometry. For towers resembling a cone or a cylinder, r will be based on the actual radius of the tower. When the tower geometry is irregular, one possibility is to use the average radius $r_{avg} = [r_C h_B + r_B(h_A + h_B) + r_A h_A](h_A + h_B)^{-1}$, where r_C is the tower top radius, r_B is the tower midsection radius, r_A is the tower base radius, h_A is the height from base to midsection, h_B is the height from midsection to top, and $(h_A + h_B)$ is the tower height, as shown in Figure 9.2(a) [2]. This expression was originally proposed by Chisholm *et al.* [46] for use in the inverted cone expression listed in Table 9.3.

Another approach to calculate tower surge impedance based on tower geometry consists in estimating the tower capacitance as $C = \varepsilon_0 C_f \sqrt{4\pi A}$, where C_f is a shape factor that depends on the tower height and radius and A is the surface area of the object assumed to represent the tower. The surge impedance is then calculated

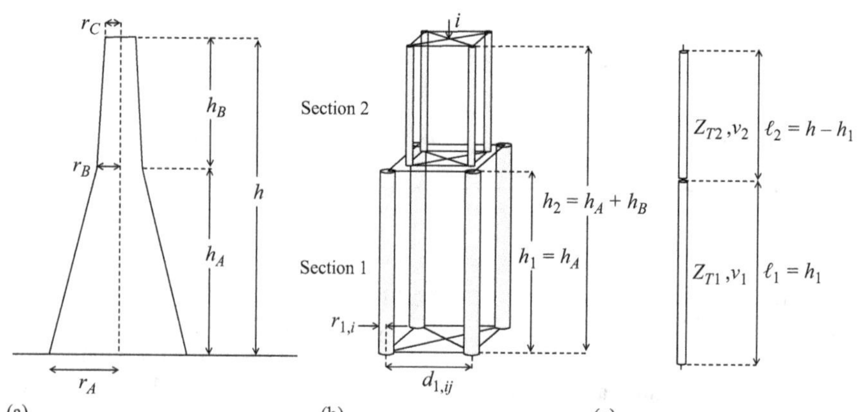

Figure 9.2 Tower modelling as a multiconductor transmission line

as $Z_T = \tau/C$ [46]. More details about this approach and geometric tower models in general can be found in Chapter 3 in Volume 2.

9.2.3.2 Non-uniform transmission line models

Determination of Z_T using geometric tower models is sometimes problematic because actual tower structures can depart significantly from the simple geometries listed in Tables 9.2 and 9.3. Another problem is that using a transmission line with a single surge impedance value as shown in Figure 9.1(b), or even in Figure 9.1(c) provided $Z_{T1} = Z_{T2} = Z_{T3} = Z_{T4}$, is unable to simulate complex tower structures in sufficient detail.

To overcome this problem, one possibility is to represent the tower as a cascade of uniform transmission lines with impedances that are calculated as a function of the tower characteristics at a given height. In Figure 9.1(c), for example, impedance Z_{T1}, which is related to the tower legs, would be different from impedance Z_{T2}, which is related to the tower body. In the same figure, additional tower sections could be used to better represent the fine structure of the tower. Lumped circuit elements could also be inserted between transmission line segments to improve the model behaviour [8,34,36,37]. Since the resulting tower model will consist of several uniform transmission lines in series, the tower as a whole can be viewed as a non-uniform transmission line, that is, a transmission line whose characteristic impedance depends on position. This type of model is also called multistorey [36] or multistair [47] tower model. Non-uniform tower models with continuously varying impedance as a function of height have also been proposed [51].

Determination of the impedances to be used in each section of a non-uniform tower model can be made using some of the expressions listed in Tables 9.2 and 9.3. Another possibility is to determine these elements from either experiments or numerical electromagnetic field analysis [8,36,37]. A compromise between accuracy and simplicity consists in representing each section of the tower as a multiconductor transmission line formed by a set of vertical conductors as shown in Figure 9.2(b). For calculating the self and mutual impedances of each tower section composed of vertical cylinders, different expressions are available. Some of them are listed in Table 9.4, where h_k is the height of the kth tower section, $d_{k,ij}$ is the distance between segments i and j of tower section k, and $r_{k,i}$ is the radius of the ith segment of section k.

The equations of Hara and Yamamoto [38] shown in Table 9.4 were obtained from experiments with vertical conductors. The equations of De Conti *et al.* [50] were derived considering a quasi-static approach. They were validated through comparisons with simulations with an electromagnetic model. Ametani *et al.* [47] and Gutiérrez *et al.* [49] also proposed equations for representing transmission towers as vertical multiconductor transmission lines. However, their equations were written in frequency domain including a correction for the finite ground conductivity considering the approximation of Deri *et al.* [22], which was later shown to lead to significant errors if applied to vertical wires [52]. For this reason, they were omitted in Table 9.4. Also, the equations proposed in [47] should be

Table 9.4 Surge impedance of multiconductor tower models

Authors	Self-impedance term $(\boldsymbol{Z}_{k,ii})$	Mutual-impedance term $(\boldsymbol{Z}_{k,ij})$
Hara and Yamamoto [38]	$60\left[\ln\left(\sqrt{2}\frac{2h_k}{r_{k,i}}\right)-2\right]$	$60\left[\ln\left(\sqrt{2}\frac{2h_k}{d_{k,ij}}\right)-2\right]$
De Conti *et al.* [50]	$60\left[\ln\left(\frac{4h_k}{r_{k,i}}\right)-1\right]$	$60\ln\frac{2h_k+\sqrt{4h_k^2+d_{k,ij}^2}}{d_{k,ij}^2}+30\frac{d_{k,ij}}{h_k}-60\sqrt{1+\frac{d_{k,ij}^2}{4h_k^2}}$

viewed with caution, since they reduce to the original Jordan's formula [42], which is in error, for the particular case of a single conductor above a lossless ground.

Distance $d_{k,ij}$ in the equations listed in Table 9.4 can be determined from the average spacing between two conductors pertaining to section k, or by using the equivalent formulas proposed by Hara and Yamamoto [38]. The value of $r_{k,i}$ is readily obtained if the tower is built with tubular conductors. If the tower is built with angle sections, $r_{k,i}$ can be determined as suggested in [38] or, preferably, by comparisons with numerical electromagnetic field simulations.

For each tower section, it is possible to obtain a single-conductor equivalent by assuming equal voltage at each line termination and considering that the sum of the currents entering a tower section is equal to the total current i shown in Figure 9.2(b). For symmetric structures,

$$Z_{Tk}=\frac{Z_{k,11}+Z_{k,12}+Z_{k,13}+\cdots+Z_{k,1n}}{n} \tag{9.9}$$

where n is the number of vertical cylinders assumed to represent the kth tower section and Z_{Tk} is the characteristic impedance of the equivalent single-conductor transmission line representing this section. The resulting equivalent circuit shown in Figure 9.2(c) can then be easily implemented in electromagnetic transient simulators for calculation of the tower top voltage. A single transmission line equivalent can also be obtained as the geometric mean of the individual values of Z_{Tk} [47]. If access to the crossarm voltages is necessary, Section 2 of the tower shown in Figure 9.2(b) can be further divided similarly as shown in Figure 9.1(c).

9.2.3.3 Propagation speed

The propagation speed associated with a lossless conductor with arbitrary orientation above the ground plane is the speed of light. One could therefore expect the same for a transmission tower model. However, results obtained in reduced- and real-scale experiments indicated in some cases a propagation speed lower than the speed of light [35,40,41]. This would be related to the time delay associated with current propagation in crossarms, shield wire supports, and slant elements that are found in actual towers [45,46]. This has motivated some authors to consider a reduced apparent propagation speed in the modelling of transmission line towers, with values ranging from $0.8c$ to $0.9c$ [15]. For a predominantly vertical structure, except when the additional propagation delay associated with shield wire supports is of importance, the travel time can be assumed to correspond to the tower height divided by the speed of light [2]. This is usually the case when the multiconductor transmission line representation discussed in Section 9.2.3.2 is used.

9.2.4 Example

As an example of tower surge impedance calculation, a 138-kV tower with dimensions $r_A = 3$ m, $r_B = 0.4$ m, $r_C = 0.3$ m, $h_A = 33$ m, and $h_B = 6.8$ m [see Figure 9.2(a)] is considered. In [33], the complete tower with slant elements and lattice structures was simulated with the hybrid electromagnetic model (HEM) [9], including a shield wire. A triangular current waveform with front time of 5 μs and time-to-half value of 50 μs was injected at the tower top by an ideal current source. Results obtained in [53] indicated that an equivalent tower impedance of 210 Ω and a propagation speed of $0.85c$ would be able to accurately reproduce the resulting voltage waveforms in all tested cases.

Table 9.5 shows tower surge impedances calculated with some of the equations listed in Tables 9.2 and 9.3 considering $h = 39.8$ m and an average radius of 2.94 m obtained from the expression proposed by Chisholm *et al.* [46]. Also included in Table 9.5 are results obtained considering a multiconductor transmission line representation in which the tower was divided into four sections (three sections of 11 m, each representing the tower legs and one additional section of 6.8 m representing the tower body), each consisting of four vertical conductors. The surge impedance of each section was calculated from (9.9) as $Z_{Tk} = \left(Z_{k,11} + 2Z_{k,12} + Z_{k,13}\right)/4$ using the expressions of De Conti *et al.* [50]. The equivalent surge impedance was calculated as the geometric mean of the individual impedances corresponding to each section. In the HEM simulations, a 3.5-cm radius was assumed to represent all conductors. In the multiconductor transmission line representation, two different values were considered for this parameter: 3.5 and 7 cm.

Table 9.5 Calculated tower surge impedance values

Geometry	**Equations**	**138-kV tower (Ω)**
Vertical cylinder	Jordan [42]	103
	Sargent and Darveniza [44]	159
	De Conti *et al.* [50]	180
Vertical cone	Sargent and Darveniza [44]	177
Inverted cone	Chisholm *et al.* [46]	198
Multiconductor transmission line	De Conti *et al.* [50]	221 ($r = 3.5$ cm)
		210 ($r = 7.0$ cm)

It is seen in Table 9.5 that the best results obtained with the geometric tower models are the ones corresponding to the inverted cone model (198 Ω) [46,49] and the revised Jordan equation (180 Ω) [50]. A good agreement with the reference value of 210 Ω is obtained with the multiconductor transmission line representation, regardless of the assumed conductor radius. The fact that a perfect match was obtained for a conductor radius of 7 cm that is twice the value assumed in the simulations with the electromagnetic model is justified by the fact that the multiconductor tower model does not include the effect of slant elements and lattice structures, which reduces the tower surge impedance [7]. Assuming a relatively larger conductor radius in the equivalent circuit is able to compensate for this effect.

9.2.5 *Discussion*

The representation of transmission towers as an equivalent transmission line should be viewed with caution in more rigorous studies, since it is an approximate representation of a complex electromagnetic problem. It is apparent that several of the requirements for the use of transmission line theory listed in Section 9.1.1 are violated even in the study of the transient response of a single vertical conductor of finite length above the ground, which can be viewed as a particular case of the more complex problem of studying the current distribution in an actual tower struck by lightning. The electromagnetic field structure associated with this type of problem is non-TEM. This poses difficulties to the definition of voltage, which becomes path dependent [54].

One of the results of modelling a transmission tower as a transmission line with constant and uniform characteristic impedance is the undistorted and unattenuated current propagation. Nevertheless, it has been demonstrated that current pulses that propagate in vertical cylinders attenuate and distort regardless of the direction of current propagation [55]. Undistorted and unattenuated current propagation is possible in conical structures if the current pulse propagates from the apex to the base, but the opposite is not true [56]. This indicates that all expressions listed in Tables 9.2–9.4 for the calculation of the tower surge impedance must be viewed with caution, because undistorted and unattenuated current propagation is implicitly assumed in most cases. Care must also be taken when comparing theoretical surge impedance values with experimentally derived ones because of possible differences in the considered voltage definitions [50].

Other critical points associated with representing a transmission tower as an equivalent transmission line are listed as follows. For example, the tower surge response is affected by the presence and orientation of the return-stroke channel [40,41,48], which is neglected in the derivation of the expressions listed in Tables 9.2–9.4. Also, time delays are considered for calculating the tower top voltage such as in (9.7), whereas static conditions are assumed in the modelling of the transmission line conductors connected to the tower with the expressions listed in Table 9.1. In addition, the TEM coupling coefficients given in (9.8) are not valid for calculating the insulator voltages for fast lightning current waveforms due to the associated non-TEM field structure [8]. The coupling coefficients are also to some extent affected by shield wire corona [10]. All these factors may lead to the incorrect estimation of the flashover rate of a transmission tower.

Most of the problems listed above are minimized if the front time t_f of the impinging lightning current waveform is larger than twice the travel time of the tower. For this condition, it is generally possible to model the tower with sufficient accuracy as a transmission line (see, e.g., [50]). The non-uniform transmission line modelling approach discussed in Section 9.2.3.2 is generally preferred because it allows a more detailed representation of the tower characteristics as a function of height. A cascade of uniform transmission lines with different surge impedances is also able to emulate, to some extent, the attenuation and distortion of tower currents. The model behaviour can be improved if suitably selected lumped-circuit parameters are inserted between the tower sections as in [8,34,36,37]. This latter type of model can be successfully used even if the condition $t_f > 2\tau$ does not necessarily hold.

To conclude, any tower model based on transmission line theory should be properly calibrated and validated before being used in the estimation of the lightning performance of transmission lines. This can be made, for example, with the use of numerical electromagnetic models and/or experimental data.

9.3 Grounding

9.3.1 Overview

Tower grounding is composed of the tower foundation plus additional grounding electrodes. In high-voltage transmission lines, the usual practice is to use counterpoise cables, which are buried horizontal wires with total length of tens of meters extending from each of the tower legs along the right of way of the line.

The transient response of a grounding system depends on electrode configuration, physical properties of the soil, and characteristics of the injected current. The grounding system is usually characterized by the harmonic impedance seen at the current injection point. This parameter, which relates the ground potential rise to the injected current at each frequency of interest, can be determined from either measurements or calculations. If the current amplitude is sufficiently large, soil ionization can occur, which also affects the grounding transient response. In any case, since the soil is a dispersive medium, the associated electromagnetic field structure is non-TEM. For this reason, the most appropriate procedure for the theoretical estimation of the grounding impedance in a wide frequency range is the use of electromagnetic models [9,57]. In some cases, however, circuit or transmission line theory can also be used with sufficient accuracy to estimate this parameter [58–60].

In this section, a discussion is presented on the determination of the grounding impedance of transmission towers for lightning studies using circuit and transmission line theory. A more detailed discussion on the lightning response of grounding electrodes is found in Chapter 7 in this volume.

9.3.2 Lumped-circuit representation

The simplest form of representing the tower grounding is to treat it as a linear resistance R_G, such as in Figure 9.1. Although this model is not able to reproduce the frequency-dependent behaviour of the grounding system [61], it is often used in analytical studies because it leads to the straightforward determination of the reflection coefficient ρ_b appearing in (9.7) [2,62]. It is also easily implemented in electromagnetic transient simulators.

The most usual choice for R_G is assuming it to correspond to R_{LF}, which is the low-frequency resistance of the electrode configuration. The resistance R_G can also be assumed to correspond to the impulse impedance Z_P, which relates the peak voltage and the peak current at the current injection point in a grounding system [61]. The use of Z_P to represent tower-footing grounding is shown by Visacro and Silveira [63] to lead to a backflashover estimation in 138- and 230-kV lines that is nearly the same as the one obtained with a rigorous representation of the grounding system. Furthermore, the use of $R_G = R_{LF}$ leads to extremely conservative results in terms of backflashover occurrence [63].

A simple linear lumped-circuit model that is able to represent the grounding impedance of a compact electrode system is shown in Figure 9.3(a) [64]. It consists of the series connection of inductance L with the parallel combination of conductance G and capacitance C. Table 9.6 presents the expressions proposed by Sunde [19] for calculating L, C, and G for both vertical and horizontal electrodes with length ℓ and radius a buried in a soil with resistivity ρ_g, permittivity ε_g, and permeability μ_0.

If L and C are neglected, the circuit shown in Figure 9.3(a) reduces to the linear resistance model discussed earlier. In this case, G obtained from Table 9.6 is the reciprocal of the low-frequency resistance R_{LF} associated with a vertical or horizontal wire buried in the ground.

Another possibility of representing the tower grounding in lightning studies consists in using a non-linear resistance to account for the occurrence of soil ionization due to high currents. A popular expression is the one recommended by CIGRÉ [2]:

$$R_G(t) = \frac{R_{LF}}{\sqrt{1 + i(t)/I_c}} \tag{9.10}$$

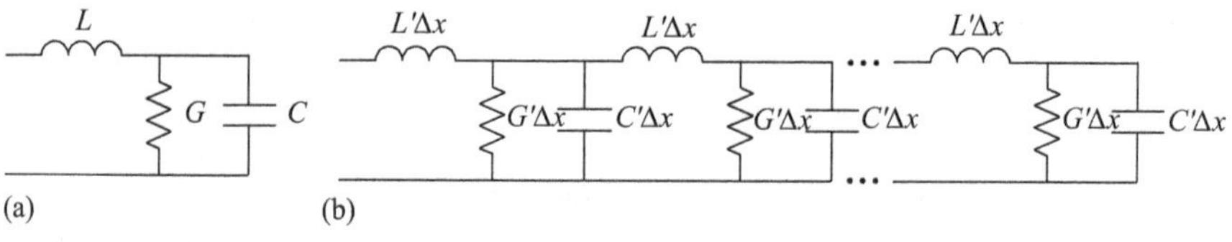

Figure 9.3 (a) Lumped- and (b) distributed-circuit models of a grounding electrode

Table 9.6 Circuit parameters for grounding electrodes [19]

Vertical wire	**Horizontal wire**
$G = R^{-1} = \left[\frac{\rho_g}{2\pi\ell}\left(\ln\frac{4\ell}{a} - 1\right)\right]^{-1}$	$G = R^{-1} = \left[\frac{\rho_g}{\pi\ell}\left(\ln\frac{2\ell}{\sqrt{2ad}} - 1\right)\right]^{-1}$
$C = \frac{\rho_g \varepsilon_g}{R} = 2\pi\varepsilon_g\ell\left(\ln\frac{4\ell}{a} - 1\right)^{-1}$	$C = \frac{\rho_g \varepsilon_g}{R} = \pi\varepsilon_g\ell\left(\ln\frac{2\ell}{\sqrt{2ad}} - 1\right)^{-1}$
$L = \frac{\mu_0 \ell}{2\pi}\left(\ln\frac{2\ell}{a} - 1\right)$	$L = \frac{\mu_0 \ell}{2\pi}\left(\ln\frac{2\ell}{a} - 1\right)$

where *i*(*t*) is the current entering the grounding electrode and $I_C = \rho_g E_0 / \left(2\pi R_{LF}^2\right)$ is the critical current, which depends both on the soil resistivity ρ_g and the critical electric field E_0. Although E_0 = 400 kV/m is recommended by CIGRÉ [2], selecting this parameter can be problematic because a wide range of values have been reported in the literature [65,66]. In fact, the reduction of R_G as a function of *i*(*t*) as predicted by (9.10) can be very sensitive to E_0 depending on the values of ρ_g and R_{LF} [67]. Also, some authors (e.g. [68,69]) suggest a dynamic behaviour for E_0, which would make the application of (9.10) difficult. Finally, this equation was derived assuming spherical field symmetry, which may restrict its applicability to compact grounding systems [2].

9.3.3 Distributed-circuit representation

9.3.3.1 Cascade of lumped-circuit elements

The equivalent circuit shown in Figure 9.3(a) is sufficiently accurate only for compact electrode configurations [58,70]. To extend this simple model to the simulation of long counterpoise cables or long vertical rods, a distributed-circuit approach can be used. One possibility is dividing the buried wire in *n* short segments of length Δx and representing each segment as a circuit in the form shown in Figure 9.3(a). The parameters of each segment will correspond to the per-unit-length inductance (L'), conductance (G'), and capacitance (C') of the wire multiplied by Δx. The resulting model will be the cascade of *n* circuit segments shown in Figure 9.3(b).

For determining the per-unit-length parameters L', G', and C', the usual practice is to divide *L*, *C*, and *G* given in Table 9.6 by the total electrode length ℓ [19], although other expressions are also available (e.g. [71–73]). By using Sunde's expressions listed in Table 9.6 and neglecting soil ionization, it is shown in [59] that this distributed-circuit model is able to represent the transient behaviour of buried vertical and horizontal wires with reasonable accuracy for high-resistivity soils and relatively slow current wavefronts. This type of model is also frequently used to account for soil ionization in the simulation of the transient response of long electrodes by gradually increasing the conductance and capacitance associated with each segment after the critical electric field is exceeded in the soil (e.g. [71,74]).

9.3.3.2 Transmission line model

Although there is an improvement over the lumped-circuit representation shown in Figure 9.3(a), the equivalent circuit of Figure 9.3(b) cannot be easily extended to account for the variation of the soil resistivity and permittivity with frequency, which has a significant effect in the transient response of a grounding system [26–29]. Also, this model is sensitive to the considered number of circuit segments.

These difficulties are overcome if $\Delta x \rightarrow 0$ is assumed. In this case, the circuit of Figure 9.3(b) can be represented using transmission line theory. For a single grounding electrode, the characteristic admittance (9.3) becomes the scalar

$$Y_C = \sqrt{YZ^{-1}} = \sqrt{(G' + j\omega C')(j\omega L')^{-1}} \tag{9.11}$$

where L', G', and C' are obtained as before and the propagation constant is given by

$$\gamma = \sqrt{YZ} = \sqrt{(G' + j\omega C')(j\omega L')} \tag{9.12}$$

The variation of ρ_g and ε_g with frequency can be easily included in the calculation of (9.11) and (9.12) simply by considering a different value for G' and C' at each frequency [60]. A series resistance could also be included in (9.11) and (9.12) to represent internal losses in the grounding electrodes, but the influence of this term is usually negligible compared to the effect of G', L', and C' [61].

By assuming the grounding system to extend from $x = 0$ to $x = \ell$, a solution in the form of (9.4) can be written to calculate terminal voltages and currents, except that now only scalars will be involved. If it is assumed that the current at $x = \ell$ is equal to zero, the grounding impedance Z_G can be calculated as $Z_G = V(0)/I(0)$, resulting in

$$Z_G = Z_C \coth(\gamma\ell) \tag{9.13}$$

where $Z_C = Y_C^{-1}$.

The equivalent grounding impedance of two parallel electrodes separated by a distance s, each carrying half of the total injected current, can be calculated using (9.9) for $n = 2$ if it is assumed that both experience the same voltage at $x = 0$. In this case,

$$Z_G = \left(Z_{self} + Z_{mutual}\right)/2 \tag{9.14}$$

where Z_{self} and Z_{mutual} are both calculated with (9.13), the former exactly as before and the latter by replacing the conductor radius a by the separation s in the calculation of the per-unit-length parameters [19]. This expression can be used, for example, for estimating the grounding impedance of a pair of parallel counterpoise cables extending from the tower legs along the right of way. If equal current sharing is assumed and the electromagnetic coupling between two pairs of counterpoise cables extending in opposite directions is neglected, the grounding impedance seen at the current injection point reads

$$Z_G = \left(Z_{self} + Z_{mutual}\right)/4 \tag{9.15}$$

For the simulation of the transient response of the tower grounding using transmission line theory, one can represent it as a multiconductor transmission line whose equations are solved either in frequency domain, with the calculation of resulting time-domain waveforms using inverse Fourier or Laplace techniques [60], or directly in time domain [30,31,75].

The modelling of vertical and horizontal grounding electrodes using transmission line theory presents a good agreement with field theory in conditions mostly related to high-resistivity soils and slow current wavefronts [59,60]. This agreement is better than the one presented by the cascade of lumped-circuit elements described in the previous section. For low-resistivity soils or fast current wavefronts, the impulse impedance of the grounding system obtained from transmission line theory is likely to be overestimated [59].

9.3.4 N-port linear circuit model based on rational approximations

The distributed-circuit approach discussed in the previous section is able to lead to sufficiently accurate results in many cases of practical interest [58–60]. However, it cannot be easily extended to simulate grounding configurations composed of electrodes with arbitrary orientation.

To overcome this difficulty, one possibility is to derive an N-port linear circuit model based on the rational approximation of grounding impedance ($\boldsymbol{Z_G}$) or admittance ($\boldsymbol{Y_G}$) matrices associated with a given grounding system [76–78]. These matrices can be used to represent a grounding system with N current-injection points (e.g. the four legs of a transmission tower) or a grounding system with Q current-injection points plus $N - Q$ observation points where the ground potential rise must be calculated (e.g. a grounding mesh in a substation), considering or not the variation of the ground resistivity and permittivity with frequency.

The determination of $\boldsymbol{Z_G}$ or $\boldsymbol{Y_G}$ over a wide frequency range is usually performed using a numerical electromagnetic model. The idea is to represent any of these matrices as a sum of rational functions of the form

$$\boldsymbol{F}(s) = \sum_{m=1}^{M} \frac{\boldsymbol{c}_m}{s - a_m} + \boldsymbol{d} + s\boldsymbol{e} \tag{9.16}$$

where a_m is the mth pole associated with either $\boldsymbol{Z_G}$ or $\boldsymbol{Y_G}$, $\boldsymbol{c}_m$ is a $N \times N$ matrix of residues associated with the mth pole, $\boldsymbol{d}$ is an $N \times N$ matrix that assures a non-zero value for $\boldsymbol{F}(s)$ at high frequencies, $\boldsymbol{e}$ is an N x N matrix that allows $\boldsymbol{F}(s)$ to increase with frequency as s approaches infinity, and s is the Laplace variable. In (9.16), a single set of poles is used to represent all matrix elements, but a more general representation is also possible. The derivation of a_m, $\boldsymbol{c}_m$, $\boldsymbol{d}$, and $\boldsymbol{e}$ from $\boldsymbol{Z_G}$ or $\boldsymbol{Y_G}$

can be performed with different fitting techniques [79–81]. An N-port linear circuit model for direct use in electromagnetic transient simulators can be readily obtained provided $\boldsymbol{F}(s) = \boldsymbol{Y}_G$ [82]. An alternative approach to include an N-port function of the form (9.16) in such simulators given that $\boldsymbol{F}(s) = \boldsymbol{Z}_G$ is proposed in [83].

If symmetry considerations are used as described in the previous section, $\boldsymbol{Z}_G$ and $\boldsymbol{Y}_G$ reduce to scalars, in which case (9.16) reads $f(s) = \sum_{m=1}^{M} c_m(s - a_m)^{-1} + d + se$, where the definition of the parameters is the same as before except that now $N = 1$. In this case, obtaining an equivalent circuit is straightforward [82]. If some loss of accuracy is admitted in the high-frequency range, Z_G can also be calculated with transmission line theory using (9.13)–(9.15), and the resulting equivalent circuit can be easily included in electromagnetic transient simulators.

The representation of $\boldsymbol{Z}_G$ or $\boldsymbol{Y}_G$ as an N-port equivalent circuit is able to reproduce the transient behaviour of a grounding system with great accuracy as long as these matrices are calculated with electromagnetic models, the fitting of the elements in (9.16) is correctly performed, and a linear model is desired.

9.4 Insulator strings

9.4.1 Introduction

Insulator string modelling for calculating the lightning performance of transmission lines is focused on determining the occurrence (or not) of insulation breakdown due to a lightning overvoltage, for which a flashover model or criterion are needed. In this context, the calculated lightning outage rate of a transmission line is directly affected by the assumed insulator string model.

The insulation breakdown process is a statistical phenomenon influenced by several parameters such as atmospheric conditions (relative humidity and pressure), waveform and polarity of the lightning overvoltage across the insulator, which have an influence on the characteristics of the resulting electric field, and the voltage withstand characteristic of the insulator, which depends on insulator material, dry-arc distance, etc. [3].

One way to predict insulation breakdown on a specific gap or insulator string configuration under a surge voltage consists in performing high-voltage laboratory tests to reproduce the specific conditions of the problem under analysis. However, from the point of view of the study of the lightning performance of transmission lines, this represents a time-consuming and costly procedure. In this context, the application of a flashover model preferably derived from experimental data acquired from this type of tests and following physical premises related to insulation breakdown process is advisable.

This section addresses the main flashover models available in the literature to estimate the lightning withstand capability of transmission line insulator strings.

9.4.2 Flashover models

Basically, four categories of models are available for determining flashover occurrence across gaps or insulator strings: (i) voltage-dependent switch; (ii) voltage–time curve; (iii) disruptive effect (DE) model; and (iv) leader progression model (LPM). A description of each approach is presented next.

9.4.2.1 Voltage-dependent switch

The simplest modelling technique consists in representing the insulator string as a voltage-dependent switch that closes when the voltage across the insulator string reaches a specific value, such as the line CFO. In spite of being computationally fast and easy to implement, which has motivated its use in analytical procedures to estimate the lightning performance of transmission lines [2], this model disregards any time-dependent characteristic of the breakdown process.

The CIGRÉ methodology to estimate the lightning performance of transmission lines [2] adopts a voltage-dependent switch governed by the so-called non-standard critical flashover overvoltage (CFO_{NS}), indicated in (9.17), where τ_t is the voltage tail time and V_{PF} is the steady-state voltage of the line. Although not explicitly shown in (9.17), such criterion aims to consider the effect of tower-footing grounding resistance and span length between adjacent towers on the flashover process, following the CIGRÉ analytical procedure.

$$CFO_{NS} = \left(0.977 + \frac{2.82}{\tau_t}\right)\left(1 - 0.2\frac{V_{PF}}{CFO}\right)CFO \tag{9.17}$$

9.4.2.2 Voltage–time curve

The voltage–time curve model is an attempt to include the time dependence of the flashover process in the modelling of the impulse withstand voltage of an insulator. This model aims to express the relationship between the overvoltage that leads the insulation to flashover and the time to flashover. Such relationship is experimentally obtained from high-voltage laboratory tests for particular gap and insulator string configurations subjected to impulse voltages of specific waveform and polarity. The test is based on varying the peak value of the applied voltage waveform and recording the instant of time at which a flashover occurs.

Equation (9.18) describes the voltage–time curve for insulator string lengths from 1 to 6 m obtained for a negative standard lightning impulse voltage wave [1,3]. In this equation, V is the flashover voltage in kV, t is the time to flashover in μs, and ℓ is the insulator string length in m.

$$V = \left(400 + \frac{710}{t^{0.75}}\right)\ell \tag{9.18}$$

Equation (9.19), which is valid from about 2 to 11 μs, describes the voltage–time curve in terms of the line CFO as proposed by [84]:

$$V = CFO\left(0.58 + \frac{1.39}{t^{0.5}}\right) \tag{9.19}$$

Similarly as in case of the voltage-dependent switch described in the previous section, a flashover is assumed to occur as soon as the voltage across the insulator string reaches V given by (9.18) or (9.19). One important flaw of this method is that the voltage–time curve does not necessarily relate voltage–time pairs referring to the same instant of time. Actually, for flashovers occurring on the tail of the applied voltage waveform, the voltage–time curve relates the crest voltage to the time to flashover. This introduces some inaccuracy on the relationship between the flashover time and the expected flashover overvoltage. Also, the voltage–time curve depends on the polarity and waveform of the impulse voltage considered in the laboratory tests. Its application for predicting flashovers due to non-standard impulse voltages is therefore prone to errors.

9.4.2.3 DE model

This section discusses a method that considers not only the peak value but also the voltage waveform on estimating the likelihood of a gap or insulator to withstand an impinging overvoltage. The DE model, also known as integration method, was originally proposed by Witzke and Bliss to estimate the impulse withstand capability of power transformers subjected to non-standard lightning waveforms [85,86]. Later, this model was extended to air gaps and insulators. The method consists in integrating the impinging voltage waveform during a time interval in which the instantaneous voltage is higher than a threshold voltage U_0. The result of this integration is expressed in terms of the variable *DE*. If *DE* reaches a base value *DEb* determined from laboratory tests, flashover occurs. The model equation is described as

$$DE = \int_{t_0}^{t} [U(t) - U_0]^k dt \tag{9.20}$$

where $U(t)$ is the applied voltage, U_0 is the voltage level below which no flashover occurs, t_0 is the time instant when $U(t)$ first exceeds U_0, and k is a constant.

This model follows two fundamental aspects [2,84]: (i) the breakdown process only starts and is maintained if a minimum threshold voltage U_0 is exceeded; (ii) flashover occurs only when the applied voltage is higher than the minimum threshold voltage U_0.

The DE model is traditionally used for estimating the insulation strength under non-standard voltage impulses. However, the constants of this model are usually derived from high-voltage laboratory experiments performed with standard voltage waveforms. Some approaches to extract these constants from experimental voltage–time curves are described next:

1. Kind [87]: It is proposed to use exponent k equal to 1, leading to the equal-area criterion.
2. Caldwell and Darveniza [88]: Two adjustment procedures are proposed, namely (i) consideration of a fixed value of U_0 to allow the calculation of k and *DEb*; (ii) assumption of $k = 1$ and variation of U_0 from 0% to 90% of the CFO. Caldwell and Darveniza [88] concluded that the best agreement between the predicted and experimental results for rod–rod gaps is obtained following the second procedure, also known as equal-area method [87].

3. Chowdhuri *et al.* [89]: The exponent k is assumed equal to $\alpha v(t)/U_0$, where α is experimentally obtained for the applied voltage waveform and tested object. The optimal value of α is the one that guarantees *DEb* with minimum mean square deviation σ_{DE} from the experimental voltage–time curve.
4. Ancajima *et al.* [90]: In this approach, *DEb* is calculated for a fixed value of U_0 for each pair of points of the voltage–time curve that best fits experimental data, along with the average and the standard deviation related to the calculated *DE*. This procedure is repeated varying U_0. The set of *DEb* and U_0 to be chosen corresponds to the average value of *DEb* that leads to the minimum standard deviation with respect to experimental data and the value of U_0 that provides the optimum value for *DEb*.
5. Gomes *et al.* [91]: Based on the procedure of Ancajima *et al.* [90], the DE parameters for $k = 1$ are obtained considering the solution of (9.20) assuming either (a) an analytical approach based on a triangular voltage waveform or (b) a numerical approach based on a double-exponential voltage waveform.

A set of DE constants for estimating the lightning performance of transmission lines is presented in [92].

9.4.2.4 Leader progression model

The LPM is a physical model that aims at representing the main stages of the breakdown process, notably the corona inception, streamer propagation and leader propagation phases [2].

The breakdown process can be summarized as follows. When the voltage across the gap exceeds the corona inception voltage, the propagation of streamers is initiated. If the applied voltage remains sufficiently high, the streamers can cross the gap. After this, the leader development phase begins. When the leader bridges the gap, flashover occurs.

Taking this sequence into account, the time-to-breakdown t_b can be considered as the sum of the times related to each phase:

$$t_b = t_i + t_s + t_l \tag{9.21}$$

where t_i is the corona inception time, t_s is the time required for the streamer to cross the gap, and t_l is the leader propagation time.

As discussed in [2], the LPM is fundamentally concerned with the propagation time related to the leader propagation speed. This parameter depends on the voltage across the gap and on the length of the gap yet to be covered by the leader. As long as the unbridged part of the gap is subjected to an average electric field larger than or equal to E_0, the breakdown gradient, leader progression continues.

LPM expressions applied for estimating flashover occurrence on high-voltage transmission line insulators assume the leader propagation speed to depend on both applied voltage and length of the unbridged part of the gap. The constants of the model are extracted from high-voltage laboratory tests considering different gap or insulator string configurations for positive and negative impulse polarity.

Table 9.7 LPM constants suggested by CIGRÉ [2,93].

Insulator type	Polarity of applied impulse	k (m^2/kV^2-µs)	E_0 (kV/m)
Cap-and-pin porcelain and glass insulator	Negative	1.3×10^{-6}	600
	Positive	1.2×10^{-6}	520
Air gaps, post-insulators, long-rod polymer insulators	Negative	1×10^{-6}	670
	Positive	0.8×10^{-6}	600

The most popular formulation of the LPM is presented as follows [2]:

$$\frac{dg}{dt} = kV(t)\left[\frac{V(t)}{l-g} - E_0\right] \tag{9.22}$$

where dg/dt is the leader propagation speed in m/s, k is a constant, $V(t)$ is the voltage across the gap in kV, g is the leader length in m, l is the air gap length in m, and E_0 is the breakdown gradient in kV/m as mentioned above. Table 9.7 summarizes the LPM constants suggested by CIGRÉ [2].

Relevant references that propose variations of the LPM are the works of Wagner and Hileman [94], Suzuki and Miyake [95], Shindo and Suzuki [96], Pigini *et al.* [97], and Motoyama [98].

Wang *et al.* [99] claim that most of the experiments carried out to obtain LPM constants consider only air gaps (as noted in Table 9.7) and that the extrapolation of these results for practical insulators must be viewed with caution. Considering typical porcelain insulators and composite insulator strings of 110–500 kV transmission lines, Wang *et al.* [99] proposed an LPM and a set of constants to describe insulator breakdown process considering both positive and negative impulse voltages, assuming both standard and short-tail lightning impulses. The results indicate 15%–30% higher breakdown voltages under short-tail impulses than under standard impulses.

9.4.3 Final remarks

This section addresses the main flashover models available in the literature. Selection of the model for the calculation of the lightning performance of transmission lines has a direct influence on the estimated outage rate and must be made with caution. To avoid inconsistent results, attention must be given both to model constraints and to the characteristics of the voltage waveform across the insulator string.

9.5 Surge arresters

9.5.1 Introduction

Transmission line surge arrester is a protective device installed in parallel with the line insulator string to limit overvoltages between the phase conductor and the tower structure, preventing them to exceed the line insulation level and, consequently, avoiding the occurrence of a flashover.

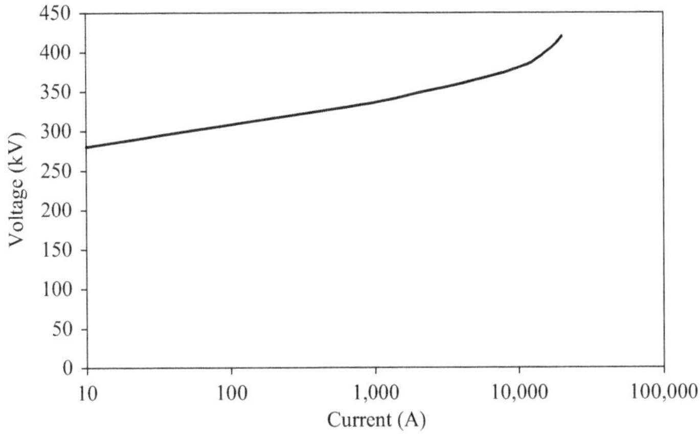

Figure 9.4 V–I characteristic of a metal-oxide surge arrester

This device presents a non-linear *V–I* characteristic that exhibits very large resistance for steady-state voltages and lower resistance for transient voltages that exceed its operating voltage. Figure 9.4 illustrates a typical *V–I* curve of a metal-oxide surge arrester.

There are several models to describe the operation of transmission line surge arresters for the estimation of the lightning performance of transmission lines. Basically, such models can be classified into two categories. The first category (i) considers the surge arrester as a non-linear resistor whose *V–I* characteristic depends only on the magnitude of the arrester current. The second category (ii) considers the frequency-dependent characteristic of surge arresters since the voltage across the arrester is a function of both the rate of rise and the magnitude of the impinging current [100].

Once a suitable model is chosen, the main difficulty is related to identifying the model parameters required to match the *V–I* characteristic of a given surge arrester. Laboratory tests and the application of trial-and-error procedures are used to determine acceptable values for such parameters.

The most common models recommended to represent transmission line surge arresters in computer simulations are the conventional model formed by a non-linear resistor [14,23], the IEEE model [100], and the Pinceti–Giannettoni model [101]. These models are briefly described in the next sections.

9.5.2 Conventional model

The conventional model used to represent gapless surge arresters in computer simulations represents the device as a non-linear resistor characterized by the following power function [14,23]:

$$i(t) = p\left[\frac{v(t)}{v_{ref}}\right]^q \tag{9.23}$$

where $i(t)$ is the arrester current, $v(t)$ is the voltage across the arrester terminals, and p, v_{ref}, and q are model constants. Specifically, v_{ref} is a reference voltage commonly defined as twice the rated voltage [23].

Since the V–I curve is characterized by two distinct regions, the application of this model commonly assumes a linear representation in the low-voltage region [23]. In the overvoltage protection region, for voltages above v_{ref}, one or more segments may be considered [14]. In this case, each segment is represented by (9.23) with a specific set of parameters.

9.5.3 IEEE model

An important drawback of the conventional model is that it does not represent the frequency-dependent behaviour of a metal-oxide surge arrester, which is important for the correct characterization of the arrester response to fast front surges.

According to IEEE working group 3.4.11 [100], the residual voltage across the arrester increases with reducing the time to peak of the arrester current for front times below 10 μs. It is indicated that the increase of the residual voltage could reach approximately 6% when the front time of the discharge current is reduced from 8 to 1.3 μs. Also, the peak voltage leads the peak current [100].

In order to consider such dynamic behaviour, IEEE working group 3.4.11 [100] proposed a surge arrester model described by the circuit illustrated in Figure 9.5. It is composed of two non-linear resistors A_0 and A_1 separated by an RL filter. Inductor L_0 represents the inductance due to the arrester leads, resistor R_0 avoids numerical instability in computer simulations, and capacitor C represents the external capacitance of the arrester. This circuit is valid for modelling gapless metal-oxide arresters only.

The proposed model matches the overall behaviour of a metal-oxide arrester as explained next. For slow-front surges, the influence of the RL filter is negligible and the two non-linear resistors are basically in parallel. For fast-front surges, the impedance of the RL filter increases, leading to an increase in the current in A_0. Since the behaviour of A_0 is characterized by a higher voltage for a given current compared to A_1 (see the curves proposed for A_0 and A_1 in [100]), the surge arrester model shown in Figure 9.5 will reproduce the increase in the residual voltage due to fast-front surges observed in actual metal-oxide surge arresters.

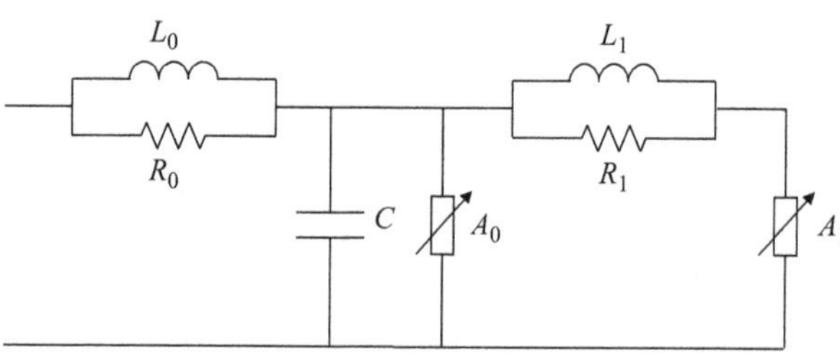

Figure 9.5 IEEE frequency-dependent model [100]

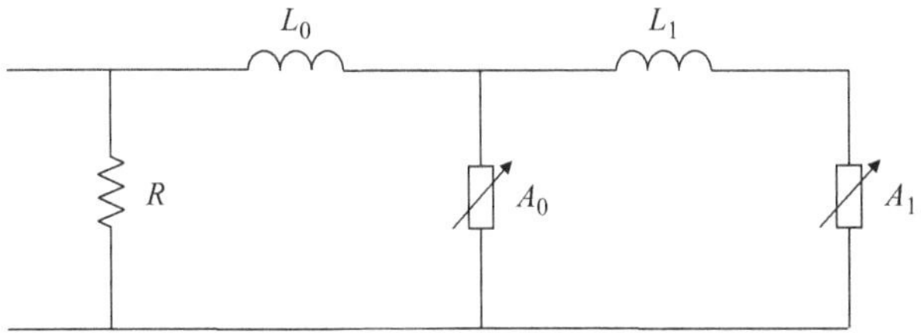

Figure 9.6 Pinceti and Giannettoni model [101]

Based on the physical dimensions of the arrester, equations to derive the initial values of L_0, R_0, L_1, R_1, C and the non-linear characteristics of the resistors A_0 and A_1 are proposed in [100]. A trial-and-error procedure may then be applied to adjust L_1 to match the residual voltages indicated in the data sheet provided by the arrester manufacturer.

This model is able to reproduce with sufficient accuracy experimental data for currents with times to crest in the range of 0.5 to 45 μs [100]. In [102], a comparison between the application of the conventional model and the IEEE model to reproduce the operation of a 120-kV surge arrester is presented, indicating that model results are essentially the same for surge front times of 0.5 μs or larger. Significant differences were observed only for surge front times below 0.5 μs.

9.5.4 Pinceti–Giannettoni model

Pinceti and Giannettoni [101] proposed a simplified model based on the IEEE model that is illustrated in Figure 9.6. Compared to the IEEE model, the capacitance is eliminated and the two resistors in parallel with the inductors are replaced by a single resistor R across the input terminals, aiming to avoid numerical instability in computer simulations.

The principle of the Pinceti–Giannettoni model is similar to that of the IEEE frequency-dependent model. However, the criterion proposed for the parameter identification is simpler, considering the estimation of model parameters directly from electrical data reported in the arrester data sheet. The physical dimensions of the surge arrester are not input data for this model.

The Pinceti–Giannettoni model considers the same A_0 and A_1 non-linear characteristics described by the IEEE model and specific formulas for determining L_1 and L_0 based on the nominal and residual voltages of the surge arrester.

9.6 Summary

In this chapter, a discussion is presented on the modelling of phase conductors, shield wires, towers, grounding systems, insulator strings, and surge arresters for the calculation of lightning overvoltages on transmission lines. It is expected that the content of this chapter will provide elements for the reader to understand the relevant aspects of the modelling of power transmission line components, with focus on the study of direct lightning strikes to high-voltage transmission lines.

References

[1] Anderson J.G. 'Lightning performance of transmission lines' in *Transmission Line Reference Book – 345 kV and above*. 2nd edn. California: Electric Power Research Institute; 1982. pp. 545–597.

[2] CIGRÉ Working Group 01. 'Guide to procedures for estimating the lightning performance of transmission lines'. 1991.

[3] IEEE Std. 1243. 'Guide for improving the lightning performance of transmission lines'. 1997.

[4] Silveira F.H., Visacro S., and Souza R.E. 'Lightning performance of transmission lines: assessing the quality of traditional methodologies to determine backflashover rate of transmission lines taking as reference results provided by an advanced approach'. *Electric Power Systems Research*. 2017;153:60–65.

[5] Martinez J.A., and Castro-Aranda F. 'Lightning performance analysis of overhead transmission lines using the EMTP'. *IEEE Transactions on Power Delivery*. 2005;20(3):2200–2210.

[6] Engelbrecht C.S., Tannemaat I., and Hesen P.L.J. 'Insulation coordination and statistical evaluation of the lightning performance with ATP/EMTP'. *Proceedings of 2015 Asia-Pacific International Conference on Lightning*. Nagoya, Japan, Jun 2015. pp. 793–798.

[7] Ishii M., and Baba Y. 'Numerical electromagnetic field analysis of tower surge response'. *IEEE Transactions on Power Delivery*. 1997;12(1):483–488.

[8] Baba Y., and Ishii M. 'Numerical electromagnetic field analysis on lightning surge response of tower with shield wire'. *IEEE Transactions on Power Delivery*. 2000;15(3):1010–1015.

[9] Visacro S., and Soares A. 'HEM: a model for simulation of lightning-related engineering problems'. *IEEE Transactions on Power Delivery*. 2005;20(2): 1206–1208.

[10] Thang T.H., Baba Y., Nagaoka N., Ametani A., Itamoto N., and Rakov V.A. 'FDTD simulation of insulator voltages at a lightning-struck tower considering ground-wire corona'. *IEEE Transactions on Power Delivery*. 2013;28(3):1635–1642.

[11] Visacro S., and Silveira F.H. 'The impact of the frequency dependence of soil parameters on the lightning performance of transmission lines'. *IEEE Transactions on Electromagnetic Compatibility*. 2015;57(3):434–441.

[12] Paul C.R. *Analysis of Multiconductor Transmission Lines*. New York: John Wiley and Sons; 1994.

[13] Rachidi F., and Tkachenko S. *Electromagnetic Field Interaction with Transmission Lines – from Classical Theory to HF Radiation Effects*. Southampton: WIT Press; 2008.

[14] Dommel H.W. *Electromagnetic Transients Program Reference Manual: EMTP Theory Book*. Portland: Boneville Power Administration; 1986.

[15] Martinez-Velasco J.A. *Power System Transients: Parameter Determination*. Boca Raton: CRC Press; 2010.

[16] Pettersson P. 'Propagation of waves on a wire above a lossy ground – different formulations with approximations'. *IEEE Transactions on Power Delivery*. 1999;14(3):1173–1180.

[17] Ametani A., Miyamoto Y., Baba Y., and Nagaoka N. 'Wave propagation on an overhead multiconductor in a high-frequency region'. *IEEE Transactions on Electromagnetic Compatibility*. 2014;56(6):1638–1648.

[18] Nakagawa M. 'Admittance correction effects of a single overhead line'. *IEEE Transactions on Power Apparatus and Systems*. 1981;100(3):1154–1161.

[19] Sunde E.D. *Earth Conduction Effects in Transmission Systems*. New York: Van Nostrand; 1949.

[20] Rachidi F. 'Transient analysis of multiconductor transmission lines above a lossy ground'. *IEEE Transactions on Power Delivery*. 1999;14(1):294–302.

[21] Carson J.R. 'Wave propagation in overhead wires with ground return'. *Bell Systems Technical Journal*. 1926;5:539–556.

[22] Deri A., Tevan G., Semlyen A., and Castanheira A. 'The complex ground return plane–a simplified model for homogeneous and multi-layer earth return'. *IEEE Transactions on Power Apparatus and Systems*. 1981;100(8):3686–3693.

[23] *ATP Alternative Transients Program Rule Book*. USA, 1997.

[24] Wise W.H. 'Potential coefficients for ground return circuits'. *Bell Systems Technical Journal*. 1948;27:365–371.

[25] De Conti A., and Emídio M.P.S. 'Extension of a modal-domain transmission line model to include frequency-dependent ground parameters'. *Electric Power Systems Research*. 2016;138:120–130.

[26] Visacro S., Alipio R., Vale M.H.M., and Pereira C. 'The response of grounding electrodes to lightning currents: the effect of frequency dependent soil resistivity and permittivity'. *IEEE Transactions on Electromagnetic Compatibility*. 2011;53(2):401–406.

[27] Visacro S., and Alipio R. 'Frequency dependence of soil parameters: experimental results, predicting formula and influence on the lightning response of grounding electrodes'. *IEEE Transactions on Power Delivery*. 2012;27(2):927–935.

[28] Alipio R., and Visacro S. 'Modeling the frequency dependence of electrical parameters of soil'. *IEEE Transactions on Electromagnetic Compatibility*. 2014;56(5):1163–1171.

[29] Cavka D., Mora N., and Rachidi F. 'A Comparison of frequency-dependent soil models: application to the analysis of grounding systems'. *IEEE Transactions on Electromagnetic Compatibility*. 2014;56(1):177–187.

[30] Marti J.R. 'Accurate modelling of frequency-dependent transmission lines in electromagnetic transient simulations'. *IEEE Transactions on Power Apparatus and Systems*. 1982;101(1):147–157.

[31] Morched B., Gustavsen B., and Tartibi M. 'A universal model for accurate calculation of electromagnetic transients on overhead lines and underground cables'. *IEEE Transactions on Power Delivery*. 1999;14(3):1032–1038.

[32] Dawalibi F.P., Ruan W., Fortin S., Ma J., and Daily W.K. 'Computation of power line structure surge impedances using the electromagnetic field

method'. *Proceedings of the IEEE/PES Transmission and Distribution Conference and Exposition*. Atlanta, USA, Nov 2001. pp. 663–668.

[33] Soares Jr. A., and Visacro S. 'Simplified expressions for tower surge impedance based on application of a field-circuit modeling approach'. *Proceedings of the 27th International Conference on Lightning Protection*. Avignon, France, Sep 2004. pp. 487–491.

[34] Hashimoto S., Baba Y., Nagaoka N., Ametani A., and Itamoto N. 'An equivalent circuit of a transmission-line tower struck by lightning'. *Proceedings of the 30th International Conference on Lightning Protection*. Cagliari, Italy, Sep 2010. pp. 936-1–936-5.

[35] Kawai M. 'Studies of the surge response on a transmission line tower'. *IEEE Transactions on Power Apparatus and Systems*. 1964;83(1):30–34.

[36] Ishii M., Kawamura T., Kouno T., Ohsaki W., Murotani K., and Higuchi, T. 'Multistory transmission tower model for lightning surge analysis'. *IEEE Transactions on Power Delivery*. 1991;6(3):1327–1335.

[37] Yamada T., Mochizuki A., Sawada J., *et al.* 'Experimental evaluation of a UHV tower model for lightning surge analysis'. *IEEE Transactions on Power Delivery*. 1995;10(1):393–402.

[38] Hara T., and Yamamoto O. 'Modelling of a transmission tower for lightning surge analysis'. *IEE Proceedings on Generation, Transmission and Distribution*. 1996;143(3):283–289.

[39] Goni O., Hossain F., Rahman M., Kaneko E., and Takahashi H. 'Simulation and experimental analyses of electromagnetic transients behaviors of lightning surge on vertical conductors'. *IEEE Transactions on Power Delivery*. 2006;21(4):1778–1786.

[40] Motoyama H., Kinoshita Y., and Nonaka K. 'Experimental study on lightning surge response of 500-kV transmission tower with overhead lines'. *IEEE Transactions on Power Delivery*. 2008;23(4):2488–2495.

[41] Motoyama H., Kinoshita Y., Nonaka K., and Baba Y. 'Experimental and analytical studies on lightning surge response of 500-kV transmission tower'. *IEEE Transactions on Power Delivery*. 2009;24(4):2232–2239.

[42] Jordan C.A. 'Lightning computations for transmission lines with overhead ground wires, part II'. *General Electric Review*. 1934;37(4):180–186.

[43] Wagner C.F., and Hileman A.R. 'A new approach to the calculation of the lightning performance of transmission line III – a simplified method: stroke to tower'. *AIEE Transactions on Power Apparatus and Systems*. 1960;79: 589–603.

[44] Sargent M.A., and Darveniza M. 'Tower surge impedance'. *IEEE Transactions on Power Apparatus and Systems*. 1969;88(5):680–687.

[45] Chisholm W., Chow Y.L., Srivastava K.D. 'Lightning surge response of transmission towers'. *IEEE Transactions on Power Apparatus and Systems*. 1983;102(9):3232–3242.

[46] Chisholm W., Chow Y.L., and Srivastava K.D. 'Travel time of transmission towers'. *IEEE Transactions on Power Apparatus and Systems*. 1985;104(10): 2922–2928.

[47] Ametani A., Kasai Y., Sawada J., Mochizuki A., and Yamada T. 'Frequency-dependent impedance of vertical conductors and a multiconductor tower model'. *IEE Proceedings on Generation, Transmission and Distribution*. 1994;141(4):339–345.

[48] Motoyama H., and Matsubara H. 'Analytical and experimental study on surge response of transmission tower'. *IEEE Transactions on Power Delivery*. 2000;15(2):812–819.

[49] Gutiérrez R.J.A., Moreno P., Naredo J.L., *et al.* 'Nonuniform transmission tower model for lightning transient studies'. *IEEE Transactions on Power Delivery*. 2004;19(2):490–496.

[50] De Conti A., Visacro S., Soares Jr. A., and Schroeder M.A.O. 'Revision, extension, and validation of Jordan's formula to calculate the surge impedance of vertical conductors'. *IEEE Transactions on Electromagnetic Compatibility*. 2006;48(3):530–536.

[51] Almeida M.E., and Correia de Barros M.T. 'Tower modelling for lightning surge analysis using Electro-Magnetic Transients Program'. *IEE Proceedings on Generation, Transmission and Distribution*. 1994;141(6):637–639.

[52] Du Y., Wang X., and Chen N. 'Circuit parameters of vertical wires above a lossy ground in PEEC models'. *IEEE Transactions on Electromagnetic Compatibility*. 2012;54(4):871–879.

[53] De Conti A., Visacro S., and Silva E.G. 'Tower surge impedance: Jordan's equations revisited'. *Przegląd Elektrotechniczny (Electrical Review)*. 2010;86(3): 22–24.

[54] Grcev L., and Rachidi F. 'On tower impedances for transient analysis'. *IEEE Transactions on Power Delivery*. 2004;19(3):1238–1244.

[55] Baba Y., and Rakov V.A. 'On the mechanism of attenuation of current waves propagating along a vertical perfectly conducting wire above ground: application to lightning'. *IEEE Transactions on Electromagnetic Compatibility*. 2005;47(3):521–532.

[56] Baba Y., and Rakov V.A. 'On the interpretation of ground reflections observed in small-scale experiments simulating lightning strikes to towers'. *IEEE Transactions on Electromagnetic Compatibility*. 2005;47(3):533–542.

[57] Grcev L., and Dawalibi F. 'An electromagnetic model for transients in grounding systems'. *IEEE Transactions on Power Delivery*. 1990;5(4):1773–1781.

[58] Grcev L., and Popov M. 'On high-frequency circuit equivalents of a vertical ground rod'. *IEEE Transactions on Electromagnetic Compatibility*. 2005;20(2): 1598–1603.

[59] Grcev L. 'Modeling of grounding electrodes under lightning currents'. *IEEE Transactions on Electromagnetic Compatibility*. 2009;51(3):559–571.

[60] Alipio R., Costa R.M., Dias R., De Conti A., and Visacro S. 'Grounding modeling using transmission line theory: extension to arrangements composed of multiple electrodes'. *Proceedings of the 33rd International Conference on Lightning Protection*. Estoril, Portugal, Sep 2016. pp. 1–5.

[61] Visacro S. 'A comprehensive approach to the grounding response to lightning currents'. *IEEE Transactions on Power Delivery*. 2007;22(1):381–386.

[62] IEEE Working Group Report. 'Estimating lightning performance of transmission lines II – updates to analytical models'. *IEEE Transactions on Power Delivery*. 1993;8(3):1254–1267.

[63] Visacro S., and Silveira F.H. 'Lightning performance of transmission lines: requirements of tower-footing electrodes consisting of long counterpoise wires'. *IEEE Transactions on Power Delivery*. 2016;31(4):1524–1532.

[64] Rudenberg R. *Electrical Shock Waves in Power Systems*. Cambridge: Harvard Univ. Press; 1968.

[65] Mousa A. 'The soil ionization gradient associated with discharge of high current into concentrated electrodes'. *IEEE Transactions on Power Delivery*. 1994;9(3):1669–1677.

[66] He J., and Zhang B. 'Progress in lightning impulse characteristics of grounding electrodes with soil ionization'. *IEEE Transactions on Industry Applications*. 2015;51(6):4924–4933.

[67] Souza R.E., Silveira F.H., and Visacro S. 'The effect of soil ionization on the lightning performance of transmission lines'. *Proceedings of the 32nd International Conference on Lightning Protection*. Shanghai, China, Sep 2014. pp. 1307–1311.

[68] Cota F.E.M., and Visacro S. 'Experimental investigation on the evolution of the soil ionization effect'. *Proceedings of XIV SIPDA – International Symposium on Lightning Protection*. Natal, Brazil, Oct 2017.

[69] Oliveira V.H.D., Pedrosa A.G., and Visacro S. 'Experimental investigation on soil ionization – first results'. *Proceedings of International Conference on Grounding and Earthing & International Conference on Lightning Physics and Effects*. Bonito, Brazil, 2012. pp. 158–161.

[70] De Conti A., and Visacro S. 'A simplified model to represent typical grounding configurations applied in medium-voltage and low-voltage distribution lines'. *Proceedings of IX SIPDA – International Symposium on Lightning Protection*. Foz do Iguaçu, Brazil, Nov 2007.

[71] He J., Gao Y., Zeng R., *et al.* 'Effective length of counterpoise wire under lightning current'. *IEEE Transactions on Power Delivery*. 2005;20(2):1585–1591.

[72] Liu Y., Theethayi N., and Thottappillil R. 'An engineering model for transient analysis of grounding system under lightning strikes: nonuniform transmission-line approach'. *IEEE Transactions on Power Delivery*. 2005;20(2):722–730.

[73] Grcev L., Markovski B., and Grceva S. 'On inductance of buried horizontal bare conductors'. *IEEE Transactions on Electromagnetic Compatibility*. 2011;53(4):1083–1087.

[74] Wu J., Zhang B., He J., and Zeng R. 'A comprehensive approach for transient performance of grounding system in the time domain'. *IEEE Transactions on Electromagnetic Compatibility*. 2015;57(2):250–256.

[75] Theethayi N., Baba Y., Rachidi F., and Thottappillil R. 'On the choice between transmission line equations and full-wave Maxwell's equations for transient analysis of buried wires'. *IEEE Transactions on Electromagnetic Compatibility*. 2008;50(2):347–357.

[76] Holdyk A., and Gustavsen B. 'Inclusion of field-solver-based tower footing grounding models in electromagnetic transient programs'. *IEEE Transactions on Industry Applications*. 2015;51(6):5101–5106.

[77] Sarajcev P., Vujevic S., and Lovric D. 'Interfacing harmonic electromagnetic models of grounding systems with the EMTP-ATP software package'. *Renewable Energy*. 2014;68:63–170.

[78] Sheshyekani K., Akbari M., Tabei B., and Kazemi R. 'Wideband modeling of large grounding systems to interface with electromagnetic transient solvers'. *IEEE Transactions on Power Delivery*. 2014;29(4):1868–1876.

[79] Gustavsen B., and Semlyen A. 'Rational approximation of frequency domain responses by vector fitting'. *IEEE Transactions on Power Delivery*. 1999;14(13): 1052–1061.

[80] Noda T. 'Identification of a multiphase network equivalent for electromagnetic transient calculations using partitioned frequency response'. *IEEE Transactions on Power Delivery*. 2005;20(2):1134–1142.

[81] Sheshyekani K., Karami H., Dehkhoda P., Paolone M., and Rachidi F. 'Application of the matrix pencil method to rational fitting of frequency-domain responses'. *IEEE Transactions on Power Delivery*. 2012;27(4): 2399–2408.

[82] Gustavsen B., and Semlyen A. 'Computer code for rational approximation of frequency-dependent admittance matrices'. *IEEE Transactions on Power Delivery*. 2002;17(4):1093–1098.

[83] Popov M., Grcev L., Hoidalen H.K., Gustavsen B., and Terzija V. 'Investigation of the overvoltage and fast transient phenomena on transformer terminals by taking into account the grounding effects'. *IEEE Transactions on Industry Applications*. 2015;51(6):5218–5227.

[84] Hileman H. *Insulation Coordination for Power Systems*. Boca Raton: CRC; 1999. pp. 627–640.

[85] Witzke R.L., and Bliss T.J. 'Surge protection of cable-connected equipment'. *Transactions of the AIEE*, 1950;69(1):527–542.

[86] Witzke R.L., and Bliss T.J. 'Co-ordination of Lightning arrester location with transformer insulation level'. *Transactions of the AIEE*. 1950;69(2):964–975.

[87] Kind D. 'Die aufbauflache bei stossspannungsbeanspruchung technischer elektrodena'. *ETZ-A*, 1958;79:65–69.

[88] Caldwell R.O., and Darveniza M. 'Experimental and analytical studies of the effect of non-standard waveshapes of the impulse strength of external insulations'. *IEEE Transactions on Power Apparatus and Systems*. 1973;PAS-92(4): 1420–1428.

[89] Chowdhuri P., Mishra A.K., and McConnell B.W. 'Volt-time characteristics of short air gaps under nonstandard lightning voltage waves'. *IEEE Trans. Power Del.* 1997;12(1):470–476.

[90] Ancajima A., Carrus A., Cinieri E., and Mazzetti C. 'Optimal selection of disruptive effect models parameters for the reproduction of MV insulators volt-time characteristics under standard and non-standard lightning impulses'. *Proceedings of IEEE Power Tech*. Lausanne, Switzerland, Jul 2007. pp. 760–765.

[91] Gomes R.M., Lima G.S., De Conti A., *et al.* 'Volt-time curve measurements of 15-kV distribution line polymeric insulators and estimation of DE parameters'. *Proceedings 2015 Asia-Pacific International Conference on Lightning*. Nagoya, Japan, Jun 2015. pp. 937–941.

[92] Chisholm W.A. 'New challenges in lightning impulse flashover modeling of air gaps and insulators'. *IEEE Electrical Insulation Magazine*. 2010;26(2): 14–24.

[93] EPRI AC Transmission Line Reference Book– 200 kV and Above, Third Edition. EPRI, Palo Alto, CA: 2005. 1011974.

[94] Wagner C.F., and Hileman A.R. 'Mechanism of breakdown of laboratory gaps'. *AIEE Transactions on Power Apparatus and Systems*. 1961;80 (3):605–618.

[95] Suzuki T., and Miyake K. 'Experimental study of breakdown voltage time characteristics of large air gaps with lightning impulses'. *IEEE Transactions on Power Apparatus and Systems*. 1977;96(1):227–233.

[96] Shindo T., and Suzuki T. 'A new calculation method of breakdown voltage–time characteristics of long air gaps'. *IEEE Transactions on Power Apparatus and Systems*. 1985;PAS-104(6):1556–1563.

[97] Pigini A., Rizzi G., Garragnati E., Porrino A., Baldo G., and Pesavento G. 'Performance of large air gaps under lightning overvoltages: experimental study and analysis of accuracy of predetermination methods'. *IEEE Transactions on Power Delivery*. 1989;4(2):1379–1392.

[98] Motoyama H. 'Experimental study and analysis of breakdown characteristics of long air gaps with short tail lightning impulse'. *IEEE Transactions on Power Delivery*. 1996;11(2):972–979.

[99] Wang X., Yu Z., and He J. 'Breakdown process experiments of 110- to 500-kV insulator strings under short tail lightning impulse'. *IEEE Transactions on Power Delivery*. 2014;29(5):2394–2401.

[100] IEEE WG 3.4.11. 'Modeling of metal oxide surge arresters'. *IEEE Transactions on Power Delivery*. 1992;7(1):302–309.

[101] Pinceti P., and Giannettoni M. 'A simplified model for zinc oxide surge arrester'. *IEEE Transactions on Power Delivery*. 1999;14(2):393–398.

[102] Miguel P.M. 'Comparison of surge arrester models'. *IEEE Transactions on Power Delivery*. 2014;29(1):21–28.

Chapter 10
Modelling of power distribution components

Alexandre Piantini[1], *Miltom Shigihara*[1]
and Acácio Silva Neto[1]

Distribution networks may be disturbed by lightning in case of either direct strikes to the line conductors or nearby strokes. Both situations may lead to overvoltages with amplitudes high enough to provoke line flashovers, damage to distribution or consumer equipment, or other power quality issues. In order to evaluate the lightning transients and their impact on a power distribution network, the knowledge of the response of the main system components when subjected to the resulting surges is of essential importance.

In this chapter, after a brief description of typical medium-voltage (MV) and low-voltage (LV) distribution networks' topologies, models of the main system components, which can be used in programs for the analysis of lightning transients, are presented and discussed. As the basis for the representation of the line conductors has already been presented in Chapter 9 in this volume, the reader is referred to that chapter for information about the transmission line equations, per-unit length parameters, surge impedance, travelling waves, and electromagnetic coupling.

The distribution transformer high-frequency behaviour plays a major role concerning both the voltages induced by indirect strokes and those transferred from the primary to the secondary side. Although the complexity of the circuit to represent the transformer depends on the phenomenon under analysis, accurate but still simple models are presented in this Chapter. These models can be used to simulate the behaviour of typical distribution transformers in the calculation of lightning-induced voltages on primary and secondary networks, as well as surges transferred to the LV side taking into account the effect of connected loads.

The analysis of the lightning performance of distribution lines entails the use of an adequate representation of the insulators. An accurate model, validated from many experiments, is presented and discussed in this chapter. Models of surge-protective devices (SPDs) used on MV and LV networks are also presented. Grounding systems, discussed in detail in Chapter 7 in this volume, have an important effect on the lightning overvoltages, and a brief summary of their

[1]Institute of Energy and Environment, Lightning and High Voltage Research Center, University of São Paulo, São Paulo, Brazil

representation is also included. Finally, as lightning transients on secondary networks are greatly affected by the characteristics of the loads, the chapter presents some models to represent the input impedances of typical LV power installations.

10.1 Typical network configurations

There are various possible configurations for power distribution grids. A brief description of the basic characteristics of typical MV and LV network topologies is presented in this section.

10.1.1 MV networks

Primary distribution systems deliver electricity from a main source substation to consumers at voltages ranging usually from 1 to 34.5 kV, but possibly at a phase-to-phase voltage of up to 69 kV [1]. They can be single- or three-phase and either overhead or underground.

Underground installations improve the aesthetics of neighbourhoods, reduce vehicular crashes with poles, reduce the exposure to electromagnetic fields, and have a better lightning performance. On the other hand, installation costs are higher and access to equipment for repairs may be more difficult. Overhead installations can operate at higher temperatures and are also advantageous in terms of service restoration, which is faster. In addition, street lights can be easily installed. However, unlike underground lines, they can be adversely affected by the weather, animals, trees, and vehicular accidents.

MV distribution networks can have different configurations depending on the density of consumer units, loads, reliability, and planning for increasing load over time. The network topology has direct bearing on its operation and performance. Most primary feeders are radially operated. The radial configuration is characterized, on the one hand, by minimal control of power flow, and, on the other hand, by straightforward circuit design, easy fault current protection, and low cost. Protection in the system is mainly relegated to fuses. A better reliability can be obtained by the use of normally open tie switches between circuits. Though the circuits operate radially, in case of a fault, the tie switches – which can be either manually or automatically operated – allow some part of the faulted circuit to be quickly restored. A more reliable service is the loop scheme. When load density is high and/or when higher reliability is desired, the mesh configuration is used, but it may require more elaborate protection and switching provisions in order to obtain the best performance. An alternative is to network the secondary system, which results in high reliability, but there are disadvantages of required excess transformer capacity and more complex protection requirements [2].

MV feeders are usually three-phase, with or without a multi-grounded neutral below the phase conductors, the former being more common. The neutral is usually 0.9–1.5 m below the phase conductors [3]. Circuit lengths vary widely, from a few kilometres to tens of kilometres. Branching from the feeders are one or more laterals, which may be three-phase, two-phase, or single-phase. For remote distribution systems in rural areas, in very sparsely populated regions, the single-phase earth return system may be applied. A non-conventional and low-cost alternative – the shield wire scheme (SWS) [4] – has been shown to be advantageous in regions such as the Amazon, where the existence of small towns and villages located relatively close to transmission lines, but far from the sub-transmission and electricity distribution systems, is not uncommon. The use of conventional MV lines for providing power supply to these communities is sometimes not feasible due to economic reasons. Loads greater than 4 MVA can be supplied with the SWS, and the insertion of this technology in the State of Rondônia, Brazil, enabled energy to be supplied to more than 40,000 people [5].

Distribution primaries may be of either conventional or compact type. Conventional lines have generally horizontal configuration and bare or covered conductors are fixed at the crossarms through ceramic or polymeric insulators. Distance between phase wires varies usually between 0.5 and 1.5 m; a typical conductor diameter is 1 cm. Compact lines – spacer cable systems – are overhead circuits with very close spacing among conductors. The conductors are covered with two or three layers of polymer and supported by a high-strength messenger, which provides mechanical support and a system neutral. The covered wires are hung beneath the messenger and supported by spacers (15–40 cm); the configuration has a fixed critical impulse flashover voltage (CFO), generally in the range of 150–200 kV [1]. Spacer cable circuits are usually installed where tree contacts are a reliability issue or in areas with right-of-way concerns, as they can be installed under existing open-wire circuits to save space.

The phase conductors are installed at a height that provides adequate electrical safety to the ground, nearby buildings, trees, highways, and railways. Regulatory agencies or power distribution companies specify the minimum clearances for each configuration, which may range, for instance, from 5.5 m, on exclusive pedestrian routes, to 12 m, on electrified railways. Span lengths vary usually from 30 m, in urban networks, up to about 120 m in rural areas, while the distance between adjacent neutral groundings is typically in the range of 100–200 m.

Detailed information of distribution power systems, including typical primary system configurations and parameters, can be found in [3].

10.1.2 LV networks

The LV network comprehends that part of the electric distribution system in which the voltage levels are up to 1,000 V. That includes the LV side of distribution transformers, the secondary circuit, and the consumers' installations.

The most common layouts are the radial, mesh, open ring, and link. The former has just one infeed point, whereas in a meshed network there are at least two possible electrical paths through which consumers can be supplied. The open ring arrangement provides at least two alternative paths to each consumer. In normal operation, each section of the ring can be treated as a radial feeder. In the event of a fault, after its isolation a normally open switch is closed and supply can be restored to the other parts of the ring. In the link arrangement, two secondary substations are interconnected. However, due to a normally open switch, the system operates as two radial feeders.

The optimum design is strongly associated with the load density, since each arrangement has its own advantages and disadvantages in terms of cost, simplicity, reliability, flexibility, voltage drop, short circuit power, and degree of protection sophistication. Link, open ring, and meshed networks are employed in densely populated urban areas, where the consumers are normally fed through underground cables. By their turn, radial overhead networks are used in both rural and urban regions.

There is a large diversity of combinations concerning the grid structure, operating criteria, and load types. Distribution transformer rating and connections, voltage level, number and type of conductors, total line length, as well as earthing practices may vary from country to country and are much dependent on the characteristics of the particular district concerned [6,7]. Nevertheless, it is possible to identify the most important network parameters required for the assessment of the basic features of the lightning surges on LV networks.

Three-phase systems may have either three or four conductors, depending on whether the neutral is present or not, whereas single-phase systems are in general composed of three conductors, two phases plus neutral. Figure 10.1 presents two typical examples of transformer connections used in secondary lines.

Both bare and separately covered conductors are commonly used at LV. Also usual is the bunched cable, in which the phase conductors are isolated and twisted around the bare neutral, that has also the mechanical function of supporting the line.

In urban areas, the LV network is commonly installed below the MV line. A typical vertical clearance between the lines is 3 m. The minimum height of the conductors above ground level is usually between about 4 and 6 m, depending on the locality. Aluminium conductors with cross sections in the range of 35–70 mm^2 are normally used [8].

The total length of a LV network may vary from 100 to about 2,000 m. In rural areas, a typical length is 1,000 m, whereas 300 m is more representative of urban networks. Average distances between loads are typically 300 and 50 m on rural and urban networks, respectively. Distances between neutral grounding points are usually in the range of 150–300 m; a common spacing is 200 m. The maximum length of the service drop from the pole to the customer's premises depends on the load, but it usually does not exceed 30 m [8].

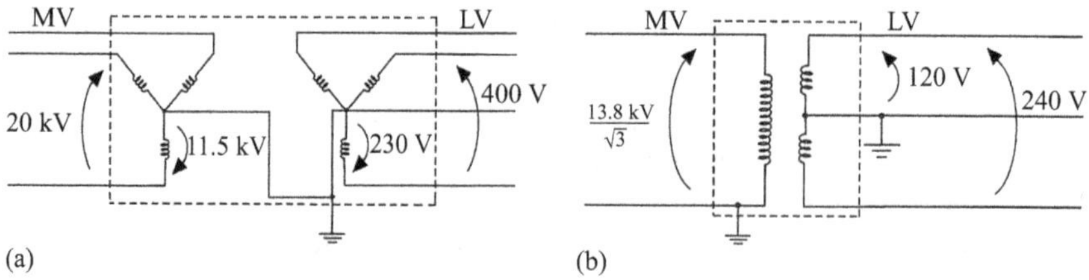

Figure 10.1 Typical examples of LV systems (adapted from [8]): (a) three-phase, four conductors; (b) single-phase, three conductors

Different types of system grounding arrangements are used in LV distribution networks. The IEC Publication 60364 series classifies these practices according to the relationships of the power system conductors and of the exposed conductive parts of components and equipment in the electrical installation with respect to the grounding system [9]. The IEC nomenclature adopts basically two letters to designate the systems. The first, which may be either 'I' or 'T', is related to the system grounding; 'I' means that all live parts of the system are isolated from earth or that points of the network are connected to ground through impedances. On the other hand, 'T' signifies a direct connection of at least one point in the network to ground. The second letter refers to the equipment grounding, and may be 'T' or 'N'. 'T' signifies that accessible conductive parts of the equipment in the installation are directly connected to ground, independently of the grounding of any point of the power system, whereas 'N' means a direct connection of accessible conductive parts of the equipment to the ground points of the power system. These connections may be made by means of a protected earth neutral (PEN) or protected earth (PE) conductor.

There are basically three types of system grounding, namely IT, TT, and TN, although the latter can be further subdivided into TN-C, TN-S, and TN-C-S depending on the way the neutral and protective functions are provided. In the TN-C system, a single conductor (PEN) performs both functions, whereas in the TN-S system separate conductors (neutral and PE) are used. The TN-C-S system is a combination of the TN-C and TN-S systems, as the neutral and protective functions are combined in one conductor from the distribution transformer to the service entrance equipment and provided by separate conductors beyond this point. The TN-C-S system is the most commonly used in public LV networks [6,10].

10.2 Modelling of distribution system components

The evaluation of lightning transients depends on the behaviour of the system components when subjected to lightning surges. In this section, models of the most important power distribution components are presented and discussed.

10.2.1 Poles

The heights of poles used in MV distribution lines are generally between 9 and 14 m. Due to the short travel time along the ground lead of the shield wire or the neutral conductor, pole effects may be either ignored [1] or represented by an inductance of about 1 μH/m.

10.2.2 Distribution transformers

Distribution transformers can be of either single or three phases and range in size from 5 to over 2,000 kVA. There are a wide variety of possibilities concerning type, connections, and winding arrangement. In rural areas, the majority are pole-mounted and rated up to 300 kVA, whereas in highly concentrated urban areas, where power ratings can exceed 2,000 kVA, transformers rated above 250 kVA are usually of the pad-mounted type [6].

There are various types of transformer connections that can be applied in three-phase systems. Both three-phase transformers and banks of single-phase units are used, the winding arrangements being related to the type of primary supply (effectively earthed, impedance-earthed, or unearthed), to the load characteristics (rated voltage, degree of unbalance), and to considerations such as primary and secondary earth faults, and susceptibility to ferroresonance. The combinations of delta and wye, as well as open-delta configurations, are among the most common ones. The grounded wye-grounded wye connection is shown in Figure 10.1(a). Guidance for application of transformer connections in three-phase distribution systems is given in [11], where all combinations of delta and wye, grounded and ungrounded, as well as other special configurations, are considered.

The evaluation of lightning transients requires the utilization of reliable models to represent all the elements involved in the phenomenon. The high frequency behaviour of the distribution transformers plays a major role concerning both the voltages induced by indirect strokes and those transferred from the primary to the secondary side. In the first case, the most important information regards the transformer impedance seen from the LV line input, whereas the analysis of transferred surges requires a more detailed representation [8].

In order to investigate the input impedances seen from the LV side, frequency domain measurements were performed by Hoidalen [12] on 15 distribution transformers with rated power from 50 to 1,250 kVA. The results showed that, in the range of 10–500 kHz, the average impedance can be represented by means of either an inductance or a capacitance, depending on the condition of the transformer neutral. If the neutral is connected to earth (TN or TT systems), the impedance can be approximated by an inductance between 4 and 40 μH, strongly correlated with the transformer rated power and voltage. On the other hand, if the neutral is isolated (IT system), a satisfactory approximation is obtained through a capacitance between 2 and 20 nF which, however, is poorly correlated with the transformer ratings [13]. Although lightning-induced voltages may have important higher-frequency components, e.g. 1 MHz, such simplified representations are usually adequate for practical purposes.

A comparison between measured and calculated input impedances seen from the LV terminals of two 13.8 kV – 220/127 V delta – earthed wye transformers with power ratings of 30 and 112.5 kVA is presented in Figure 10.2. The differences among the impedances of the terminals were not significant and therefore only one curve is shown for each transformer. It can be seen that up to about 1 MHz the impedances of both transformers can be reasonably approximated by a simple RLC parallel circuit. The values of the parameters to represent the impedance of each terminal are $R = 1{,}100\ \Omega$, $L = 48\ \mu\text{H}$, and $C = 0.76$ nF for the 30-kVA transformer and $R = 760\ \Omega$, $L = 12.4\ \mu\text{H}$, and $C = 1.0$ nF for the 112.5-kVA transformer [8]. The influence of the terminations at the high-voltage (HV) side on the measured input impedance was found to be negligible. In the tests, voltages were applied between the secondary terminals short-circuited and the transformer tank. This procedure is valid as long as the voltages induced on the phase conductors have approximately the same amplitude and waveform. This is indeed the normal

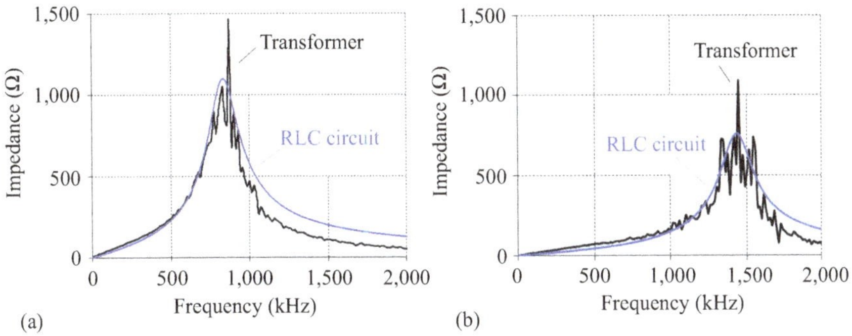

Figure 10.2 Comparison of the magnitudes of the transformer input impedance seen from each LV terminal (1) and the impedance of an RLC parallel circuit (adapted from [8]: (a) 30-kVA transformer ($R = 1{,}100\ \Omega$, $L = 48\ \mu H$, and $C = 0.76\ nF$); (b) 112.5-kVA transformer ($R = 760\ \Omega$, $L = 12.4\ \mu H$, and $C = 1.0\ nF$)

situation as far as nearby strokes are concerned, especially in case of a bunched cable, where the twisted conductors are very close to each other and can be assumed to be at the same height.

The evaluation of the voltage surges transferred from the primary to the secondary windings requires the knowledge of the voltages induced at the HV side as well as of the transformer behaviour with respect to high-frequency signals. Several models for calculating transferred voltages in case of fast transients have been proposed in the literature.

The model presented by Morched *et al.* in [14] can be used to simulate the high-frequency behaviour of any type of multi-phase, multi-winding transformer by means of combinations of RLC networks that match the frequency response of the transformer at its terminals. The frequency response can be obtained either from measurements or from models based on the physical layout of the transformer. The accuracy and numerical stability of the model were verified from tests and its validation was confirmed through comparisons between measured and simulated – using the Electromagnetic Transients Program (EMTP) [15] – step responses of a 125-MVA, 215/44-kV, wye-wye-connected transformer.

The voltages transferred to the LV transformer terminals, when lightning strikes close to a distribution line, were calculated by Borghetti *et al.* [16] for some simple configurations considering two different transformer models: the well-known capacitive PI-circuit and the model proposed by Vaessen [17], which is valid for unloaded transformers. The results showed that the voltage waveforms can be completely different depending on the model used, and that in some cases the ratio between the amplitudes of the transferred voltages is about 20:1.

The model developed by Piantini and Malagodi [18] for unloaded transformers was shown to satisfy the requirements concerning accuracy and simplicity and was applied in [19] and [20] to analyse voltages transferred to the secondary in case of nearby lightning strokes. In [21], it was modified to enable the representation of a specific transformer under the loaded condition as well. Based on the transfer characteristics of nine typical distribution transformers, delta–earthed wye connected and with power ratings ranging from 15 to 225 kVA, a further improvement was proposed by Piantini and Kanashiro [22]. The model was used in [23] to study the effect of the secondary loads on transferred voltages. The transformer was treated as a quadrupole and its transfer function and input, output, and transfer impedances were obtained through the application of impulse voltages to the short-circuited primary terminals with the secondary ones open, and vice versa [18]. Details of the modelling procedure, which does not account for saturation effects of the transformer core, are described in [24], whereas the circuit parameters corresponding to four typical distribution transformers, together with the test results that validated the model, are presented in [22].

The influence of different loads on the secondary side of an MV distribution transformer was analysed by Richter and Zeller [25]. Three cases were considered regarding the transformer termination: open, open cable, and cable with a resistive load. A simple single-phase model was presented and a relatively good agreement was obtained between measurements and computations for the situations considered. A simple linear distribution transformer model, capable of predicting transient voltages transferred from primary to secondary and vice versa, was presented by Manyahi and Thottappillil [26]. The model was based on the admittance network representation proposed in [14], with the assumption that the transient voltages at the transformer terminals are the same in all phases. Tests were performed in the open circuit condition, and a good agreement was found between the measured and simulated transferred voltages.

The voltages transferred from the MV to the LV distribution system were studied by Montaño and Cooray [27] using the Agrawal *et al.* coupling model [28] for the calculation of voltages induced on the primary line and the transformer models presented in [22,29]. From the simulations results, the authors concluded that the transformer has a significant impact on the signature of the overvoltages transferred to the secondary network. The studies conducted in [30–33] dealt with the voltages transferred to the LV network in case of direct strikes to the MV line. The investigations were performed using the model described in [22], applied however to different transformers. In the calculations done in [34] for the case of indirect strokes, the transformer was simulated by the circuit adopted in [30].

The investigation carried out by Borghetti *et al.* [35] emphasized the importance of an adequate representation of the distribution transformer in the simulation of transferred voltages. The calculations were done with the LIOV-EMTP code [36–38] considering two transformer models: the pure capacitive PI-circuit and the more complex and accurate model presented in [14]. The parameters of both models were obtained experimentally and the results showed that the capacitive PI-circuit model overestimates the transferred voltages by about one order of magnitude in comparison with the high-frequency model proposed in [14]. The main conclusion of the paper regards the crucial role played by the transformer in the simulation of transferred overvoltages.

A model of core-type distribution transformers that can be used to evaluate transferred surges was presented by Noda *et al.* [39]. It consists of the equivalent circuit of the transformer with circuit blocks representing winding-to-winding and winding-to-enclosure capacitance, skin effects of winding conductors and iron core, and multiple resonances due to the combination of winding inductance and turn-to-turn capacitance. The model parameters can be determined by measurements using an impedance analyser. Transient simulations using the EMTP [15], considering a 10-kVA unloaded transformer modelled with the proposed method, agreed well with laboratory test results. However, it was later observed that the model gives inaccurate results for a specific type of transformer because the skin effect of the secondary windings was neglected. Therefore, an improved model taking the skin effect into account was proposed by Honda *et al.* [40] and field tests were performed using an actual-scale distribution line. The primary line was three-phase, three-wire, 430 m long and matched at both ends, whereas the LV line was one-phase, three-wire, 164 m long, and open-ended. Through an impulse generator, currents were injected at different points of the system, simulating direct and indirect lightning hits. Comparisons of measured and calculated voltages at the secondary side of a 10 kVA transformer confirmed the accuracy of the model.

An adequate representation of the high-frequency behaviour of the power transformers is essential for the appraisal of the surges transferred to the LV network. Models differ remarkably and the desired compromise between accuracy and simplicity depends on the specific application. Figure 10.3 depicts the simplified model obtained in [22,23] for various three-phase, 13.8 kV – 220/127 V delta – grounded wye connected distribution transformers. The parameters to represent each phase of two of these transformers, rated 30 and 112.5 kVA, are given in Table 10.1. Figure 10.4 presents comparisons between measured and calculated transferred voltages in case of secondary both in open circuit and connected to a balanced load, for impulses with different waveforms applied to the short-circuited primary terminals of the 30-kVA transformer. The load was simulated by three resistors of the same value connected in a grounded wye configuration, and the voltages were measured across one of the resistors. Similar comparisons are presented in Figure 10.5 for the transformer rated 112.5 kVA. The relatively good

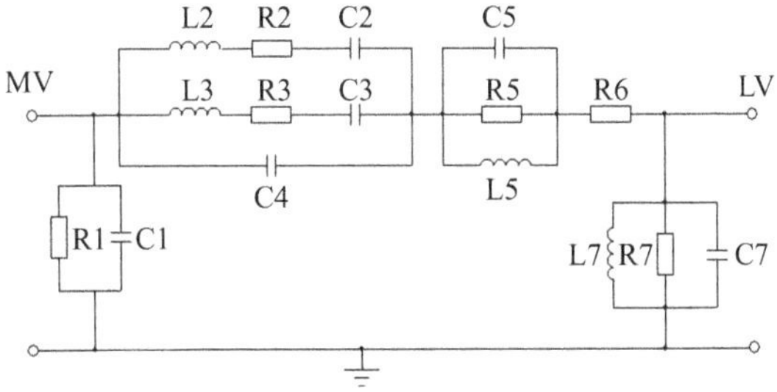

Figure 10.3 Simplified distribution transformer model for the evaluation of transferred voltages (adapted from [18] and [19])

Table 10.1 Parameters of the model shown in Figure 10.3 for the transformers rated 30 and 112.5 kVA [8,22,23]

Parameters	Transformer	
	30 kVA	**112.5 kVA**
R1 (kΩ)	–	–
R2 (kΩ)	14	3
R3 (kΩ)	0.8	5
R5 (kΩ)	–	–
R6 (kΩ)	1.1	0.35
R7 (kΩ)	1.615	1.5
C1 (pF)	493	600
C2 (pF)	94.8	1,126
C3 (pF)	21.5	146
C4 (pF)	50	600
C5 (pF)	–	400
C7 (pF)	760	850
L2 (mH)	16	35
L3 (mH)	1.84	15
L5 (mH)	–	–
L7 (mH)	0.05	0.0124

agreement observed, also in comparisons involving other transformers and load conditions [22,23], indicates the suitability of the model for the evaluation of transferred lightning surges.

In [41], Gustavsen and Semlyen presented an efficient methodology for transient modelling of power transformers based on measured or calculated frequency responses. The responses are approximated with rational functions by means of the method of vector fitting [42], which is shown to be well suited for this purpose. Applications of the methodology can be found, e.g., in [43–51].

10.2.3 Insulators

The assessment of the lightning dielectric strength of power equipment is based on tests performed using the standard lightning impulse voltage (1.2/50-μs waveshape). However, lightning overvoltages may vary widely as they depend on the line configuration, soil characteristics, and on many lightning parameters [52–59]. The occurrence of a line flashover depends not only on the surge magnitude, but also on its polarity and waveshape. Therefore, the analysis of the insulator behaviour under non-standard impulses requires a model to estimate the corresponding volt–time curves. Physical-model formulas that may be used to produce volt–time curves for various non-standard impulses fall into the following three general categories [60–62]:

(a) methods that model the breakdown phenomena directly, such as the leader progression model [63–65];
(b) integration methods such as the disruptive effect (DE) model [66–72];
(c) methods that use the standard volt–time curve directly during the period before the waveshape becomes non-standard [73,74].

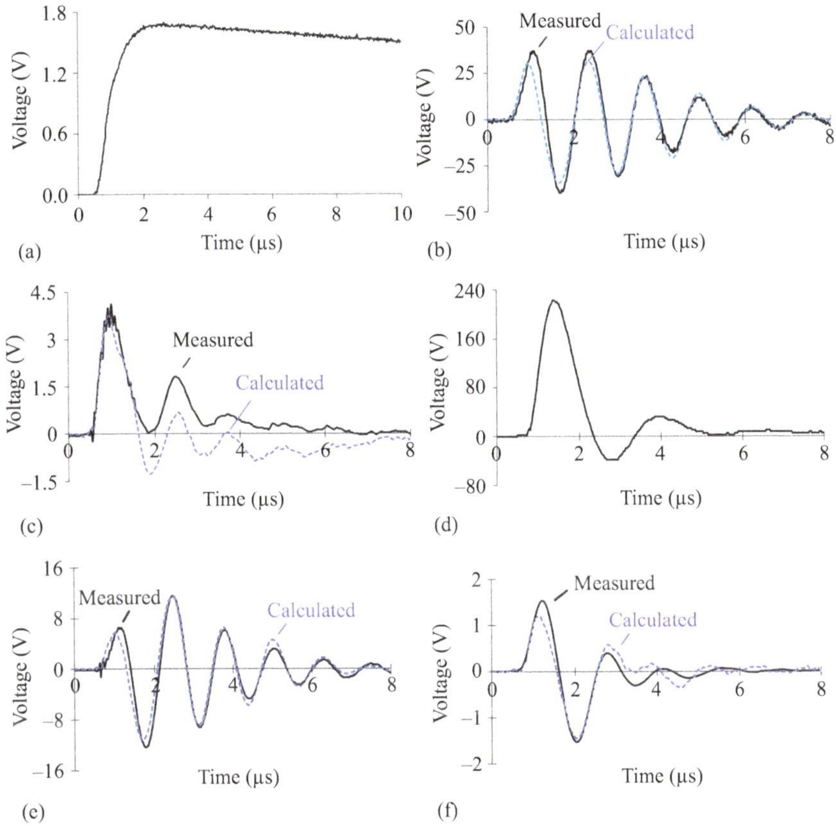

Figure 10.4 Measured and calculated voltages transferred to the secondary of the 30-kVA transformer under different load conditions (adapted from [8]): (a) voltage applied to the HV terminals (standard 1.2/50 μs waveform, with amplitude of 1.70 kV for Panel b and 0.95 kV for Panel c); (b) transferred voltages (no-load condition) corresponding to Panel a; (c) transferred voltages (balanced load of 50 Ω) corresponding to Panel a with amplitude of 0.95 kV; (d) voltage applied to the HV terminals (Panels e and f); (e) transferred voltages (no-load condition) corresponding to Panel d; (f) transferred voltages (balanced load of 50 Ω) corresponding to Panel d

All these methods are in current use for the estimation of the dielectric strength of transmission line insulators under lightning conditions, but none is universally accepted [74].

Distribution insulators are commonly simulated as a switch which closes when the instantaneous voltage across its terminals exceeds a certain value, usually the CFO [75–81]. However, for the analysis of the line performance against indirect strokes, the value of 1.5× CFO is recommended by [1]. As the overvoltages induced by nearby strokes tend to have short pulse width compared to the time to

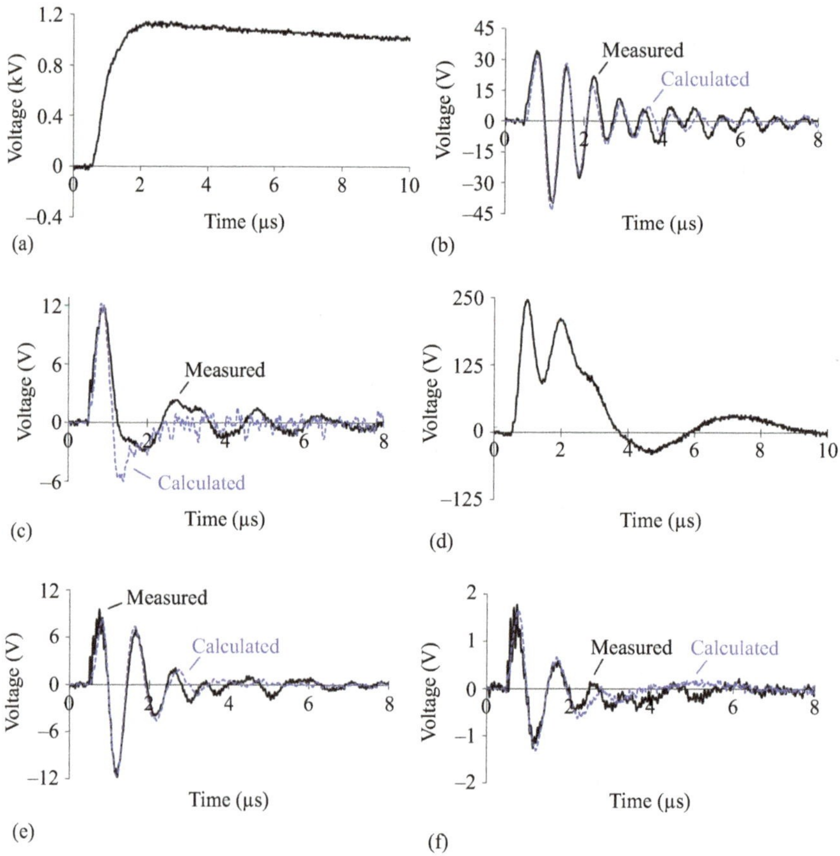

Figure 10.5 Measured and calculated voltages transferred to the secondary of the 112.5-kVA transformer under different load conditions (adapted from [8]): (a) voltage applied to the HV terminals (standard 1.2/50 μs waveform, Panels b and c); (b) transferred voltages (no-load condition) corresponding to Panel a; (c) transferred voltages (balanced load of 50 Ω) corresponding to Panel a; (d) voltage applied to the HV terminals (Panels e and f); (e) transferred voltages (balanced load of 100 Ω) corresponding to Panel d; (f) transferred voltages (balanced load of 10 Ω) corresponding to Panel d

half value of a typical stroke [1], this approximation accounts for the turn-up in the insulation volt–time curve. The same 1.5 factor is used in the analysis of direct strokes to lines protected with shield wires or surge arresters [1], as it is assumed that reflections from adjacent poles will quickly reduce the tail and, if flashover does not occur at the time to crest assumed for the stroke current (2 μs), it will not occur at all. The 1.5 factor approximates the turn-up in the insulation volt–time curve at 2 μs, as illustrated in Figure 10.6.

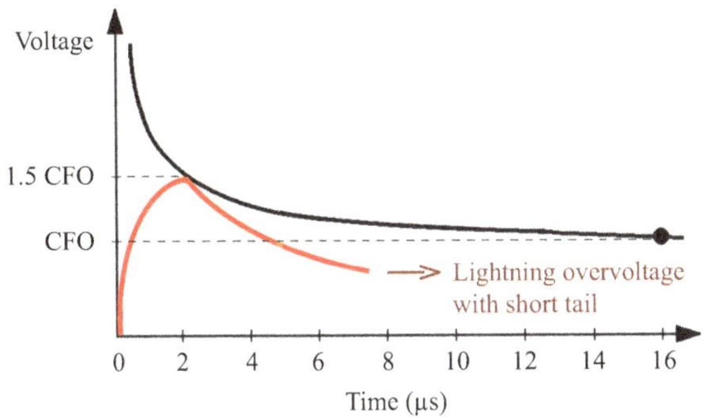

Figure 10.6 Turn-up in the insulation volt–time curve at 2 µs

The DE method [66–72] has also been extensively used to model the lightning impulse flashover process [21,31–33,59,82–84], as it provides a more suitable criterion to indicate the occurrence of flashovers, i.e. to close the switch that represents the insulator. The model can be generally expressed by the following equation:

$$DE = \int_{t_0}^{t_b} [v(t) - V_0]^k dt \qquad (10.1)$$

where $v(t)$ is the applied impulse voltage stressing the dielectric at time t, V_0 is the onset voltage, t_0 is the time at which $v(t_0) = V_0$, and t_b is the time to breakdown. *DE* is assumed to be a function of both voltage amplitude and time, but these factors do not necessarily have the same importance. The exponent K allows varying the relative weight given to these two quantities. Kind [68] and Darveniza and Vlastos [69] suggest that $K = 1$, whereas Hileman proposes $K = 1.36$ and Ancajima *et al.* [72] and Shigihara *et al.* [82] adopt the equation proposed by Chowdhuri *et al.* [70]:

$$K = \alpha \frac{v(t)}{V_0} \qquad (10.2)$$

where α depends on the type of air gap and waveshape.

According to the method, flashover occurs if *DE*, obtained from (10.1), exceeds a certain critical value *DEc*.

Different procedures exist for the estimation of the parameters V_0, K, and *DEc* required for the application of the method. In [85] and [86], respectively, the method by Shigihara *et al.* [82] was applied to estimate the parameters of the DE model for the prediction of the volt–time curves of typical 15- and 24-kV pin-type porcelain insulators. The corresponding parameters are presented in Table 10.2.

In [85,86], the measured volt–time characteristics corresponding to the standard and to four non-standard lightning impulse waveshapes (1.2/4, 1.2/10, 3/10, and 7.5/30 µs) were compared with those predicted by three different procedures

Table 10.2 Parameters for the application of the DE model to distribution insulators according to Shigihara et al. (adapted from [82]). The CFO can be obtained either from tests or from manufacturers' catalogues

Voltage class (kV)	Polarity	CFO (kV)	V_0 (kV)	a	*DEc*
15	Negative	133.2	0.63 × CFO	0.28	0.00033
24		164.4	0.50 × CFO	0.20	0.00030
15	Positive	112.8	0.70 × CFO	0.22	0.00085
24		142.9	0.56 × CFO	0.19	0.00022

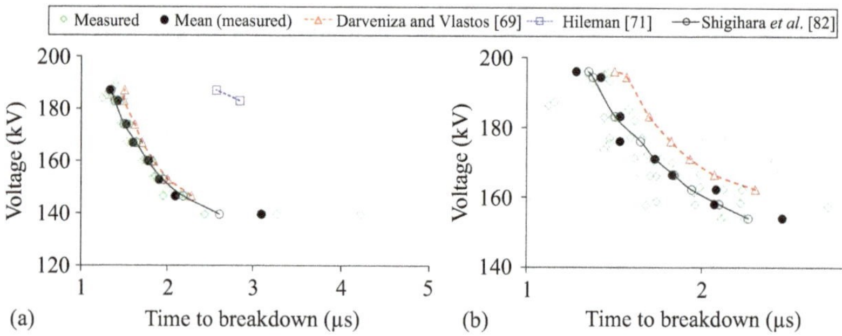

Figure 10.7 Volt–time curves for the 1.2/4-μs waveshape (15-kV insulator). Adapted from [85]. (a) positive polarity; (b) negative polarity

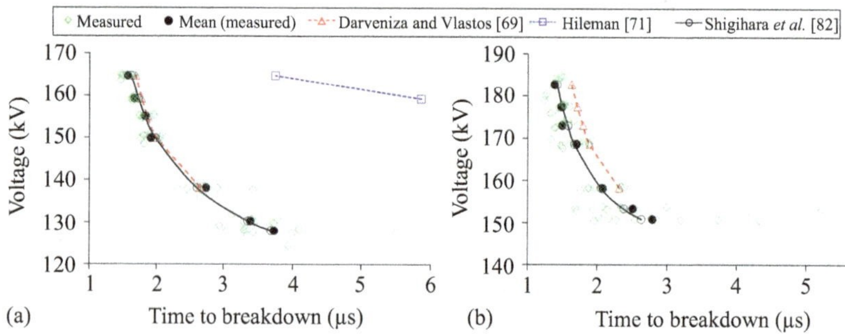

Figure 10.8 Volt–time curves for the 1.2/10-μs waveshape (15-kV insulator). Adapted from [85]. (a) positive polarity; (b) negative polarity

related to the DE model, namely Darveniza and Vlastos [69], Hileman [71], and Shigihara *et al.* [82]. For illustration, some of these comparisons are presented in Figures 10.7–10.11, which show all the measured points (i.e. the points corresponding to all the voltage applications), as well as the mean values corresponding

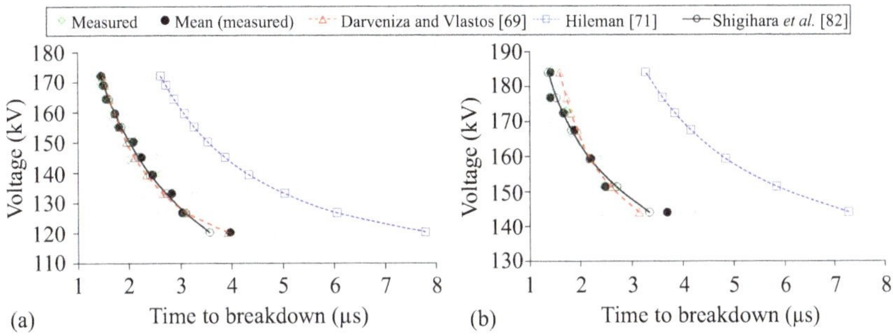

Figure 10.9 Volt–time curves for the 1.2/50-μs waveshape (15-kV insulator). Adapted from [85]. (a) positive polarity; (b) negative polarity

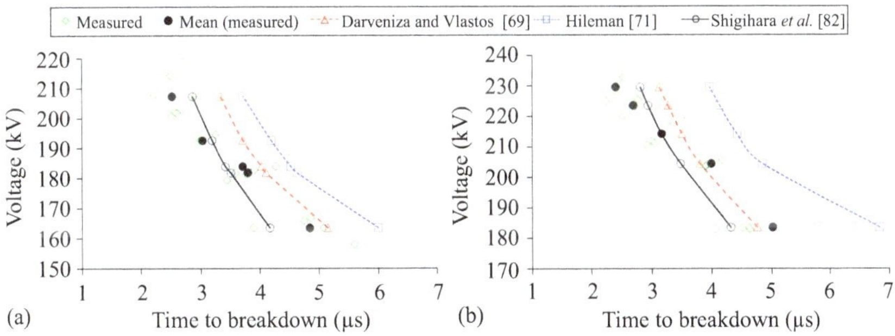

Figure 10.10 Volt–time curves for the 3/10-μs waveshape (24-kV insulator). Adapted from [86]. (a) positive polarity; (b) negative polarity

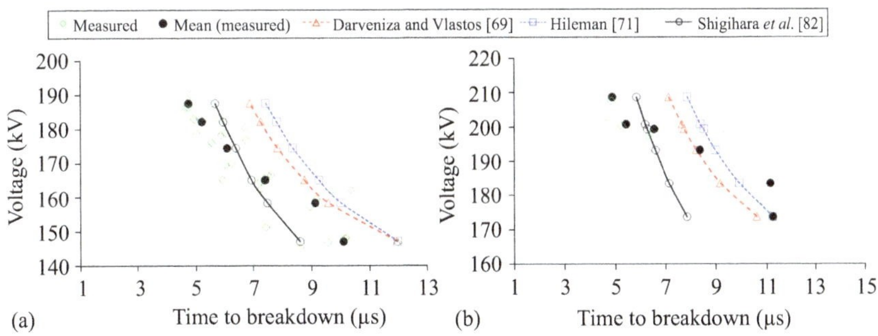

Figure 10.11 Volt–time curves for the 7.5/30-μs waveshape (24-kV insulator). Adapted from [86]. (a) positive polarity; (b) negative polarity

to each voltage level. The dispersion of the times to breakdown tends to increase as the peak value of the applied voltage diminishes.

The results show that, for some impulse waveshapes and polarities, the method by Darveniza and Vlastos [69] does not predict insulator breakdown for the lower voltage levels. This can be observed, e.g., in Figures 10.7 and 10.8. The values of the parameters relevant to the procedure by Hileman [71] may lead to large errors when applied to distribution insulators – flashovers are not predicted, e.g., for the cases shown in Figures 10.7(b) and 10.8(b), but actually it was formulated for application to transmission systems, for which it yields satisfactory results. On the other hand, a relatively good agreement was found between experimental results and calculations performed using the method proposed by Shigihara *et al.* [82] for the 15 and 24 kV insulators and the five impulse waveshapes considered.

10.2.4 Surge arresters and LV SPDs

Models used for transmission line arresters in transient simulations can also be applied to MV distribution surge arresters. The equations to derive the initial values of the parameters required for the IEEE and the Pinceti–Giannettoni models described in Chapter 9 in this volume are presented in this subsection, as well as some typical volt–current (*V–I*) characteristics of distribution arresters. A brief discussion about modelling the behaviour of LV SPDs is also presented.

10.2.4.1 IEEE model

Obtaining the parameters of the circuit shown in Figure 9.5 in this volume – reproduced in Figure 10.12 for convenience – is the main difficulty in using the IEEE model. However, some formulas based on the height of the arrester and the number of parallel columns of metal oxide discs can be used for an initial estimation [87].

The inductance L_1 and the resistance R_1 comprising the filter between the two non-linear resistors can be estimated as

$$L_1 = \frac{15d}{n}\ (\mu\text{H}) \tag{10.3}$$

$$R_1 = \frac{65d}{n}\ (\Omega) \tag{10.4}$$

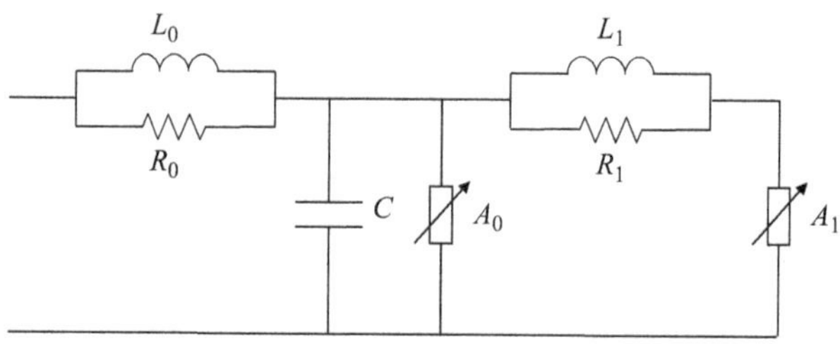

Figure 10.12 IEEE frequency-dependent model (adapted from [87])

where d is the estimated height of the arrester in metres (use overall dimensions from catalogue data) and n is the number of parallel columns of metal oxide in the arrester.

The values of L_0, R_0, and C can be estimated according to the following expressions:

$$L_0 = 0.2\frac{d}{n}\ (\mu\text{H}) \tag{10.5}$$

$$R_0 = 100\frac{d}{n}\ (\Omega) \tag{10.6}$$

$$C = 100\frac{n}{d}\ (\text{pF}) \tag{10.7}$$

Regarding the non-linear V–I characteristics of the resistors A_0 and A_1, they can be estimated from the per unitized curves given in Figure 10.13, where the reference voltage V_{10} is the resulting peak voltage corresponding to a test current impulse with peak of 10 kA and 8/20 μs waveshape.

10.2.4.2 Pinceti–Giannettoni model

The circuit corresponding to the Pinceti–Giannettoni model [88], which is derived from the IEEE model [87], is shown in Figure 9.6 in this volume, reproduced in Figure 10.14 for convenience. Its main innovation regards the simplicity of the criteria proposed for the identification of the parameters, which are obtained as follows:

- the V–I characteristics of the non-linear resistors A_0 and A_1 are derived from those shown in Figure 10.13;
- $R = 1$ MΩ, and it aims only at avoiding numerical instabilities;

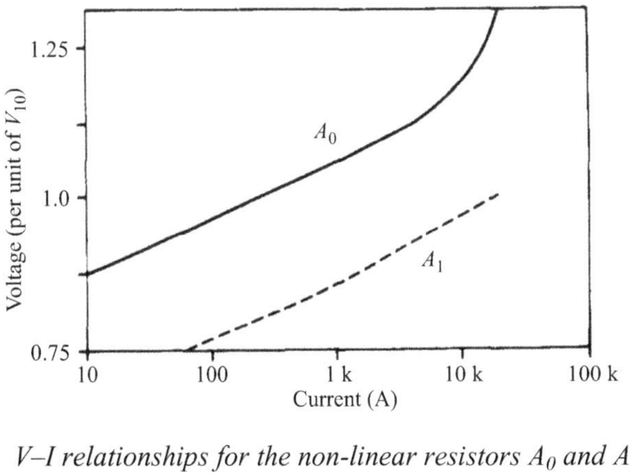

Figure 10.13 V–I relationships for the non-linear resistors A_0 and A_1 of the IEEE model (adapted from [87])

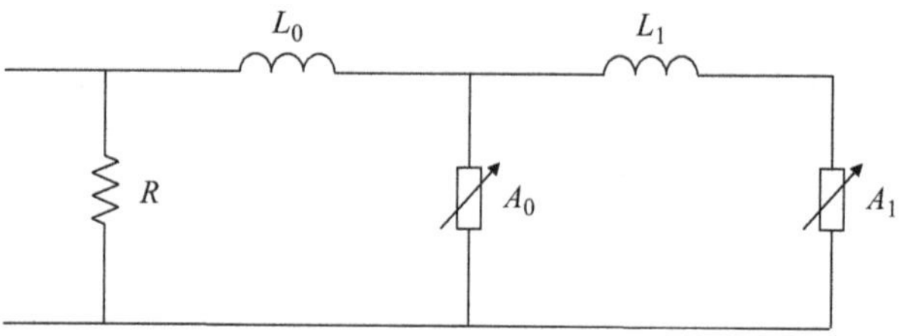

Figure 10.14 Pinceti–Giannettoni model (adapted from [88])

- the inductances L_0 and L_1 are calculated from

$$L_0 = \frac{1}{12}\frac{V_{r(1/T_2)} - V_{r(8/20)}}{V_{r(8/20)}} V_n \text{ (μH)} \tag{10.8}$$

$$L_1 = \frac{1}{4}\frac{V_{r(1/T_2)} - V_{r(8/20)}}{V_{r(8/20)}} V_n \text{ (μH)} \tag{10.9}$$

where V_n is the arrester-rated voltage, $V_{r(1/T_2)}$ is the residual voltage at 10 kA for fast front current surges ($1/T_2$ μs), and $V_{r(8/20)}$ is the residual voltage for 10 kA surge with 8/20-μs waveshape. As mentioned in [88], although different values may be used by different manufacturers for the tail time T_2, this fact does not cause any trouble because the residual voltage reaches its peak value on the rising front of the impulse current.

10.2.4.3 Other models

The required iterative procedures and the need of data not reported on manufacturers' datasheets might bring some difficulties in the calculation and adjustment of the parameters required for the IEEE [87] and the Pinceti–Giannettoni [88] models. In [89], a simplified model, derived from [87], was proposed by Fernández and Díaz, who presented a straightforward approach for obtaining its parameters based on manufacturer's data. The results obtained with the model, whose equivalent circuit is shown in Figure 10.15, were verified through simulations and measurements.

Besides the previous models, gapless distribution arresters are often represented, in lightning transient studies, by their V–I characteristics [58,90–92], while

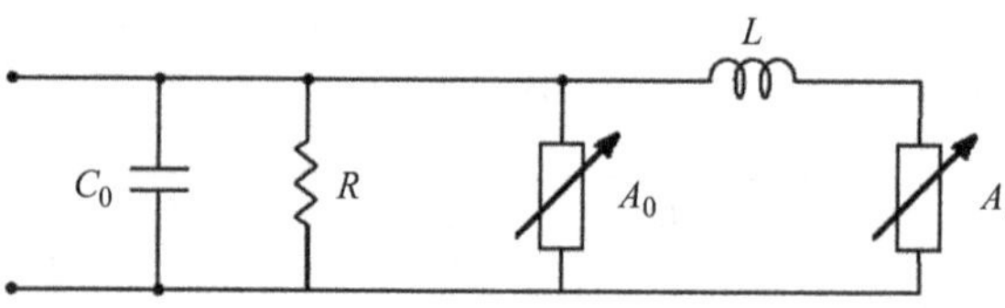

Figure 10.15 Equivalent circuit of the Fernández–Díaz model (adapted from [89])

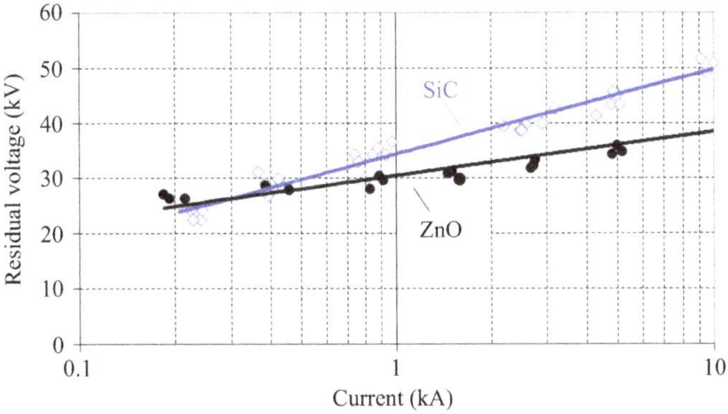

Figure 10.16 V–I characteristics of SiC and ZnO surge arresters (12 kV, 5 kA). Adapted from [57]

gapped arresters are commonly simulated by means of a gap in series with the non-linear resistor [93]; see also Matsuda *et al.*'s work [80].

A comparison between the curves of typical silicon-carbide (SiC) and zinc-oxide (ZnO) arresters is presented in Figure 10.16. The rated voltage and rated discharge current of the nine arresters (3 ZnO and 6 SiC), of different manufacturers, are 12 kV and 5 kA, respectively. As the scatters among the results relative to surge arresters of the same type were small, the results corresponding to each type were grouped. The curves depicted in Figure 10.16 were used as reference for simulating surge arresters in scale model experiments [57,90,94].

10.2.4.4 Low-voltage SPDs

As mentioned in Chapter 8 in this volume, varistors can have a quite complex operation characteristic. Figure 8.49 in this volume shows that, besides a non-linear resistance, the equivalent circuit is composed of a stray capacitance, a stray inductance, the resistances of intergranular boundaries, and the bulk resistance of the varistor. The circuit is reproduced in Figure 10.17, for convenience. Information

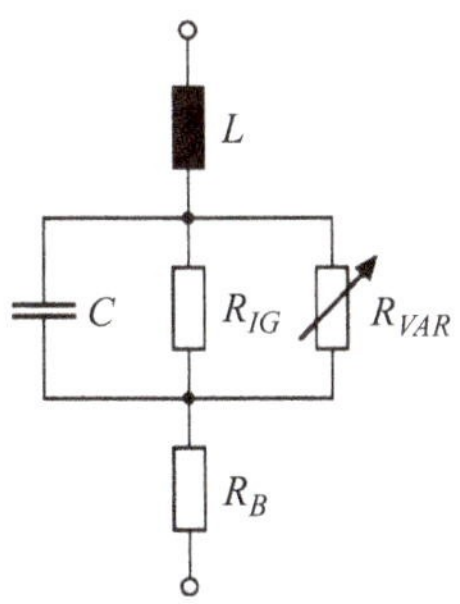

Figure 10.17 Equivalent circuit of a varistor. L: inductance of leads; C: capacitance; R_{IG}: resistance of intergranular boundaries; R_{VAR}: non-linear resistance (ideal varistor); R_B: bulk resistance

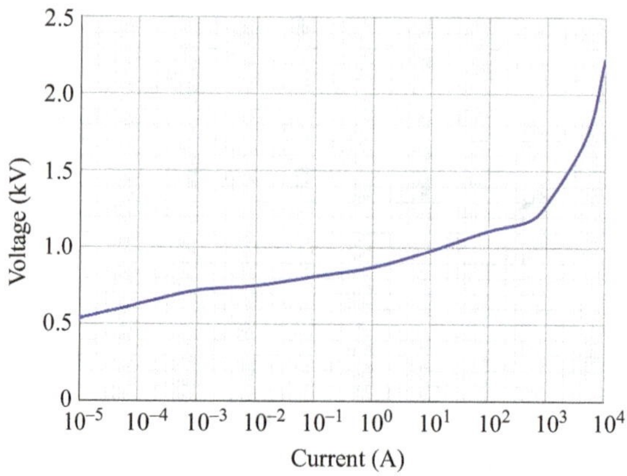

Figure 10.18 V–I characteristic of the non-linear resistor which, in parallel with a 780-pF capacitor, constitutes the equivalent circuit of the LV SPDs considered in Chapter 5 in Volume 2 (adapted from [8])

about the circuit components, as well as examples of *V–I* curves of different varistors, is given in Section 8.3.7.3 in this volume.

Simpler models are commonly used to represent the behaviour of LV SPDs. In [80], varistors are simulated by their *V–I* curves, whereas in [8] – and in Chapter 5 in Volume 2 – a capacitor of 780 pF is considered in parallel with a non-linear resistor with the *V–I* characteristic depicted in Figure 10.18, which is representative of a typical 440-V varistor.

10.2.5 Grounding

Distribution grounding systems are characterized by small dimensions – usually one or a few interconnected grounding rods – due both to the lack of space for installation and high costs resulting from the large number of grounding structures. In order not to impair the performance of the overcurrent protection system, high values of ground resistances observed in critical, high-resistivity soil conditions are compensated through an increase in the number of grounding points [95]. However, this approach is not enough to prevent flashovers caused by lightning overvoltages. Due to the inherent high-frequency content of the lightning surges, the performance of distribution systems is strongly affected by the ground resistance locally [96–98] and high values cannot be compensated by increasing the number of grounding points. Thus, to achieve a proper protection, it is essential to implement technical procedures to ensure not only that the ground resistance values measured during the implementation of the grounding system are within the limits set by the standards, but also that they will remain within these limits in all seasons of the year.

IEEE Std. 81 [99] points out the significant influence of moisture content, temperature, and soil compaction on the resistivity, but it does not suggest measurement procedures to minimize the problem. In case of substations, IEEE Std. 80

[100] recommends performing periodic measurements for the verification of maximum values of the ground resistance. However, this procedure is not economically feasible for distribution systems due to the high number of grounding points. Aiming at finding solutions to this problem, an investigation on the influence of the seasonal moisture on soil resistivity and its effect on distribution grounding systems was performed by Coelho *et al.* [95]. It was shown that the soil moisture has great influence on resistivity and that the influence of seasonal moisture on the ground resistance depends not only on the characteristics of the topsoil, but also on the resistivity of the deeper layers and grounding system topology. From measurements considering the most representative soils of the southern region of Brazil, an extrapolation procedure was proposed to determine the critical values for resistivity and ground resistance of small grounding structures.

The influence of the soil moisture on the apparent resistivity depends both on the size of the grounding system and on the soil characteristics. For small grounding systems, such as those usually used in power distribution systems, the apparent soil resistivity depends significantly on the precipitation conditions. Such influence tends to decrease as the system dimension increases, but the degree of variation depends on the soil type. In general, the apparent resistivity of compact or shallow soils is only slightly affected by seasonality in case of grounding systems with equivalent radius larger than about 6 m. For extremely porous and very deep soils, a similar condition is observed only in case of systems of much higher dimensions [95].

Due to the small dimensions of distribution grounding systems, they can be reasonably well modelled by the impulse impedance, which is defined as the ratio of the peak values of the ground potential rise (GPR) and impressed current. As shown in Chapter 7 in this volume, such representation allows promptly determining the maximum GPR value for any given peak current.

10.2.6 Loads

Lightning transients on secondary networks are greatly affected by the characteristics of the loads. The input impedance of a LV power installation, i.e. the impedance seen by the power line at the service entrance, varies according to the grounding system arrangement, number and configuration of the circuits, type, size and length of the conductors, and frequency response of the connected electric appliances. As a consequence, remarkable differences may be observed between the impedances of distinct installations. This justifies, to a certain extent, the use of simple models in the analysis of lightning transients, where loads are often simulated as lumped resistors [101–103], inductors [31], or capacitors [104]. A more accurate representation of the load frequency-dependent behaviour, however, may allow for a more precise evaluation of the main features of the overvoltages [12].

Although the connected electric appliances have a major influence on the input impedance of TN systems, for frequencies above 100 kHz such impedance can be approximated by an inductance in the range of 2–20 μH, as indicated by measurements in the range of 5 kHz to 2 MHz performed by Hoidalen on several residential units [12,13]. The inductance decreases with the number of circuits and branches, so that lower values correspond to larger installations.

By its turn, the impedance of an IT system is affected to a lesser degree by the connected consumer loads. Its behaviour is typically capacitive up to about 100 kHz, turning into inductive for higher frequencies. The equivalent circuit can be modelled by a capacitance between 2 and 200 nF in series with an inductance in the range of 2–20 μH. Larger capacitance values are associated with larger installations. Figure 10.19 presents the frequency responses of the equivalent circuits of the input impedances of a 127 m^2 residential apartment [13]. The measurements were performed at the meter cabinet between the phase conductors and the PE conductor, with the installation disconnected from the distribution network. Data relative to the IT system were obtained by connecting the neutral to the phase wires, whereas the tests related to the TN system were carried out with the neutral connected to the PE conductor.

LV consumers use microelectronic components which are sensitive to short circuit faults, large load variations, power interruptions, and lightning surges. Immunity test requirements for residential, commercial, and industrial apparatuses in relation to continuous and transient, conducted, and radiated disturbances,

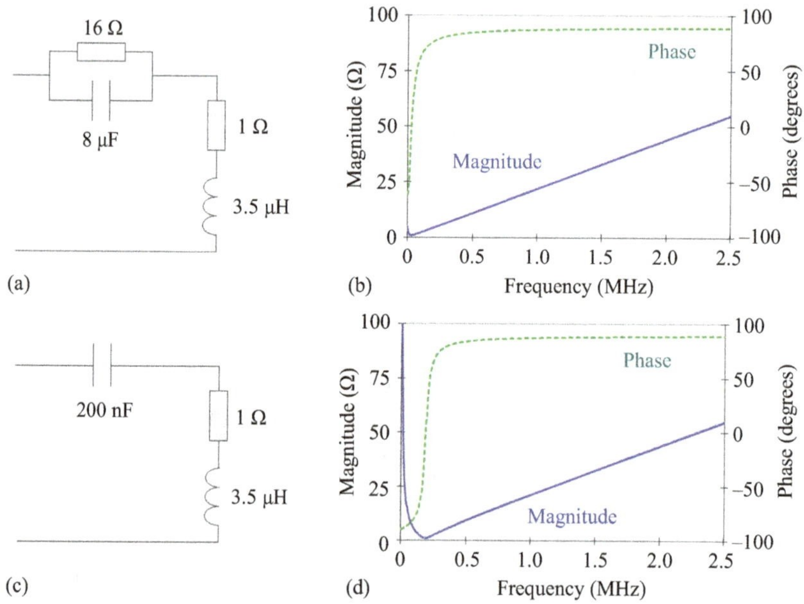

Figure 10.19 Equivalent circuits and corresponding frequency responses of the input impedances of typical LV power installations: (a) equivalent circuit (TN system); (b) frequency response (TN system); (c) equivalent circuit (IT system); (d) frequency response (IT system) (adapted from [13])

including electrostatic discharges, are specified by IEC publication 61000-6 series [105,106]. Immunity test requirements for electrical fast transients and surges are specified by IEC publication 61000-4 series [107,108]. Regarding withstanding capability, IEC 61000-6-1 [105] specifies immunity levels (line-to-earth: ±2 kV; line-to-line: ± 1kV) for surges in the supply voltage for residential, commercial, and light-industrial environments.

According to IEC 60364-4-44 [109], equipment shall be selected so that its rated impulse withstand voltage is not less than the required impulse withstand voltage as specified in Table 10.3. It is the responsibility of each product committee to include the rated impulse withstand voltage in their relevant standards [110].

The Information Technology Industry Council (ITI) [111] curve describes the magnitude and duration of steady-state and transient voltage disturbances which typically can be tolerated, with no interruption in function, by most information technology equipment connected to 120 V, 60 Hz systems. It replaces the curves formerly developed by the Computer Business Equipment Manufacturers Association (CBEMA), the ITI's forerunner organization, and describes conditions for seven types of events, as well as a no-damage region and a prohibited region [110].

The withstand levels of many pieces of household equipment were determined in [112] through the use of a combination wave generator (i.e. a generator which applies a 1.2/50-μs voltage wave when driving an open circuit and a 8/20-μs current wave when connected to a short-circuit) and a ring wave generator. The pieces of equipment were tested by supplying terminals connected to the source in the following conditions: connected (operating), stand-by (connected to supply, but in OFF status), and disconnected (by a mechanical switch). Tests were carried out by gradually increasing surge severity until the withstand level was reached. The severity was increased until the equipment was damaged. Table 10.4 shows the results, in which the values are generally higher than the immunity values specified by IEC 61000-6-1 [105].

Table 10.3 Required rated impulse withstand voltage of LV equipment (adapted from [109])

Nominal voltage of the installation (V)		**Required impulse withstand voltage (kV) between live conductors and PE for**			
Three-phase systems*	**Single-phase systems with middle point**	**Equipment at the origin of the installation (overvoltage category IV)**	**Equipment of distribution and final circuits (overvoltage category III)**	**Appliances and current using equipment (overvoltage category II)**	**Specially protected equipment (overvoltage category I)**
–	120–240	4	2.5	1.5	0.8
230/400* 277/480*	–	6	4	2.5	1.5
400/690	–	8	6	4	2.5

*In Canada and USA, for voltages to ground higher than 300 V, the impulse withstand voltage corresponding to the next highest voltage in column one applies

Table 10.4 Surge withstand voltage of some appliances based on tests [112]

Appliance	Surge withstand voltage (kV)	
	Common mode	Differential mode
VCR*	2§	1
DVD Player†	≥6§	2
Television†	≥6§	2
PC*	2	3.5
Multifunctional printer†	≥6	2
Microwave oven*	4	
Audio micro system*	≥6§	2
Fax machine*	2	4
Air conditioner*	≥6	2
Refrigerator‡	≥6	≥6
UPS*	2	≥6

*Fixed voltage input
†Automatically regulated voltage input
‡Refrigerator without electronic control
§Equipment without grounding terminal (voltage applied between active conductors and a metallic plate under the equipment)

10.3 Concluding remarks

This chapter aimed at presenting and discussing the characteristics of typical power distribution networks, along with models of the main system components, including poles, transformers, insulators, surge arresters, SPDs, grounding systems, and the equivalent input impedance of LV power installations. The models are very useful as they are reliable, relatively simple, and can be used in programs for the evaluation of lightning electromagnetic transients in MV and LV networks.

References

[1] IEEE Std 1410. 'IEEE Guide for improving the lightning performance of electric power overhead distribution lines'. New York, 2011.

[2] Heydt G.T. 'The next generation of power distribution systems'. *IEEE Transactions on Smart Grid*, 2010;**1**(3):225–35.

[3] Short T.A. *Electric power distribution handbook*. CRC Press LLC, Boca Raton, FL, 2004.

[4] Iliceto F., Cinieri E., Casely-Hayford L. and Dokyi G.O. 'New concepts on MV distribution from insulated shield wires of HV lines: operation results of an experimental system and applications in Ghana'. *IEEE Transactions on Power Delivery*, 1989;**4**(4):2130–44.

[5] Ramos J.E., Piantini A., Pires V.A. and D'Ajuz A. 'The Brazilian experience with the use of the shield wire line technology (SWL) for energy distribution'. *IEEE Latin America Transactions*, 2009;**7**(6):650–6.

[6] Lakervi E. and Holmes E.J. 'Electricity distribution network design' (*IEE Power Engineering Series*, London, **21**, 1996, 2nd edn.).

[7] Silvestro F., Massuco S., Fornari F., Burt G. and Foote, C. 'Classification of low-voltage grids based on energy flows and grid structures', Project Dispower, Contract No. ENK5-CT-2001-00522, 2004.

[8] Piantini A. 'Lightning protection of low-voltage networks'. In: Cooray V. *Lightning protection*, Chap. 12, pp. 553–634. (*IET Power and Energy Series*, London, **58**, 2010, 1st edn.).

[9] IEC 60364-4-41. 'Low-voltage electrical installations – Part 4-41: Protection for safety – Protection against electric shock', Geneva, 2005.

[10] IEEE Std. C62.41.1. 'IEEE Guide on the surge environment in low-voltage (1000 V and less) AC power circuits', New York, 2002.

[11] IEEE C57.105. 'IEEE Guide for application of transformer connections in three-phase distribution systems', New York, 1978.

[12] Hoidalen H.K. *Lightning-induced overvoltages in low-voltage systems*. PhD Thesis, Norwegian University of Science and Technology, 1997.

[13] Hoidalen H.K. 'Lightning-induced voltages in low-voltage systems and its dependency on overhead line terminations'. *Proceedings of the 24th International Conference on Lightning Protection (ICLP)*, Birmingham, Sep. 1998, pp. 287–92.

[14] Morched A., Martí L. and Ottevangers J. 'A high frequency transformer model for the EMTP'. *IEEE Transactions on Power Delivery*, 1993;**8**(3): 1615–26.

[15] Dommel H.W. *Electromagnetic transients program reference manual (EMTP Theory Book)*. The University of British Columbia, Vancouver, 1986.

[16] Borghetti A., Iorio R., Nucci C.A. and Pelacchi P. 'Calculation of voltages induced by nearby lightning on overhead lines terminated on distribution transformers'. *Proceedings of the International Conference on Power Systems Transients (IPST)*, Lisbon, September 1995, pp. 311–6.

[17] Vaessen P.T.M. 'Transformer model for high frequencies'. *IEEE Transactions on Power Delivery*, 1988;**3**(4):1761–8.

[18] Piantini A. and Malagodi C.V.S. 'Modelling of three-phase distribution transformers for calculating lightning induced voltages transferred to the secondary'. *Proceedings of the 5th International Symposium on Lightning Protection (SIPDA)*, São Paulo, May 1999, pp. 59–64.

[19] Piantini A. and Malagodi C.V.S. 'Voltage surges transferred to the secondary of distribution transformers'. *Proceedings of the 11th International Symposium on High Voltage Engineering (ISH)*, London, 1, Aug. 1999, pp. 1365–8.

[20] Piantini A. and Malagodi C.V.S. 'Voltages transferred to the low-voltage side of distribution transformers due to lightning discharges close to overhead lines'. *Proceedings of the 5th International Symposium on Lightning Protection (SIPDA)*, São Paulo, May 1999, pp. 201–5.

[21] Piantini A., Bassi W., Janiszewski J.M. and Matsuo N.M. 'A simple transformer model for analysis of transferred lightning surges from MV to LV

lines'. *Proceedings of the 15th International Conference on Electricity distribution (CIRED)*, Nice, 1999, paper 2-18/1-6.

[22] Piantini A. and Kanashiro A.G. 'A distribution transformer model for calculating transferred voltages'. *Proceedings of the 26th International Conference on Lightning Protection (ICLP)*, Cracow, 2, Sep. 2002, pp. 429–34.

[23] Kanashiro A.G. and Piantini A. 'The effect of the secondary loads on the voltage surges transferred through distribution transformers'. *Proceedings of the 13th International Symposium on High Voltage Engineering (ISH)*, Rotterdam, 2003.

[24] Kanashiro A.G., Piantini A. and Burani G.F. 'A methodology for transformer modelling concerning high frequency surges'. *Proceedings of the 6th International Symposium on Lightning Protection (SIPDA)*, São Paulo, Nov. 2001, pp. 275–80.

[25] Richter B. and Zeller P. 'Lightning and leakage currents through MO-surge arresters in distribution systems'. *Proceedings of the 26th International Conference on Lightning Protection (ICLP)*, Cracow, 2, Sep. 2002, pp. 527–30.

[26] Manyahi M.J. and Thottappillil R. 'Transfer of lightning transients through distribution transformer circuits'. *Proceedings of the 26th International Conference on Lightning Protection (ICLP)*, Cracow, 2, Sep. 2002, pp. 435–40.

[27] Montanõ R. and Cooray V. 'Penetration of lightning induced transient from high voltage to low-voltage power system across distribution transformers'. *Proceedings of the 27th International Conference on Lightning Protection (ICLP)*, Avignon, 2, Sep. 2004, pp. 303–8.

[28] Agrawal A.K., Price H.J. and Gurbaxani S.H. 'Transient response of a multiconductor transmission line excited by a nonuniform electromagnetic field'. *IEEE Transactions on Electromagnetic Compatibility*, 1980:**EMC-22**(2): 119–29.

[29] Bachega R., De Araujo J., De Souza A. and Martinez M.L.B. 'Digital models of transformer for application on electromagnetic transient studies'. *Proceedings of the International Conference on Grounding and Earthing & 3rd Brazilian Workshop on Atmospheric Electricity*, Rio de Janeiro, Nov. 2002, pp. 279–84.

[30] Piantini A., Kanashiro A.G. and Obase P.F. 'Lightning surges transferred to the low-voltage network'. *Proceedings of the 7th International Symposium on Lightning Protection (SIPDA)*, Curitiba, Nov. 2003, pp. 216–21.

[31] Obase P.F., Piantini A. and Kanashiro A.G. 'Overvoltages transferred to low-voltage networks due to direct strokes on the primary'. *Proceedings of the 8th International Symposium on Lightning Protection (SIPDA)*, São Paulo, Nov. 2005, pp. 489–94.

[32] Obase P.F., Piantini A. and Kanashiro A.G. 'Overvoltages on LV networks associated with direct strokes on the primary line'. *Proceedings of the 28th International Conference on Lightning Protection (ICLP)*, Kanazawa, 1, Sep. 2006, pp. 479–84.

[33] Obase P.F., Piantini A. and Kanashiro A.G. 'Overvoltages on distribution networks caused by direct lightning hits on the primary line'. *Proceedings of*

the International Conference on Grounding and Earthing and International Conference on Lightning Physics and Effects, Belo Horizonte, Nov. 2004, pp. 198–203.

[34] De Conti A. and Visacro S. 'Calculation of lightning-induced voltages on low-voltage distribution networks'. *Proceedings of the 7th International Symposium on Lightning Protection (SIPDA)*, São Paulo, Nov. 2005, pp. 483–8.

[35] Borghetti A., Morched A.S., Napolitano F., Nucci C.A. and Paolone M. 'Lightning-induced overvoltages transferred from medium-voltage to low-voltage networks'. *Proceedings of the IEEE Power Tech*, St. Petersburg, June 2005, pp. 1–7.

[36] Nucci C.A., Bardazzi V., Iorio R., Mansoldo A. and Porrino A. 'A code for the calculation of lightning-induced overvoltages and its interface with the Electromagnetic Transient program'. *Proceedings of the 22nd International Conference on Lightning Protection (ICLP)*, Budapest, Sep. 1994, pp. 19–23.

[37] Paolone M., Nucci C.A. and Rachidi F. 'A new finite difference time domain scheme for the evaluation of lightning induced overvoltages on multi-conductor overhead lines'. *Proceedings of the 5th International Conference on Power System Transients (IPST)*, Rio de Janeiro, June 2001.

[38] Borghetti A., Gutierrez J.A., Nucci C.A., Paolone M., Petrache E. and Rachidi F. 'Lightning-induced voltages on complex distribution systems: models, advanced software tools and experimental validation'. *Journal of Electrostatics*, 2004;**60**(2–4):163–74.

[39] Noda T., Nakamoto H. and Yokoyama S. 'Accurate modeling of core-type distribution transformers for electromagnetic transient studies'. *IEEE Transactions on Power Delivery*, 2002;**17**(4):969–76.

[40] Honda H., Noda T., Asakawa A. and Yokoyama S. 'Improvements to a pole-mounted distribution transformer model for electromagnetic transient studies and validation by field tests'. *Proceedings of the 28th International Conference on Lightning Protection (ICLP)*, Kanazawa, 2, Sep. 2006, pp. 789–94.

[41] Gustavsen B. and Semlyen A. 'Application of vector fitting to state equation representation of transformers for simulation of electromagnetic transients'. *IEEE Transactions on Power Delivery*, 1998;**13**(3):834–42.

[42] Gustavsen B. and Semlyen A. 'Simulation of transmission line transients using vector fitting and modal decomposition'. *IEEE Transactions on Power Delivery*, 1998;**13**(2):605–14.

[43] Gustavsen B. 'Wide band modeling of power transformers'. *IEEE Transactions Power Delivery*, 2004;**19**(10):414–22.

[44] Liang G., Sun H., Zhang X. and Cui X. 'Modeling of transformer windings under very fast transient overvoltages'. *IEEE Transactions on Electromagnetic Compatibility*, 2006;**48**:621–27.

[45] Alvarez D.L., Rosero J.A. and Mombello E.E. 'Circuit model of transformers windings using vector fitting, for frequency response analysis (FRA)'. *2013 Workshop on Power Electronics and Power Quality Applications (PEPQA)*, 2013, pp. 1–6.

[46] Alvarez D.L., Rosero J.A. and Mombello E.E. 'Circuit model of transformers windings using vector fitting, for frequency response analysis (FRA) PART II: Core influence'. *2013 Workshop on Power Electronics and Power Quality Applications (PEPQA)*, 2013, pp. 1–5.

[47] Eslamian M. and Vahidi B. 'New equivalent circuit of transformer winding for the calculation of resonance transients considering frequency-dependent losses'. *IEEE Transactions on Power Delivery*, 2015;**30**:1743–51.

[48] Samarawickrama K., Jacob N.D., Gole A.M. and Kordi B. 'Impulse generator optimum setup for transient testing of transformers using frequency-response analysis and genetic algorithm'. *IEEE Transactions on Power Delivery*, 2015;**30**:1949–57.

[49] Jurisic B., Uglesic I., Xemard A. and Paladian F. 'Difficulties in high frequency transformer modeling'. *Electric Power Systems Research*, 2016;**138**:25–32.

[50] De Conti A., Oliveira V.C., Rodrigues P.R., Silveira F.H., Silvino J.L. and Alipio R. 'Effect of a lossy dispersive ground on lightning overvoltages transferred to the low-voltage side of a single-phase distribution transformer'. *Electric Power Systems Research*, 2017;**153**:104–10.

[51] Tavares P., Villibor J., Lopes G., Faria G., Pereira M. and Wanderley Neto E. 'Evaluation of vector fitting methodology on distribution transformers modelling based on lightning impulse test results'. *2018 IEEE Electrical Insulation Conference (EIC)*, 2018, pp. 331–35.

[52] Cooray V. and De La Rosa F. 'Shapes and amplitudes of the initial peaks of lightning-induced voltage in power lines over finitely conducting earth: theory and comparison with experiment'. *IEEE Transaction on Antennas and Propagation*, 1986;**34**(1):88–92.

[53] Yokoyama S., Miyake K. and Fukui S. 'Advanced observations of lightning induced voltage on power distribution lines (II)'. *IEEE Transaction on Power Delivery*, 1989;**4**(4):2196–203.

[54] Nucci C.A., Rachidi F., Ianoz M. and Mazzetti C. 'Lightning-induced overvoltages on overhead lines'. *IEEE Transactions on Electromagnetic Compatibility*, 1993;**35**(1):75–86.

[55] Rachidi F., Nucci C.A., Ianoz M. and Mazzetti C. 'Influence of a lossy ground on lightning-induced voltages on overhead lines'. *IEEE Transactions Electromagnetic Compatibility*, 1996;**38**(3):250–64.

[56] Piantini A., Carvalho T.O., Silva Neto A., Janiszewski J.M., Altafim R.A.C. and Nogueira A.L.T. 'A system for simultaneous measurements of lightning induced voltages on lines with and without arresters'. *Proceedings of the 27th Int. Conf. Lightning Protection (ICLP)*, Avignon, France, 2004, vol. 1, pp. 297–302.

[57] Piantini A., Janiszewski J.M., Borghetti A., Nucci C.A. and Paolone M. 'A scale model for the study of the LEMP response of complex power distribution networks'. *IEEE Transactions Power Delivery*, 2007;**22**(1):710–20.

[58] Piantini A. and Janiszewski J.M. 'Lightning-induced voltages on overhead lines – application of the Extended Rusck Model'. *IEEE Transactions on Electromagnetic Compatibility*, 2009;**51**(3):548–58.

[59] Silva A.R., Piantini A., Da Silva D.A. and Shigihara M. 'A method for the evaluation of the behavior of a 15 kV insulator under bipolar oscillating impulse voltages'. *International Journal of Electrical Power & Energy Systems*, 2019;**109**:307–13.

[60] CIGRE WG 33.01. 'Guide to procedures for estimating the lightning performance of transmission lines', Paris, 1991.

[61] Task Force 15.09 on Nonstandard Lightning Voltage Waves. 'Review of research on nonstandard lightning voltage waves'. *IEEE Transactions on Power Delivery*, 1994;**9**:1972–81.

[62] Chisholm W.A. 'New challenges in lightning impulse flashover modeling of air gaps and insulators'. *IEEE Electrical Insulation Magazine*, 2010;**26**(2): 14–25.

[63] Pigini A., Rizzi G., Garbagnati E., Porrino A., Baldo G. and Pesavento G. 'Performance of large air gaps under lightning overvoltages: experimental study and analysis of accuracy of predetermination methods'. *IEEE Transactions on Power Delivery*, 1989;**4**(2):1379–92.

[64] Dellera L. and Garbagnati E. 'Lightning stroke simulation by means of the leader progression model. I. Description of the model and evaluation of exposure of free-standing structures'. *IEEE Transactions on Power Delivery*, 1990;**5**(4):2009–22.

[65] Dellera L. and Garbagnati E. 'Lightning stroke simulation by means of the leader progression model. II. Exposure and shielding failure evaluation of overhead lines with assessment of application graphs'. *IEEE Transactions on Power Delivery*, 1990;**5**(4):2023–9

[66] Witzke R.L. and Bliss T.J. 'Surge protection of cable-connected equipment'. *AIEE Transactions*, 1950;**69**:527–42.

[67] Witzke R.L. and Bliss T.J. 'Coordination of lightning arrester location with transformer insulation level'. *AIEE Transactions*, 1950;**69**:964–75.

[68] Kind D. 'Die aufbaufläche bei sto*β*spannungsbeanspruchung technischer elektrodenanordnungen'. *Elektrotechnische Zeitschrift*, 1958;**79**:65–9.

[69] Darveniza M. and Vlastos A.E. 'The generalized integration method for predicting impulse volt-time characteristics for non-standard wave shapes – a theoretical basis'. *IEEE Transactions on Electrical Insulation*, 1988;**23**(3): 373–81.

[70] Chowdhuri P., Mishra A.K. and McConnel B.W. 'Volt-time characteristics of short air gaps under nonstandard lightning voltage waves'. *IEEE Transactions on Power Delivery*, 1997;**12**(1):470–6.

[71] Hileman A.R. *Insulation coordination for power systems*. 1999, CRC Press, Boca Raton, FL, pp. 627–40.

[72] Ancajima A., Carrus A., Cinieri E. and Mazzetti C. 'Behavior of MV insulators under lightning-induced overvoltages: experimental results and reproduction of volt–time characteristics by disruptive effect models'. *IEEE Transactions on Power Delivery*, 2010;**25**(1):221–30.

[73] Cortina R., Garbagnati E., Serravalli W., Dellera L., Pigini A. and Thione L. 'Some aspects of the evaluation of the lightning performance of electrical

systems'. *International Conference on High Voltage Electric Systems*, Sep. 1980, CIGRÉ, pp 21.

[74] IEEE Std 1243. 'IEEE guide for improving the lightning performance of transmission lines', New York, 1997.

[75] Bassi W. and Janiszewski J.M. 'Evaluation of currents and charges in low-voltage surge arresters due to lightning strikes'. *IEEE Transactions on Power Delivery*, 2003;**18**(1):90–4.

[76] De Conti A. and Visacro S. 'Evaluation of lightning surges transferred from medium voltage to low-voltage networks'. *IEE Proceedings – Generation, Transmission and Distribution*, 2005;**152**(3):351–56.

[77] De Conti A., Perez E., Soto E., Silveira F.H., Visacro S. and Torres H. 'Calculation of lightning-induced voltages on overhead distribution lines including insulation breakdown'. *IEEE Transactions on Power Delivery*, 2010;**25**(4):3078–84.

[78] De Conti A., Campos A.F.M., Silveira F.H., Lima J.L.C. and Costa S.E. 'Calculation of lightning flashovers on distribution lines'. *Proceedings of the 2011 International Symposium on Lightning Protection (SIPDA)*, 2011, Fortaleza, Brazil, pp. 205–10.

[79] Chen J. and Zhu M. 'Calculation of lightning flashover rates of overhead distribution lines considering direct and indirect strokes'. *IEEE Transactions on Electromagnetic Compatibility*, 2014;**56**(3):668–74.

[80] Matsuda Y., Yokoyama S., Michishita K. and Sato T. 'Analysis of surge arrester damage accident on distribution line due to winter lightning'. *Proceedings of the 11th Asia-Pacific International Conference on Lightning (APL 2019)*, Hong Kong, 2019.

[81] Tanida M., Nakagawa M., Yasui S., Uenishi H., Kunii Y. and Kanazawa Y. 'Analytical study of lightning protection performance by excluding the current-limiting arcing horns of a phase'. *Proceedings of the 11th Asia-Pacific International Conference on Lightning (APL 2019)*, Hong Kong, 2019.

[82] Shigihara M., Piantini A., Ramos M.C.E.S., Braz C.P., Mazzetti C. and Ancajima A. 'Evaluation of the performance of different methods for estimating the parameters of the disruptive effect model – application to 15 kV insulators'. *IEEE Transactions on Dielectrics and Electrical Insulation*, 2016;**23**(2):1030–7.

[83] Rodrigues A.R., Guimarães G.C., Boaventura W.C., Lima J.L.C., Chaves M.L.R. and Silva A.M.B. 'Volt–time curve prediction of distribution insulators under standard and typical lightning overvoltages using the disruptive effect method'. *Journal of Control, Automation and Electrical Systems*, 2017;**28**(2):259–70.

[84] Lopes G.P., Pedroso J.A.D. and Martinez M.L.B. 'Evaluation of CFO for medium voltage insulators submitted to non-standard impulse shapes experimental results'. *2013 IEEE Electrical Insulation Conference (EIC)*, 2013, pp. 419–23.

[85] Shigihara M., Piantini A., Ramos M.C.E.S., Braz C.P., Mazzetti C. and Ancajima A. 'Comparison of different procedures to predict the

volt-time curves of 15 kV insulators'. *Electric Power Systems Research*, 2017;**153**:82–7.

[86] Shigihara M. and Piantini A. 'Volt-time curves of 24 kV porcelain insulators under non-standard impulse waveshapes'. *Proceedings of the 33rd International Conference on Lightning Protection (ICLP)*, Estoril, 2016, ID 254.

[87] IEEE Working Group 3.4.11. 'Modeling of metal oxide surge arresters'. *IEEE Transactions on Power Delivery*, 1992;**7**(1):302–9.

[88] Pinceti P. and Giannettoni M. 'A simplified model for zinc oxide surge arresters'. *IEEE Transactions on Power Delivery*, 1999;**15**(2):393–8.

[89] Fernandez F. and Diaz R. 'Metal oxide surge arrester model for fast transient simulations'. *Proceedings of the 4th International Conference on Power System Transients*, Rio de Janeiro, Brazil, June 2001, pp. 20–4.

[90] Piantini A. 'Lightning transients on LV networks caused by indirect strokes'. *Journal of Lightning Research*, 2007;**1**:111–31.

[91] Michishita K., Ishii M. and Hongo Y. 'Flashover rate of distribution line due to indirect negative lightning return strokes'. *IEEE Transactions on Power Delivery*, 2009;**24**(1):472–79.

[92] Paolone M., Nucci C.A., Petrache E. and Rachidi F. 'Mitigation of lightning-induced overvoltages in medium voltage distribution lines by means of periodical grounding of shielding wires and of surge arresters, modeling and experimental validation'. *IEEE Transactions on Power Delivery*, 2004; **19**(1):423–31.

[93] Nakada K., Yokoyama S., Yokota T., Asakawa A. and Kawabata T. 'Analytical study on prevention methods for distribution arrester outages caused by winter lightning'. *IEEE Transactions on Power Delivery*, 1998; **13**(4):1399–04.

[94] Piantini A. and Janiszewski J.M. 'Use of surge arresters for protection of overhead lines against nearby lightning'. *Proceedings of the 10th International Symposium on High Voltage Engineering*, Montreal, 1997, pp. 213–16.

[95] Coelho V.L., Piantini A., Almaguer H.A.D., Coelho R.A., Boaventura W.C. and Paulino J.O.S. 'The influence of seasonal soil moisture on the behavior of soil resistivity and power distribution grounding systems'. *Electric Power Systems Research* (Print), 2014;**118**:76–82.

[96] Piantini A. 'Analysis of the effectiveness of shield wires in mitigating lightning-induced voltages on power distribution lines'. *Electric Power Systems Research*, 2018;**159**:9–16.

[97] Piantini A. 'Lightning protection of overhead power distribution lines'. *Proceedings of the 29th International Conference on Lightning Protection (ICLP)*, Uppsala, June 2008, pp. 1–29.

[98] Yokoyama S. 'Designing concept on lightning protection of overhead power distribution line'. *Proceedings of the 9th International Symposium on Lightning Protection (SIPDA)*, Foz do Iguaçu, Nov. 2007, pp. 647–62.

[99] IEEE Std 81. 'IEEE Power Engineering Society. IEEE guide for measuring earth resistivity, ground impedance, and earth surface potentials of a grounding system', New York, 2012.

[100] IEEE Std 80. 'IEEE Power Engineering Society. IEEE guide for safety in AC substation grounding', New York, 2000.

[101] Goedde G.L., Dugan R.C. and Rowe L.D. 'Full scale lightning surge tests of distribution transformers and secondary systems'. *IEEE Transactions on Power Delivery*, 1992;**7**(3):1592–600.

[102] Smith D.R. and Puri J.L. 'A simplified lumped parameter model for finding distribution transformer and secondary system responses to lightning'. *IEEE Transactions on Power Delivery*, 1988;**4**(3):1921–36.

[103] Puri J.L., Abi-Samra N.C., Dionise T.J. and Smith D.R. 'Lightning induced failures in distribution transformers'. *IEEE Transactions on Power Delivery*, 1988;**3**(4):1784–801.

[104] Mirra C., Porrino A., Ardito A. and Nucci C.A. 'Lightning overvoltages in low voltage networks'. *Proceedings of the International Conference on Electricity Distribution (CIRED)*, Birmingham, Conference Publication No. 438, June 1997, pp. 2.19.1–6.

[105] IEC 61000-6-1. 'Electromagnetic Compatibility (EMC) – Part 6-1: Generic standards – Immunity for residential, commercial and light-industrial environments', Geneva, 2016.

[106] IEC 61000-6-2. 'Electromagnetic Compatibility (EMC) – Part 6-2: Generic standards – Immunity for industrial environments', Geneva, 2016.

[107] IEC 61000-4-4. 'Electromagnetic Compatibility (EMC) – Part 4-4: Testing and measurement techniques – Electrical fast transients/burst immunity test', Geneva, 2012.

[108] IEC 61000-4-5. 'Electromagnetic Compatibility (EMC) – Part 4-5: Testing measurement techniques – Surge immunity test', Geneva, 2014.

[109] IEC 60364-4-44. 'Low voltage electrical installations – Part 4-44: Protection for safety – Protection against voltage disturbances and electromagnetic disturbances', Geneva, 2018.

[110] Piantini A., Hermoso Alameda B., *et al. Lightning protection of low-voltage networks*. Paris: CIGRÉ, 2013, 81 pp. CIGRE Working Group C4.408. Technical Brochure 550.

[111] Information Technology Industry Council (ITI), 1250 Eye Street NW, Suite 200, Washington, D.C. (http://www.itic.org).

[112] Matsuo N.M., Kagan N., Domingues I.T., Jesus N.C., Silva M.H.I. and Takauti N.H. 'Methodology for the assessment of possible damages in low voltage equipment due to lightning surges primary'. *Proceedings of the 9th International Symposium on Lightning Protection (SIPDA)*, Foz do Iguaçu, November 2007, pp. 329–34.

Index

D

M

R

S

T

www.ingramcontent.com/pod-product-compliance
Lightning Source LLC
LaVergne TN
LVHW011558200926
848484LV00001B/329